Graph Algorithms and Applications 2

Graph Algorithms and Applications 2

EDITORS

Giuseppe Liotta
University of Perugia, Italy

Roberto Tamassia
Brown University, USA

Ioannis G Tollis
*University of Crete - ICS-FORTH, Greece, and
The University of Texas at Dallas, USA*

 World Scientific

NEW JERSEY · LONDON · SINGAPORE · BEIJING · SHANGHAI · HONG KONG · TAIPEI · CHENNAI

Published by

World Scientific Publishing Co. Pte. Ltd.

5 Toh Tuck Link, Singapore 596224

USA office: Suite 202, 1060 Main Street, River Edge, NJ 07661

UK office: 57 Shelton Street, Covent Garden, London WC2H 9HE

British Library Cataloguing-in-Publication Data
A catalogue record for this book is available from the British Library.

GRAPH ALGORITHMS AND APPLICATIONS 2

ISBN 981-238-855-9 (pbk)

Printed in Singapore.

Preface

This book contains volumes 4–5 of the *Journal of Graph Algorithms and Applications (JGAA)*. Topics of interest for *JGAA* include:

Design and analysis of graph algorithms: exact and approximation graph algorithms; centralized and distributed graph algorithms; static and dynamic graph algorithms; internal- and external-memory graph algorithms; sequential and parallel graph algorithms; deterministic and randomized graph algorithms.

Experiences with graph algorithms: animations; experimentations; implementations.

Applications of graph algorithms: computational biology; computational geometry; computer graphics; computer-aided design; computer and interconnection networks; constraint systems; databases; graph drawing; graph embedding and layout; knowledge representation; multimedia; software engineering; telecommunication networks; user interfaces and visualization; VLSI circuits.

JGAA is supported by distinguished advisory and editorial boards, has high scientific standards, and takes advantage of current electronic document technology. The electronic version of *JGAA* is available on the Web at

http://jgaa.info/

We would like to express our gratitude to the members of the advisory board for their encouragement and support of the journal, to the members of the editorial board and guest editors for their invaluable service and to the many anonymous referees for their essential role in the selection process. Finally, we would like to thank all the authors who have submitted papers to *JGAA*.

Giuseppe Liotta
Roberto Tamassia
Ioannis G. Tollis

Journal of Graph Algorithms and Applications

Managing Editor:

Giuseppe Liotta, University of Perugia

Editors-in-Chief:

Roberto Tamassia, Brown University
Ioannis G. Tollis, University of Crete – ICS-FORTH, Greece, and
The University of Texas at Dallas, USA

Advisory Board:

I. Chlamtac, University of Texas at Dallas
S. Even, Technion
G. N. Frederickson, Purdue University
T. C. Hu, University of California at San Diego
D. E. Knuth, Stanford University
C. L. Liu, University of Illinois
K. Mehlhorn, Max-Planck-Institut für Informatik
T. Nishizeki, Tohoku University
F. P. Preparata, Brown University
I. H. Sudborough, University of Texas at Dallas
R. E. Tarjan, Princeton University
M. Yannakakis, Stanford University

Editorial Board:

G. Di Battista, University of Roma Tre
P. Eades, University of Sydney
D. Eppstein, University of California at Irvine
M. Fürer, Pennsylvania State University
A. Gibbons, King's College
M. T. Goodrich, University of California at Irvine
X. He, State University of New York at Buffalo
A. Itai, Technion
Y. Kajitani, University of Kitakyushu
M. Kaufmann, Universität Tübingen
S. Khuller, University of Maryland
E. W. Mayr, Technischen Universität München
J. S. B. Mitchell, State University of New York at Stony Brook
B. Raghavachari, University of Texas at Dallas
D. Wagner, University of Karlsruhe
T. Warnow, University of Texas at Austin

Contents

Volume 4:1 (2000)

Journal of Graph Algorithms and Applications

http://www.cs.brown.edu/publications/jgaa/

vol. 4, no. 1, pp. 1–16 (2000)

Approximations of Weighted Independent Set and Hereditary Subset Problems

Magnús M. Halldórsson

Science Institute
University of Iceland
IS-107 Reykjavik, Iceland
http://www.hi.is/~mmh
mmh@hi.is

Abstract

The focus of this study is to clarify the approximability of weighted versions of the maximum independent set problem. In particular, we report improved performance ratios in bounded-degree graphs, inductive graphs, and general graphs, as well as for the unweighted problem in sparse graphs. Where possible, the techniques are applied to related hereditary subgraph and subset problem, obtaining ratios better than previously reported for e.g. Weighted Set Packing, Longest Common Subsequence, and Independent Set in hypergraphs.

Communicated by S. Khuller: submitted August 1999; revised April 2000.

Earlier version appears in COCOON '99 [12]. Work done in part at School of Informatics, Kyoto University, Japan.

Halldórsson, *Weighted Independent Set*, JGAA, 4(1) 1–16 (2000)

1   Introduction

An *independent set*, or a stable set, in a graph is a set of mutually nonadjacent vertices. The problem of finding a maximum independent set in a graph, INDSET, is one of most fundamental combinatorial NP-hard problem. It serves also as the primary representative for the family of subgraph problems that are hereditary under vertex deletions. We are interested in finding approximation algorithms that yield good performance ratios, or guarantees on the quality of the solution they find vis-a-vis the optimal solution.

The focus of this paper is to present improved performance ratios for three major versions of the independent set problem: in weighted graphs, bounded-degree graphs and sparse graphs. We also apply some of the methods to a number of related (or not-so related) problems that obey certain hereditariness property, most of which had not been approximated before.

A considerable amount of research has been done on the approximability of INDSET in the last decade. It has been shown to be hard to approximate through advances in the study of interactive proof systems. In particular, Håstad [19] showed it hard to approximate within $n^{1-\epsilon}$, for any $\epsilon > 0$, unless NP-hard problems have randomized polynomial algorithms. The best performance ratio known is $O(n/\log^2 n)$, due to Boppana and Halldórsson [4].

For bounded-degree graphs, Halldórsson and Radhakrishnan [17] gave the first asymptotic improvement over maximal solutions, obtaining a ratio of $O(\Delta/\log\log\Delta)$. For small values of Δ, an algorithm of Berman and Fujito [3] attains the best bound known of $(\Delta + 3)/5$. See the survey [13] for a more complete description of earlier results. The best asymptotic bound known is $O(\Delta\log\log\Delta/\log\Delta)$ due to Vishwanathan [31] (first recorded in [13]), combining two results on semi-definite programming due to Karger, Motwani and Sudan [22] and Alon and Kahale [2].

The current paper is divided into four independent section, each of which treats a different technique for finding independent sets. They are ordered both in chronological order of inquiry, as well as the depth of the solution technique. We first study in Section 2 an elementary general partitioning technique that yields nontrivial performance ratio for a large class of problems satisfying a property that we call *semi-heredity*. All results holds for for weighted versions of the problems. We obtain a $O(n/\log n)$ approximation for INDEPENDENT SET IN HYPERGRAPHS, LONGEST COMMON SUBSEQUENCE, MAX SATISFY-ING LINEAR SUBSYSTEM, and MAX INDEPENDENT SEQUENCE. We strengthen the ratio for problems that do not contain a forbidden clique, obtaining a $O(n(\log\log n/\log n)^2)$ performance ratio for INDSET and MAX HEREDITARY SUBGRAPH. (All problems are defined in their respective sections.)

In Section 3, we consider another elementary strategy, partitioning the vertices into weight classes. It easily yields that weighted versions of semi-hereditary problems on any class of graphs are approximable within $O(\log n)$ of the respective unweighted case. However, this overhead factor reduces to a constant in the case of ratios in the currently achievable range, giving a $O(n/\log^2 n)$ ratio for WIS.

Halldórsson, *Weighted Independent Set*, JGAA, 4(1) 1–16 (2000)   3

We consider in Section 3.1 the approximation of the weighted set packing problem (WSP), in terms of m, the number of base elements. We match the best ratio known for the unweighted case of $O(\sqrt{m})$. We also describe a simplified argument of Lehmann [23] with a better constant factor.

In Section 4, we consider approximations based on semi-definite programming (SDP) relaxations. We generalize the result of Vishwanathan [31] in two ways. First, we apply it to the weighted case, obtaining a $O(\Delta \log \log \Delta / \log \Delta)$ ratio for WIS. This improves on the previous best ratio of $(\Delta + 2)/3$ due to Halldórsson and Lau [15]. Halperin [18] has independently obtained the same ratio, using different techniques. Our ratio also holds in terms of another parameter, $\delta(G)$, the inductiveness of the graph, giving a $O(\delta \log \log \delta / \log \delta)$ approximation of WIS. This improves on the previous best ratio known of $(\delta + 1)/2$ due to Hochbaum [20]. For the other direction, we apply the technique to sparse unweighted graphs, obtaining a ratio of $O(\overline{d} \log \log \overline{d} / \log \overline{d})$, the first asymptotic improvement on Turán's bound [6, 20].

Notation   Let $G = (V, E)$ be a graph, let n denote its number of vertices and let Δ $(\overline{d})$ denote its maximum (average) degree. WIS takes as input instance (G, w), where G is a graph and $w : V \mapsto \mathbf{R}$ is a vector of vertex weights, and asks for a set of independent vertices whose sum of weights is maximized. The maximum weight of an independent set in instance (G, w) denoted by $\alpha(G, w)$, or $\alpha(G)$ on unweighted graphs. Let $|S|$ denote the cardinality of a set S, and let $w(S)$ denote the sum of the weights of the elements of S. Let $w(G)$ denote $w(V(G))$.

We say that a problem is *approximable within* $f(n)$, if there is a polynomial time algorithm which on any instance with n distinguished elements returns a feasible solution within a $f(n)$ factor from optimal. We let OPT denote some optimal solution of the given problem instance and HEU the output of the algorithm under study on that same instance. We also overload those term to refer to the weight of those solutions.

2   Partitioning into easy subproblems

We consider a collection of problems that involve finding a feasible subset of the input of maximum weight. The input contains a collection of n *distinguished elements*, each carrying an associated nonnegative rational weight. Each set of distinguished elements uniquely induces a candidate for a *solution*, which we assume is efficiently computable from the set. The weight of a solution is the sum of the weights of the distinguished elements in the solution.

A property is said to be *hereditary* if whenever a set S of distinguished elements corresponds to a feasible solution, any subset of S also corresponds to a feasible solution. A property is *semi-hereditary* if under the same circumstances, any subset S' of S uniquely induces a feasible solution, possibly corresponding to a superset of S'.

Halldórsson, *Weighted Independent Set*, JGAA, 4(1) 1–16 (2000)   4

To illustrate the concept of semi-hereditarity, consider the problem Maximum Common Subtree [1]. Given is a collection of n free trees, and we are to find a tree that is isomorphic to a subtree (i.e. connected induced subgraph) of each input tree. Verifying if a particular tree is isomorphic to a subtree of another tree is polynomial solvable. Consider the vertices of the first input tree as the distinguished elements. A given subset of these vertices is not necessarily a proper solution, but it uniquely induces a tree that minimally connects the vertices of the subset. Thus, the additional power of the semi-hereditary property is necessary to capture this problem.

Hereditary graph properties are special cases of these definitions. A property of graphs is *hereditary* if whenever it holds for a graph it also holds for its induced subgraphs. For a hereditary graph property, the associated subgraph problem is that of finding a subgraph of maximum vertex-weight satisfying the property. Here, the vertices form the distinguished elements.

Our key tool is a simple partitioning idea, that has been used in various contexts before.

Proposition 2.1 *Let Π be a semi-hereditary subset property. Suppose that given an instance I, we can produce t instances $I_1, I_2, \ldots, I_t$ that cover the set of distinguished elements (i.e. each distinguished element is contained in at least one I_i). Further, suppose we can solve exactly the maximum Π-subset problem on each I_i. Then, the largest of these t solutions yields an approximation of the maximum Π-subset of I within t.*

In the remainder of this section we describe applications of this approach to a number of particular problems.

2.1   Partition into small subsets

Proposition 2.2 *Let Π be a semi-hereditary property for which feasibility can be decided in time at most polynomial in the size of the input and at most simply exponential in the number of distinguished elements. Then, the maximum weighted Π-subgraph can be approximated within $n/\log n$.*

We achieve this by arbitrarily partitioning the set of distinguished elements into $n/\log n$ sets each with $\log n$ elements. For each subset of each set, obtain the candidate solution for this subset and determine feasibility. By our assumptions, each step can be done in polynomial time, and in total at most $2^{\log n} \cdot n/\log n = n^2/\log n$ sets are generated and tested. By this procedure, we find optimal solutions within each of the $n/\log n$ sets. Since the optimal solution of the whole is divided among these sets, the performance ratio is at most $n/\log n$.

Surprisingly, this $n/\log n$-approximation appears to be the best that is known for most such problems. A property is *nontrivial* if it holds for some graphs and fails for others. It is known that, the subgraph problem for any nontrivial hereditary property cannot be approximated within any constant unless $P = NP$, and stronger results hold for properties that fail for some clique or some independent set [25].

Halldórsson, *Weighted Independent Set*, JGAA, 4(1) 1–16 (2000)   5

We apply Proposition 2.2 to several problems featured in the compendium on optimization problems [5]:

Weighted Independent Sets in Hypergraphs   Given a hypergraph, or a set system, $(S, \mathcal{C})$ where S is a set of weighted base elements (vertices) and $\mathcal{C} = \{C_1, C_2, \ldots, C_n\}$ is a collection of subsets of S, find a maximum weight subset S' of vertices such that no subset C_i is fully contained in S'.

Hofmeister and Lefmann [21] analyzed a Ramsey-theoretic algorithm generalizing that of [4], and showed its performance ratio to be $O(n/(\log^{(r-1)} n))$ for the case of r-uniform hypergraphs. It is straightforward to verify the heredity thus a $O(n/\log n)$ performance ratio holds by Proposition 2.1.

Longest Common Subsequence   Given a finite set R of strings from a finite alphabet Σ, find a longest possible string w that is a subsequence of each string x in R. The problem is clearly hereditary, and feasibility can be tested for each string x in R separately via dynamic programming. Hence, by applying Proposition 2.2, partitioning the smallest string in the input, we obtain a performance ratio of $O(m/\log m)$, where m is the size of the smallest string.

Max Satisfying Linear Subsystem   Given a system $Ax = b$ of linear equations, with A an integer $m \times n$ matrix and b an integer m vector, find a rational vector $x \in \mathcal{Q}^n$ that satisfies the maximum number of equations.

This problem is clearly hereditary, since any subset of a feasible collection of equations is also feasible. Feasibility of a given system can be solved in polynomial time via linear programming. Hence, $O(m/\log m)$ approximation follows from Proposition 2.2. This holds equally if equality is replaced by inequalities $(>, \geq)$. It also holds if a particular set of constraints/equations are required to be satisfied by a solution.

Max Independent Sequence   Given a graph, find a maximum length sequence $v_1, v_2, \ldots, v_m$ of independent vertices such that, for all $i < m$, a vertex v_i' exists which is adjacent to v_{i+1} but is not adjacent to any v_j for $j \leq i$. This problem was introduced by Blundo (see [5]).

First observe that solutions to the problem are hereditary: if $v_1, v_2, \ldots, v_m$ is an independent sequence, then so is any subsequence $v_{a_1}, v_{a_2}, \ldots, v_{a_x}$. This is because, for all $i < x$, there exists a node v_i' that is adjacent to $v_{a_{i+1}}$ but not adjacent to any v_j for $j < a_{i+1}$ and hence not to any v_{a_j} for $j \leq i$. Feasibility of a solution can be tested in time polynomial in the size of the input. Independence is easily tested by testing all pairs in the proposed solution. A valid set can be turned into a valid sequence by inductively finding the element adjacent to a vertex outside the set that is adjacent to no other unselected vertex.

Thus, we obtain an $O(n/\log n)$ approximation via Proposition 2.2. We can also argue strong approximation hardness bounds.

Proposition 2.3 MAX INDEPENDENT SEQUENCE *is no easier than* INDSET, *within 2. Thus, it is hard to approximate within* $n^{1-\epsilon}$, *for any* $\epsilon > 0$, *unless* $NP = ZPP$.

Proof.   Given a graph G on vertices $v_1, v_2, \ldots, v_n$, the graph H_G consists of G and n additional vertices $\{w_1, w_2, \ldots, w_n\}$ connected into a clique, with $(v_i, w_j) \in E(H_G)$ iff $i \geq j$. Then, any independent set in G corresponds to an independent sequence in H_G. The converse is also true, with the possible exclusion of one w_i vertex; in that case, we can replace that w_i vertex with some v_j vertex that must exist and be independent of the other v-vertices in the set. Hence, we get a size-preserving reduction. The new graph contains twice as many vertices, thus the performance ratio lower bound is weaker for MAX INDEPENDENT SEQUENCE by a factor of 2. The hardness now follows from the result of Håstad [19] on INDSET. $\qquad\square$

Theorem 2.4 *Weighted versions of* INDSET IN HYPERGRAPHS, MAX HEREDITARY SUBGRAPH *and* MAX INDEPENDENT SEQUENCE *can be approximated within* $O(n/\log n)$.

2.2   Weighted Independent Sets and Other Hereditary Graph Properties

A theorem of Erdős and Szekeres [7] on Ramsey numbers yields an efficient algorithm [4] for finding either cliques or independent sets of nontrivial size.

Fact 2.5 (Erdős, Szekeres) *Any graph on n vertices contains a clique on s vertices or an independent set on t vertices such that* $\binom{s+t-2}{s-1} \geq n$.

We use this theorem to approximate a large class of hereditary subgraph problems.

Theorem 2.6 MAX WEIGHTED HEREDITARY SUBGRAPH *can be approximated within* $O(n(\log\log n/\log n)^2)$, *for properties that fail for some cliques or some independent set.*

Proof.   Let n denote here the size of the input graph G to the MAX WEIGHTED HEREDITARY SUBGRAPH problem. We say that a graph is *amenable* if it is either an independent set or consists of at most $\log n/\log\log n$ disjoint cliques. Theorem 2.5 implies that we can find in G either an independent set of size at least $\log^2 n$, or a clique of size at least $\log n/2\log\log n$. Thus we can find an amenable subgraph of size $X = \log^2 n/3\log\log n$, by at most $\log n$ applications of Theorem 2.5.

We then pull these amenable subgraphs one by one from G, obtaining a partition of G into amenable subgraphs. The number of subgraphs in the partition will be at most $3n/X$. Namely, at most $n/(\log^2(n/X)/3\log\log n) = n/X(1+o(1))$ subgraphs are found before the size of G drops below n/X and the remainder is at most another n/X.

We can solve WIS on an amenable subgraph by exhaustively checking all $(\log n/\log\log n)^{\log n/\log\log n} = O(n)$ possible combinations of selecting up to one vertex from each clique. More generally, assume without loss of generality that our hereditary subgraph property fails for cliques of size s. We can solve it optimally on an amenable subgraph by exhaustively checking all combinations of selecting at most $s-1$ vertices from each clique. That number is still at most $(\log n/\log\log n)^{s\log n/\log\log n}$, which is $poly(n)$ for fixed s. In the case that the property fails for some independent set, we exchange the roles of independent sets and cliques in our partitioning routine with no change in the results. $\square$

Examples of such properties include: bipartite, k-colorable, k-clique free, planar.

2.3 Limitations of partitioning

The wide applicability of this partitioning technique might offer a glimmer of hope for approximating the independent set problem in general graphs within $n^{1-\epsilon}$, for some $\epsilon > 0$. The following observation casts a shade on that proposal.

For a property Π, the Π-chromatic number of a graph is the minimum number of classes that the vertex set can be partitioned into such that the graph induced by each class satisfies Π. Scheinerman [29] has shown that for any nontrivial hereditary property Π, the Π-chromatic number of a random graph approaches $\theta(n/\log n)$. This indicates that our results are essentially the best possible.

3 Partitioning into weight classes

We now consider a simple general strategy for obtaining approximations to weighted subgraph problems, that always comes within a $\log n$ factor from the unweighted case and often within less.

Theorem 3.1 *Let Π be a hereditary subgraph problem. Suppose Π can be approximated within ρ on unweighted graphs (or on a subclass thereof). Then, the vertex-weighted version can be approximated within $O(\rho \cdot \log n)$.*

Proof. Consider the following strategy. Let W be the maximum vertex weight. Delete all vertices of weight at most W/n. Let V_i be the set of vertices whose weight lies in $(W/2^i, W/2^{i-1}]$, for $i = 1, 2, \ldots, \lg n$. Run the ρ-approximate algorithm on the V_i, ignoring the weights. Output the maximum weight solution, denoted by HEU.

We claim that the performance ratio of this method is at most $2\rho\lg n + 1$. First, note that the set of vertices of small weight adds up to at most W, or less than that of HEU. Second, if G' is the graph induced by vertices of weight more than W/n,

$$OPT(G') \le \sum_{i=1}^{\lg n} OPT(V_i) \le \sum_{i=1}^{\lg n} 2\rho\,HEU(V_i) = 2\rho\,HEU(G),$$

Halldórsson, *Weighted Independent Set*, JGAA, 4(1) 1–16 (2000)   8

where the additional factor of 2 comes from the rounding of the weights.   □

We note that the logarithmic loss in approximation is caused by a logarithmic decrease in subgraph sizes. However, when the performance function is close to linear, as is the case today, decrease in subgraph size affects performance only slightly. We illustrate this with WIS, matching the known approximation for unweighted graphs.

Theorem 3.2 *If a hereditary subgraph problem can be approximated within* $g(n) = n^{1-\Omega(1/\log\log n)}$, *then its weighted version can also be approximated within* $O(g(n))$. *In particular, WIS can be approximated within* $O(n/\log^2 n)$.

Proof. Let G be a graph partitioned into subgraphs $V_1, \ldots, V_{\log n}$ as in Theorem 3.1, let OPT be an optimal solution and HEU the heuristic solution found. Observe that the function g satisfies $g(N) = O(g(n) \cdot N/n)$ when $N \geq n/\lg n$, and $g(N) = O(g(n)/\log n)$ when $N \leq n/\lg n$,

Let L be the set of indices ℓ that satisfy

$$w(V_\ell \cap OPT) \geq w(OPT)/2\lg n, \tag{1}$$

and note that $\sum_{i \in L} w(V_i \cap OPT) \geq w(OPT)/2$.

Suppose that for some $\ell' \in L$, $|V_{\ell'}| < n/\lg n$. By (1), $w(V_i \cap OPT) \leq w(OPT) \leq (2\lg n)w(V_{\ell'} \cap OPT)$, for all i. Thus,

$$\rho \leq \frac{w(OPT)}{w(HEU)} \leq 4\lg n \frac{w(V_{\ell'} \cap OPT)}{w(HEU)} \leq 4\lg n \cdot g(|V_{\ell'}|) = O(g(n)).$$

Otherwise, $g(|V_\ell|) = O(g(n) \cdot |V_\ell|/n)$ for all $\ell \in L$. Then,

$$\rho \;\leq\; \sum_i \frac{w(V_i \cap OPT)}{w(HEU)} \leq 2\frac{\sum_{\ell \in L} w(V_\ell \cap OPT)}{w(HEU)}$$

$$\leq\; 2\sum_i g(|V_i|) = \frac{g(n)}{n} \sum_{\ell \in L} O(|V_\ell|) = O(g(n)).$$

The $O(n/\log^2 n)$ ratio for WIS now follows from the result of [4] for the unweighted case.   □

3.1   Weighted set packing

The WSP problem is as follows. Given a set S of m base elements, and a collection $\mathcal{C} = \{C_1, C_2, \ldots, C_n\}$ of weighted subsets of S, find a subcollection $\mathcal{C}' \subseteq \mathcal{C}$ of disjoint sets of maximum total weight $\sum_{C_i' \in \mathcal{C}'} w(C_i')$. A variety of applications of this problem to practical optimization problems is surveyed in [30]. It has recently been used to model multi-unit combinatorial auctions [27, 10] and and in the formation of coalitions in multiagent systems [28].

By forming the intersection graph of the given hypergraph (with a vertex for each set, and two vertices being adjacent if the corresponding sets intersect),

Halldórsson, *Weighted Independent Set*, JGAA, 4(1) 1–16 (2000)   9

a weighted set packing instance can be transformed to a weighted independent set instance on n vertices. Hence, approximations of WIS — as a function of n — carry over to WSP.

For approximations of unweighted set packing *as a function of* m $(= |S|)$, Halldórsson, Kratochvíl, and Telle [14] gave a simple $\sqrt{m}$-approximate greedy algorithm, and noted that $m^{1/2-\epsilon}$-approximation is hard via [19]. We observe that the positive results hold also for the weighted case, by a simple variant of the greedy method.

Theorem 3.3 WSP *can be approximated within* $2\sqrt{m}$ *in time proportional to the time it takes to sort the weights.*

Proof. The algorithm initially removes all sets of cardinality $\sqrt{m}$ or more. It then greedily selects sets of maximum weight that are disjoint from the previously selected sets.

```
SetPackingApprox(S,C)
   Max ← the set in C of maximum weight
   C ← {C ∈ C : |C| ≤ √m}
   Output the larger of GreedySP(S,C) and Max
end
GreedySP(S,C)
   t ← 0, Cₜ ← C
   repeat
     t ← t + 1
     Xₜ ← C ∈ Cₜ₋₁ of maximum weight
     Zₜ ← {C ∈ Cₜ₋₁ : X ∩ C ≠ ∅}
     Cₜ ← Cₜ₋₁ − Zₜ
   until C = ∅
   return {X₁, X₂, ..., Xₜ}
end
```

Figure 1: Greedy set packing algorithm

Consider Z_t, the sets eliminated in some iteration i. Observe that the optimal solution contains at most $\sqrt{m}$ sets from Z_t (since sets in Z_t have an element in common with X_t which is of cardinality at most $\sqrt{m}$), all of which are of weight at most that of X_t, the set chosen by the algorithm. Hence, in every iteration, the contribution added to the algorithm's solution is at least $\sqrt{m}$-th fraction of what the optimal solution could get.

Also, the optimal solution contains at most $\sqrt{m}$ sets among those eliminated in the second line of SetPackingApprox, since each of them is of cardinality at least $\sqrt{m}$. Since the algorithm contains at least the weight of the maximum weight set, this is at most $\sqrt{m}$ times the algorithm's solution. Combined, the optimal solution is of weight at most $2\sqrt{m}$ times the algorithm's solution. $\square$

Halldórsson, *Weighted Independent Set*, JGAA, 4(1) 1–16 (2000)

We now describe an improvement due to Lehmann [23] that shows that the greedy algorithm can be modified to give a slightly better ratio of $\sqrt{m}$ by itself. The modification to GREEDYSP is to change line 4 to

$$X_t \leftarrow C \in \mathcal{C}_{t-1} \text{ that maximizes } w(C)/\sqrt{|C|}.$$

Let OPT be some optimal set packing solution. Consider any iteration t of the algorithm, and let OPT_t be the sets in $OPT \cap Z_t$. Note first, that for any set $C \in \mathcal{C}_{t-1}$,

$$w(C) \leq \sqrt{|C|}\frac{w(X_t)}{\sqrt{|X_t|}},$$

because of how X_t was chosen. Thus,

$$w(OPT_t) = \sum_{C \in OPT_t} w(C) \leq \frac{w(X_t)}{\sqrt{|X_t|}} \sum_{C \in OPT_t} \sqrt{|C|}.$$

Since the sets in OPT_t must be disjoint and of total cardinality at most m, the sum on the right hand side is maximized when all the sets are of equal size. This gives

$$w(OPT_t) \leq \frac{w(X_t)}{\sqrt{|X_t|}}\sqrt{|OPT_t| \cdot m}.$$

Note that OPT_t contains at most one set for each element of X_t, so $|OPT_t| \leq |X_t|$. Hence, $w(OPT_t) \leq \sqrt{m}\, w(X_t)$. Since this holds for each iteration, a ratio of $\sqrt{m}$ follows. Gonen and Lehmann [10] show that no greedy algorithm can obtain a better ratio.

One can also observe that the constant factor can be arbitrarily improved, if one can afford a commensurate increase in the polynomial complexity. Modify SETPACKINGAPPROX to set Max as the maximum weight set packing in $(S, \mathcal{C})$ containing at most s sets. Also, change the upper bound on the cardinality of sets to be included in $\mathcal{C}$ from $\sqrt{m}$ to $q = \sqrt{m/s}$. To analyze this, let us split the optimal packing into a packing of sets of size greater than q and that of sets at most q. A packing of the former can contain at most $m/q = \sqrt{sm}$ sets, hence Max approximates it within $\sqrt{m/s}$ factor. Also, we know that GREEDYSP approximates the latter within the same factor. The better of the two solutions now yields a $2\sqrt{m/s}$ approximation.

4 Semi-definite programming

A fascinating polynomial-time computable function $\vartheta(G)$ introduced by Lovász [24] has the remarkable "sandwiching" property that it always lies between two NP-hard functions, $\alpha(G) \leq \vartheta(G) \leq \overline{\chi}(G)$. This property suggests that it may be particularly suited for obtaining good approximations to either function. While some of those hopes have been dashed [8], a number of fruitful applications have been found and it remains the most promising candidate for obtaining improved approximations [9].

Karger, Motwani and Sudan [22] gave improved approximations for k-colorable graphs via the theta function, followed by Alon and Kahale [2] that obtained improved approximations for INDSET in the case of linear-sized independent sets. Mahajan and Ramesh [26] showed how these and related algorithms can be derandomized. Vishwanathan [31] observed that an improved performance ratio for the independent set problem of $O(\Delta \log \log \Delta / \log \Delta)$ could be obtained by combining together theorems of [22, 2].

We illustrate here how the improved $O(\Delta \log \log \Delta / \log \Delta)$ also applies to weighted independent sets. For this purpose, we give straightforward generalizations of the results of [22] and [2]. Halperin [18] has independently obtained the same bound, using a different rounding procedure.

We actually prove a stronger bound. A graph is said to be δ-*inductive* if, there is a linear ordering of the vertices such that each vertex has at most δ neighbors ordered after itself. We obtain a $O(\delta \log \log \delta / \log \delta)$ approximation of WIS, improving on the previous best $(\delta + 1)/2$ [20].

The Lovász number $\vartheta(G)$ of a graph G is the least number k such that there exists a representation of unit vectors v_i to each vertex $i \in V$, such that for any two nonadjacent vertices i and j the dot product of their vectors satisfies the equality

$$(v_i \cdot v_j) = -\frac{1}{k}.$$

Such a representation can be computed in polynomial time, or more precisely, one can obtain a representation that is arbitrarily close. We shall use several other definitions of this measure later, that were introduced in the original paper of Lovász [24]; the current one was defined in [22] as the *strict vector chromatic number*.

We first give a weighted version of a result of [22]. For our purposes, it suffices to use the simpler method of "rounding by hyperplanes", as the constant in the exponent is not important.

Proposition 4.1 *Let G be a weighted δ-inductive graph satisfying $\vartheta(\overline{G}) \leq k$. Then an independent set in G of weight $\Omega(w(G)/\delta^{1-1/2k})$ can be constructed with high probability in polynomial time.*

Proof. The inductiveness of the graph implies that its edges can be directed so that each vertex has outdegree at most d, and thus we say it has at most d *out-neighbors*. We assume such a direction on the edges.

¿From the bound on $\vartheta(\overline{G})$, we can represent the vertices as Euclidean vectors, such that for *adjacent* vertices i and j the corresponding vectors v_i and v_j satisfy $(v_i \cdot v_j) = -\frac{1}{k}$. Given such a representation, the algorithm selects r hyperplanes at random (by choosing uniformly random vectors on the unit sphere around the origin as normals), dividing $\mathbf{R}^n$ into 2^r partitions. The algorithm examines each of the partitions, collects the set of vertices with no out-neighbors in the same partition, and outputs the set of maximum weight. We give a lower bound on the expected weight of this set.

Let $q = \frac{1}{2} + \frac{1}{\pi k}$ and let $r = 2 + \lceil \log_{1/(1-q)} \delta \rceil$. Note that $(1-q)^r \leq 1/4\delta$. Also, it can be shown that $1/\lg 1/(1-q) \leq 1 - 1/2k$, using that $1/(1-q) = 2(1 + 1/(\pi k/2 - 1))$ and that $\ln(1+x) \geq x/(1+x)$. That means that

$$2^r \leq 8\delta^{1/\lg 1/(1-q)} = O(\delta^{1-1/2k}).$$

The probability that a random hyperplane separates the vectors associated with two vertices is ϕ/π, where ϕ is the angle between the vectors. When the vertices are adjacent, this probability is

$$\frac{arccos(-1/k)}{\pi} \leq \frac{1}{2} + \frac{1}{\pi k} = q,$$

where the inequality is obtained from the Taylor expansion of $arccos(x)$. Hence, the probability that none of the r hyperplanes cuts a given edge is at most

$$p = (1-q)^r \leq \frac{1}{4\delta}.$$

The probability, for a given vertex v, that all of v's outedges are cut is then at least $1 - \delta p \geq 1 - 1/4$. Consider, for a given partition P, the independent set I of vertices with no out-neighbors (i.e. outdegree zero) within its partition. If $w(P)$ denotes the weight of a given partition P, the expected weight of I is at least $w(P)(1 - 1/4)$. Averaging over the 2^r partitions, the expected weight of the set output is at least

$$\frac{w(G)(1 - 1/4)}{2^r} = \Omega(w(G)/\delta^{1-1/2k}).$$

$\square$

We now consider a generalization of the Lovász number to weighted graphs. An *orthonormal representation* of a graph $G = (V, E)$ is an assignment of a unit vector b_v in Euclidean space to each vertex v of G, such that $b_u \cdot b_v = 0$ if $u \neq v$ and $(u, v) \notin E$. The (weighted) theta function $\vartheta(G, w)$ [11] equals the minimum over all unit vectors d and all orthonormal labelings b_v of

$$\max_{v \in V} \frac{w(v)}{(d \cdot b_v)^2}.$$

An equivalent dual characterization is to define it as the maximum over all unit vectors d and all orthonormal representations b_v of the complement graph $\overline{G}$ of $\sum_{v \in V} (d \cdot b_v)^2 w(v)$. The Lovász number $\vartheta(G)$ is $\vartheta(G, \mathbf{1})$, the theta function on the unit-weighted graph.

Proposition 4.2 *If $\vartheta(G, w) \geq 2w(G)/k$ (e.g. if $\alpha(G, w) \geq 2w(G)/k$), then we can find an induced subgraph K in G such that $\vartheta(\overline{K}) \leq k$ and $w(K) \geq w(G)/k$.*

Halldórsson, *Weighted Independent Set*, JGAA, 4(1) 1–16 (2000)   13

Proof. We emulate [2]. Let d be a unit vector and b_v a representation such that $\vartheta(G, w) = \sum_{v \in V} (d \cdot b_v)^2 w(v)$. For each v, let a_v denote $(d \cdot b_v)^2$. We can then split the sum for $\vartheta(G, w)$ into two parts: those where a_v is small or at most $1/k$ for some breakpoint k, and those where a_v is large. Thus,

$$\vartheta(G, w) = \sum_{a_v \leq 1/k} a_v w(v) + \sum_{a_v \geq 1/k} a_v w(v) \leq w(G)/k + \sum_{a_v \geq 1/k} a_v w(v).$$

Let K be the subgraph induced by vertices v with $a_v \geq 1/k$. If $\vartheta(G, w) \geq 2w(G)/k$, we have that since $a_v \leq 1$ for each vertex v,

$$w(K) = \sum_{v \in V(K)} w(v) \geq \sum_{a_v \geq 1/k} a_v w(v) \geq \vartheta(G, w) - w(G)/k \geq w(G)/k.$$

Also,

$$\max_{v \in K} \frac{1}{(d \cdot b_v)^2} \leq k,$$

hence the Lovász number of K is at most k by its definition. $\qquad\square$

Theorem 4.3 WIS *can be approximated within* $O(\delta \log \log \delta / \log \delta)$.

Proof. Let (G, w) be an instance with $\alpha(G, w) = 2w(G)/k$, for some k. We find via Proposition 4.2 a subgraph K_k with $\vartheta(\overline{K_k}) \leq k$ and $w(K_k) \geq w(G)/k$. We then find via Proposition 4.1 an independent set in $K_k \subset G$ of weight at least $\frac{w(G)/k}{\delta^{1-1/2k}}$. The approximation ratio is then at most $2\delta^{1-1/2k}$.

Alternatively, δ-inductive graphs are well-known to be $\delta+1$-colorable. Thus, the heaviest color class is an independent set of weight at least $w(G)/(\delta + 1)$, for a $2(\delta + 1)/k$ approximation. Observe that first ratio is increasing with k and the latter decreasing, with breakpoint achieved when $k = \frac{1}{2} \log \delta / \log \log \delta$, in which case both ratios are $O(\delta \log \log \delta / \log \delta)$. $\qquad\square$

4.1   Sparse graphs

Inductive graphs can be thought of as being "everywhere sparse". For reasons of padding, it is not possible to get similar ratios for WIS on all sparse graphs. However, we can obtain this for the unweighted INDSET problem, improving on the previous best ratio known of $(2\overline{d} + 3)/5$ [16].

Theorem 4.4 IS *can be approximated within* $O(\overline{d} \log \log \overline{d} / \log \overline{d})$.

Proof. For a graph G of average degree $\overline{d}$, let t denote $n/\alpha(G)$, i.e. $\alpha(G) = n/t$. Consider the subgraph H induced by vertices of degree at most $2t\overline{d}$. Then, $\Delta(H) \leq 2t\overline{d}(G)$. At least $t\overline{d}(n - |V(H)|)$ edges are removed, while G contained only $\frac{1}{2}\overline{d}n$ edges. Hence, at most $\frac{1}{2t}n$ vertices are removed, and thus $\alpha(H) \geq \alpha(G)/2$.

Halldórsson, *Weighted Independent Set*, JGAA, 4(1) 1–16 (2000)   14

Apply Propositions 4.1 and 4.2 on H to obtain a subgraph $K \subset H \subseteq G$ with $\vartheta(\overline{K}) \leq k = 4t$ and at least n/k vertices, We then obtain an independent set in K with at least

$$\Omega\left(\frac{|V(H)|/k}{\Delta(H)^{1-1/2k}}\right) = \Omega\left(\frac{n/t}{(2t\overline{d}(G))^{1-1/8t}}\right) = \Omega\left(\frac{n/t^2}{\overline{d}(G)^{1-1/8t}}\right)$$

vertices, for a performance ratio of $O(\overline{d}(G)^{1-1/8t}t^2)$. Recall that a minimum-degree greedy algorithm attains the Turán bound of $n/(\overline{d}+1)$ [6] for a $O(\overline{d}/t)$ approximation [16]. The two functions cross when t is about $\frac{1}{24}\log\overline{d}/\log\log\overline{d}$, for the desired ratio. $\qquad\square$

References

[1] T. Akutsu and M. M. Halldórsson. On the approximation of largest common point sets and largest common subtrees. *Theoretical Comput. Sci.*, 233:33–50, Dec. 1999.

[2] N. Alon and N. Kahale. Approximating the independence number via the θ function. *Math. Programming*, 80:253–264, 1998.

[3] P. Berman and T. Fujito. On the approximation properties of independent set problem in degree 3 graphs. In *Proc. Fourth Workshop on Algorithms and Data Structures*, pages 449–460. Springer LNCS #955, Aug. 1995.

[4] R. B. Boppana and M. M. Halldórsson. Approximating maximum independent sets by excluding subgraphs. *BIT*, 32(2):180–196, June 1992.

[5] P. Crescenzi and V. Kann. A compendium of NP optimization problems. Dynamic online survey at http://www.nada.kth.se/theory/problemlist.html, 1999.

[6] P. Erdős. On the graph theorem of Turán (in Hungarian). *Mat. Lapok*, 21:249–251, 1970.

[7] P. Erdős and G. Szekeres. A combinatorial problem in geometry. *Compositio Math.*, 2:463–470, 1935.

[8] U. Feige. Randomized graph products, chromatic numbers, and the Lovász ϑ-function. *Combinatorica*, 17(1):79–90, 1997.

[9] U. Feige and J. Kilian. Heuristics for finding large independent sets, with applications to coloring semi-random graphs. In *Proc. 39th Ann. IEEE Symp. on Found. of Comp. Sci.*, 1998.

[10] R. Gonen and D. Lehmann. Optimal solutions for multi-unit combinatorial auctions: Branch and bound heuristics. Unpublished manuscript, May 2000.

Halldórsson, *Weighted Independent Set*, JGAA, 4(1) 1–16 (2000)   15

[11] M. Grötschel, L. Lovász, and A. Schrijver. *Geometric Algorithms and Combinatorial Optimization.* Springer-Verlag, 1988.

[12] M. M. Halldórsson. Approximations of weighted independent set and hereditary subset problems. In *Computing and Combinatorics, Proceedings of the 5th Annual International Conference (COCOON)*, volume 1627 of *Springer Lecture Notes in Computer Science*, pages 261–270, Tokyo, Japan, July 1998.

[13] M. M. Halldórsson. A survey on independent set approximations. In *Proc. 1st Intl. Wkshop on Approxim. Algor. (APPROX)*, volume 1444 of *Springer Lecture Notes in Computer Science*, pages 1–14, Aalborg, Denmark, July 1998.

[14] M. M. Halldórsson, J. Kratochvíl, and J. A. Telle. Independent sets with domination constraints. *Disc. Appl. Math.*, 99:39–54, Dec. 1999. URL: http://www.elsevier.nl/locate/jnlnr/05267.

[15] M. M. Halldórsson and H. C. Lau. Low-degree graph partitioning via local search with applications to constraint satisfaction, max cut, and 3-coloring. *Journal of Graph Algorithms and Applications*, 1(3):1–13, 1997.

[16] M. M. Halldórsson and J. Radhakrishnan. Greed is good: Approximating independent sets in sparse and bounded-degree graphs. *Algorithmica*, 18:145–163, 1997.

[17] M. M. Halldórsson and J. Radhakrishnan. Improved approximations of independent sets in bounded-degree graphs via subgraph removal. *Nordic J. Computing*, 1(4):275–292, Winter 1994.

[18] E. Halperin. Improved approximation algorithms for the vertex cover problem in graphs and hypergraphs. In *Proc. Eleventh ACM-SIAM Symp. on Discrete Algorithms*, pages 329–337, 2000.

[19] J. Håstad. Clique is hard to approximate within $n^{1-\epsilon}$. *Acta Mathematica*, 182:105–142, 1999.

[20] D. S. Hochbaum. Efficient bounds for the stable set, vertex cover, and set packing problems. *Disc. Applied Math.*, 6:243–254, 1983.

[21] T. Hofmeister and H. Lefmann. Approximating maximum independent sets in uniform hypergraphs. In *Proc. 23rd Intl. Symp. Math. Found. of Comp. Sci. (MFCS)*, volume 1450 of *Springer Lecture Notes in Computer Science*, pages 562–570, Brno, Czech Republic, Aug. 1998.

[22] D. Karger, R. Motwani, and M. Sudan. Approximate graph coloring by semi-definite programming. *J. ACM*, 45(2):246–265, Mar. 1998.

[23] D. Lehmann. Personal communication, May 1999.

[24] L. Lovász. On the Shannon capacity of a graph. *IEEE Trans. Inform. Theory*, IT-25(1):1–7, Jan. 1979.

[25] C. Lund and M. Yannakakis. The approximation of maximum subgraph problems. In *Proceedings of the 20th International Conference on Automata, Languages, and Programming (ICALP)*, Springer Lecture Notes in Computer Science, 1993.

[26] S. Mahajan and H. Ramesh. Derandomizing semidefinite programming based approximation algorithms. *SIAM J. Comput.*, 28(5):1641–1663, 1999.

[27] T. Sandholm. An algorithm for optimal winner determination in combinatorial auctions. In *Proc. International Joint Conference on Artificial Intelligence (IJCAI)*, Stockholm, Sweden, 1999.

[28] T. Sandholm, K. Larson, M. Andersson, O. Shehory, and F. Tohmé. Coalition structure generation with worst case guarantees. *Artificial Intelligence*, 111(1-2):209–238, 1999.

[29] E. R. Scheinerman. Generalized chromatic numbers of random graphs. *SIAM J. Disc. Math.*, 5(1):74–80, Feb. 1992.

[30] R. R. Vemuganti. Applications of set covering, set packing and set partitioning models: A survey. In D.-Z. Du and P. M. P. (Eds.), editors, *Handbook of Combinatorial Optimization*, volume 1, pages 573–746. Kluwer Academic Publishers, 1998.

[31] S. Vishwanathan. Personal communication, 1996.

Volume 4:2 (2000)

Journal of Graph Algorithms and Applications

http://www.cs.brown.edu/publications/jgaa/

vol. 4, no. 2, pp. 1–11 (2000)

Approximation Algorithms for Some Graph Partitioning Problems

G. He

Battelle, Pacific Northwest National Lab
P. O. Box 999 / MS K9-55, Richland, WA 99352 USA
http://www.pnl.gov/remote/expertise/ggh.htm
George.He@pnl.gov

J. Liu[1]

Department of Mathematics and Computer Sciences
University of Lethbridge, Lethbridge, Alberta, Canada, T1K 3M4
http://www.cs.uleth.ca/
liu@cs.uleth.ca

C. Zhao

Department of Mathematics and Computer Sciences
Indiana State University, Terre Haute, IN 47809 USA
http://math.indstate.edu/zhao.html
cheng@laurel.indstate.edu

Abstract

This paper considers problems of the following type: given an edge-weighted k-colored input graph with maximum color class size c, find a minimum or maximum c-way cut such that each color class is totally partitioned. Equivalently, given a weighted complete k-partite graph, cover its vertices with a minimum number of disjoint cliques in such a way that the total weight of the cliques is maximized or minimized. Our study was motivated by some work called the index domain alignment problem [6], which shows its relevance to optimization of distributed computation. Solutions of these problems also have applications in logistics [3] and manufacturing systems [10]. In this paper, we design some approximation algorithms by extending the matching algorithms to these problems. Both theoretical and experimental results show that the algorithms we designed produce good approximations.

Communicated by D. Eppstein.
Submitted: July 2000. Revised: August 2000.

---

[1]Research by this author was partially supported by the Natural Sciences and Engineering Research Council of Canada.

G. He et al., *Approximation algorithms*, JGAA, 4(2) 1–11 (2000)

1   Introduction

Distributed memory architectures are becoming increasingly popular since the promised scalability at reduced costs, and the availability of high performance microprocessors. This architecture requires that data associated with a given computation be partitioned and distributed to the local storage of each individual processor. How this is done will affect the program performance. However, programming of distributed memory multiprocessors is difficult and error-prone due to the lack of a single uniform global address space. The recent research (Li and Chen [6], Banerjee, Eigenmann, Nicolau and Padua [1], Gupta and Banerjee [4], Knoble, Lukas and Steele [5], and Ramanujam and Sadayappan [9]) has concentrated on automating this process. In [6], Li and Chen modeled this process as the (primary) index domain alignment problem and proved that the problem is NP-complete for the class of graphs with alignment dimension 2.

In [7], we have further generalized the index domain alignment problem into the following four graph partitioning problems, and proved that these problems are polynomially equivalent and NP-complete. We follow the standard definitions and notations in [2].

Let G be a graph. An edge weight on G is a map w from $E(G)$ to $Z^+ \cup \{0\}$, the set of nonnegative integers. If H is a subgraph of G, we denote $w(H) = \sum_{e \in E(H)} w(e)$.

Let $G = (V(G), E(G), w)$ be an edge weighted k-vertex colorable graph with a list of vertex color classes $U_1, \ldots, U_k$, where U_i is the set of vertices with color i. Let $c = max\{|U_1|, |U_2|, \cdots, |U_k|\}$. G is called a *c-way k-colored graph* with color classes $U_1, \ldots, U_k$. If $|U_i| = c$ for $1 \leq i \leq k$, we say that G is a *c-way k-colored balanced graph*.

Let $G = (V(G), E(G), w)$ be a c-way k-colored graph with color classes $U_1, \ldots, U_k$. An *orthogonal partition* of G is a vertex partition of $V(G)$ into $V_1, \ldots, V_c$ such that $|V_i \cap U_j| \leq 1$ for any U_i and V_j. That is, each U_i contains vertices of different colors.

The maximum (minimum) orthogonal partition problem

Let $G = (V(G), E(G), w)$ be an edge weighted, c-way k-colored graph with color classes $U_1, \ldots, U_k$. Find an orthogonal partition $V_1, \ldots, V_c$ of G so that

$$\sum_{i \neq j}^{c} \sum_{e \in E([V_i, V_j])} w(e)$$

is the maximum (minimum) among all the orthogonal partitions of G.

We call such a partition $V_1, \ldots, V_c$ an *optimum partition* of G. Let S be the spanning subgraph of G such that e is an edge of S if and only if e is an edge joining V_i and V_j for some $i \neq j$. S is called a maximum (minimum) *orthogonal subgraph* of G for the maximum (minimum) orthogonal partition problem of G.

Note that a k-vertex colorable graph is a k-partite graph. We have the following variation of the above problems, which are easier to deal with.

G. He et al., *Approximation algorithms*, JGAA, 4(2) 1–11 (2000) 3

The maximum (minimum) disjoint k-clique problem

Let $K_{c,\ldots,c}$ be an edge weighted complete k-partite graph with c vertices in each part. Find a set of c vertex-disjoint k-cliques of $K_{c,\ldots,c}$ that has the maximum (minimum) weight.

We would like to point out that both the maximum and minimum vertex disjoint $k(\geq 2)$-clique problems have applications in logistics (see [3]) and manufacturing systems in industry (see [10]). The logistics application arises in the context of making weekly assignments of sets of drivers to loads. Often we are left with a driver assignment problem that is an instance of c vertex disjoint 3-cliques: we are given an edge weighted complete 3-partite graph with a vertex set V_1 of drivers, a vertex set V_2 of beginning-of-the-week loads and a vertex set V_3 of midweek loads. We are looking for a set of vertex-disjoint triangles, say $\{\Delta_{uvw}\}$ so we can maximize the total revenue or minimize the total cost. The manufacturing system application occurs in the selection of quality tools and assignment of the tools and quality operations to machine centers or inspection stations. This case can also be modeled as an instance of c vertex disjoint 3 (or higher)-cliques.

Since the maximum (minimum) orthogonal partition problem and the maximum (minimum) clique problem are all NP-complete, it is worthwhile to find approximation algorithms for these problems, and this is the goal of our paper.

In Section 2, we extend the optimum matching algorithms to these problems to obtain approximation algorithms, and analyze these algorithms. Section 3 provides experimental results and comparisons with other algorithms, for example, the random search algorithm. Section 4 shows that in some sense these approximations are tight. Section 5 gives concluding remarks.

2 Approximation algorithms

First, we define a useful graph *merge* operation.

Let $G = (V(G), E(G), w)$ be a c-way k-colored graph with color classes $U_1, \ldots, U_k$. Consider the induced subgraph $G_{i,j} = G[U_i \cup U_j]$. Let $F_{i,j} = \{u_1 v_1, u_2 v_2, \ldots, u_l v_l\}$ be a matching of $G_{i,j}$ where $u_k \in U_i$ and $v_k \in U_j$. Merge U_j with U_i by identifying the vertex v_k with u_k, also identifying an unsaturated vertex v under $F_{i,j}$ in U_i with an unsaturated vertex u under $F_{i,j}$ in U_j in an obvious way, and leaving the vertices which are not $F_{i,j}$-saturated and cannot be paired off unchanged. The new vertex set is denoted by $U_{i(j)}$. Remove the loops and edges within $U_{i(j)}$, and replace multiple edges by a single edge with the sum of weights of the multiple edges as the new weight of the edge. The new weighted graph $(G_{i(j)}, w')$ is called the *merge of U_j to U_i from G along $F_{i,j}$*. We note that $(G_{i(j)}, w')$ is an edge weighted $(k-1)$-colored graph.

Next, we extend G to a complete k-partite balanced graph $C(G)$ with c vertices in each part by adding in some new vertices and edges. Let w' be the extension of w to $C(G)$ such that $w'(e) = 0$ if e is a new edge. Note that

G. He et al., *Approximation algorithms*, JGAA, 4(2) 1–11 (2000)   4

$C(G) = G$ if G is a complete k-colored balanced graph with c vertices in each part.

Now we consider some approximate solutions based on the idea of optimum matching algorithms.

Algorithm A1. (For the maximum orthogonal partition and the minimum k-clique problems)

Begin
1 Input G with color classes $U_1, \ldots, U_k$.
2 If $C(G) \neq G$, construct $C(G)$ with columns $U'_1, \ldots U'_k$;
   label the new vertices with symbol $'$;
   assign the new edges weight 0 to obtain the weight function w'
3 Set $i = k$, and $G^i = C(G)$.
4 Repeat
   Construct $G^i_{1,i} = [U'_1, U'_i]$.
   Find a minimum weighted 1-factor F_i of $G^i_{1,i}$.
   Relabel the vertices in U'_i so that $F_i = \{v_1 u^i_1, v_2 u^i_2, \ldots, v_c u^i_c\}$.
   Construct a merge $G^i_{1(i)}$ from G^i along with F_i.
   Set i to $i - 1$, $G^i = G^i_{1(i)}$.
   Until $i = 2$.
5 Find a minimum weighted 1-factor F_2 of G^2.
6 Set $V'_1 = \{v_1, u^2_1, u^3_1, \ldots, u^k_1\}, \ldots, V'_c = \{v_c, u^2_c, u^3_c, \ldots, u^k_c\}$.
7 Set $V_i = V'_i - \{$ vertices with label $' \}$.
8 Output $K = G[V_1] \cup \cdots \cup G[V_c]$ and $X = G \backslash E(K)$.
End.

Theorem 1 *Let G be an edge weighted complete k-colored graph with c vertices in each part. Algorithm A1 computes in $O(kc^3)$ time an approximation solution K for the minimum disjoint k-clique problem satisfying*

$$w(K) \leq \frac{1}{c} w(G).$$

If G is an edge weighted c-way k-colored graph, then the subgraph X constructed by Algorithm A1 is an orthogonal subgraph for the maximum partition problem with $w(X) \geq \frac{c-1}{c} w(G)$.

Proof: For the clique problem, $G = C(G)$ and $U_i = U'_i$. We prove the theorem by induction on k. If $k = 2$, G is a complete bipartite graph $K_{c,c}$, $E(G)$ has a decomposition into c edge-disjoint 1-factors. Therefore, Algorithm A1 produces an optimum 1-factor with the minimum weight. Being an optimum one, we must have $w(X) \leq \frac{1}{n} w(G)$. The theorem is true for $k = 2$.

Suppose that the theorem is true for any weighted complete $(k-1)$-partite graphs with c vertices in each part. Now let $k > 2$. We are going to prove that the theorem is true for k.

Let G be a weighted complete k-partite graph with n vertices in each part. In Algorithm A1, first at step 4, we consider the subgraph $G_{1,k} = [U_1, U_k]$.

G. He et al., *Approximation algorithms*, JGAA, 4(2) 1–11 (2000)   5

Then Algorithm A1 delivers a minimum weighted 1-factor F_k of $G_{1,k}$. We have that $w(F_k) \leq \frac{1}{c} w(G_{1,k})$. Now we let G' be a merge graph $(G_{1(k)}, w')$ from G along the 1-factor F_m. Let K' be the result obtained by applying Algorithm A1 to G'. Then $w'(K') \leq \frac{1}{c} w'(G')$ by the induction hypothesis. We note that K' consists of c disjoint $(k-1)$-cliques. At step 6, we construct K^* from K' as follows: if a clique C in K' contains the vertex u_i, then we include v_i to C to obtain a k-clique. It is clear that K^* is a feasible solution to G for the minimum clique problem. Then we have

$$
\begin{aligned}
w(K) = w(K^*) \;&=\; w'(K') + w(F_k) \\
&\leq\; \frac{1}{c} w'(G') + w(F_k) \\
&=\; \frac{1}{c}\Big(\sum_{e \notin [U_1, U_k]} w(e) \Big) + w(F_k) \\
&\leq\; \frac{1}{c}\Big(\sum_{e \notin [U_1, U_k]} w(e) \Big) + \frac{1}{c}\Big(\sum_{e \in [U_1, U_k]} w(e) \Big) \\
&=\; \frac{1}{c} w(G).
\end{aligned}
$$

For the second part of the theorem, we have that $w(X) = w'(C(G)) - w(K) = w(G) - w(K) \geq w(G) - \frac{1}{c} w(G) = \frac{c-1}{c} w(G)$.

As for the running time, we see that the main operation in Algorithm A1 is finding a minimum weighted 1-factor which takes $O(c^3)$ steps (see [2] and [8]). There are $k-1$ such operations, and other operations can be performed in $O(c^3)$ steps, hence the above algorithm has time complexity $O(kc^3)$.

This completes the proof. ∎

Algorithm A2. (For the minimum orthogonal partition
                and the maximum clique problems)

Begin
1 Input G with color classes $U_1, \ldots, U_k$, where $|U_1| = c$.
2 Set $i = k$, and $G^i = G$.
3 Repeat until $i = 2$.
     Construct $G^i_{1,i} = [U_1, U_i]$.
     Find a maximum weighted 1-factor F_i of $G^i_{1,i}$.
     Relabel the vertices in U_i so that $F_i = \{v_1 u_1^i, \ldots, v_l u_l^i\}$.
     Construct a merge $G^i_{1(i)}$ from G^i along with F_i.
     Set i to $i-1$, $G^i = G^i_{1(i)}$.
4 Find a maximum weighted 1-factor F_2 of G^2.
5 Let $V_1 = \{v_1, u_1^2, u_1^3, \ldots, u_1^{k_1}\}, \ldots, V_c = \{v_c, u_c^2, u_c^3, \ldots, u_c^{k_c}\}$.
     Output $K = G[V_1] \cup \cdots \cup G[V_c]$ and $X = G - E(K)$.
End.

Similarly, we can prove the following:

G. He et al., *Approximation algorithms*, JGAA, 4(2) 1–11 (2000)   6

Theorem 2 *Let G be an edge weighted complete k-partite graph with c vertices in each part. Algorithm A2 computes in $O(kc^3)$ time an approximation solution K for the maximum disjoint k-clique problem satisfying*

$$w(K) \geq \frac{1}{c}w(G).$$

If G is a c-way k-colored edge weighted graph, then the subgraph X constructed by Algorithm A2 is an orthogonal solution for the minimum orthogonal partition problem with $w(X) \leq \frac{c-1}{c}w(G)$.

We say that an algorithm A is a δ-*approximation algorithm* for a problem P if there is a number δ such that for every instance I of P, the approximate solution $S_A(I)$ given by A is related to the exact solution on $S(I)$ by

$$\left|\frac{S_A(I) - S(I)}{S(I)}\right| \leq \delta.$$

It is desirable to design approximation algorithm with a small δ.

Theorem 3 *Algorithm A1 is a $\frac{1}{c}$-approximation algorithm for the maximum orthogonal partition problem.*

Proof: Given a c-way k-colored edge weighted graph G, let X^* be an optimum solution and X be a solution obtained by Algorithm A1 for the maximum orthogonal partition problem. Then $w(X^*) \geq w(X) \geq \frac{c-1}{c}w(G)$. Hence $\frac{w(X)}{w(X^*)} \geq \frac{w(X)}{w(G)} \geq \frac{c-1}{c}$, and $0 \leq 1 - \frac{w(X)}{w(X^*)} \leq 1 - \frac{c-1}{c} = \frac{1}{c}$. ■

Remark 4 *We note that Algorithm A1 is significant when c is large.*

Similarly, we can prove the following.

Theorem 5 *Algorithm A2 is a $\frac{c-1}{c}$-approximation algorithm for the maximum k-clique problem.*

Remark 6 *We point out that if G is a complete k-partite graph with c vertices in each part and with constant weight for each edge, then both of our algorithms produce optimum solutions. One natural question is, for which input graphs, will Algorithms A1 and A2 generate optimum solutions?*

To investigate the above question, let G be an edge weighted c-way k-colored graph with color classes $\{U_i : i = 1, \ldots, k\}$. We construct a new graph $T(G)$ from G with vertex set $\{U_i : i = 1, \ldots, k\}$ and U_iU_j is an edge of $T(G)$ if and only if there is an edge in G which joins U_i and U_j. In other words, $T(G)$ is obtained from G by shrinking each of U_i into one vertex and deleting the multiple edges.

We easily have the following observation.

G. He et al., *Approximation algorithms*, JGAA, 4(2) 1–11 (2000)   7

Theorem 7 *Let $G = (V, E, w)$ be a c-way k-colored edge weighted graph with color classes $U_1, \ldots, U_k$. If $T(G)$ is acyclic, then we can obtain optimum solutions for the maximum and minimum orthogonal partition problems using Algorithms A1 and A2.*

Proof: We only prove the case of minimum orthogonal partition. We note that if $T(G)$ is acyclic, then in each merge operation of Algorithm A2, there are no multiple edges. The matching produced in step 4 in Algorithm A2 is actually a maximum matching between U_i and U_{i+1} in G for some i. Let X be a solution produced by Algorithm A2 with the orthogonal partition $V_1, V_2, \ldots, V_c$ and let X^* be an optimum solution to the problem. Then $K = C(G) - E(X)$ is the solution for the minimum clique problem. Let $K^* = C(G) - E(X^*)$. Then $K^* \cap [U_i, U_{i+1}]$ is a matching, hence $w(K^* \cap [U_i, U_{i+1}]) \leq w(K \cap [U_i, U_{i+1}])$. Therefore, $w(K^*) = \sum_i w(K^* \cap [U_i, U_{i+1}]) \leq \sum_i w(K \cap [U_i, U_{i+1}]) = w(K)$. It follows that $w(X) = w(G) - w(K) \leq w(G) - w(K^*) = w(X^*)$. Therefore, $w(X)$ is an optimum solution. ∎

3 Experimental comparison

In this section, we compare Algorithms A1 and A2 with some simple algorithms when k is small. We first describe two simple algorithms for the four partition problems when $k = 3$. The input consists of an edge weighted graph G and a list of color classes U_1, U_2, U_3.

Random Selection Algorithm

This procedure randomly picks three vertices v_1, v_2 and v_3 from U_1, U_2 and U_3 respectively to form a triangle (3-clique). Remove this triangle and repeat this procedure until a set of c vertex-disjoint 3-cliques is found.

The Cubic (3-clique) Greedy Algorithm

This algorithm searches for an optimum (maximum or minimum) weighted triangle containing a starting vertex $v_1 \in U_1$. If v_1, v_2, v_3 are the vertices of this triangle, we remove them from the sets: U_1 , U_2 and U_3, and apply the algorithm again.

Our comparison results are based on 1000 tests of graphs with 30 vertices in each set and with integer weights uniformly in the range from 0 to 9. The tables below give the mean values and the standard deviations of the 1000 approximation solutions obtained by the algorithms.

Method	Random	Greedy	Algorithm A1
Mean	405	218	60
Standard Deviation	26	18	6.6

Comparison results for minimum 3-cliques

G. He et al., *Approximation algorithms*, JGAA, 4(2) 1–11 (2000)　　　8

Method	Random	Greedy	Algorithm A2
Mean	405	736	749
Standard Deviation	26	8.8	6.6

Comparison results for maximum 3-cliques

We see that Cubic Greedy Algorithms are better than the Random select Algorithm and Algorithms A1 and A2 are better than Cubic Greedy Algorithms.

In the following we will compare only Algorithm A2 with the (4-clique) Greedy Algorithm when $k = 4$. The results are also based on 1000 trials of graphs with 100 vertices in each set and with integer weights uniformly in the range from 1 to 100.

Method	Greedy	Algorithm A2
Mean	41806	52801
Standard Deviation	391	173

Comparison results for maximum 4-cliques

To summarize, Algorithms A1 and A2 appear to be the better performers both in terms of speed and quality of the solution. We are interested in studying how often these methods find an optimum solution. By looking at some smaller instances G of the problem (where we could find an optimum solution by exhaustive searching) with $k = 3$, $c = 5$ (k is the number of sets and c is the number of vertices in each set) and uniform weight, the matching method found an optimum solution about 40% of the time and the cubic greedy method found it about 30% of the time.

4　How good are these algorithms?

We consider the following decision problem.

Problem *Let W be any positive integer and G an edge weighted c-way k-colored graph with color classes $U_1, \ldots, U_k$. Is there an orthogonal subgraph X of G such that $w(X) \geq W$ ($w(X) \leq W$)?*

We note that these two problems are actually other versions of the optimum orthogonal partition problems. Therefore, they are NP-complete. We recall that our algorithms A1 (A2) can produce an approximate solution X with $w(X) \geq \frac{c-1}{c}w(G)$ ($w(X) \leq \frac{c-1}{c}w(G)$). The following theorem shows that there is no efficient algorithm for finding an approximation solution X with a better ratio $\frac{w(X)}{w(G)}$.

Theorem 8 *For every real $0 < \epsilon < \frac{1}{c}$, it is NP-complete to decide whether a given edge weighted, c-way, k-colored graph $G = (V, E, w)$ has an orthogonal subgraph X such that $w(X) \geq (\frac{c-1}{c} + \epsilon)w(G)$, (or $w(C) \leq (\frac{c-1}{c} - \epsilon)w(G)$).*

Proof: We will prove the case for maximum orthogonal partition problem only. For each pair of input instance G and W in the decision problem, we are going to

G. He et al., *Approximation algorithms*, JGAA, 4(2) 1–11 (2000)   9

construct, in polynomial time, a c-way k'-colored graph G' with weight function w' such that $w(X) \geq W$ is equivalent to $w'(X') \geq (\frac{c-1}{c} + \epsilon)w'(G')$, where X' is an orthogonal subgraph of G'.

(I) If $W < \lceil (\frac{c-1}{c} + \epsilon)w(G) \rceil$, then add in one pendant edge to G with weight W' to obtain G' (the new vertex forms one color class in G'), where W' satisfies $\lceil (\frac{c-1}{c} + \epsilon)(w(G) + W') \rceil - W' = W$ (the smallest integer W' satisfying $\frac{W+W'}{w(G)+W'} \geq \frac{c-1}{c} + \epsilon$ will do). Let the weight function on $E(G')$ be w'. Then $w'(G') = w(G) + W'$ and $w'(X') = w(X) + W'$. Therefore,

$$w'(X') \geq \lceil (\frac{c-1}{c} + \epsilon)w'(G') \rceil$$

is equivalent to

$$w(X) + W' \geq \lceil (\frac{n-1}{n} + \epsilon)(w(G) + W') \rceil,$$

which is equivalent to

$$w(X) \geq \lceil (\frac{c-1}{c} + \epsilon)(w(G) + W') \rceil - W' = W.$$

(II) Let $W > \lceil (\frac{c-1}{c} + \epsilon)w(G) \rceil$. We note that there exists a c-way, W-colored, edge weighted graph (H, w_1) such that $w_1(X_H) = \frac{c-1}{c}w_1(H)$, where X_H is an orthogonal subgraph with the maximum weight (the complete W-partite graph with c-vertices in each part and with a constant weight for each edge will do). Now we choose such a graph H satisfying (let the weight of $w_1(H)$ be big enough)

$$W + \frac{c-1}{c}w_1(H) \leq \lceil (\frac{c-1}{c} + \epsilon)(w(G) + w_1(H)) \rceil.$$

Two cases arise:

Case (a). If $W = \lceil (\frac{c-1}{c} + \epsilon)(w(G) + w_1(H)) \rceil - \frac{c-1}{c}w_1(H)$, then let G' be the disjoint union of G and H. The weight function w' of G' is defined as follows: $w'(e) = w(e)$ if $e \in E(G)$ and $w'(e) = w_1(e)$ if $e \in E(H)$. It is easy to see that $w'(G') = w(G) + w_1(H)$, $w'(X') = w(X) + \frac{c-1}{c}w_1(H)$. Therefore,

$$w'(X') \geq \lceil (\frac{c-1}{c} + \epsilon)w'(G') \rceil,$$

which is equivalent to

$$w(C) + \frac{c-1}{c}w_1(H) \geq \lceil (\frac{c-1}{c} + \epsilon)(w(G) + w_1(H)) \rceil,$$

which is equivalent to

$$w(C) \geq \lceil (\frac{c-1}{c} + \epsilon)(w(G) + w_1(H)) \rceil - \frac{c-1}{c}w_1(H) = W.$$

G. He et al., *Approximation algorithms*, JGAA, 4(2) 1–11 (2000)   10

Case (b). If Case (a) does not hold, then $W_H = W + \frac{c-1}{c}w(H) < \lceil(\frac{c-1}{c} + \epsilon)(w(G) + w_1(H))\rceil$. Let G' and w' be defined as in Case (a). Since $w'(G') = w(G) + w_1(H)$, $w'(X') = w(X) + \frac{c-1}{c}w_1(H)$, it follows that

$$W_H < \lceil(\frac{c-1}{c} + \epsilon)(w'(G'))\rceil.$$

We modify G' to obtain G'' by the method used in (I) [that is, we add in one pendant edge with weight W'_H to G' (the weight W'_H is determined by the method in (I)), the resulting graph is denoted by G'' and its weight function is w'']. Let X'' be an orthogonal subgraph of G''. Then

$$w''(X'') \geq \lceil(\frac{c-1}{c} + \epsilon)w''(G'')\rceil$$

is equivalent to

$$w'(X') + W'_H \geq \lceil(\frac{c-1}{c} + \epsilon)(w'(G') + W'_H)\rceil,$$

which is equivalent to

$$w'(X') \geq \lceil(\frac{c-1}{c} + \epsilon)(w'(G') + W'_H)\rceil - W'_H = W_H,$$

which is equivalent to

$$w(X) + \frac{c-1}{c}w_1(H) \geq W + \frac{c-1}{c}w_1(H),$$

which is equivalent to
$$w(X) \geq W.$$

In any case, we have constructed, in polynomial time, a c-way k'-colored graph G' with the weight function w' such that $w(X) \geq W$ is equivalent to $w'(X') \geq (\frac{c-1}{c} + \epsilon)w'(G')$, where X' is an orthogonal subgraph of G'. Therefore, it is NP-complete to decide whether or not $w(X) \geq (\frac{c-1}{c} + \epsilon)w(G)$. This completes the proof. ∎

5   Conclusion

We have provided efficient and deterministic algorithms for four graph partition problems. Our algorithms produce good approximate solutions, both in theory, as shown by our mathematical analysis, and in practice, as confirmed by our experimental results. We have also shown that there is no efficient algorithm is likely to yield a better ratio of the weight of the approximation solution over the weight of the input graph.

G. He et al., *Approximation algorithms*, JGAA, 4(2) 1–11 (2000)   11

Acknowledgments

We thank the anonymous referees for their valuable comments.

References

[1] U. Banerjee, R. Eigenmann, A. Nicolau and D. A. Padua, Automatic Program Parallelization, Proceedings of the IEEE, Vol. 81, No. 2, 1993, pp. 211–243.

[2] J. A. Bondy and U. S. R. Murty, Graph Theory with Applications, Elsevier, New York, 1976.

[3] J. Exoo, J. Jiang and C. Zhao, An Application of Multi-Dimensional Matching to Logistics, Utilitas Mathematica 57(2000), pp. 227–235.

[4] M. Gupta and P. Banerjee, Demonstration of Automatic Data Partitioning Techniques for Parallelizing Compilers on Multicomputers, IEEE Transactions on Parallel and Distributed Systems, No. 2, 3 (1992), pp. 179–193.

[5] K. Knoble, J. Lukas and G. L. Steele, Data Optimization, Allocation of Arrays to Reduce Communication in SIMD Machines, Journal of Parallel and Distributed Computing, 8 (1990), pp., 102–118.

[6] J. Li and M. Chen, Index Domain Alignment: Minimizing Cost of Cross-reference between Distributed Arrays, in Proceedings of the third Symposium on Frontiers of Massively Parallel Computation, College Park, October, 1990, pp. 424–433.

[7] G. He, Jiping Liu and Cheng Zhao, Extremal n-colorable Subgraphs with Index Domain Alignment Restrictions, Submitted for publication.

[8] L. Lovász and M. Plummer, Matching Theory, Annals of Discrete Mathematics 29, North-Holland, 1986.

[9] J. Ramanujam and P. Sadayappan, Compile-Time Techniques for Data Distribution Memory Machines, IEEE Transactions on Parallel and Distributed Systems, No.4, 2 (1991), pp. 472–483.

[10] M. Zhou and C. Zhao, Quality Planning in Manufacturing Systems, In the Proceedings, 8th Annual Industrial Engineering Research Conference (IERC), May 23–26, 1999, Pheonix, AZ.

Volume 4:3 (2000)

Journal of Graph Algorithms and Applications

http://www.cs.brown.edu/publications/jgaa/

vol. 4, no. 3, pp. 1–3 (2000)

Advances in Graph Drawing

Special Issue on Selected Papers from the Sixth International Symposium on Graph Drawing, GD'98

Guest Editors' Foreword

Giuseppe Liotta

Dipartimento di Ingegneria Elettronica e dell'Informazione
Università di Perugia
via G. Duranti 93, 06125 Perugia, Italy
http://www.diei.unipg.it/PAG_PERS/liotta/liotta.htm
liotta@diei.unipg.it

Sue H. Whitesides

School of Computer Science
McGill University
Montreal, PQ H3A2A7, Canada
http://www.cs.mcgill.ca/~sue/
sue@cs.mcgill.ca

G.Liotta and S.H. Whitesides, *Guest Editors' Foreword*, JGAA, 4(3) 1–3 (2000)2

Introduction

This Special Issue on the Sixth International Symposium on Graph Drawing (GD'98)[1] brings together eight papers based on work presented at the conference. As editors, we chose to invite papers from GD'98 that would reflect the broad nature of the annual symposia on graph drawing, which showcase theoretical contributions as well as experimental work and visualization systems.

We have also included a contribution based on the GD'98 Graph Drawing Contest. This contest is an annual conference tradition. Long before the conference takes place, the contest committee posts at the conference web site textual descriptions of one or more non-trivial graphs, typically with a set of desired features or requirements for "good" visualizations. Contestants from all over the world compete for cash prizes, awarded at the conference banquet. The contest is known for pushing graph drawing technologies to their limits and for stimulating new research directions.

One of our goals as editors has been to explore the potential of an electronic journal to convey information that cannot be captured easily by a static medium such as paper. Hence we are pleased to include two contributions accompanied by web sites. Since the web sites are maintained by the authors rather than by the journal, their evolution over time is at the authors' discretion. An impression of the original web site, which went through the refereeing process together with its associated paper, can be gotten either by looking at a "snapshot" version maintained by the journal or by looking at the paper itself.

All contributions in this Special Issue have been through a rigorous review process, whether they follow the style of traditional journal papers or whether they describe working systems or a contest entry. We thank the authors, the referees, and the editorial board of the journal for their careful work and for their patience, generosity, and support of our endeavor to explore the potential of electronic journal publication. We hope that we have captured a bit of the dynamic quality that the range of research interests presented at the graph drawing symposia imparts.

Scanning the Issue

Among theoretical contributions, the paper by M. Dillencourt, D. Eppstein, and D.S. Hirschberg introduces and studies the *geometric thickness* of a graph, a notion that lies between the graph-theoretical thickness and the book thickness.

A variant of the well-known binary space partition decomposition defined in the context of computational geometry is described in the paper by C. A. Duncan, M. T. Goodrich, and S. G. Kobourov, for the purpose of drawing huge graphs. They present the *balanced aspect ratio (BAR) tree*, which supports

---

[1]GD'98 was held August 13-15, 1998, at McGill University, Montreal, Canada. Proceedings published in the Springer-Verlag Lecture Notes in Computer Science (LNCS) series, vol. 1547, available on-line at http://link.springer.de/link/service/series/0558/tocs/t1547.htm .

G.Liotta and S.H. Whitesides, *Guest Editors' Foreword*, JGAA, 4(3) 1–3 (2000)3

recursive division of a graph into subgraphs of roughly equal size, such that the drawing of each subgraph has a balanced aspect ratio.

The problem of measuring similarities between different drawings of the same graph is studied in the paper by S. Bridgeman and R. Tamassia, who define a general framework for quantifying how much a change in a drawing affects the user's mental map. This type of study provides basic principles for designing interactive algorithms for drawing graphs when preserving the mental map is a priority.

Two of the papers with a strong experimental component deal with *orthogonal* drawings, i.e. drawings where the vertices are placed at grid points and the edges are chains of horizontal and vertical segments.

The paper by J. M. Six, K. G. Kakoulis, and I. G. Tollis gives a postprocessing technique called *refinement*, whose purpose is to improve the readability of orthogonal drawings in the plane. Their experimental analysis shows significant improvements in such measurable graph drawing aesthetics as area, bends, crossings, and total edge length.

The paper by G. Di Battista, M. Patrignani, and F. Vargiu studies orthogonal drawings in $3D$ space and presents a new approach, called *split&push*, to compute orthogonal drawings of graphs of maximum degree six. A final drawing is produced through a sequence of steps: starting from a degenerate drawing, each step splits the current drawing into two pieces and finds a structure closer to the final version. The experimental analysis compares the resulting algorithm with other existing algorithms, taking into account computational time, volume, average edge length, and average number of bends.

The paper by U. Brandes and D. Wagner is motivated by a very practical application: visualizing timetables of train schedules. They do this with drawings of graphs whose vertices have fixed positions determined by the geographic locations of the railway stations and whose edges join pairs of stations connected by non-stop train service. To avoid visual clutter, some edges are drawn as straight lines, while others are drawn as Bezier curves. Force-directed methods compute the final visualization.

The paper by P. Eades and Mao Lin Huang presents a strategy intended for visualizing and navigating in huge graphs. Using this strategy, a user views an *abridgment* of a graph, that is, a small part of the graph that is currently of interest. By changing the abridgment, the user may travel through the graph. The changes use animation to transform smoothly from one view to the next. The strategy has been implemented in a prototype system called DA-TU, which is accessible through the web.

Finally, the paper by U. Brandes, V. Kääb, A. Löh, D. Wagner, and T. Willhalm, chosen to represent the Graph Drawing Contest, shows an example of dynamic $3D$ straight-line drawing of directed graphs. The contestants were given a dynamic graph of links between World Wide Web pages. The goal was to depict the content as it evolved. The authors use a force directed approach to create an animation of the given graph. The animation is accessible through the web.

Journal of Graph Algorithms and Applications

http://www.cs.brown.edu/publications/jgaa/

vol. 4, no. 3, pp. 5–17 (2000)

Geometric Thickness of Complete Graphs

Michael B. Dillencourt   *David Eppstein*   *Daniel S. Hirschberg*

Information and Computer Science
University of California
Irvine, CA 92697-3425, USA
http://www.ics.uci.edu/~{dillenco,eppstein,dan}
{dillenco,eppstein,dan}@ics.uci.edu

Abstract

We define the *geometric thickness* of a graph to be the smallest number of layers such that we can draw the graph in the plane with straight-line edges and assign each edge to a layer so that no two edges on the same layer cross. The geometric thickness lies between two previously studied quantities, the (graph-theoretical) thickness and the book thickness. We investigate the geometric thickness of the family of complete graphs, $\{K_n\}$. We show that the geometric thickness of K_n lies between $\lceil (n/5.646) + 0.342 \rceil$ and $\lceil n/4 \rceil$, and we give exact values of the geometric thickness of K_n for $n \leq 12$ and $n \in \{15, 16\}$. We also consider the geometric thickness of the family of complete bipartite graphs. In particular, we show that, unlike the case of complete graphs, there are complete bipartite graphs with arbitrarily large numbers of vertices for which the geometric thickness coincides with the standard graph-theoretical thickness.

Communicated by G. Liotta and S. H. Whitesides; submitted November 1998; revised November 1999.

Research supported in part by NSF Grants CDA-9617349, CCR-9703572, CCR-9258355, and matching funds from Xerox Corp. A preliminary version of this paper appeared in the *Sixth Symposium on Graph Drawing, GD '98*, (Montréal, Canada, August 1998), Springer-Verlag Lecture Notes in Computer Science 1547, 102–110.

M. B. Dillencourt et al., *Geometric Thickness*, JGAA, 4(3) 5–17 (2000)   6

1   Introduction

Suppose we wish to display a nonplanar graph on a color terminal in a way that minimizes the apparent complexity to a user viewing the graph. One possible approach would be to use straight-line edges, color each edge, and require that two intersecting edges have distinct colors. A natural question then arises: for a given graph, what is the minimum number of colors required?

Or suppose we wish to print a circuit onto a circuit board, using uninsulated wires, so that if two wires cross, they must be on different layers, and that we wish to minimize the number of layers required. If we allow each wire to bend arbitrarily, this problem has been studied previously; indeed, it reduces to the graph-theoretical thickness of a graph, defined below. However, suppose that we wish to further reduce the complexity of the layout by restricting the number of bends in each wire. In particular, if we do not allow any bends, then the question becomes: for a given circuit, what is the minimum number of layers required to print the circuit using straight-line wires?

These two problems motivate the subject of this paper, namely the *geometric thickness* of a graph. We define $\bar{\theta}(G)$, the geometric thickness of a graph G, to be the smallest value of k such that we can assign planar point locations to the vertices of G, represent each edge of G as a line segment, and assign each edge to one of k layers so that no two edges on the same layer cross. This corresponds to the notion of "real linear thickness" introduced by Kainen [15]. Graphs with geometric thickness 2 (called "doubly-linear graphs) have been studied by Hutchinson *et al.* [13], where the connection with certain types of visibility graphs was explored.

A notion related to geometrical thickness is that of (graph-theoretical) thickness of a graph, $\theta(G)$, which has been studied extensively [1, 3, 8, 9, 10, 14, 16] and has been defined as the minimum number of planar graphs into which a graph can be decomposed. The key difference between geometric thickness and graph-theoretical thickness is that geometric thickness requires that the vertex placements be consistent at all layers and that straight-line edges be used, whereas graph-theoretical thickness imposes no consistency requirement between layers.

Alternatively, the graph-theoretical thickness can be defined as the minimum number of planar layers required to embed a graph such that the vertex placements agree on all layers but the edges can be arbitrary curves [15]. The equivalence of the two definitions follows from the observation that, given any planar embedding of a graph, the vertex locations can be reassigned arbitrarily in the plane without altering the topology of the planar embedding provided we are allowed to bend the edges at will [15]. This observation is easily verified by induction, moving one vertex at a time.

The (graph-theoretical) thickness is now known for all complete graphs [1,

M. B. Dillencourt et al., *Geometric Thickness*, JGAA, 4(3) 5–17 (2000)   7

2, 4, 17, 19], and is given by the following formula:

$$\theta(K_n) = \begin{cases} 1, & 1 \leq n \leq 4 \\ 2, & 5 \leq n \leq 8 \\ 3, & 9 \leq n \leq 10 \\ \left\lceil \frac{n+2}{6} \right\rceil, & n > 10 \end{cases} \tag{1.1}$$

Another notion related to geometric thickness is the *book thickness* of a graph G, $bt(G)$, defined as follows [5]. A *book with k pages* or a *k-book*, is a line L (called the *spine*) in 3-space together with k distinct half-planes (called *pages*) having L as their common boundary. A *k-book embedding* of G is an embedding of G in a k-book such that each vertex is on the spine, each edge either lies entirely in the spine or is a curve lying in a single page, and no two edges intersect except at their endpoints. The book thickness of G is then the smallest k such that G has a k-book embedding.

It is not hard to see that the book thickness of a graph is equivalent to a restricted version of the geometric thickness where the vertices are required to form the vertices of a convex n-gon. This is essentially Lemma 2.1, page 321 of [5]. It follows that $\theta(G) \leq \overline{\theta}(G) \leq bt(G)$. It is shown in [5] that $bt(K_n) = \lceil n/2 \rceil$.

In this paper, we focus on the geometric thickness of complete graphs. In Section 2 we provide an upper bound, $\overline{\theta}(K_n) \leq \lceil n/4 \rceil$. In Section 3 we provide a lower bound. In particular, we show that $\overline{\theta}(K_n) \geq \left\lceil \frac{3-\sqrt{7}}{2}(n+1) \right\rceil \geq \left\lceil \frac{n+1}{5.646} \right\rceil$. This follows from a more precise expression which gives a slightly better lower bound for certain values of n.

These lower and upper bounds do not match in general. The smallest values for which they do not match are $n \in \{13, 14, 15\}$. For these values of n, the upper bound on $\overline{\theta}(K_n)$ from Section 2 is 4, and the lower bound from Section 3 is 3. In Section 4, we resolve one of these three cases by showing that $\overline{\theta}(K_{15}) = 4$. For $n = 16$ the two bounds match again, but they are distinct for all larger n.

Section 5 briefly addresses the geometric thickness of complete bipartite graphs; we show that

$$\left\lceil \frac{ab}{2a + 2b - 4} \right\rceil \leq \theta(K_{a,b}) \leq \overline{\theta}(K_{a,b}) \leq \left\lceil \frac{\min(a, b)}{2} \right\rceil.$$

When a is much greater than b, the leftmost and rightmost quantities in the above inequality are equal. Hence there are complete bipartite graphs with arbitrarily many vertices for which the standard thickness and geometric thickness coincide. We also show that the bounds on geometric thickness of complete bipartite graphs given above are not tight, by showing that $\overline{\theta}(K_{6,6}) = 2$ and $\overline{\theta}(K_{6,8}) = 3$.

Section 6 contains a table of the lower and upper bounds on $\overline{\theta}(K_n)$ established in this paper for $n \leq 100$ and lists a few open problems.

M. B. Dillencourt et al., *Geometric Thickness*, JGAA, 4(3) 5–17 (2000)   8

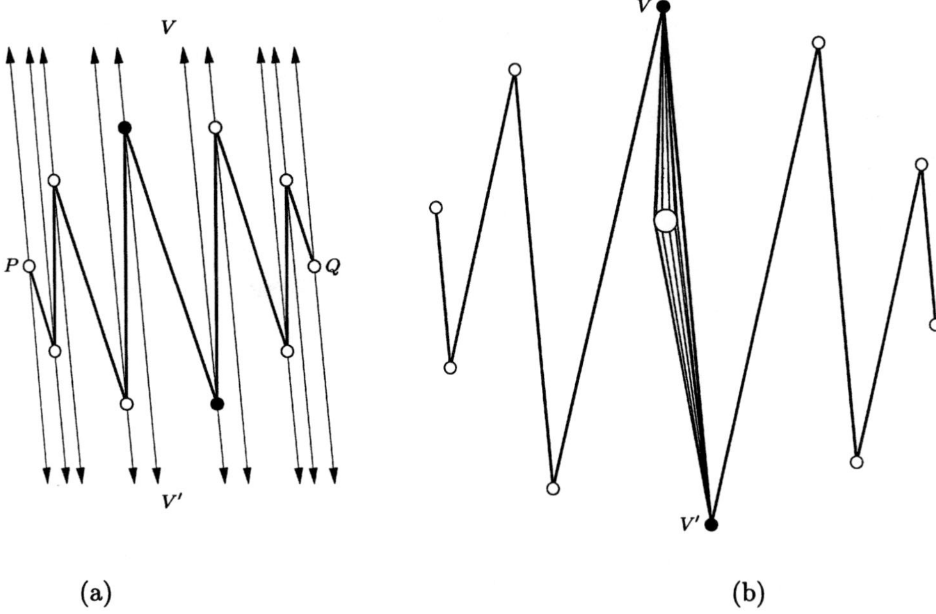

Figure 1: Construction for embedding K_{2k} with geometric thickness of $k/2$, illustrated for $k = 10$. (a) The inner ring. (b) The outer ring. The circle in the center of (b) represents the inner ring shown in (a).

2   Upper Bounds

Theorem 2.1 $\overline{\theta}(K_n) \leq \lceil n/4 \rceil$.

Proof   Assume that n is a multiple of 4, and let $n = 2k$ (so, in particular, k is even). We show that n vertices can be arranged in two *rings* of k vertices each, an *outer ring* and an *inner ring*, so that K_n can be embedded using only $k/2$ layers and with no edges on the same layer crossing.

The vertices of the inner ring are arranged to form a regular k-gon. For each pair of diametrically opposite vertices P and Q, consider the zigzag path as illustrated by the thicker lines in Figure 1(a). This path has exactly one diag-

M. B. Dillencourt et al., *Geometric Thickness*, JGAA, 4(3) 5–17 (2000)   9

onal connecting diametrically opposite points (namely, the diagonal connecting the two dark points in the figure.) Note that the union of these zigzag paths, taken over all $k/2$ pairs of diametrically opposite vertices, contains all $\binom{k}{2}$ edges connecting vertices on the inner ring. Note also that for each choice of diametrically opposite vertices, parallel rays can be drawn through each vertex, in two opposite directions, so that none of the rays crosses any edge of the zigzag path. These rays are also illustrated in Figure 1(a).

By continuity, if the infinite endpoints of a collection of parallel rays (e.g., the family of rays pointing "upwards" in Figure 1(a)) are replaced by a suitably chosen common endpoint (so that the rays become segments), the common endpoint can be chosen so that none of the segments cross any of the edges of the zigzag path. We do this for each collection of parallel rays, thus forming an outer ring of k vertices. This can be done in such a way that the vertices on the outer ring also form a regular k-gon. By further stretching the outer ring if necessary, and by moving the inner ring slightly, the figure can be perturbed so that none of the diagonals of the polygon comprising the outer ring intersect the polygon comprising the inner ring. The outer ring constructed in this fashion is illustrated in Figure 1(b).

Once the $2k$ vertices have been placed as described above, the edges of the complete graph can be decomposed into $k/2$ layers. Each layer consists of:

1. A zigzag path through the outer ring, as shown in Figure 1(b).

2. All edges connecting V and V' to vertices of the inner ring, where V and V' are the (unique) pair of diametrically opposite points joined by an edge in the zigzag path through the outer ring. (These edges are shown as edges connecting the circle with V and V' in Figure 1(b), and as arrows in Figure 1(a)).

3. The zigzag path through the inner ring that does not intersect any of the edges connecting V and V' with inner-ring vertices. (These are the heavier lines in Figure 1(a).)

It is straightforward to verify that this is indeed a decomposition of the edges of K_n into $k/2 = n/4$ layers.   ■

3 Lower Bounds

Theorem 3.1 *For all $n \geq 1$,*

$$\bar{\theta}(K_n) \geq \max_{1 \leq x \leq n/2} \frac{\binom{n}{2} - 2\binom{x}{2} - 3}{3n - 2x - 7}. \tag{3.1}$$

In particular, for $n \geq 12$,

$$\bar{\theta}(K_n) \geq \left\lceil \frac{3 - \sqrt{7}}{2} n + 0.342 \right\rceil \geq \left\lceil \frac{n}{5.646} + 0.342 \right\rceil. \tag{3.2}$$

M. B. Dillencourt et al., *Geometric Thickness*, JGAA, 4(3) 5–17 (2000)   10

Proof   We first prove a slightly less precise bound, namely

$$\overline{\theta}(K_n) \geq \frac{3 - \sqrt{7}}{2}n - O(1).$$
(3.3)

For graph G and vertex set X, let $G[X]$ denote the subgraph of G induced by X. Let S be any planar point set, and let $T_1, \ldots T_k$ be a set of straight-line planar triangulations of S such that every segment connecting two points in S is an edge of at least one of the T_i. Find two parallel lines that cut S into three subsets A, B, and C (with B the middle set), with $|A| = |C| = x$, where x is a value to be chosen later. For any T_i, the subgraph $T_i[A]$ is connected, because any line joining two vertices of A can be retracted onto a path through $T_i[A]$ by moving it away from the line separating A from B. Similarly, $T_i[C]$ is connected, and hence each of the subgraphs $T_i[A]$ and $T_i[C]$ has at least $x - 1$ edges.

By Euler's formula, each T_i has at most $3n - 6$ edges, so the number of edges of T_i not belonging to $T_i[A] \cup T_i[C]$ is at most $3n - 6 - 2(x - 1) = 3n - 2x - 4$. Hence

$$\binom{n}{2} \leq 2\binom{x}{2} + k(3n - 2x - 4).$$
(3.4)

Solving for k, we have

$$k \geq \frac{\binom{n}{2} - 2\binom{x}{2}}{3n - 2x - 4},$$

and hence

$$k \geq \frac{n^2 - 2x^2}{6n - 4x} - O(1).$$
(3.5)

If $x = cn$ for some constant c, then the fraction in (3.5) is of the form $n(1 - 2c^2)/(6 - 4c)$. This is maximized when $c = (3 - \sqrt{7})/2$. Substituting the value $x = (3 - \sqrt{7})n/2$ into (3.5) yields (3.3).

To obtain the sharper conclusion of the theorem, observe that by choosing the direction of the two parallel lines appropriately, we can force at least one point of the convex hull of S to lie in B. Hence, of the edges of T_i that do not belong to $T_i[A] \cup T_i[C]$, at least three are on the convex hull. If we do not count these three edges, then each T_i has at most $3n - 2x - 7$ edges not belonging to $T_i[A] \cup T_i[C]$, and we can strengthen (3.4) to

$$\binom{n}{2} - 3 \leq 2\binom{x}{2} + k(3n - 2x - 7),$$

or

$$k \geq \frac{\binom{n}{2} - 2\binom{x}{2} - 3}{3n - 2x - 7}.$$
(3.6)

Since (3.6) holds for any x, (3.1) follows.

M. B. Dillencourt et al., *Geometric Thickness*, JGAA, 4(3) 5–17 (2000)   11

To prove (3.2), let $f(x)$ be the expression on the right-hand side of (3.6). Consider the inequality $f(x) \geq x_0$, where x_0 is a constant to be specified later. After cross-multiplication, this inequality becomes

$$-x^2 + x + \frac{n^2}{2} - \frac{n}{2} - 3 - (3n - 7 - 2x)x_0 \geq 0. \tag{3.7}$$

The expression in the left-hand side of (3.7) represents an inverted parabola in x. If we let $x = x_0$, we obtain

$$x_0^2 + (8 - 3n)x_0 + \frac{n^2}{2} - \frac{n}{2} - 3 \geq 0, \tag{3.8}$$

and if we let $x = x_0 + 1$ we obtain the same inequality. Now, consider x_0 of the form $An + B - \epsilon$. Choose A and B so that if $\epsilon = 0$, the terms involving n^2 and n vanish in (3.8). This gives the values $A = (3 - \sqrt{7})/2$ and $B = \sqrt{7}(23/14) - 4$. Substituting $x_0 = An + B - \epsilon$ with these values of A and B into (3.8), we obtain

$$\sqrt{7} \cdot \epsilon \cdot n + (\epsilon^2 - \frac{23\epsilon}{\sqrt{7}} - 3/28) \geq 0. \tag{3.9}$$

For $\epsilon = 0.0045$, (3.9) will be true when $n \geq 12$. Therefore, for all $x \in [x_0, x_0 + 1]$, $f(x) \geq x_0$, when $\epsilon = 0.0045$ and $n \geq 12$. In particular, $f(\lceil x_0 \rceil) \geq x_0$. Since k is an integer, (3.2) follows from (3.6). ∎

4   The Geometric Thickness of K_{15}

The lower bounds on geometric thickness provided by equation (3.1) of Theorem 3.1 are asymptotically larger than the lower bounds on graph-theoretical thickness provided by equation (1.1), and they are in fact at least as large for all values of $n \geq 12$. However, they are not tight. In particular, we show that $\bar{\theta}(K_{15}) = 4$, even though (3.1) only gives a lower bound of 3.

Theorem 4.1 $\bar{\theta}(K_{15}) = 4$.

To prove this theorem, we first note that the upper bound, $\bar{\theta}(K_{15}) \leq 4$, follows immediately from Theorem 2.1.

To prove the lower bound, assume that we are given a planar point set S, with $|S| = 15$. We show that there cannot exist a set of three triangulations of S that cover all $\binom{15}{2} = 105$ line segments joining pairs of points in S. We use the following two facts: (1) A planar triangulation with n vertices and b convex hull vertices contains $3n - 3 - b$ edges; and (2) Any planar triangulation of a given point set necessarily contains all convex hull edges. There are several cases, depending on how many points of S lie on the convex hull.

Case 1: 3 points on convex hull. Let the convex hull points be A, B and C. Let A_1 (respectively, B_1, C_1) be the point furthest from edge BC (respectively AC, AB) within triangle ABC. Let A_2 (respectively, B_2, C_2) be the point next furthest from edge BC (respectively AC, AB) within triangle ABC.

M. B. Dillencourt et al., *Geometric Thickness*, JGAA, 4(3) 5–17 (2000)   12

Lemma 4.2 *The edge AA_1 will appear in every triangulation of S.*

Proof   Orient triangle ABC so that edge BC is on the x-axis and point A is above the x-axis. For an edge PQ to intersect AA_1, at least one of P or Q must lie above the line parallel to BC that passes through A_1. But there is only one such point, namely A.   ■

Lemma 4.3 *At least one of the edges A_1A_2 or AA_2 will appear in every triangulation of S.*

Proof   Orient triangle ABC so that edge BC is on the x-axis and point A is above the x-axis. For an edge PQ to intersect A_1A_2 or AA_2, at least one of P or Q must lie above the line parallel to BC that passes through A_2. There are only two such points, A and A_1. Hence an edge intersecting A_1A_2 must necessarily be AX and an edge intersecting AA_2 must necessarily be A_1Y, for some points X and Y outside triangle AA_1A_2. Since edges AX and A_1Y both split triangle AA_1A_2, they intersect, so both edges cannot be present in a triangulation. It follows that either A_1A_2 or AA_2 must be present.   ■

Now let Z be the set of 12 edges consisting of the three convex hull edges and the nine edges pp_1, pp_2, p_1p_2 (where $p \in \{A, B, C\}$). Each triangulation of S contains 39 edges, and since any triangulation contains all three convex hull edges, it follows from Lemmas 4.2 and 4.3 that at least 9 edges of any triangulation must belong to Z. Hence a triangulation contains at most 30 edges not in Z. Thus three triangulations can contain at most $30 \cdot 3 + 12 = 102$ edges, and hence cannot contain all 105 edges joining pairs of points in S.

Case 2: 4 points on convex hull. Let A,B,C,D be the four convex hull vertices. Assume triangle DAB has at least one point of S in its interior (if not, switch A and C). Let A_1 be the point inside triangle DAB furthest from the line DB. By Lemma 4.2, the edge AA_1 must appear in every triangulation of S, as must the 4 convex hull edges. Since any triangulation of S has 38 edges, three triangulations can account for at most $3 \cdot 33 + 5 = 104$ edges.

Case 3: 5 or more points on convex hull. Let h be the number of points on the convex hull. A triangulation of S will have $42 - h$ edges, and all h hull edges must be in each triangulation. So the total number of edges in three triangulations is at most $3(42 - 2h) + h = 126 - 5h$, which is at most 101 for $h \geq 5$.

This completes the proof of Theorem 4.1.

5   Geometric Thickness of Complete Bipartite Graphs

In this section we consider the geometric thickness of complete bipartite graphs, $K_{a,b}$. We first give an upper bound, (Theorem 5.1); it is convenient to state this bound in conjunction with the obvious lower bound on standard thickness

M. B. Dillencourt et al., *Geometric Thickness*, JGAA, 4(3) 5–17 (2000)   13

that follows from Euler's formula. It follows from this theorem that, for any b, $\overline{\theta}(K_{a,b}) = \theta(K_{a,b})$ provided a is sufficiently large (Corollary 5.2). Hence, unlike the situation with complete graphs, there are complete bipartite graphs with arbitrarily many vertices for which the standard thickness and geometric thickness coincide. We show that the lower bound in Theorem 5.1 is not a tight bound for geometric thickness by showing that $\overline{\theta}(K_{6,8}) = 3$. A pair of planar drawings demonstrating that $\theta(K_{6,8}) = 2$ can be found in [16]. Finally we show that the upper bound in Theorem 5.1 is also not tight, since $\overline{\theta}(K_{6,6}) = 2$ while Theorem 5.1 only implies that $\overline{\theta}(K_{6,6}) \leq 3$.

Theorem 5.1 *For the complete bipartite graph $K_{a,b}$,*

$$\left\lceil \frac{ab}{2a + 2b - 4} \right\rceil \leq \theta(K_{a,b}) \leq \overline{\theta}(K_{a,b}) \leq \left\lceil \frac{\min(a, b)}{2} \right\rceil. \tag{5.1}$$

Proof   The first inequality follows from Euler's formula, since a planar bipartite graph with $a + b$ vertices can have at most $2a + 2b - 4$ edges. To establish the final inequality, assume that $a \leq b$ and a is even. Draw b blue vertices in a horizontal line, with $a/2$ red vertices above the line and $a/2$ red vertices below. Each layer consists of all edges connecting the blue vertices with one red vertex from above the line and one red vertex from below. ∎

Corollary 5.2 *For any integer b, $\overline{\theta}(K_{a,b}) = \theta(K_{a,b})$ provided*

$$a > \begin{cases} \frac{(b-2)^2}{2}, & \text{if } b \text{ is even} \\ (b-1)(b-2), & \text{if } b \text{ is odd} \end{cases} \tag{5.2}$$

Proof   If $a > b$, the leftmost and rightmost quantities in (5.1) will be equal provided $ab/(2a + 2b - 4) > (b - 2)/2$ if b is even, or provided $ab/(2a + 2b - 4) > (b - 1)/2$ if b is odd. By clearing fractions and simplifying, we see that this happens when (5.2) holds. ∎

Theorem 5.3 $\overline{\theta}(K_{6,8}) = 3$.

Proof   It follows from the second inequality in Theorem 5.1 that $\overline{\theta}(K_{6,8}) \leq 3$, so we need only show that $\overline{\theta}(K_{6,8}) > 2$. Suppose that we did have an embedding of $K_{6,8}$ with geometric thickness 2, with underlying points set S. Since $K_{6,8}$ has 14 vertices and 48 edges, and since Euler's formula implies that a planar bipartite graph with 14 vertices has at most 24 edges, it follows that each layer has exactly 24 edges and that each face of each layer is a quadrilateral.

Two-color the points of S according to the bipartition of $K_{6,8}$. We claim that there must be at least one red vertex and one blue vertex on the convex hull of S. Suppose, to the contrary, that all convex hull vertices are the same color (say red). Then because each layer is bipartite and because the convex hull contains at least three vertices, the outer face in either layer would consist of at least 6

M. B. Dillencourt et al., *Geometric Thickness*, JGAA, 4(3) 5–17 (2000)   14

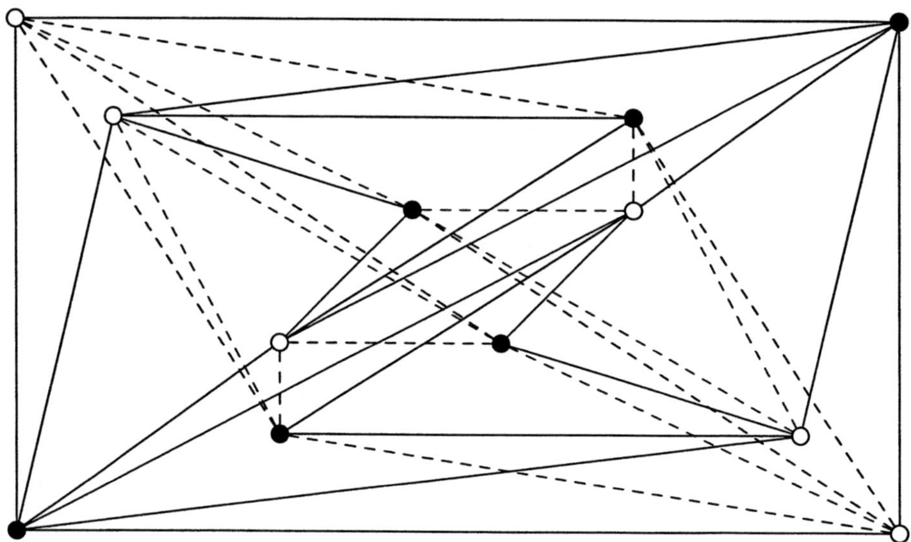

Figure 2: A drawing showing that $\overline{\theta}(K_{6,6}) = 2$. The solid lines represent one layer, the dashed lines the other.

vertices (namely the convex hull vertices and three intermediate blue vertices), which is impossible because each face is bounded by a quadrilateral. The claim implies that one of the layers (say the first) must contain a convex hull edge. But then this edge could be added to the second layer without destroying either planarity or bipartiteness. Since the second layer already has 14 vertices and 24 edges, this is impossible.   ∎

Figure 2 establishes the final claim of the introduction to this section, namely that $\overline{\theta}(K_{6,6}) = 2$.

6   Final Remarks

In this paper we have defined the geometric thickness, $\overline{\theta}$, of a graph, a measure of approximate planarity that we believe is a natural notion. We have established upper bounds and lower bounds on the geometric thickness of complete graphs. Table 1 contains the upper and lower bounds on $\overline{\theta}(K_n)$ for $n \leq 100$.

Many open questions remain about geometric thickness. Here we mention several.

1. Find exact values for $\overline{\theta}(K_n)$ (i.e., remove the gap between upper and lower bounds in Table 1). In particular, what are the values for K_{13} and K_{14}?

M. B. Dillencourt et al., *Geometric Thickness*, JGAA, 4(3) 5–17 (2000)   15

Table 1: Upper and lower bounds on $\bar{\theta}(K_n)$ established in this paper.

n	LB	UB
1- 4	1	1
5- 8	2	2
9-12	3	3
13-14	3	4
15-16	4	4
17-20	4	5
21-24	5	6
25-26	5	7
27-28	6	7
29-31	6	8
32	7	8
33-36	7	9
37	7	10

n	LB	UB
38-40	8	10
41-43	8	11
44	9	11
45-48	9	12
49-52	10	13
53-54	10	14
55-56	11	14
57-60	11	15
61-64	12	16
65	12	17
66-68	13	17
69-71	13	18
72	14	18

n	LB	UB
73-76	14	19
77	14	20
78-80	15	20
81-82	15	21
83-84	16	21
85-88	16	22
89-92	17	23
93-94	17	24
95-96	18	24
97-99	18	25
100	19	25

Note: Upper bounds are from Theorem 2.1. The lower bounds for $n \geq 12$ are from Theorem 3.1, with the exception of the lower bound for $n = 15$ which is from Theorem 4.1. Lower bounds for $n < 12$ are from (1.1).

2. What is the smallest graph G for which $\bar{\theta}(G) > \theta(G)$? We note that the existence of a graph G such that $\bar{\theta}(G) > \theta(G)$ (e.g., K_{15}) establishes Conjecture 2.4 of [15].

3. Is it true that $\bar{\theta}(G) = O\left(\theta(G)\right)$ for all graphs G? It follows from Theorem 2.1 that this is true for complete graphs. For the crossing number [11, 18], which like the thickness is a measure of how far a graph is from being planar, the analogous question is known to have a negative answer.

Bienstock and Dean [7] have described families of graphs which have crossing number 4 but arbitrarily high rectilinear crossing number (where the rectilinear crossing number is the crossing number restricted to drawings in which all edges are line segments).

4. What is the complexity of computing $\overline{\theta}(G)$ for a given graph G? Computing $\theta(G)$ is known to be NP-complete [16], and it certainly seems plausible to conjecture that the same holds for computing $\overline{\theta}(G)$. Since the proof in [16] relies heavily on the fact that $\theta(K_{6,8}) = 2$, Theorem 5.3 of this paper shows that this proof cannot be immediately adapted to geometric thickness. Bienstock [6] has shown that it is NP-complete to compute the rectilinear crossing number of a graph, and that it is NP-hard to determine whether the rectilinear crossing number of a given graph equals the crossing number.

References

[1] V. B. Alekseev and V. S. Gončakov. The thickness of an arbitrary complete graph. *Math USSR Sbornik*, 30(2):187–202, 1976.

[2] L. W. Beineke. The decomposition of complete graphs into planar subgraphs. In F. Harary, editor, *Graph Theory and Theoretical Physics*, chapter 4, pages 139–153. Academic Press, London, UK, 1967.

[3] L. W. Beineke. Biplanar graphs: a survey. *Computers & Mathematics with Applications*, 34(11):1–8, December 1997.

[4] L. W. Beineke and F. Harary. The thickness of the complete graph. *Canadian Journal of Mathematics*, 17:850–859, 1965.

[5] F. Bernhart and P. C. Kainen. The book thickness of a graph. *Journal of Combinatorial Theory Series B*, 27:320–331, 1979.

[6] D. Bienstock. Some provably hard crossing number problems. *Discrete & Computational Geometry*, 6(5):443–459, 1991.

[7] D. Bienstock and N. Dean. Bounds for rectilinear crossing numbers. *Journal of Graph Theory*, 17(3):333–348, 1993.

[8] R. Cimikowski. On heuristics for determining the thickness of a graph. *Information Sciences*, 85:87–98, 1995.

[9] A. M. Dean, J. P. Hutchinson, and E. R. Scheinerman. On the thickness and arboricity of a graph. *Journal of Combinatorial Theory Series B*, 52:147–151, 1991.

M. B. Dillencourt et al., *Geometric Thickness*, JGAA, 4(3) 5–17 (2000)   17

[10] J. H. Halton. On the thickness of graphs of given degree. *Information Sciences*, 54:219–238, 1991.

[11] F. Harary and A. Hill. On the number of crossings in a complete graph. *Proc. Edinburgh Math Soc.*, 13(2):333–338, 1962/1963.

[12] N. Hartsfield and G. Ringel. *Pearls in Graph Theory*. Academic Press, Boston, MA, 1990.

[13] J. P. Hutchinson, T. Shermer, and A. Vince. On representation of some thickness-two graphs. In F. J. Brandenburg, editor, *Symposium on Graph Drawing (GD '95)*, pages 324–332, Passau, Germany, September 1995. Springer-Verlag Lecture Notes in Computer Science 1027.

[14] B. Jackson and G. Ringel. Plane constructions for graphs, networks, and maps: Measurements of planarity. In G. Hammer and P. D, editors, *Selected Topics in Operations Research and Mathematical Economics: Proceedings of the 8th Symposium on Operations Research*, pages 315–324, Karlsruhe, West Germany, August 1983. Springer-Verlag Lecture Notes in Economics and Mathematical Systems 226.

[15] P. C. Kainen. Thickness and coarseness of graphs. *Abhandlungen aus dem Mathematischen Seminar der Universität Hamburg*, 39:88–95, 1973.

[16] A. Mansfield. Determining the thickness of a graph is NP-hard. *Mathematical Proceedings of the Cambridge Philosophical Society*, 93(9):9–23, 1983.

[17] J. Mayer. Decomposition de K_{16} en trois graphes planaires. *Journal of Combinatorial Theory Series B*, 13:71, 1972.

[18] J. Pach and G. Tóth. Which crossing number is it, anyway? In *Proceedings of the 39th Annual IEEE Symposium on the Foundations of Computer Science*, pages 617–626, Palo Alto, CA, November 1998.

[19] J. Vasak. The thickness of the complete graph having $6m + 4$ points. Manuscript. Cited in [12, 14].

Journal of Graph Algorithms and Applications

http://www.cs.brown.edu/publications/jgaa/

vol. 4, no. 3, pp. 19–46 (2000)

Balanced Aspect Ratio Trees and Their Use for Drawing Large Graphs

Christian A. Duncan

Max-Planck-Institut für Informatik
Saarbrücken, Germany
http://www.mpi-sb.mpg.de/~ duncan
christian.duncan@mpi-sb.mpg.de

Michael T. Goodrich   *Stephen G. Kobourov*

Center for Geometric Computing
The Johns Hopkins University
Baltimore, MD 21218
http://www.cs.jhu.edu/labs/cgc/
goodrich@cs.jhu.edu   kobourov@cs.jhu.edu

Abstract

We describe a new approach for cluster-based drawing of large graphs, which obtains clusters by using binary space partition (BSP) trees. We also introduce a novel BSP-type decomposition, called the *balanced aspect ratio* (BAR) tree, which guarantees that the cells produced are convex and have bounded aspect ratios. In addition, the tree depth is $O(\log n)$, and its construction takes $O(n \log n)$ time, where n is the number of points. We show that the BAR tree can be used to recursively divide a graph embedded in the plane into subgraphs of roughly equal size, such that the drawing of each subgraph has a *balanced aspect ratio*. As a result, we obtain a representation of a graph as a collection of $O(\log n)$ layers, where each succeeding layer represents the graph in an increasing level of detail. The overall running time of the algorithm is $O(n \log n + m + D_0(G))$, where n and m are the number of vertices and edges of the graph G, and $D_0(G)$ is the time it takes to obtain an initial embedding of G in the plane. In particular, if the graph is planar each layer is a graph drawn with straight lines and without crossings on the $n \times n$ grid and the running time reduces to $O(n \log n)$.

Communicated by G. Liotta and S. H. Whitesides: submitted November 1998; revised November 1999.

Research supported in part by ARO grant DAAH04–96–1–0013 and NSF grant CCR-9732300.

Duncan, Goodrich, and Kobourov, *BAR Trees*, JGAA, 4(3) 19–46 (2000)   20

1 Introduction

In the past decade hundreds of graph drawing algorithms have been developed (e.g., see [7, 8]), and research in methods for visually representing graphical information is now a thriving area with several different emphases. One general emphasis in graph drawing research is directed at algorithms that display an entire graph, with each vertex and edge explicitly depicted. Such drawings have the advantage of showing the global structure of the graph. A disadvantage, however, is that they can be cluttered for drawings of large graphs, where details are typically hard to discern. For example, such drawings are inappropriate for display on a computer screen any time the number of vertices is more than the number of pixels on the screen. For this reason, there is a growing emphasis in graph drawing research on algorithms that do not draw an entire graph, but instead partially draw a graph, either by showing high-level structures and allowing users to "zoom in" on areas of interest, or by showing substructures of the graph and allowing users to "scroll" from one area of the graph to another. Such approaches are well suited for displaying large graphs, such as significant portions of the world wide web graph, where every web page is a vertex and every hyper-link is an edge.

A common technique used for scrolling viewpoints is the *fish-eye view* [16, 18, 27], which shows an area of interest quite large and detailed (such as nodes representing a user's web pages) and shows other areas successively smaller and in less detail (such as nodes representing a user's department and organization web pages). Fish-eye views allow a user to understand the structure of a graph near a specific set of nodes, but they often do not display global structures.

An alternate technique displays the global structure present in a graph by clustering smaller subgraphs and drawing these subgraphs as single nodes or filled-in regions. By grouping vertices together into *clusters*, we can recursively divide a given graph into layers of increasing detail. These layers can then be viewed in a top-down fashion or even in fish-eye view by following a single path in a cluster-based recursion tree. If clusters of a graph are given as input along with the graph itself, then several authors give various algorithms for displaying these clusters in two or three dimensions [10, 11, 13, 14, 24, 31]. If, as will often be the case, clusters of a graph are not given *a priori*, then various heuristics can be applied for finding clusters using properties such as connectivity, cluster size, geometric proximity, or statistical variation [1, 17, 23, 25]. Once a clustering has been determined, we can generate the layers in a hierarchical drawing of the graph, with the layer depth (i.e., number of layers) being determined by the depth of the recursive clustering hierarchy. This approach allows the graph to be represented by a sequence of drawings of increasing detail. As illustrated by Eades and Feng [10], this hierarchical approach to drawing large graphs can be very effective. Thus, our interest in this paper is to further the study of methods for producing good graph clusterings that can be used for graph drawing purposes.

We feel that a good clustering algorithm and its associated drawing method should come as close as possible to achieving the following goals:

Duncan, Goodrich, and Kobourov, *BAR Trees*, JGAA, 4(3) 19–46 (2000)   21

1. *Balanced clustering*: in each level of the hierarchy the size of the clusters should be about the same.

2. *Small cluster depth*: there should be a small number of layers in the recursive decomposition.

3. *Convex cluster drawings*: the drawing of each cluster should fit in a simple convex region, which we call the *cluster region* for that subgraph.

4. *Balanced aspect ratio*: cluster regions should not be too "skinny".

5. *Efficiency*: computing the clustering and its associated drawing should not take too long.

In this paper we study how well we can achieve these goals for large graph drawings using clustering. Previous algorithms optimize one or more of the above criteria at the expense of some of the rest. Our goal is to simultaneously satisfy all of them. Our approach relies on creating the clusters using binary space partition (BSP) trees, defined by recursively cutting regions with straight lines.

1.1   BSP Tree Based Clustered Graph Drawing

The main idea behind the use of a BSP tree in $\mathbb{R}^2$ to define clusters is very simple. Given a graph $G = (V, E)$, where $n = |V|$ and $m = |E|$, we can use any existing method to embed it in the plane, provided that method places vertices at distinct points in the plane (e.g., see [7, 20, 32]). For example, if G is planar we can use any existing method for embedding G in the plane such that vertices are at grid points, and edges of the graph are straight lines that do not cross [6, 12, 28, 30, 33]. Once the graph drawing is defined, we build a binary space partition tree on the vertices of this drawing. Each node v in this tree corresponds to a convex region R of the plane, and associated with v is a line that separates R into two regions, each of which are associated with a child of v. Thus, any such BSP tree defined on the points corresponding to vertices of G naturally defines a hierarchical clustering of the nodes of G. Such a clustering could then be used, for example, with an algorithm like that of Eades and Feng [10], who present a technique for drawing a 3-dimensional representation of a clustered graph.

The main problem with using BSP trees to define clusters for a graph drawing algorithm is that previous methods for constructing BSP trees do not give rise to clustered drawings that achieve the design goals listed above. For example, the standard k-d tree and its variants (e.g., see [15, 26]), which use axis-parallel lines to recursively divide the number of points in a region in half, maintain every criteria but the balanced aspect ratio. Likewise, quad-trees and fair-split trees (e.g., see [4, 26]), which always split by a line parallel to a coordinate axis to recursively divide the area of a region more or less in half, maintain balanced aspect ratio but can have a depth that is $\Theta(n)$.

In graph drawing, aesthetics are very important, and while "fat" regions appear rounder, a series of skinny regions can be distracting. But depth is also

Duncan, Goodrich, and Kobourov, *BAR Trees*, JGAA, 4(3) 19–46 (2000)   22

important, for a deep hierarchy of clusterings would be computationally expensive to traverse and would not provide very balanced clusters. The balanced box-decomposition tree of Arya et al. [3, 2] has $O(\log n)$ depth and has regions with good aspect ratio, but it sacrifices convexity by introducing holes into the middle of regions, which makes this data structure less attractive for use in clustering for graph drawing applications. Indeed, to our knowledge, there is no previous BSP-type hierarchical decomposition tree that achieves all of the above design goals.

1.2   The Balanced Aspect Ratio (BAR) Tree

In this paper we present a new type of binary space partition tree that is better suited for the application of defining clusters in a large graph. Our data structure, which we call the *balanced aspect ratio* (BAR) tree, is a BSP-type decomposition tree that has $O(\log n)$ depth and creates convex regions with bounded aspect ratio (also called "fat" regions). In this paper we present the BAR tree in $\mathbb{R}^2$. The generalized BAR tree in $\mathbb{R}^d$ is presented in [9]. The construction of the BAR tree is very similar to that of a k-d tree, except for two important differences:

1. In addition to axis-aligned cuts, the BAR tree allows for one more cut direction: a 45°-angled cut.

2. Rather than insisting that the number of points in a region be cut in half at every level, the BAR tree guarantees that the number of points is cut roughly in half every two levels, which is something that does not seem possible to do with either a k-d tree or a quadtree (or even a hybrid of the two) while guaranteeing regions with bounded aspect ratios.

In short, the BAR tree is an $O(\log n)$-depth BSP-type data structure that creates fat, convex regions. Thus, the BAR tree is "balanced" in two ways: on the one hand, clusters on the same level have roughly the same number of points, and, on the other hand, each cluster region has a bounded aspect ratio.

We show that a BAR tree achieves this combined set of goals by proving the existence of a cut, which we call a *two-cut*. A two-cut might not reduce the point size by any amount but maintains balanced aspect ratio and ensures the existence of a subsequent cut, which we call a *one-cut*, that both maintains good aspect ratio *and* reduces the point size by at least two-thirds. In Section 3, we formally define one- and two-cuts and describe how to construct a BAR tree.

1.3   Our Results for Cluster-Based Graph Drawing

In Section 4, we show how to use the BAR tree in a cluster-based graph drawing algorithm. The Large Graph Drawing (LGD) algorithm runs in $O(n \log n + m + D_0(G))$ time, where n and m are the number of vertices and edges in the graph G and $D_0(G)$ is the time to embed G in the plane. If the graph is planar,

Duncan, Goodrich, and Kobourov, *BAR Trees*, JGAA, 4(3) 19–46 (2000)   23

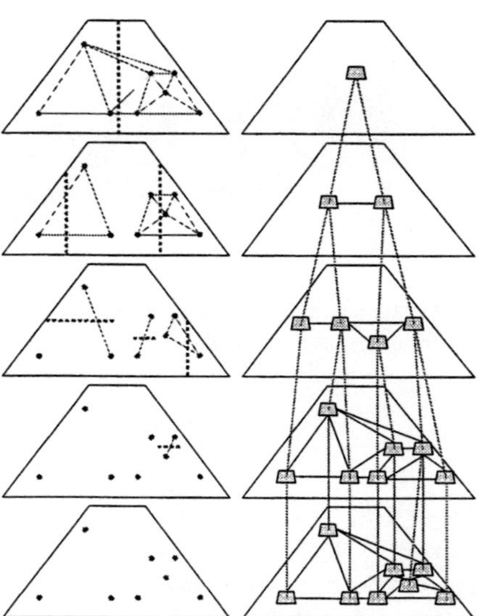

Figure 1: A clustered graph $C = (G, T)$. The underlying graph G is at the lowest level on the right. The clustering of G on the right is obtained from the BSP cuts on the left. Each cluster is represented by a single node. Edges between layers on the right are edges of the tree T.

the algorithm introduces no edge crossings and the running time reduces to $O(n \log n)$.

The algorithm creates a hierarchical cluster representation of a graph, with balanced clusters at each layer and with cluster depth $O(\log n)$. Each cluster region has a *balanced aspect ratio*, guaranteed by the BAR tree data structure. In the actual display of the clustered graph we represent the clusters either by their convex hulls, or by a larger region defined by the BSP tree, or simply by a single node, see Figure 1.

2   Using a BSP Tree for Cluster Drawing

Let $G = (V, E)$ be the graph that we want to draw, where $|V| = n$ and $|E| = m$. Note that graph G is given combinatorially, i.e., defined by the order of the neighbors around each vertex. An embedding of G also assigns distinct coordinates in $\mathbb{R}^2$ for every vertex $v \in V(G)$. The edges of the graph are drawn as straight lines. For the rest of this paper, we assume that the vertices of G have integer coordinates, that is, the graph is embedded on the integer grid.

The goal of our LGD algorithm is to produce a representation of the graph G given a BSP tree T, see Figure 1. Similar to [10] we define the *clustered graph* $C = (G, T)$ to be the graph G, and the BSP tree T, such that the vertices of G coincide with the leaves of T. An internal node of T represents a cluster, which

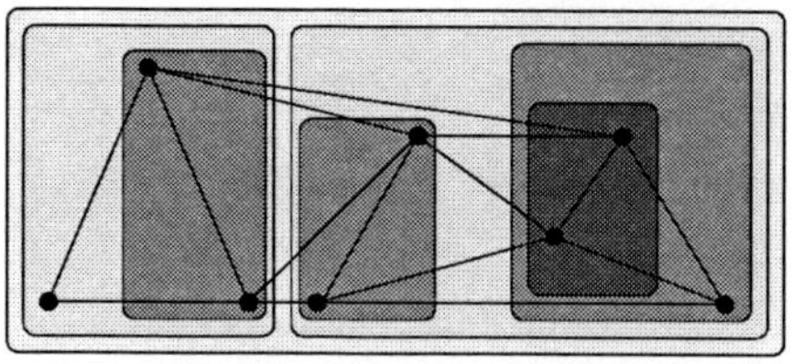

Duncan, Goodrich, and Kobourov, *BAR Trees*, JGAA, 4(3) 19–46 (2000)   24

Figure 2: A 2-dimensional representation of a clustered graph $C = (G,T)$. The underlying graph G and the clustering are the same as in Figure 1. a simple closed curve.

consists of all the vertices in its subtree. All the nodes of T at a given depth i represent the clusters of that level.

A *view at level i*, $G_i = (V_i, E_i)$, consists of the nodes of depth i in T and a set of representative edges, for $0 \leq i \leq \texttt{depth}(T)$. An edge (u, v) belongs to E_i if there is an edge between a and b in G, where a is in the subtree of u and b is in the subtree of v. In addition, each node $u \in T$ has an associated region, corresponding to the partition given by T. In Figure 1 we show an example of a 3-dimensional representation of a graph G and in Figure 2 we show a 2-dimensional representation of the same graph.

We create the graphs G_i in a bottom-up fashion, starting with G_k and going all the way up to G_0, where $k = \texttt{depth}(T)$. Define the combinatorial graph $H = (V(H), E(H))$, where initially $V(H) = \{u \in T : \texttt{depth}(u) = k\}$ and $E(H) = E(G)$. Notice that H is well defined since the leaves of T are exactly the vertices of G.

At each new level i we perform a *shrinking* of H. Suppose $u, v \in V(H)$, and $\texttt{parent}(u) = \texttt{parent}(v)$. We replace the pair by their parent and remove the edge (u, v) if it exists. We also remove any multiple edges that this operation may have created and maintain for each surviving edge a pointer to the original edge in G. Thus a *shrinking* of the graph H consists of all such operations, necessary to transform H into a representation of G at one higher level in the tree T.

At each level G_i is a subgraph of G with certain edges removed. Since we are producing a representation of G in 3-dimensions, every vertex must have three coordinates. The first two coordinates correspond to the location of the vertex on the integer grid. The third coordinate of a vertex $v \in V_i$ is equal to i, that is, all the vertices in G_i are embedded in the plane given by $z = i$. To obtain G_i from G_{i+1}, for $i = 0, \ldots, k-1$, we use the combinatorial graph H from level $i + 1$. Initially $E_i = E_{i+1}$. We then perform a shrinking of H and while removing an edge from H we remove its associated edge from E_i.

Thus the algorithm on Figure 3 runs in $O(n \cdot \texttt{depth}(T) + m)$ time. Using any of the previous known types of BSP trees, we can maintain most but never all of the desired properties. For example, if T is a k-d tree the cluster regions do not have balanced aspect ratios. We next describe how to construct a BSP tree which satisfies all of our goal criteria.

Duncan, Goodrich, and Kobourov, *BAR Trees*, JGAA, 4(3) 19–46 (2000)   25

```
create_clustered_graph(T, G)
    H ← G
    k ← depth(T)
    for i = k downto 0
        obtain G_i from H
        shrink H
    return C
```

Figure 3: Given graph G embedded in the plane and BSP tree T create clustered graph C. Here H is a combinatorial graph initially the same as G. The operations of obtaining G_i from H and shrinking of H are defined in Section 2.

3   The BAR Tree

Let us now discuss in detail the definition of our particular BSP-type decomposition tree, the BAR tree, and its construction. We begin with some general definitions.

Definition 1 *The following terms relate to various potential cuts:*

- *A canonical cut direction is any of the following three vectors:*

$$\vec{v}_x = (1, 0), \vec{v}_y = (0, 1), \vec{v}_z = (1, -1).$$

- *A canonical cut is any line whose normal is a canonical cut direction. For example, the line $x - y = 3$ has normal $\vec{v}_z$.*

- *A canonical region is any convex polygon such that each side is a segment of a canonical cut.*

Since there are three cut directions[1], a canonical region can have at most six sides. For convenience, we define six labels representing the six sides of the polygon. Notice that some of these sides may have zero length. For a canonical region R, we let x^l and x^r represent the corresponding left and right sides of R with normal $\vec{v}_x$. Similarly, we define y^l, y^r, z^l, and z^r, see Figure 4.

Definition 2 *For a canonical region R, let $\mathrm{diam}_i(R)$ be the L_m metric distance between the two sides of R with normal $\vec{v}_i$. For a side l in R, we define $|l|$ to be the length of the line segment l measured in the L_m metric.*

For simplicity in our arguments and notation, we use the L_∞ metric although any of the standard L_m metrics is acceptable. In the L_∞ metric the distance between two lines normal to $\vec{v}_z$ and the length of a line segment normal to $\vec{v}_z$ are

---

[1]Note the assymetry of not having the canonical direction $\vec{v}_w = (1, 1)$. The arguments that rely on the three canonical directions above also hold if we add this fourth direction, or any others.

Duncan, Goodrich, and Kobourov, *BAR Trees*, JGAA, 4(3) 19–46 (2000)   26

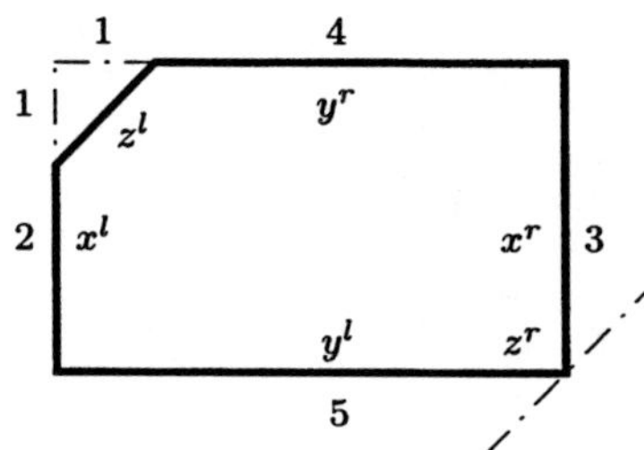

Figure 4: A labelling of the various sides of a canonical region R.

defined differently than in the L_2 metric. In particular, for a canonical region R with sides z^l and z^r, the length $|z^l|$ (or $|z^r|$) is the vertical distance between the two endpoints. The distance between the lines associated with z^l and z^r is one half the vertical distance between the two lines.

Definition 3 *The aspect ratio of a canonical region R is*

$$\mathtt{ar}(R) = \max(\mathtt{diam}_i(R))/\min(\mathtt{diam}_j(R)), \forall i, j \in \{x, y, z\}.$$

Given an aspect ratio parameter α, a region R is α-balanced if $\mathtt{ar}(R) \leq \alpha$.

This definition is valid only for canonical regions. Since all of the regions that appear in this section are canonical regions, whenever we refer to any region we mean a canonical region. When the term α is understood, we refer to α-balanced regions as simply *balanced* regions and refer to non-α-balanced regions as *unbalanced* regions. Throughout the paper, we also call balanced and unbalanced regions, respectively, *fat* and *skinny* regions.

To understand the various notions of a canonical region, let us look at one specific canonical region R in Figure 4. Here we see the various sides of R, x^l, x^r, y^l, y^r, z^l, z^r. In particular, although not actually a true side of R, we still represent the side z^r. It is tangent to R and has zero length. From the figure, we see the various lengths of each side:

$$\begin{aligned}
|x^l| &= 2, \quad |y^l| = 5, \quad |z^l| = 1, \\
|x^r| &= 3, \quad |y^r| = 4, \quad |z^r| = 0.
\end{aligned}$$

Since we are using the L_∞ metric, the length of z^l is 1 rather than $\sqrt{2}$ as would be the case in the L_2 metric. We can also compute $\mathtt{diam}_i(R)$ for each of the three canonical directions as well as the aspect ratio of R.

- $\mathtt{diam}_x(R) = 5$,

- $\mathtt{diam}_y(R) = 3$,

- $\mathtt{diam}_z(R) = (2 + 5)/2 = 3.5$,

- $\mathtt{ar}(R) = \max(\mathtt{diam}_i(R))/\min(\mathtt{diam}_j(R)) = \mathtt{diam}_x(R)/\mathtt{diam}_y(R) = 2.$

3.1   Constructing the BAR Tree

We now introduce the BAR tree data structure. Suppose we are given a point set S in the plane, $|S| = n$, and an initially square region R containing S. We construct a BAR tree T on S recursively dividing R into cells such that the following properties are guaranteed:

- Every cell in the tree is convex.

- Every cell in the tree has balanced aspect ratio.

- Every leaf cell contains at most a constant number of points of S.

- The tree has $O(n)$ nodes.

- The depth of the tree is $O(\log n)$.

The structure is straightforward and reminiscent of the original k-d tree. Recall that in a k-d tree, every node u in the tree represents a cell region $u.\texttt{region}$ and an axis-parallel cut $u.\texttt{cut}$ partitioning that region into two sub-regions, $u.\texttt{left}$ and $u.\texttt{right}$. The leaves of the tree are cells with a constant number of points. In general, each cut divides the region into two roughly equal halves, and thus the tree has $O(\log n)$ depth and uses $O(n)$ space. However, if the vast majority of the points is concentrated close to any particular corner of the region, no constant number of axis-parallel cuts can effectively reduce the size of the point set and maintain good aspect ratio. This is a serious concern for many applications and for ours in particular. As a result, an extensive amount of research has been dedicated to improving and analyzing the performance of k-d trees and its derivatives, often concentrating on trying to maintain some form of balanced aspect ratio [5, 19, 29].

We now show how to construct a BAR tree T from a point set S using an aspect ratio parameter α and a balance parameter β. We prove that any α-balanced region can be divided by a sequence of one or two cuts into at most three subregions. We also guarantee that each subregion is α-balanced and the number of points in each of the three subregions is less than β times the number of points in the original region. We begin by defining the notions of a one-cut and a two-cut.

Definition 4 *Let R be an α-balanced canonical region containing n points. Let β be a given balance parameter. A one-cut is any canonical cut dividing R into two subregions R_1 and R_2 such that:*

1. *R_1 and R_2 are both α-balanced canonical regions.*

2. *R_1 and R_2 contain at most βn points.*

If there exists a one-cut for R, we say R is one-cuttable.

Definition 5 *Let R be an α-balanced canonical region containing n points. Let β be a given balance parameter. A two-cut is any canonical cut dividing R into two subregions R_1 and R_2 such that:*

Duncan, Goodrich, and Kobourov, *BAR Trees*, JGAA, 4(3) 19–46 (2000)   28

```
create_BAR_tree(R, α, β)
    create node u
    u.region ← R
    if number of points in R ≤ c,
        return u
    if an (α, β)-balanced one-cut s, exists in R
        u.cut ← s
        (R₁, R₂) ← s(R)
    else let s be an (α, β)-balanced two-cut in R
        u.cut ← s
        (R₁, R₂) ← s(R)
    u.left ← create_BAR_tree(R₁, α, β)
    u.right ← create_BAR_tree(R₂, α, β)
    return u
```

Figure 5: Creating the BAR tree. The recursion stops when a cell has a constant number of points, $c \geq 1$.

1. R_1 and R_2 are both α-balanced canonical regions.

2. R_2 contains at most βn points.

3. R_1 is one-cuttable.

If there exists a two-cut for R, we say R is two-cuttable.

For an α-balanced region R which is two-cuttable, let s represent the two-cut dividing R into two regions R_1 and R_2, and let s' represent the one-cut dividing R_1. In other words, the sequence of two cuts, s and s', results in three α-balanced regions each containing at most βn points. To make it clear that α and β are parameters, we often refer to one-cuts (resp. two-cuts) of a region R as (α, β)-balanced one-cuts (resp. two-cuts).

Figure 5 shows the pseudo-code for the construction of a BAR tree. Here we use the notation $(R_1, R_2) \leftarrow s(R)$ as a shorthand for cutting the region R with a cut s resulting in subregions R_1 and R_2. We prove in the next section that every α-balanced region is either one-cuttable or two-cuttable for sufficiently large constant values of α and β. Since the algorithm only uses one-cuts and two-cuts, the regions produced are all α-balanced regions. The algorithm stops the recursion when a leaf cell has a constant number of points from S. Because at least every other cut used is a one-cut, the depth of the tree is $O(log_{1/\beta}n)$ and the size is $O(n)$. Therefore, the algorithm correctly creates a tree which satisfies the properties for a BAR tree.

Duncan, Goodrich, and Kobourov, *BAR Trees*, JGAA, 4(3) 19–46 (2000)   29

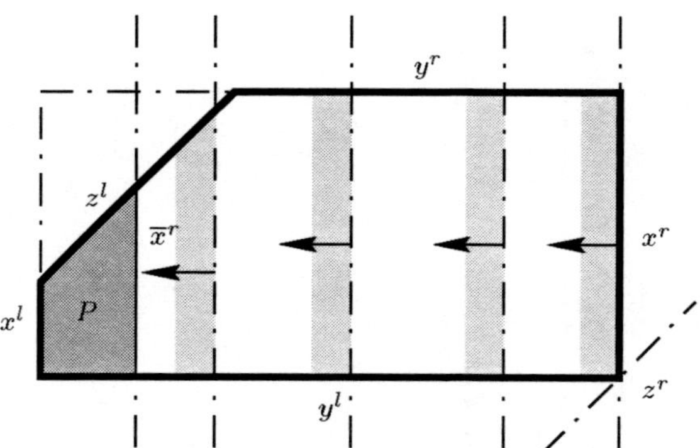

Figure 6: The shaded region P represents the region between x^l and a maximal cut of x^r for a region R.

3.2   Two-Cut Existence Theorem

Since the correctness of the previous algorithm relies on the existence of a two-cut for a region, we prove that every region R is either one-cuttable or two-cuttable. Before we do this, we need to describe some basic terminology relating to cutting a region R into two subregions.

Definition 6 *Suppose we are given an α-balanced canonical region R and a canonical direction $\vec{v}_i$. Let i^l and i^r be the two (possibly zero length) sides of R normal to $\vec{v}_i$. Let $\overline{i}^l$ be the line containing i^l and let P be the region between i^r and $\overline{i}^l$ (at first P is the same as R). Sweep $\overline{i}^l$ towards i^r until either P is empty or just before P becomes unbalanced. We call this final region $R_{i,r} = P$ maximized in the direction from i^l. Similarly, we call $\overline{i}^l$ the maximal cut of i^l. $R_{i,l}$ is similarly defined.*

Definition 7 *For a region R with n points and a canonical direction $\vec{v}_i$, let $R_{i,l}$ (resp. $R_{i,r}$) represent the region maximized in the direction from i^r (resp. i^l), If $R_{i,l} \cap R_{i,r} = \emptyset$ define R_i to be the region $R_{i,l}$ or $R_{i,r}$ with the larger number of points. Otherwise if $R_{i,l} \cap R_{i,r} \neq \emptyset$, define R_i to be R.*

Since the change in aspect ratio during the sweep is continuous, the region $R_{i,r}$ has aspect ratio equal to α. Figure 6 illustrates a maximal cut of x^r for a canonical region R using the parameter $\alpha = 2$. The region $R_{i,r}$ maximized in the direction from x^r has aspect ratio $\mathrm{ar}(R_{i,r}) = 2$. Figure 7 shows a few more examples of regions with their respective maximal cuts and associated subregions. The following lemma follows from a straightforward geometric argument.

Lemma 1 *Given regions R and $R_{i,r}$ and lines i^l and $\overline{i}^l$ as defined above, if $R_{i,r}$ is not empty and we continue sweeping in the same direction, the region between $\overline{i}^l$ and i^r will be unbalanced until it becomes empty.*

Duncan, Goodrich, and Kobourov, *BAR Trees*, JGAA, 4(3) 19–46 (2000)   30

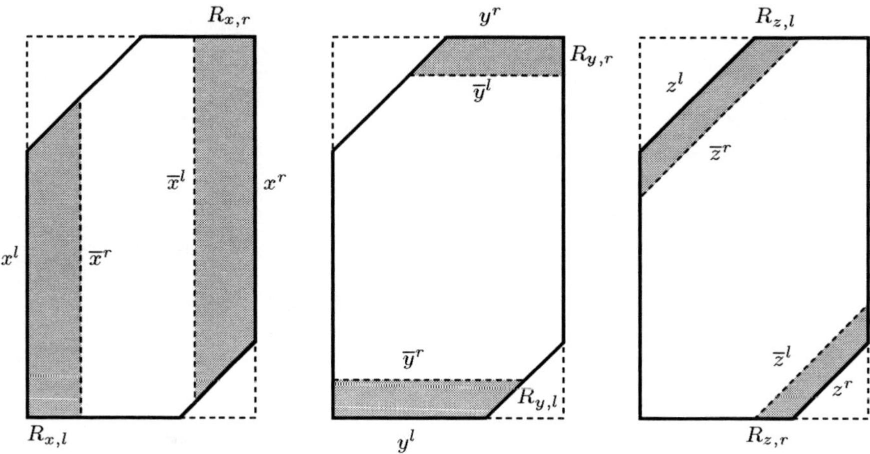

Figure 7: The labels on the sides of a general canonical region and the maximizing cuts from the respective directions.

Corollary 1 *For an α-balanced region R, if the region $R_{i,r}$ is maximized in the direction from i^l, then $\min\{\mathtt{diam}_x(R_{i,r}), \mathtt{diam}_y(R_{i,r}), \mathtt{diam}_z(R_{i,r})\} = \mathtt{diam}_i(R_{i,r})$.*

Corollary 2 *For an α-balanced region R and direction $\vec{v}_i$, if $R_{i,l} \cap R_{i,r} = \emptyset$, then any cut i^m with a normal $\vec{v}_i$ and lying between $\overline{i}^l$ and $\overline{i}^r$ produces two α-balanced subregions R_1 and R_2.*

Lemma 2 *Suppose we are given a region R with n points, a balance parameter $\beta \geq 1/2$ and two parallel lines c^l and c^r. Without loss of generality, let us orient these lines so that c^l lies to the left of c^r. Then one of the following must be true:*

- *The number of points from R to the left of c^l (i.e., away from c^r) is more than βn;*

- *The number of points from R to the right of c^r (i.e., away from c^l) is more than βn;*

- *There exists a line c' parallel and between c^l and c^r dividing R into two subregions R_1 and R_2 such that the number of points in either subregion is less than βn.*

Proof: Assume the first two conditions do not hold. Thus, we only need to prove that the last condition must hold. Let n_1 be the number of points to the left of c^l and let n_2 be the number of points to the left of c^r. We know then that $n_1 > \beta n \geq n/2$. Similarly, we know that $(n - n_2) > \beta n \geq n/2$. It follows

Duncan, Goodrich, and Kobourov, *BAR Trees*, JGAA, 4(3) 19–46 (2000)   31

then that $n_2 < n/2$. Sweep a line c' from c^l to c^r letting n_3 be the number of points to the left of c'. Since the sweep is continuous, n_3 varies from $n_1 > n/2$ to $n_2 < n/2$. In particular, there is a point where $n_3 = n/2$. This cut divides R into two subregions each with less than $n/2$ points. $\qquad\square$

Corollary 3 *For an α-balanced region R with n points, a direction $\vec{v}_i$, and $\beta \geq 1/2$, either R is one-cuttable or R_i contains more than βn points.*

Proof: If the two subregions $R_{i,l}$ and $R_{i,r}$ intersect each other, then by definition $R_i = R$ and thus contains n points. If R is one-cuttable, then the statement is trivially true. Otherwise, we have two cuts $\vec{i}^{\,r}$ and $\vec{i}^{\,l}$ associated with $R_{i,l}$ and $R_{i,r}$ respectively. From Lemma 2, either $R_{i,l}$ or $R_{i,r}$ contains more than βn points or there exists a line c' parallel and between $\vec{i}^{\,r}$ and $\vec{i}^{\,l}$ dividing R into two subregions R_1 and R_2 such that the number of points in either subregion is less than βn. However, this implies that R is one-cuttable. $\qquad\square$

The above corollary is quite useful in proving that certain regions are one-cuttable. For instance, let R be an α-balanced region such that, for some canonical direction $\vec{v}_i$, both $R_{i,l}$ and $R_{i,r}$ are empty. Since neither of these two subregions can contain any points, R must be one-cuttable. In fact, this notion can be extended to include multiple canonical directions.

Lemma 3 *Let R be an α-balanced region R with n points and $\beta \geq 2/3$. If $R_x \cap R_y \cap R_z = \emptyset$, then R is one-cuttable.*

Proof: This is a standard extension from set theory. For a set of points S, it is impossible to have three subsets of S each contain more than $2/3$ of S without their intersection containing at least one point. $\qquad\square$

If we can prove that there exist regions such that no possible assignment for the R_i's allows for a non-empty intersection, then the region R is always one-cuttable. Do there exist regions which are guaranteed to be one-cuttable? We describe two such regions which we will use to argue that every α-balanced region is inevitably two-cuttable.

Definition 8 *For a given aspect ratio parameter α we define two special canonical regions with aspect ratio α as follows:*

- *Canonical isosceles trapezoidal (CIT) regions are trapezoids which have z^l and z^r as the two opposing parallel base sides, see Figure 8a.*

- *Canonical right-angle trapezoidal (CRT) regions are trapezoids which have their two opposing parallel base sides normal to either $\vec{v}_x$ or $\vec{v}_y$, see Figure 8b.*

Lemma 4 *For $\alpha > 4$ and $\beta \geq 2/3$, canonical isosceles trapezoidal (CIT) regions are one-cuttable.*

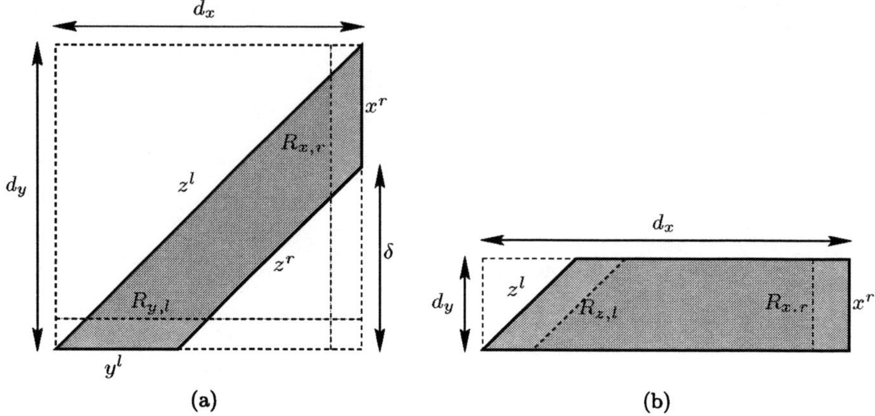

Figure 8: Examples of (a) CIT and (b) CRT regions.

Proof: Without loss of generality, we can analyze the region R in Figure 8a, since the other possible CIT regions are symmetrical. Let $d_i = \mathtt{diam}_i(R)$ for $i \in \{x, y, z\}$. Define $\delta = |z^r| = d_x - |x^r|$. Since the trapezoid's two parallel sides are z^l and z^r, we know that $d_x = d_y$ and $|x^r| = |y^l|$. Recall that in the L_∞ metric, $d_z = (|x^l| + |y^l|)/2 = |y^l|/2$. Similarly, we get $d_z = |x^r|/2$. Since the region has aspect ratio α, we have $\mathtt{ar}(R) = \alpha = d_x/d_z$. It follows that

$$
\begin{aligned}
d_x &= \alpha d_z \\
&= \alpha |x^r|/2 \\
&= \alpha(d_x - \delta)/2 \\
&= \alpha\delta/(\alpha - 2)
\end{aligned}
\tag{1}
$$

Let us examine the possible intersections of $R_x \cap R_y \cap R_z$. Since $R_{x,l}$ is empty, we know that $R_x = R_{x,r}$. Since by definition, $R_{x,r}$ is maximized from x^l, we know that $\mathtt{diam}_x(R_x) \le d_y/\alpha = d_x/\alpha$. From Equation 1 and from $\alpha > 4$, it follows that $\mathtt{diam}_x(R_x) < \delta/2$. Similarly, we know that $R_y = R_{y,l}$ and $\mathtt{diam}_y(R_y) < \delta/2$. This implies that $R_x \cap R_y = \emptyset$. From Lemma 3, R must be one-cuttable. $\square$

Lemma 5 *For $\alpha > 4$ and $\beta \ge 1/2$, canonical right-angle trapezoidal (CRT) regions are one-cuttable.*

Proof: Without loss of generality, we can again analyze the region R in Figure 8b, since the other possible CRT regions are symmetrical. Let $d_i = \mathtt{diam}_i(R)$ for $i \in \{x, y, z\}$. We know that $\max_{i \in \{x,y,z\}}(d_i) = d_x$ and $\min_{i \in \{x,y,z\}}(d_i) = d_y$ from the definition of the region. Therefore, we know that $\mathtt{ar}(R) = \alpha = d_x/d_y$. Observing that $|y^r| = d_x - d_y$, we obtain:

$$
d_y = d_x - |y^r|
$$

Duncan, Goodrich, and Kobourov, *BAR Trees*, JGAA, 4(3) 19–46 (2000)   33

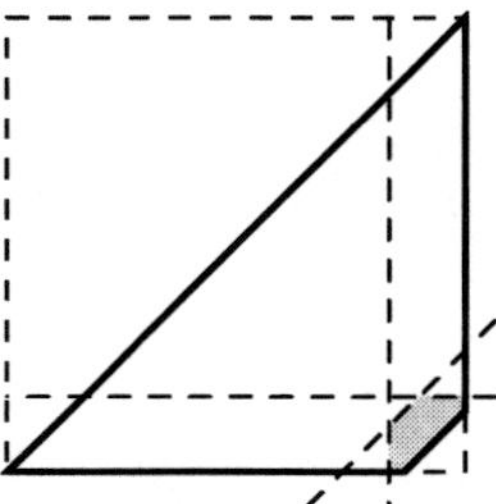

Figure 9: A region R which is not one-cuttable if the points are densely concentrated in the highlighted corner. Notice that no canonical cut can divide this region without creating a region that is too skinny.

$$
\begin{aligned}
&= \alpha d_y - |y^r| \\
&= |y^r|/(\alpha - 1)
\end{aligned}
\tag{2}
$$

Let us examine the possible intersections of $R_x \cap R_y \cap R_z$. Since $R_{x,l}$ is empty, we know that $R_x = R_{x,r}$. Since by definition, $R_{x,r}$ is maximized from x^l, we know that $\mathtt{diam}_x(R_x) \leq d_y/\alpha$. From Equation 2 and from $\alpha > 4$, it follows that $\mathtt{diam}_x(R_x) < |y^r|/12$. Similarly, we can see that $R_z = R_{z,l}$ and $\mathtt{diam}_z(R_z) < |y^r|/6$. This implies that $R_x \cap R_z = \emptyset$. From Lemma 3 it follows that R must be one-cuttable. $\qquad\square$

It is easy to construct examples where a region R is not one-cuttable for a given a point set, see Figure 9. However, the following theorem shows that by making a two-cut followed by a one-cut we can in fact divide an α-balanced region into at most three α-balanced subregions each containing less than a constant fraction of the points in R.

Theorem 1 (Two-Cut Existence Theorem) *Any α-balanced region R is either one-cuttable or two-cuttable for $\alpha \geq 6$ and $\beta \geq 2/3$.*

Proof: We can assume that R is not one-cuttable, and thus only prove that it must be two-cuttable. Again let $d_i = \mathtt{diam}_i(R)$ for $i \in \{x, y, z\}$. Without loss of generality, assume $d_y \geq d_x$. Consider the two parallel sides, z^l and z^r. We call a cut, z^i, $i \in l, r$, **small** if

$$
|z^i| \leq \min(d_x, d_y)\frac{\alpha - 2}{\alpha} = d_x \frac{\alpha - 2}{\alpha},
$$

and **large** otherwise. We now break the analysis into three cases based on the size of these two sides. Each case follows roughly the same argument. If a region is not one-cuttable, the three subregions R_x, R_y, and R_z must all intersect each other since $\beta \geq 2/3$. If one of these regions is one-cuttable, in particular either a CIT or CRT region, then R is two-cuttable. Therefore, we prove in each case that if all three subregions are not CIT or CRT regions, they cannot simultaneously intersect.

Duncan, Goodrich, and Kobourov, *BAR Trees*, JGAA, 4(3) 19–46 (2000)   34

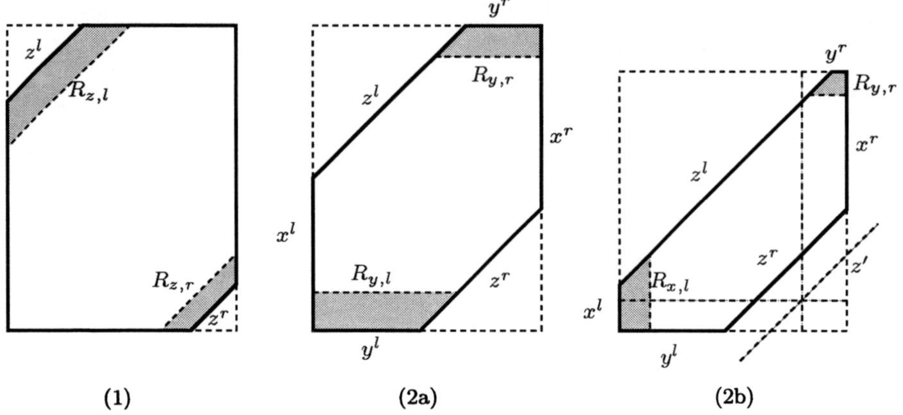

Figure 10: Case 1: both z^l, z^r are *small*. Case 2a: both sides are *large* and $|y^l| \le |x^l|$, which guarantees that $R_{y,l}$ and $R_{y,r}$ are both CRT regions. Case 2b: both sides are *large* and $|y^l| > |x^l|$.

Case 1. (z^l and z^r are both small):
Let both z^l and z^r be small, see Figure 10.1. From Equation (1) and because z^l is small, we know that $\mathtt{diam}_x(R_{z,l}) = \alpha|z^l|/(\alpha - 2) \le d_x$. The same holds for the region $\mathtt{diam}_x(R_{z,r})$. Thus these two CIT regions are disjoint. Since there was no one-cut, particularly in the z-direction, one of the two regions has more than βn points. By Lemma 4, both CIT regions are one-cuttable. Therefore, R has a two-cut, namely the one creating the CIT region with maximum points, R_z.

Case 2. (z^l and z^r are both large):
Let both z^l and z^r be large. Without loss of generality, let the larger of the two cuts be z^l. Notice that,

$$d_x(\alpha - 2)/\alpha < |z^r| \le |z^l| \le d_x.$$

Because $|z^l| \ge |z^r|$ and $d_x \le d_y$, we know that $|y^r| \le |x^r|$. Therefore, $R_{y,r}$ is a CRT region, and is one-cuttable.

If $|y^l| \le |x^l|$, then $R_{y,l}$ is also a CRT region, see Figure 10.2a. From Lemma 5, R_y is always one-cuttable. Therefore, R is two-cuttable, the two-cut being either $\overline{y}^l$ or $\overline{y}^r$.

Otherwise, we have the situation in Figure 10.2b:

$$
\begin{aligned}
|x^l| \;&<\; |y^l| \\
&=\; d_x - |z^r| \\
&\le\; d_x - d_x(\alpha - 2)/\alpha \\
&=\; d_x(1 - (\alpha - 2)/\alpha) \\
&=\; 2d_x/\alpha.
\end{aligned}
\tag{3}
$$

Duncan, Goodrich, and Kobourov, *BAR Trees*, JGAA, 4(3) 19–46 (2000)   35

We now have bounds on $|x^l|$, $|y^l|$, and $|y^r|$. Let us now bound $|x^r|$. Using Equation 3, we see that

$$
\begin{aligned}
d_y &\le d_x + |x^l| \\
&\le d_x + 2d_x/\alpha \\
&\le d_x(1 + 2/\alpha).
\end{aligned}
$$

$$
\begin{aligned}
|x^r| &= d_y - |z^r| \\
&\le d_x(1 + 2/\alpha) - d_x(1 - 2/\alpha) \\
&= 4d_x/\alpha
\end{aligned}
\tag{4}
$$

Using arguments similar to those used in proving Equation 2, we know that

$$
\begin{aligned}
\mathrm{diam}_x(R_{x,r}) &\le |x^r|/(\alpha - 1) \\
&\le 4d_x/\alpha(\alpha - 1), \text{ and}
\end{aligned}
$$

$$
\begin{aligned}
\mathrm{diam}_y(R_{y,l}) &\le |y^l|/(\alpha - 1) \\
&\le 2d_x/\alpha(\alpha - 1).
\end{aligned}
$$

Consider the intersection of $\overline{y}^r$ and $\overline{x}^l$ and the cut z' which passes through this point, see Figure 10.2b. If z' lies inside R, we can bound the size of the intersection of this cut with R by

$$
\begin{aligned}
|z'| &= (\mathrm{diam}_x(R_{x,r}) + \mathrm{diam}_y(R_{y,l})) \\
&\le 6d_x/\alpha(\alpha - 1) \\
&\le d_x/5 \\
&< |z^r|.
\end{aligned}
$$

However, this implies that z' does not intersect R. Consequently, $R_{x,r} \cap R_{y,l} = \emptyset$, and either $R_x = R_{x,l}$ or $R_y = R_{y,r}$. Since either of these subregions is one-cuttable, R is two-cuttable.

Case 3. (only one of the two cuts is large):
Without loss of generality, let the larger of the two cuts be z^l. In other words, $|z^l| > d_x(\alpha - 2)/\alpha$. Here we need to consider two subcases.

- **3i.** (long rectangle) If $d_y \ge d_x \frac{\alpha+1}{\alpha-2}$, we cannot necessarily cut the region using the direction $\vec{v}_x$. Using the same argument as in Case 2, we see that $R_{y,r}$ is a CRT region. Thus, if $R_y = R_{y,r}$, we are done. Similarly, using the argument for Case 1, we see that $R_{z,r}$ is a CIT region, see Figure 11a. Therefore, we can assume that $R_y = R_{y,l}$ and $R_z = R_{z,l}$ as in Figure 11b.

  From Equation 1, $\mathrm{diam}_y(R_{z,l}) \le \alpha d_x/(\alpha - 2)$. Similarly, from Equation 2, we know that $\mathrm{diam}_y(R_{y,l}) \le d_x/\alpha$. Thus, combining the two yields,

$$
\mathrm{diam}_y(R_{z,l}) + \mathrm{diam}_y(R_{y,l}) \le d_x \frac{\alpha}{\alpha - 2} + d_x/\alpha
$$

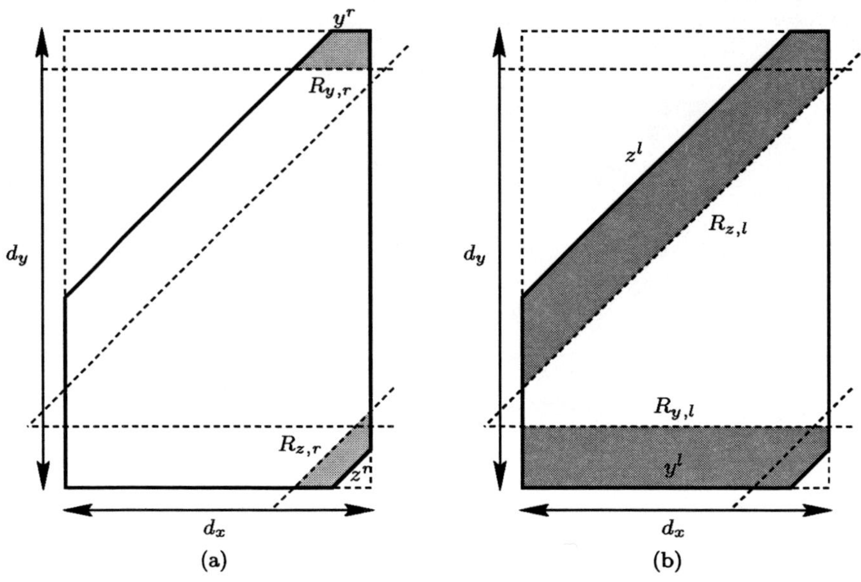

Figure 11: Case 3i, for a long rectangle. (a) Two one-cuttable subregions, $R_{y,r}$ and $R_{z,r}$. (b) Opposing not necessarily one-cuttable subregions, $R_{y,l}$ and $R_{z,l}$, but they cannot intersect.

$$\begin{aligned}
&= d_x\left(\frac{\alpha}{\alpha-2}+\frac{1}{\alpha}\right)\\
&\le d_y\frac{\alpha-2}{\alpha+1}\left(\frac{\alpha}{\alpha-2}+\frac{1}{\alpha}\right)\\
&= d_y\frac{1}{\alpha+1}\left(\alpha+1-\frac{2}{\alpha}\right)\\
&< d_y.
\end{aligned}$$

From this, we know that $R_{z,l}$ and $R_{y,l}$ cannot intersect. Therefore, either $R_z = R_{z,r}$ or $R_y = R_{y,r}$ and the region is two-cuttable.

- **3ii.** (squat rectangles) Now, we have $d_y < d_x\frac{\alpha+1}{\alpha-2}$. Since z^l is large, we know that $R_{y,r}$ is a CRT region. Since the rectangle is squat, we know that $R_{x,l}$ is also a CRT region, see Figure 12a. Since z^r is small, either $R_{z,l}$ is a CIT region or $R_{z,l} = R$. The latter case arises if maximizing from z^r and z^l produces regions which intersect each other. Notice, because of the dimensions of the region, this is not possible in either the $\vec{v}_x$ or $\vec{v}_y$ direction. Since $d_y \ge d_x$, $R_{y,l}$ cannot intersect $\cap R_{y,r}$. Notice also that, for $\alpha > 5$,

$$\begin{aligned}
\mathtt{diam}_x(R_{x,l}) &\le d_y/\alpha\\
&< d_x\frac{\alpha+1}{\alpha(\alpha-2)}
\end{aligned}$$

Duncan, Goodrich, and Kobourov, *BAR Trees*, JGAA, 4(3) 19–46 (2000)   37

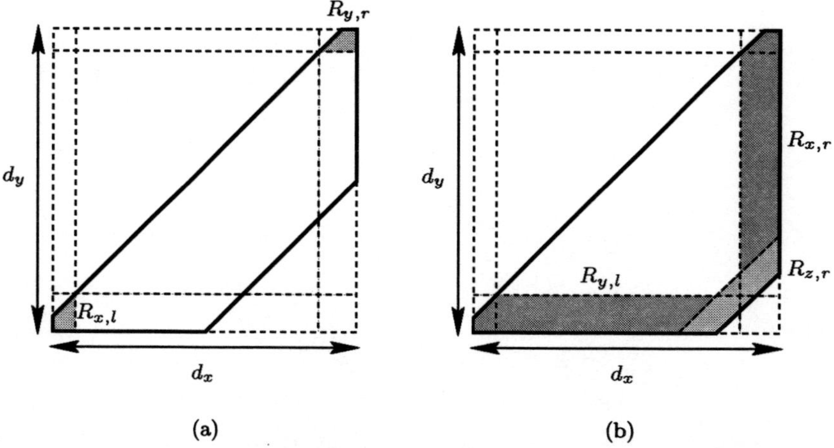

Figure 12: Case 3ii, for a short rectangle. (a) Two one-cuttable subregions, $R_{x,l}$ and $R_{y,r}$. (b) Opposing not necessarily one-cuttable subregions, $R_{x,r}$ and $R_{y,l}$. If they intersect, $R_z = R_{z,r}$ is a one-cuttable region.

$$< \quad d_x/2.$$

The same is true for $R_{x,r}$. So, $R_{x,l}$ cannot intersect $R_{x,r}$.

We only need to consider the case when $R_x = R_{x,r}$ and $R_y = R_{y,l}$. Since both regions contain more than βn points, they must intersect, see Figure 12b. It follows then that $|z^r| \leq 2d_x/\alpha$. We also know that $|z^l| \leq d_x$. Recalling that $\alpha \geq 6$, we can bound $\mathtt{diam}_z(R)$, $\mathtt{diam}_z(R_{z,r})$, and $\mathtt{diam}_z(R_{z,l})$ by

$$
\begin{aligned}
\mathtt{diam}_z(R) &\geq d_x/2 - |z^r|/2 \\
&\geq d_x/2 - d_x/\alpha \\
&\geq d_x/2 - d_x/6 \\
&= \frac{d_x}{3} \\[6pt]
\mathtt{diam}_z(R_{z,l}) &\leq \frac{d_x}{\alpha} \\
&\leq \frac{d_x}{6} \\[6pt]
\mathtt{diam}_z(R_{z,r}) &\leq \frac{|z^r|}{\alpha - 2} \\
&\leq \frac{2d_x}{\alpha^2 - 2\alpha} \\
&\leq \frac{2d_x}{24}
\end{aligned}
$$

Duncan, Goodrich, and Kobourov, *BAR Trees*, JGAA, 4(3) 19–46 (2000)   38

$$= \frac{d_x}{12}$$

$$\mathtt{diam}_z(R_{z,r}) + \mathtt{diam}_z(R_{z,r}) \;\leq\; \frac{d_x}{6} + \frac{d_x}{12}$$

$$= \frac{d_x}{4}$$

$$< \frac{d_x}{3}$$

$$\leq \mathtt{diam}_z(R).$$

This implies that $R_{z,l}$ does not intersect $R_{z,r}$ and similarly cannot intersect $R_{x,r} \cap R_{y,l}$. Therefore, we know that $R_z = R_{z,r}$. Since $R_{z,r}$ is a one-cuttable CIT region, we know that R must be two-cuttable.

This completes the proof of the two-cut existence theorem. $\qquad\qquad\square$

Theorem 2 *Given a point set S in the plane, we can construct a BAR tree representing a decomposition of the plane into "fat" regions in $O(n \log n)$ time.*

Proof: To prove this, it suffices to note that a one-cut or a two-cut in any of the three canonical directions can be found in $O(n)$ time and that the depth of the tree is $O(\log n)$. $\qquad\qquad\square$

4   Using a BAR tree for Cluster Based Drawing

Let $G = (V, E)$ be the graph that we want to draw. Once we obtain the embedding of G, using whatever algorithm is most appropriate for the graph, we associate with the graph the smallest bounding square, R, which we call G's *cluster region*. Using the embedding and its cluster region, we create the BAR tree T, as described above. Each node $u \in T$ maintains $u.\mathtt{region}$, $u.\mathtt{cluster}$, and $u.\mathtt{depth}$. Here $u.\mathtt{cluster}$ is the subgraph of G which is properly contained in $u.\mathtt{region}$. Recall that the depth of the tree T is $k = O(\log n)$. In our application of the tree structure to cluster-based graph drawing, we want every leaf to be at the same depth. Therefore, we propagate any leaf not at the maximum depth down the tree until the desired depth is reached. This is merely conceptual and does not require any additional storage space or change in the tree structure.

Using the tree T, we create the clustered graph C, which consists of k layers. Each layer is an embedded subgraph of G along with the regions and clusters obtained from T. The layers are connected with vertical edges which are simply the edges in T. The other inputs to LGD are the aspect ratio parameter α and the balance parameter, β. Here, α determines the maximal aspect ratio of a cluster region in C, and β determines the cluster balance, the ratio of a cluster's size to its parent's. For a summary of the operations, see Figure 13.

Lemma 6 *A call to LGD(G, α, β) for $\alpha = 6$, $\beta = 2/3$ results in 2/3-balanced clustering with aspect ratio less than or equal to 6 and cluster depth $O(\log n)$.*

Duncan, Goodrich, and Kobourov, *BAR Trees*, JGAA, 4(3) 19–46 (2000)   39

```
LGD(G, α, β)
   embed(G)
   T ← create_BAR_tree(G, α, β)
   C ← create_clustered_graph(T, G)
   display(C)
```

Figure 13: Main algorithm. The inputs to the algorithm are graph G along with the aspect ratio parameter α and the balance parameter β. Graph G is embedded in the plane, after which the BAR tree T is created. Finally, the clustered graph C is created and displayed.

Proof: By construction, the clusters are β-balanced and the cluster depth is equivalent to the depth of T. Thus, for $\alpha \geq 6$ and $\beta \geq 2/3$ the depth is $O(\log_{1/\beta} n)$. $\qquad\square$

Theorem 3 *For $\alpha \geq 6$, $\beta \geq 2/3$, algorithm LGD creates a 2/3-balanced clustered graph C in $O(n \log n + m + D_0(G))$ time.*

Proof: The proof follows directly from the construction of the algorithm and previous statements about the running time of each component. $\qquad\square$

Once we obtain the clustered graph C, we can display it as a 3-dimensional multi-layer graph representing each cluster by either the the convex hull of its vertices or by its associated region in the BAR tree. Along with the clustered graph C we can display a particular cluster with more details. Thus we provide the global structure using the clustered graph and the local detail using the individual clusters.

4.1   Planar Graphs

When the graph G is planar, we are able to show a few special properties of our clustered drawings.

Theorem 4 *If G is planar, for $\alpha \geq 6$, $\beta \geq 2/3$, algorithm LGD creates a 2/3-balanced clustered graph C in $O(n \log n)$ time. Moreover, C is embedded with straight lines and no crossings on the $n \times n \times k$ grid, where $k = O(\log n)$.*

Proof: We begin with a planar grid embedding with straight-line edges [6, 12, 28] and then the original layer, G_k, is planar. Since each successive layer is a proper subgraph of the previous layer, it too must be planar and drawn without edge crossings. $\qquad\square$

In Figure 14 we can see a clustered graph $C = (G, T)$ in which the clusters are represented by the partitions of the plane obtained from the BAR tree. Note that in this case there is no need to select a representative vertex for a cluster.

Duncan, Goodrich, and Kobourov, *BAR Trees*, JGAA, 4(3) 19–46 (2000)   40

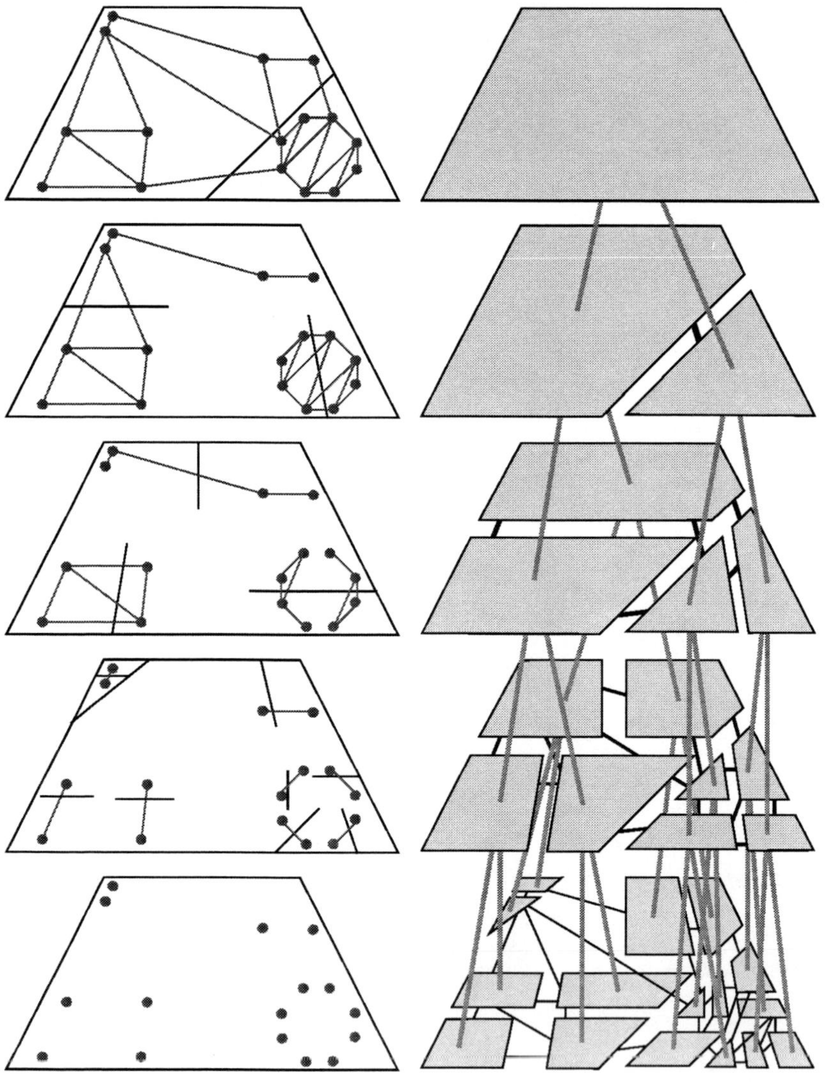

Figure 14: A clustered graph $C = (G, T)$. The clustering of G on the right is obtained from the BAR tree cuts on the left. Each cluster is represented by the region defined by the BAR tree cuts. Note the edge-region crossings at the last two levels.

Duncan, Goodrich, and Kobourov, *BAR Trees*, JGAA, 4(3) 19–46 (2000)   41

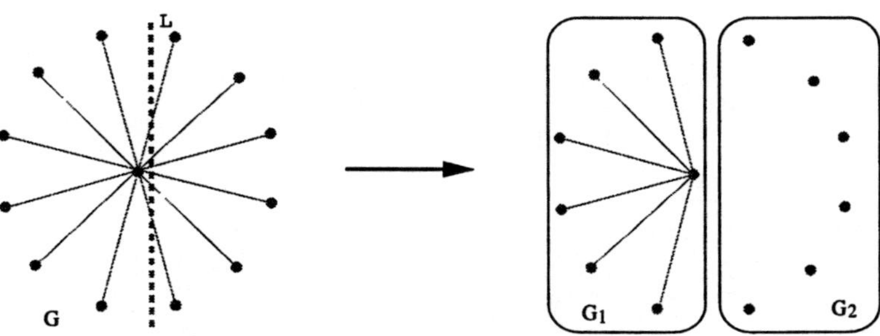

Figure 15: Graph G with an inherently large cut. Any cut L which maintains a β-balance between the clusters, where $1/2 \leq \beta < 1$, cuts $O(n)$ edges.

For such drawings it is possible to have an edge cross a region that it does not belong to. Moreover, it is possible to have an edge cross the convex hull of a cluster that it does not belong to. If we represent a cluster by the convex hulls of its connected components, however, there will be no such crossings. Thus, if we could guarantee that each cluster is connected or has a small number of connected components, the display of the graph can be improved even further. Alternatively, we can redefine the clusters at each level to be the connected components of vertices inside each cluster region of the BAR tree. With this definition of clusters we could then use the algorithm of Eades and Feng [10] to produce a new clustered embedding of the planar graph so as to have no edge or region crossings.

4.2   Extensions

Throughout this paper we do not discuss the cut sizes produced by our algorithm, that is the number of edges intersected by a cut line in the BAR tree. In some applications it is important that the number of such edges cut be as small as possible. There exist graphs, however, that do not allow for "nice" cuts of small size. Consider the *star* graph G on Figure 15. Any cut, which maintains a β-balance between the two subgraphs it produces, intersects $O(n)$ edges. If the balance parameter is $\beta = 1/2$, the cut contains $\lfloor \frac{n}{2} \rfloor$ edges. As this example shows, we cannot hope to guarantee cut sizes better than $O(n)$. Still, if the given graph has a small cut then we would like to find a small cut as well.

Minimizing the cut size violates two of our five criteria, namely, speed and convexity. First of all, looking for the best β-balanced cut is a computationally expensive operation, and while it can be done in polynomial time, it is not hard to see that it cannot be done in linear time. In addition, the best β-balanced cut may not preserve the convex cluster drawing property that LGD maintains. As shown in Figure 16, this may result in new edge crossings in our clustered graph.

Our algorithm does not guarantee that it will find the optimum β-balanced

Duncan, Goodrich, and Kobourov, *BAR Trees*, JGAA, 4(3) 19–46 (2000)   42

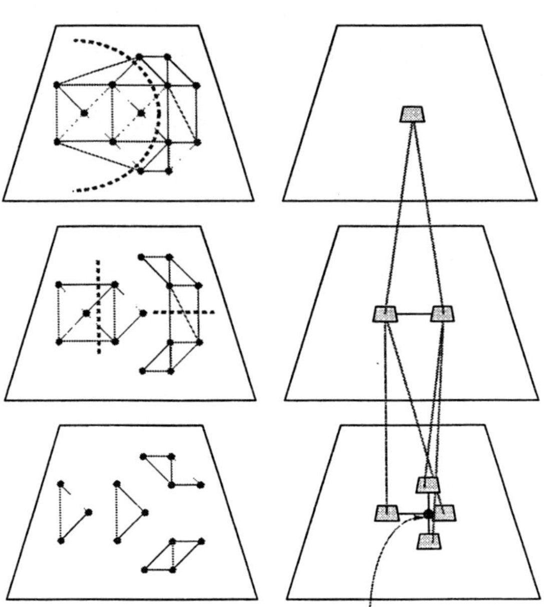

Figure 16: An example of a graph in which each cluster is represented by a single node. Note that the non-straight line cut produces a crossing in the multi-level graph.

cut but we can modify the BAR tree construction so that we find locally optimal cuts. Here are some of the possible criteria that we can use in choosing among the potential cuts: minimize cut size, minimize connected components resulting from a given cut, minimize aspect ratio, maximize β-balance.

These criteria can also be combined in various ways to produce desired optimization functions. In finding such optimal cuts, it is important to note that a one-cut, if available, might not always be a better choice over a potential two-cut. Yet again, a two-cut that minimizes the cut size may have no subsequent one-cut that does not cut many more edges. Thus, it may be reasonable to go two levels in evaluating possible scores instead of choosing greedily.

5   Conclusion and Open Problems

In this paper we present a straightforward and efficient algorithm for displaying large graphs. The LGD algorithm optimizes cluster balance, cluster depth, aspect ratio and convexity. Our algorithm does not rely on any specific graph properties, although various properties can aid in performance, and produces the clustered graph in a very efficient $O(n \log n + m + D_0(G))$ time.

The embedding of the cluster graph is determined in the very first step of our algorithm. Unfortunately, it is possible that the initial embedding is not the best one (for example, in terms of the size of the cuts produced by our

Duncan, Goodrich, and Kobourov, *BAR Trees*, JGAA, 4(3) 19–46 (2000)   43

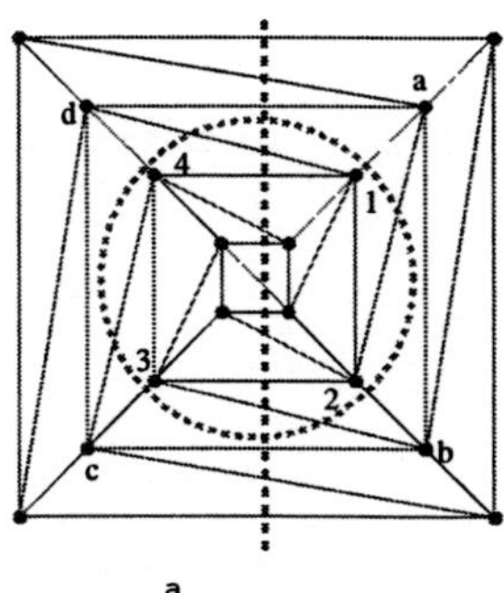
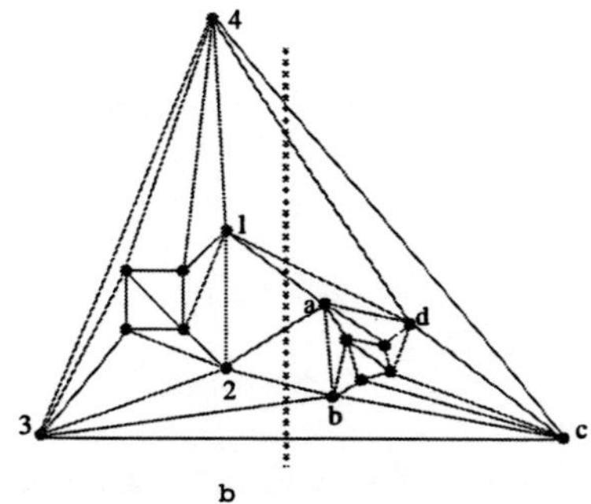

Figure 17: The graph in part (a) has no β-balanced line cut of size better than $O(n)$ but it does have a cycle cut (the dotted circle) of size $O(1)$. We can transform the graph in (a) to the graph in (b) by taking one of the faces crossed by the cycle as the outer face. Note that in (b) the cycle cut has become a line and its size is $O(1)$.

algorithm). In fact, as shown on Figure 17, G may have a minimum β-balanced cut of size $O(n)$ or $O(1)$, depending on the embedding. While it is still true that some graphs may always have cuts of size $O(n)$ (for example, the star graph, Figure 15), we would like to minimize the cut whenever we can. It is an open question whether it is possible to determine the optimal embedding, one that yields the minimum β-balanced cuts.

Another open question is related to the separator theorems of Lipton and Tarjan [21] and Miller [22]. Is it possible given a 2-connected planar graph G to always produce $O(\sqrt{dn})$ β-balanced cuts, where d is its maximum degree, and n is the number of vertices? If so, can we find an embedding for the resulting clustered graph which preserves efficiency, cluster balance, cluster depth, convexity, and guarantees good aspect ratio and straight-line drawings without crossings?

Acknowledgements

We would like to thank Rao Kosaraju and David Mount for their helpful comments regarding the balanced aspect ratio tree.

Duncan, Goodrich, and Kobourov, *BAR Trees*, JGAA, 4(3) 19–46 (2000)   44

References

[1] M. R. Anderberg. *Cluster Analysis for Applications*. Academic Press, New York, 1973.

[2] S. Arya and D. M. Mount. Approximate range searching. In *Proceedings of the 11th Annual ACM Symposium on Computational Geometry*, pages 172–181, 1995.

[3] S. Arya, D. M. Mount, N. S. Netanyahu, R. Silverman, and A. Wu. An optimal algorithm for approximate nearest neighbor searching. In *Proceedings of the 5th Annual ACM-SIAM Symposium on Discrete Algorithms*, pages 573–582, 1994.

[4] P. B. Callahan and S. R. Kosaraju. A decomposition of multidimensional point sets with applications to k-nearest-neighbors and n-body potential fields. *Journal of the ACM*, 42:67–90, 1995.

[5] S. P. Dandamudi and P. G. Sorenson. An empirical performance comparison of some variations of the k-d tree and BD tree. *International Journal of Computer and Information Sciences*, 14(3):135–159, June 1985.

[6] H. de Fraysseix, J. Pach, and R. Pollack. Small sets supporting Fary embeddings of planar graphs. In *Proceedings of the 20th Annual ACM Symposium on Theory of Computing (STOC)*, pages 426–433, 1988.

[7] G. Di Battista, P. Eades, R. Tamassia, and I. G. Tollis. Algorithms for drawing graphs: an annotated bibliography. *Computational Geometry: Theory and Applications*, 4:235–282, 1994.

[8] G. Di Battista, P. Eades, R. Tamassia, and I. G. Tollis. *Graph Drawing: Algorithms for the Visualization of Graphs*. Prentice Hall, Englewood Cliffs, NJ, 1999.

[9] C. A. Duncan, M. T. Goodrich, and S. G. Kobourov. Balanced aspect ratio trees: Combining the advantages of k-d trees and octrees. In *Proceedings of the 10th ACM-SIAM Symposium on Discrete Algorithms (SODA)*, pages 300–309, 1999.

[10] P. Eades and Q.-W. Feng. Multilevel visualization of clustered graphs. In *Proceedings of the 4th Symposium on Graph Drawing (GD '96)*, pages 101–112, 1996.

[11] P. Eades, Q.-W. Feng, and X. Lin. Straight-line drawing algorithms for hierarchical graphs and clustered graphs. In *Proceedings of the 4th Symposium on Graph Drawing (GD '96)*, pages 113–128, 1996.

[12] I. Fáry. On straight lines representation of planar graphs. *Acta Scientiarum Mathematicarum*, 11:229–233, 1948.

Duncan, Goodrich, and Kobourov, *BAR Trees*, JGAA, 4(3) 19–46 (2000)   45

[13] Q.-W. Feng, R. F. Cohen, and P. Eades. How to draw a planar clustered graph. In *Proceedings of the 1st Annual International Conference on Computing and Combinatorics (COCOON '95)*, pages 21–31, 1995.

[14] Q.-W. Feng, R. F. Cohen, and P. Eades. Planarity for clustered graphs. In *Third Annual European Symposium on Algorithms (ESA '95)*, pages 213–226, 1995.

[15] J. H. Friedman, J. L. Bentley, and R. A. Finkel. An algorithm for finding best matches in logarithmic expected time. *ACM Transactions on Mathematical Software*, 3:209–226, 1977.

[16] G. W. Furnas. Generalized fisheye views. In *Proceedings of ACM Conference on Human Factors in Computing Systems (CHI '86)*, pages 16–23, 1986.

[17] J. A. Hartigan. *Clustering Algorithms*. John Wiley & Sons, New York, 1975.

[18] K. Kaugars, J. Reinfelds, and A. Brazma. A simple algorithm for drawing large graphs on small screens. In *Graph Drawing (GD '94)*, pages 278–281, 1995.

[19] D. T. Lee and C. K. Wong. Worst-case analysis for region and partial region searches in multidimensional binary search trees and balanced quad trees. *Acta Informatica*, 9(2):23–29, Apr. 1978.

[20] R. J. Lipton, S. C. North, and J. S. Sandberg. A method for drawing graphs. In *Proceedings of the 1st Annual ACM Symposium on Computational Geometry*, pages 153–160, 1985.

[21] R. J. Lipton and R. E. Tarjan. Applications of a planar separator theorem. *SIAM Journal of Computing*, 9:615–627, 1980.

[22] G. L. Miller. Finding small simple cycle separators for 2-connected planar graphs. *Journal of Computer and System Sciences*, 32(3):265–279, 1986.

[23] F. J. Newbery. Edge concentration: A method for clustering directed graphs. In *Proceedings of the 2nd International Workshop on Software Configuration Management*, pages 76–85, 1989.

[24] S. C. North. Drawing ranked digraphs with recursive clusters. In *Graph Drawing, ALCOM International Workshop PARIS on Graph Drawing and Topological Graph Algorithms (GD '93)*, Sept. 1993.

[25] R. Sablowski and A. Frick. Automatic graph clustering. *Proc. of 4th Symposium on Graph Drawing (GD '96)*, LNCS 1190:395–400, 1996.

[26] H. Samet. *The Design and Analysis of Spatial Data Structures*. Addison-Wesley, Reading, MA, 1990.

Duncan, Goodrich, and Kobourov, *BAR Trees*, JGAA, 4(3) 19–46 (2000)   46

[27] M. Sarkar and M. H. Brown. Graphical fisheye views. *Communications of the ACM*, 37(12):73–84, 1994.

[28] W. Schnyder. Embedding planar graphs on the grid. In *Proceedings of the 1st ACM-SIAM Symposium on Discrete Algorithms (SODA)*, pages 138–148, 1990.

[29] Y. V. Silva-Filho. Average case analysis of region search in balanced k-d trees. *Information Processing Letters*, 8(5):219–223, June 1979.

[30] S. K. Stein. Convex maps. *Proceedings of the American Mathematical Society*, 2(3):464–466, 1951.

[31] K. Sugiyama and K. Misue. Visualization of structural information: Automatic drawing of compou nd digraphs. *IEEE Transactions on Systems, Man, and Cybernetics*, 21(4):876–892, 1991.

[32] W. T. Tutte. How to draw a graph. *Proceedings London Mathematical Society*, 13(52):743–768, 1963.

[33] K. Wagner. Bemerkungen zum vierfarbenproblem. *Jahresbericht der Deutschen Mathematiker-Vereinigung*, 46:26–32, 1936.

Journal of Graph Algorithms and Applications

http://www.cs.brown.edu/publications/jgaa/

vol. 4, no. 3, pp. 47–74 (2000)

Difference Metrics for Interactive Orthogonal Graph Drawing Algorithms

Stina Bridgeman   *Roberto Tamassia*

Center for Geometric Computing
Department of Computer Science
Brown University
Providence, Rhode Island 02912–1910
{ssb,rt}@cs.brown.edu

Abstract

Preserving the "mental map" is a major goal of interactive graph drawing algorithms. Several models have been proposed for formalizing the notion of mental map. Additional work needs to be done to formulate and validate "difference" metrics which can be used in practice. This paper introduces a framework for defining and validating metrics to measure the difference between two drawings of the same graph, and gives a preliminary experimental analysis of several simple metrics.

This version of the paper is suitable for color printing.

Communicated by G. Liotta and S. H. Whitesides: submitted February 1999; revised October 1999.

Research supported in part by the U.S. Army Research Office under grant DAAH04-96-1-0013, by the National Science Foundation under grants CCR-9732327 and CDA-9703080, and by a National Science Foundation Graduate Fellowship.

S. Bridgeman and R. Tamassia, *Difference Metrics*, JGAA, 4(3) 47–74 (2000)   48

1   Introduction

Graph drawing algorithms have traditionally been developed using a batch model, where the graph is redrawn from scratch every time a drawing is desired. These algorithms, however, are not well suited for interactive applications, where the user repeatedly makes modifications to the graph and requests a new drawing. When the graph is redrawn it is important to preserve the look (the user's "mental map" [19]) of the original drawing as much as possible, so the user does not need to spend a lot of time relearning the graph.

The problems of incremental graph drawing, where vertices are added one at a time, and the more general case of interactive graph drawing, where any combination of vertex/edge deletion and insertion is allowed at each step, have been starting to receive more attention. See, for example, the work of Biedl and Kaufmann [3], Brandes and Wagner [4], Bridgeman et. al. [6], Cohen et. al. [8], Fößmeier [12], Miriyala, Hornick, and Tamassia [18], Moen [20], North [21], Papakostas, Six, and Tollis [22], Papakostas and Tollis [23], and Ryall, Marks, and Shieber [26]. However, while the algorithms themselves have been motivated by the need to preserve the user's mental map, much of the evaluation of the algorithms has so far focused on traditional optimization criteria such as the area and the number of bends and crossings (see, for example, the analysis in Biedl and Kaufmann [3], Fößmeier [12], Papakostas, Six, and Tollis [22], and Papakostas and Tollis [23]). Mental map preservation is often achieved by attempting to minimize the change between drawings — typically by allowing only very limited modifications (if any) to the position of vertices and edge bends in the existing drawing — making it important to be able to measure precisely how much the look of the drawing changes. Animation can be used to provide a smooth transition between the drawings and can help compensate for greater changes in the drawing, though it is still important to limit, if not minimize, the difference between the drawings because if there is a very large change it can become difficult to generate a clear, useful animation. It is thus still important to have a measure of how the look of the drawing changes.

Studying "difference" metrics to measure how much a drawing algorithm changes the user's mental map has a number of benefits, including

- providing a basis for studying the behavior of constraint-based interactive drawing algorithms like InteractiveGiotto [6], where meaningful bounds on the movement of any given part of the drawing are difficult to obtain,

- providing a technique to compare the results of different interactive drawing algorithms, and

- providing a goal for the design of new drawing algorithms by identifying which qualities of the drawing are the most important to preserve.

Finding a good difference metric also has an immediate practical benefit, namely solving the "rotation problem" of InteractiveGiotto. Giotto [27], the core of InteractiveGiotto, does not take into account the coordinates of vertices and bends in the

S. Bridgeman and R. Tamassia, *Difference Metrics*, JGAA, 4(3) 47–74 (2000)   49

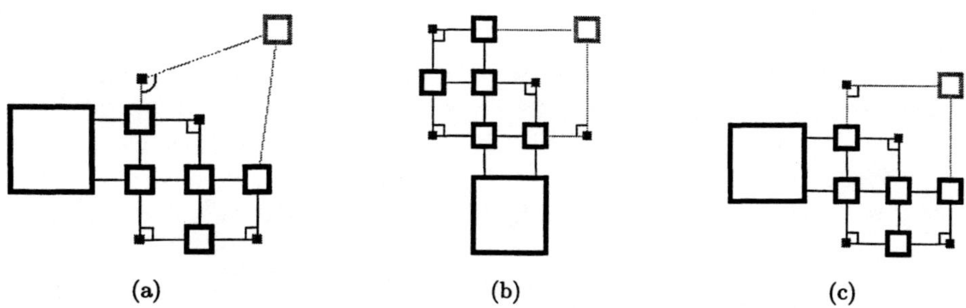

Figure 1: The rotation problem of InteractiveGiotto. (a) is the user-modified graph (the user's changes are shown in orange); (b) and (c) show the uncorrected and corrected outputs, respectively. While (b) and (c) are both clearly drawings of the graph shown in (a), the resemblance is more readily seen in the properly rotated and reflected drawing (c).

original drawing when constructing a new drawing — and, as a result, the output of InteractiveGiotto may be rotated by a multiple of 90 degrees and/or be a mirror-image reflection of the original drawing (Figure 1). The problem can be solved by computing the value of the metric for each of the possible rotations and choosing the rotation with the smallest value.

Eades et. al. [11], Lyons, Meijer, and Rappaport [16], and Misue et. al. [19] have proposed several models for formalizing the notion of the mental map, though more work needs to be done to formally define potential difference metrics and then experimentally validate (or invalidate) them. Validation can be via user studies similar to those done by Purchase, Cohen, and James [24, 25] to evaluate the impact of various graph drawing aesthetics on human understanding.

Motivated by applications to InteractiveGiotto, this paper will focus primarily on difference metrics for orthogonal drawings, though many of the metrics can be used without modification for arbitrary drawings. Section 3 describes several potential metrics, Section 4 presents a framework for evaluating the suitability of the metrics along with a preliminary evaluation, and Section 5 outlines plans for future work.

2 Preliminaries

Paired Sets of Objects Every metric presented in this paper compares two drawings D and D' of the same graph G. Each object of G has a representation in both D and D'. For example, each vertex of G has a representation in both drawings — if vertices are drawn as rectangles, this representation consists of the position, size, color, line style, etc. of the rectangle. Similarly, each edge of G has a representation in both drawings, and if edges are drawn as polylines, this

S. Bridgeman and R. Tamassia, *Difference Metrics*, JGAA, 4(3) 47–74 (2000)   50

representation consists of the positions of the bends and endpoints, plus the color, line style, and so forth.

A *paired set* of objects is a set of pairs describing the representation of each object in the two drawings. The paired set of vertices of G is the set of pairs (r_v, r'_v), where r_v and r'_v are the representations of v in D and D', respectively, for all vertices v of G. The paired set of edges is defined similarly. Referring to a paired set is simply a way of matching up the elements of each drawing according to the underlying object of G that they represent.

It should be noted that the only features of the representations that are considered here are the geometric features such as position and size; other features like vertex color and line style are also very important in preserving the look of the drawing and may be able to at least partially compensate for geometric changes but are not considered further at this stage.

Point Set Selection   Most of the metrics are based on point sets, working with paired sets of points derived from the edges and vertices of the graph rather than the edges and vertices themselves. Once derived, each point is independent from the others — there is no notion of a group of points being related because they were derived from the same vertex, for example.

Points can be selected in a number of ways. North [21] suggests that vertex positions are a significant visual feature of the drawing, and two vertex-centered methods — *centers* and *corners* — are used here to reflect that idea. "Centers" consists of the center points of each vertex; this captures how vertices move. "Corners" uses the four corners of each vertex, taking into account both vertex motion (the movement of the center) and changes in the vertex dimension. It seems important to take into account changes in vertex dimension because a vertex with a large or distinctive shape can act as a landmark to orient the user to the drawing; loss of that landmark makes orientation more difficult. Other choices of points can include edge bends and endpoints.

Points can also be derived from groups of graph objects. For example, the vertices of the graph can be partitioned, and points derived from the centroids or convex hull of the partitions. Point sets based on partitioning can be used to capture information about larger units of the graph, such as groups of vertices representing related objects.

Drawing Alignment   The key features of a graph object's representation are coordinates, which means metrics may be sensitive to the particular values of those coordinates — the scaling and translation of one point set relative to the other can make a large difference in the value of the metric (Figure 2).

To eliminate this effect, the drawings are *aligned* before coordinate-sensitive metrics are computed. This is done by extracting a (paired) set of points from the drawings and applying a point set matching algorithm to obtain the best fit. In

S. Bridgeman and R. Tamassia, *Difference Metrics*, JGAA, 4(3) 47–74 (2000)   51

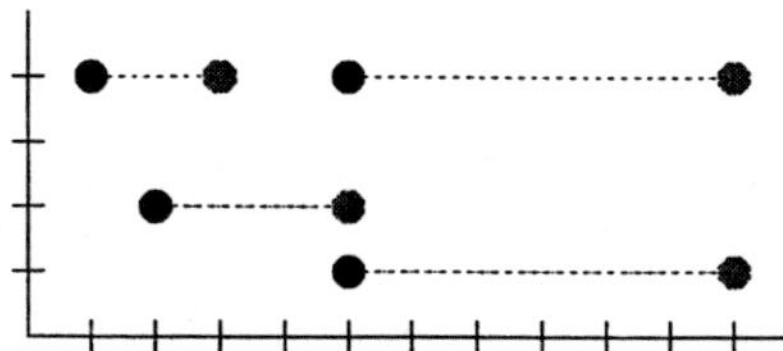

Figure 2: Two point sets (black and gray) superimposed. (Corresponding points in the two sets are connected with dotted lines.) As shown, the Euclidean distance metric (Section 3.1) would report a distance of 4.25. However, translating the gray points one unit to the left and then scaling by 1/2 in the x direction allows the point sets to be matched exactly, for a distance of 0. It should be noted that exact matches are not possible in general.

general the matching algorithm should take into account scaling, translation, and rotation, though it may be possible to eliminate one or more of the transformations for certain metrics or if something is known about the relationship between the two drawings. For example, interactive drawing algorithms often preserve the rotation of the drawing (see Biedl and Kaufmann [3] and Papakostas and Tollis [23] for examples), eliminating the need to consider rotation in the alignment stage. Even if the algorithm does not preserve the rotation, for orthogonal drawings there are only eight possible rotations for the second drawing relative to the first — four multiples of $\pi/2$, applied to the original drawing and its reflection about the x axis — which can be handled by computing the metric separately for each rotation and taking the minimum value instead of incorporating rotation into the alignment process.

A great deal of work has been done on point set matchings; see Alt, Aichholzer, and Rote [1], Chew et. al. [7], and Goodrich, Mitchell, and Orletsky [14] for several methods of obtaining both optimal and approximate matchings. Different methods can be applied when the correspondence between points is known as it is here; Imai, Sumino, and Imai [15] provide an algorithm that minimizes the maximum distance between corresponding points under translation, rotation, and scaling. In the implementation used in Section 4, the alignment is performed by using gradient search to minimize the distance squared between points.

3   Metrics

The difference metrics being considered fall into six categories:

- **distance:** metrics based on the distance between points or the distance points move between drawings

- **proximity:** metrics based on the nearness of points and the clustering of points according to the distance between them

S. Bridgeman and R. Tamassia, *Difference Metrics*, JGAA, 4(3) 47–74 (2000)   52

- **partitioning**: metrics based on partitioning points according to measures other than proximity
- **orthogonal ordering**: metrics based on the relative angle between pairs of points
- **shape**: metrics based on the sequence of horizontal and vertical segments of the graph's edges
- **topology**: metrics based on the embedding of the graph

Proximity, ordering, and topology are suggested by Eades et. al. [11], Lyons, Meijer, and Rappaport [16], and Misue et. al. [19] as qualities which should be preserved; distance (also suggested by Lyons, Meijer, and Rappaport [16]), shape, and partitioning reflect intuition about what causes drawings to look different. Within each category specific metrics were chosen to capture intuition about what qualities of the drawing are important to preserve.

An alternative taxonomy is given by Biedl et. al. [2]. This taxonomy is similar to the one given above, with the main distinctions being the inclusion of "feature similarity" based on the appearance of regions of the drawing, and the grouping of all measures based on the comparison of point sets into a single "metric similarity" category.

In the following, let P denote the (paired) set of points, and p_i and p_i' be the coordinates of point i in drawings D and D', respectively. Also let $d(p, q)$ be the Euclidean distance between points p and q.

3.1   Distance

The distance metrics reflect the simple observation that drawings that look very different cannot be aligned very well, and vice versa. Since the alignment is based on distance minimization, these metrics essentially measure the quality of the alignment.

In order to make the value of the distance metrics comparable between pairs of drawings, they are scaled by the graph's *unit length u*. For orthogonal drawings the unit length can be computed by taking the greatest common divisor of the Manhattan distances between vertex centers and bend points on edges. Non-orthogonal portions of the drawing, such as modifications of an orthogonal drawing made by the user, can be ignored during the computation. While the determination of the unit length will be unreliable if only a small portion of the drawing is orthogonal, scaling by the unit length is not necessary in some applications (e.g., solving the rotation problem of InteractiveGiotto) and can often be supplied manually if it is required (e.g., the drawing algorithm is known to place vertices on a unit grid).

Hausdorff Distance   The *undirected Hausdorff distance* is a standard metric for determining the distance between two point sets, and measures the largest distance

S. Bridgeman and R. Tamassia, *Difference Metrics*, JGAA, 4(3) 47–74 (2000)   53

between a point in one set and its nearest neighbor in the other.

$$\text{haus}(D, D') = \frac{1}{u} \max\{\max_i \min_j d(p_i, p_j'), \max_i \min_j d(p_i', p_j)\}$$

where $1 \leq i, j \leq |P|$ and $j \neq i$.

Euclidean Distance   *Euclidean distance*, used by Lyons, Meijer, and Rappaport [16], is a simple metric measuring the average distance moved by each point from the first drawing to the second; it is motivated by the notion that if points move a long way from their locations in the first drawing, the second drawing will look very different.

$$\text{dist}(D, D') = \frac{1}{u\,|P|} \sum_{1 \leq i \leq |P|} d(p_i, p_i')$$

Relative Distance   *Relative distance* measures the average change in the distance between each pair of points between the first drawing and the second. This measures how much the points in each drawing move relative to each other; it is similar in some respects to the orthogonal ordering metrics in Section 3.4.

$$\text{rdist}(D, D') = \frac{1}{|P|\,(|P| - 1)} \sum_{1 \leq i,j \leq |P|} \left| d(p_i, p_j) - d(p_i', p_j') \right|$$

3.2   Proximity

The proximity metrics reflect the idea that points near each other in the first drawing should remain near each other in the second drawing. This is stronger than the distance metrics because it captures the idea that if an entire subgraph moves relative to another (and there are only small changes within each subgraph), the distance should be less than if each point in one of the subgraphs moves in a different direction (Figure 3).

Three different metrics are used to try to capture this idea: nearest neighbor within (nn-within), nearest neighbor between (nn-between), and ϵ-clustering.

Nearest Neighbor Within   *Nearest neighbor within* is based on the reasoning that if p_j is the closest point to p_i in D, then p_j' should be closest point to p_i' in D'. Considering only distances within a single drawing means that nn-within is alignment-independent and thus not subject alignment errors, but means that it is not suitable for solving the rotation problem of InteractiveGiotto.

This metric has two versions, *weighted* and *unweighted*. In the weighted version the number of points closer to p_i' than p_j' is considered, whereas in the unweighted version only whether or not p_j' is the closest point matters. The reasoning behind the weighted version is that if there are more points between p_i' and p_j', the visual

S. Bridgeman and R. Tamassia, *Difference Metrics*, JGAA, 4(3) 47–74 (2000)   54

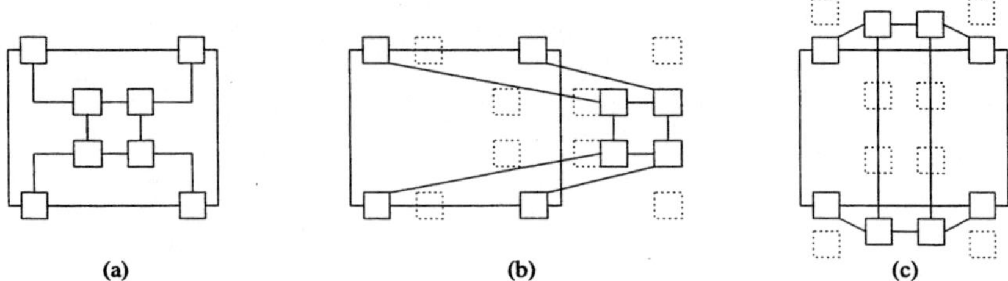

Figure 3: Proximity: (b) looks more like (a) than (c) does because the relative shape of both the inner and outer squares are preserved even though the distance (using the Euclidean distance metric) between (c) and (a) is smaller. An aligned version of the vertices of (a), used in the computation of the distance metric, is shown with dotted lines in (b) and (c).

linkage between p'_i and p'_j has been disrupted to a greater degree and the drawing looks more different.

In both cases the distance is scaled by the number of points being considered and W, the maximum weight contributed by a single point, so that the metric's value is always in the range $[0, 1]$.

$$\text{nn-w}(D, D') = \frac{1}{W\,|P|} \sum_{1 \le i \le |P|} \text{closer}(p'_i, p'_j)$$

where p_j is the closest point to p_i in D and

$$\text{closer}(p'_i, p'_j) = \left|\{k \mid d(p'_i, p'_k) < d(p'_i, p'_j)\}\right|, \qquad\qquad W = |P| - 2 \qquad \text{(weighted)}$$

$$\text{closer}(p'_i, p'_j) = \begin{cases} 0 & \text{if } d(p'_i, p'_j) \le d(p'_i, p'_k), k \ne i \\ 1 & \text{otherwise} \end{cases}, \quad W = 1 \qquad \text{(unweighted)}$$

Nearest Neighbor Between   *Nearest neighbor between* is similar to nn-within but instead measures whether or not p'_i is the closest of the points in D' to p_i when the two drawings are aligned. The idea that a point should remain nearer to its original position than any other is also the force behind layout adjustment algorithms based on the Voronoi diagram [16].

$$\text{nn-b}(D, D') = \frac{1}{W\,|P|} \sum_{1 \le i \le |P|} \text{closer}(p_i, p'_i)$$

where

$$\text{closer}(p_i, p'_i) = \left|\{j \mid d(p_i, p'_j) < d(p_i, p'_i)\}\right|, \qquad\qquad W = |P| - 1 \qquad \text{(weighted)}$$

$$\text{closer}(p_i, p'_i) = \begin{cases} 0 & \text{if } d(p_i, p'_i) \le d(p_i, p'_j), j \ne i \\ 1 & \text{otherwise} \end{cases}, \quad W = 1 \qquad \text{(unweighted)}$$

S. Bridgeman and R. Tamassia, *Difference Metrics*, JGAA, 4(3) 47–74 (2000)   55

Unlike nn-within, nn-between is not alignment- and rotation-independent and thus is suitable for solving the rotation problem.

ϵ**-Clustering**   The definition for an ϵ-cluster is from Eades et. al [11]: An ϵ-cluster for a point p_i is the set of points p_j such that $d(p_i, p_j) \leq \epsilon$, where a reasonable value to use for ϵ is

$$\epsilon = \max_i \min_{j \neq i} d(p_i, p_j)$$

The ϵ-*cluster* metric measures how the ϵ-cluster for p_i compares to that for p_i'. Let $C_D = \{(i, j) \mid d(p_i, p_j) \leq \epsilon_D\}$ and $C_{D'} = \{(i, j) \mid d(p_i', p_j') \leq \epsilon_{D'}\}$. Then

$$\text{clust}(D, D') = 1 - \frac{|C_D \cap C_{D'}|}{|C_D \cup C_{D'}|}$$

The idea is that points should be in the same ϵ-cluster in both drawings.

3.3   Partitioning

The partitioning metrics are based on dividing the point set into subsets according to some criteria, and measuring qualities of these partitions. The motivation for this is to capture "visual units" that the user may use for orientation when learning the new drawing.

Fixed Relative Position Partitioning   A variety of partitioning methods are possible. A simple one is to divide the point set so that the points in each group have the same relative position in both drawings. This identifies blocks of the graph that are the same in both drawings — the larger the partitions, the more unchanged parts and the more similar the drawings.

Two metrics are computed, the average partition size and the number of partitions. Both are scaled to have values between 0 and 1:

$$\text{alsz}(D, D') = 1 - \frac{\left(\dfrac{1}{k} \sum_{1 \leq i \leq k} |S_i|\right) - 1}{|P| - 1} \qquad \text{[average size]}$$

$$\text{alct}(D, D') = \frac{k - 1}{|P| - 1} \qquad \text{[number of partitions]}$$

where the set of partitions is $\{S_1, \ldots, S_k\}$. The idea is that as the drawings become more different, the partition size will decrease and the number of partitions will increase.

The partitioning method and the metrics computed are obviously quite simple, and can be made much more sophisticated. For example, the partitions can be

S. Bridgeman and R. Tamassia, *Difference Metrics*, JGAA, 4(3) 47–74 (2000)   56

adjusted to only include points that are also close physically; while it tends not to be the case that points from widely separated regions of the drawing are in the same partition, it can occur. A large distance between points interferes with their grouping as a single visual unit.

Additional sophistications can address things such as how visually separate two adjacent partitions are, since if they are very near or entertwined in one drawing, it is more difficult for the observer to distinguish them and use them as landmarks for the other drawing. Partitions can also be weighted to take into account how well the partition reflects a distinct unit of the graph, so that partitions containing only points derived from a connected subgraph are better than those containing points from several unconnected portions of the graph, even if those subgraphs are physically close together.

3.4   Orthogonal Ordering

The *orthogonal ordering* metric reflects the desire to preserve the relative ordering of every pair of points — if p_i is northeast of p_j in D, p_i' should remain to the northeast of p_j' in D' (Eades et. al. [11] and Misue et. al. [19]). The simplest measurement of difference in the orthogonal ordering is to take the angle between the vectors $p_j - p_i$ and $p_j' - p_i'$ (*constant-weighted* orthogonal ordering). This has the desirable feature that if p_j is far from p_i, $d(p_j, p_j')$ must be larger to result in the same angular move, which reflects the intuition that the relative position of points near each other is more important that the relative position of points that are far apart.

However, simply using the angular change fails to take into account situations such as that shown in Figure 4. This problem can be addressed by introducing a weight that depends on the particular angles involved in the move in addition to size of the move (*linear-weighted* orthogonal ordering).

$$\text{order}(D, D') = \frac{1}{W|P|} \sum_{1 \le i,j \le |P|} n \min(\text{order}(\theta_{ij}, \theta_{ij}'), \text{order}(\theta_{ij}', \theta_{ij}))$$

where θ_{ij} is the angle from the positive x axis to the vector $p_j - p_i$, θ_{ij}' is the angle from the positive x axis to the vector $p_j' - p_i'$, and

$$\text{order}(\theta_{ij}, \theta_{ij}') = \int_{\theta_{ij}}^{\theta_{ij}'} \text{weight}(\theta)\, d\theta, \quad W = \min\left\{ \int_0^\pi \text{weight}(\theta)\, d\theta, \int_\pi^{2\pi} \text{weight}(\theta)\, d\theta \right\}$$

The weight functions are

$$\text{weight}(\theta) = \begin{cases} \frac{\frac{\pi}{2} - (\theta \bmod \frac{\pi}{2})}{\frac{\pi}{4}} & \text{if } (\theta \bmod \frac{\pi}{2}) > \frac{\pi}{4} \\ \frac{\theta \bmod \frac{\pi}{2}}{\frac{\pi}{4}} & \text{if } (\theta \bmod \frac{\pi}{2}) \le \frac{\pi}{4} \end{cases} \qquad \text{(linear-weighted)}$$

$$\text{weight}(\theta) = 1 \qquad\qquad\qquad\qquad\qquad \text{(constant-weighted)}$$

S. Bridgeman and R. Tamassia, *Difference Metrics*, JGAA, 4(3) 47–74 (2000)   57

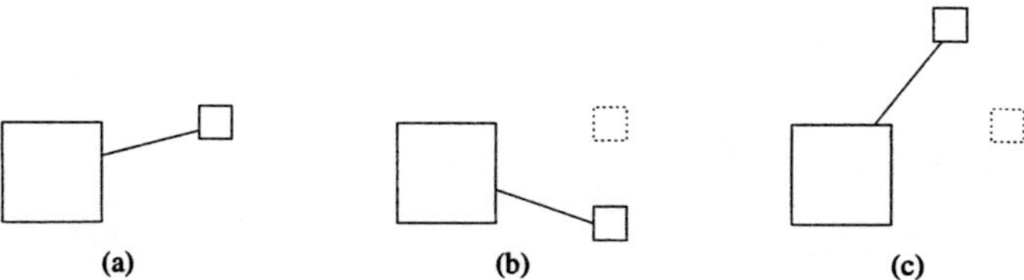

Figure 4: Orthogonal ordering: Even though the angle the vertex moves relative to the center of the large vertex is the same from (a) to (b) and from (a) to (c), the perceptual difference between (a) and (c) is much greater. The original location of the vertex is shown with a dotted box in (b) and (c) for comparison purposes.

The λ-matrix model for measuring the difference of two point sets, used by Lyons, Meijer, and Rappaport [16], is based on the concept of order type of a point set, from Goodman and Pollack [13]. This model tries to capture the notion of the relative position of vertices in a straight-line drawing and is thus related to the orthogonal ordering metric.

3.5 Shape

The *shape* metric is motivated by the reasoning that edge routing may have an effect on the overall look of the graph (Figure 5). The shape of an edge is the sequence of directions (north, south, east, and west) traveled when traversing the edge; writing the shape as a string of N, S, E, and W characters yields the *shape string* of the edge. For non-orthogonal edges the direction is taken to be the most prominent direction; for example, if the edge goes from (1,1) to (4,2) the most prominent direction is east. For each pair of edges (e_i, e_i'), the edit distance between the corresponding shape strings is computed. Two methods are used for determining the edit distance. One uses dynamic programming to compute the minimum number of insertions, deletions, or replacements of characters needed to transform one string into the other. The other is similar, but normalizes the measure according to the length of the strings; the algorithm is given by Marzal and Vidal [17]. The value of the shape metric is the average edit distance over the graph's edges.

$$\text{shape}(D, D') = \frac{1}{|E|} \sum_{1 \leq i \leq |E|} \text{edits}(e_i, e_i')$$

Shape is scale- and translation-independent.

The shape metric is similar in spirit to the cost function used by Brandes and Wagner [5] in their dynamic extension of **Giotto** [27]. Their cost function counts the number of changes in angles at vertices and edge bends; the shape metric takes this in account to some degree by noting changes in the direction of an edge segment.

S. Bridgeman and R. Tamassia, *Difference Metrics*, JGAA, 4(3) 47–74 (2000)   58

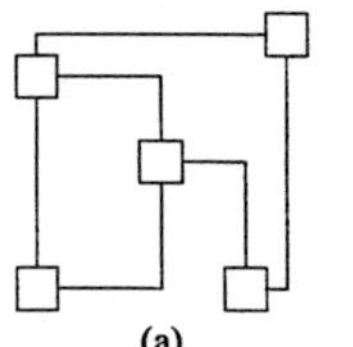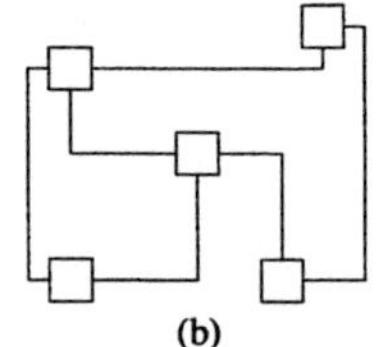

(a)   (b)

Figure 5: Shape: (a) and (b) look different even though the graphs are the same and the vertices have the same coordinates.

3.6 Topology

The topology metric reflects the idea that preserving the order of edges around a vertex is important in preserving the mental map (Eades et. al. [11] and Misue et. al. [19]) — comparing the drawing produced by **Giotto** in Figures 6 and 7 to the user's input illustrates this. However, since most interactive orthogonal drawing algorithms always preserve topology, it is not useful as a means of comparing these algorithms. (See, for example, the algorithms of Bridgeman et. al. [6], Biedl and Kaufmann [3], Fößmeier [12], Papakostas, Six, and Tollis [22], and Papakostas and Tollis [23].) Topology is also alignment-independent and so can not be used to solve the rotation problem of **InteractiveGiotto**. As a result, it is not discussed in any more detail here.

4 Analyzing the Metrics

Once defined, the suitability of the metrics must be evaluated. A good metric for measuring the difference between drawings should satisfy the following three requirements:

- it should *qualitatively* reflect the visual difference between two drawings, i.e. the value increases as the drawings diverge;

- it should *quantitatively* reflect the visual difference so that the magnitude of the difference in the metric is proportional to the perceived difference; and

- in the rotation problem of **InteractiveGiotto**, the metric should have the smallest value for the correct rotation, though this requirement can be relaxed when the difference between drawings is high since in that case there is no clear "correct" rotation.

The third point is the easiest to satisfy — in fact, most of the metrics defined in the previous section can be used to solve the rotation problem — but is still important worth considering since the problem was one of the factors that first inspired this work.

The running time of the metrics is not considered at this point. While efficiency is clearly a concern, the goal at this stage is to identify the type of measure that

S. Bridgeman and R. Tamassia, *Difference Metrics*, JGAA, 4(3) 47–74 (2000)   59

best captures visual similarity. Once this has been done, efficient implementations and/or approximations can be considered.

Some preliminary work has been done on evaluating the proposed metrics with respect to the first and third criteria. Evaluating the qualitative behavior of potential metrics requires a human-generated master ordering of pairs of drawings based on the visual difference between the existing drawing and the new drawing in each pair. This is very difficult to do when each pair of drawings is of a different graph, and most interactive drawing algorithms only produce a single drawing of a particular user-modified graph. InteractiveGiotto, however, makes it possible to obtain a series of drawings of the same input by relaxing the constraints preserving the layout. By default InteractiveGiotto preserves edge crossings, the direction (left or right) and number of bends on an edge, and the angles between consecutive edges leaving a vertex. Recent modifications allow the user to turn off the last two constraints on an edge-by-edge or vertex-by-vertex basis, making it possible to produce a series of drawings of the same graph by relaxing different sets of constraints. A smooth way of relaxing the constraints is to use a breadth-first ordering, expanding outward from the user's modifications. In the first step all of the constraints are applied, in the second step the bend and angle constraints are relaxed for all of the modified objects, in the third step the angle constraints are relaxed for all vertices adjacent to edges whose bends constraints have been relaxed, in the fourth step the bend constraints are relaxed for all edges adjacent to vertices whose angle constraints have been relaxed, and so on, alternating between angle and bend constraints until all of the constraints have been relaxed. This relaxation method is based on the idea that the user is most willing to allow restructuring of the graph near where her changes were made and so the drawings produced resemble drawings that an actual user might encounter. The result is a series of drawings of the same graph — a *relaxation sequence* — bearing varying degrees of similarity to the original.

4.1 An Example

Figures 6 and 7 show portions of two relaxation sequences produced by InteractiveGiotto; the base graphs and user modifications are those used in the first two steps of Figure 2 in Bridgeman et. al. [6]. Giotto's redraw-from-scratch drawing of the graph is also included for comparison. Figures 8 and 9 show the results of several metric-and-pointset combinations for each sequence of drawings. Since Giotto and InteractiveGiotto do not preserve the orientation of the original drawing, each metric is computed for the eight possible rotations (four multiples of $\pi/2$, with or without a reflection around the x-axis) and the lowest value chosen. The color of each column indicates the *confidence*, a measure of how much lower the metric's value is for the best rotation as compared to the second best; red is the most confident and purple is the least confident, with white indicating that the metric is rotation-independent (and so confidence is meaningless) and black indicating that two rotations had the same lowest value. The shading of the column indicates

S. Bridgeman and R. Tamassia, *Difference Metrics*, JGAA, 4(3) 47–74 (2000)   60

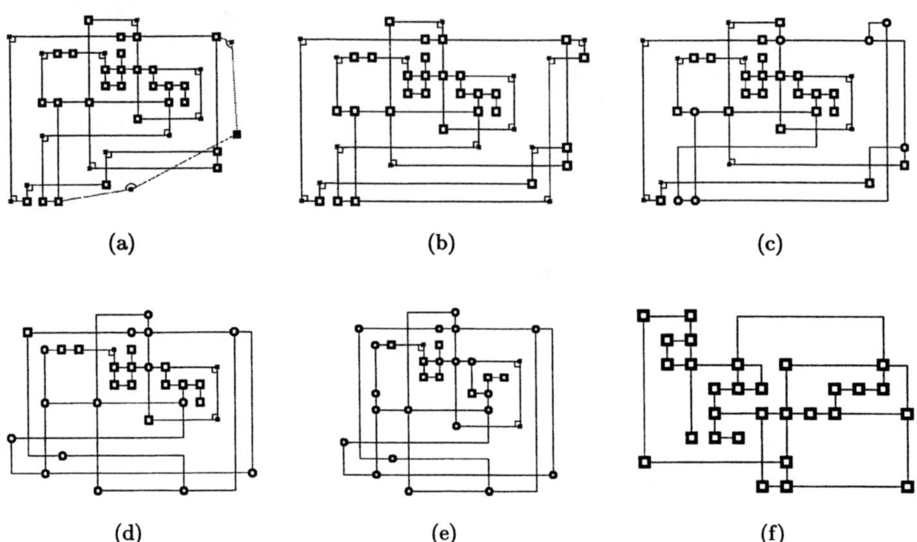

Figure 6: Relaxations for stage 1. (a) shows the user's modifications; (b)–(e) show the output from InteractiveGiotto for relaxation steps 0, 4, 6, and 7; (f) is the output from Giotto. (Intermediate relaxation steps produced drawings identical to those shown and have been left out.) Rounded vertices and bends without markers indicate vertices and bends for which the constraints have been relaxed.

whether or not the metric chose the right rotation — hashing means that the wrong rotation had the lowest value. The correct rotation is defined to be the rotation chosen by the metric/pointset combination with the highest confidence; in practive this is nearly always the rotation a human would pick as the correct answer.

The point sets shown in Figures 8 and 9 show use the following abbreviations:

- **centers**: vertex centers

- **corners**: vertex corners

- **hulls (cen)**: the vertex centers point set is partitioned using fixed relative partitioning; the points used are the points on the convex hull of each partition

- **hulls (cor)**: the vertex corners point set is partitioned using fixed relative partitioning; the points used are the points on the convex hull of each partition

- **centroids (cen)**: the vertex centers point set is partitioned using fixed relative partitioning; the points used are the centroids of each partition

- **centroids (cor)**: the vertex corners point set is partitioned using fixed relative partitioning; the points used are the centroids of each partition

The t or f in brackets following the point set name indicates whether or not points

S. Bridgeman and R. Tamassia, *Difference Metrics*, JGAA, 4(3) 47–74 (2000)   61

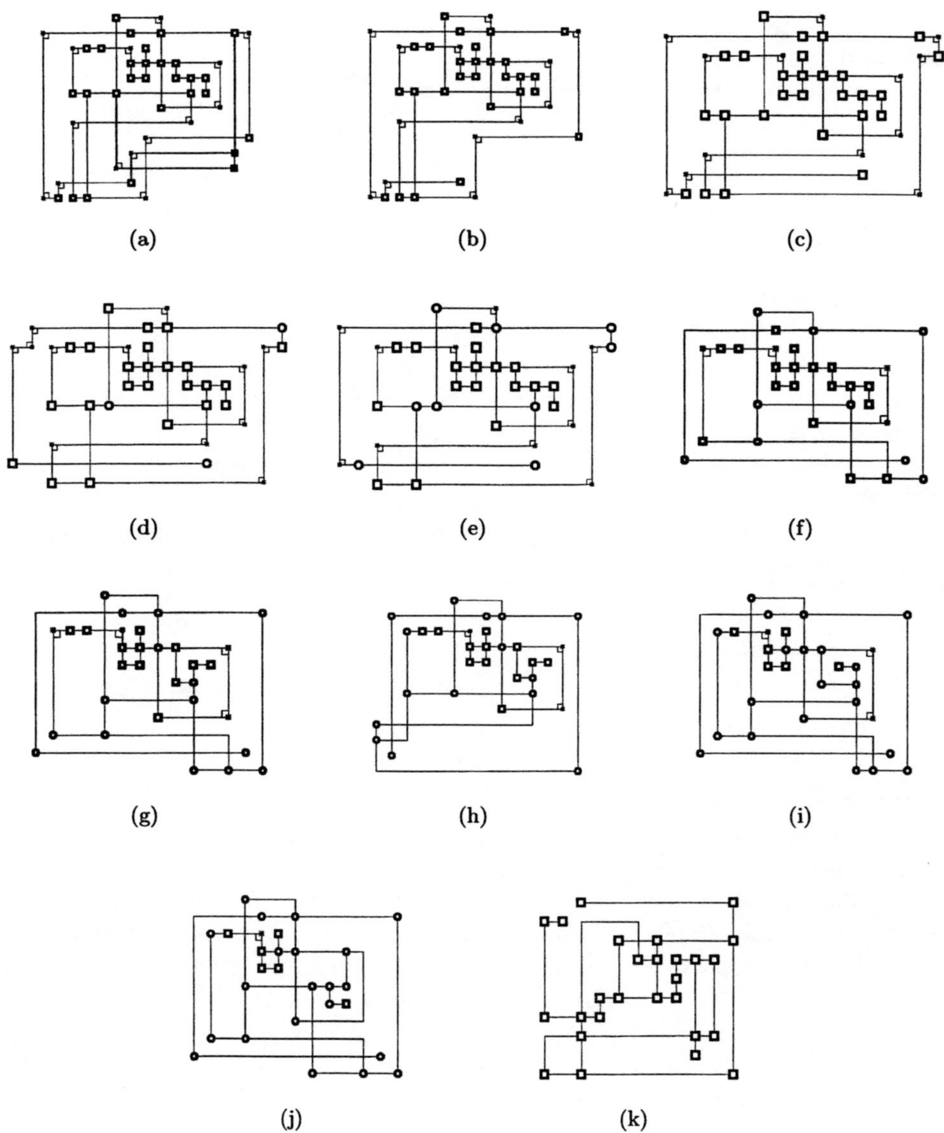

Figure 7: Relaxations for stage 2. (a) shows the starting graph, (b) shows the user's modifications (two vertices and their adjacent edges deleted), (c) is the fully-constrained output from InteractiveGiotto (step 0), (d)–(j) show the output from InteractiveGiotto for relaxation steps 2–8 (step 1 produced the same drawing as step 0), and (k) is the output from Giotto. Rounded vertices and bends without markers indicate vertices and bends for which the constraints have been relaxed.

S. Bridgeman and R. Tamassia, *Difference Metrics*, JGAA, 4(3) 47–74 (2000)   62

derived from parts of the graph modified by the user are included in the point set. The first position indicates the value for the point set used for the computation of the metric, the second the point set used for drawing alignment.

The relaxed drawings are labelled as follows:

- ni: InteractiveGiotto's drawing of the nth relaxation step

- xg: Giotto's output

4.2   Experimental Setup

For the experimental study, a set of 14 graphs with 30 vertices each were extracted from the 11,582-graph test suite used in the experimental studies of Di Battista et. al. [9, 10]. 30-vertex graphs were chosen as being small enough to work with easily, but large enough so that one modification does not affect the entire graph. Future experiments will consider graphs of different sizes.

A random modification was applied to each graph to simulate a user's modification. A modification is one of:

- *edge insert*: insertion of a single edge between two randomly chosen vertices

- *edge delete*: deletion of a single edge

- *edge split*: insertion of a single vertex at the midpoint of an existing edge, splitting the edge into two new edges

- *vertex insert*: insertion of a vertex along with a number of adjacent edges; both the number of edges and their endpoints is chosen randomly

- *vertex delete*: deletion of a vertex and its incident edges

Operations were not allowed to disconnect the graph. Since InteractiveGiotto preserves the embedding of the graph and the edge crossings and bends, new edges were routed so as to mimic how an actual user might draw the edge, instead of simply connecting the endpoints with a straight line (and potentially introducing many edge crossings). Only a single modification was made in each graph because as the user's changes affect a larger portion of the graph, the difference between the new drawings and the original rapidly becomes high and it is difficult to determine an ordering of the relaxation sequence.

Modifications of each type were applied to each of the 14 original graphs, though due to some difficulties with Giotto, the total number of modified graphs generated was only 63.

For each modified graph, a series of progressively relaxed drawings was obtained by incrementally relaxing the constraints given to InteractiveGiotto. The number of relaxed drawings for each modified graph ranged from 8 to 17, but was generally around 11. Each sequence of drawings was then ordered by a human according to increasing visual distance from the original drawing and this ordering was compared to the orderings produced by sorting the drawings in each relaxation sequence according to the computed values of the metric. The orderings were compared using

S. Bridgeman and R. Tamassia, *Difference Metrics*, JGAA, 4(3) 47–74 (2000)   63

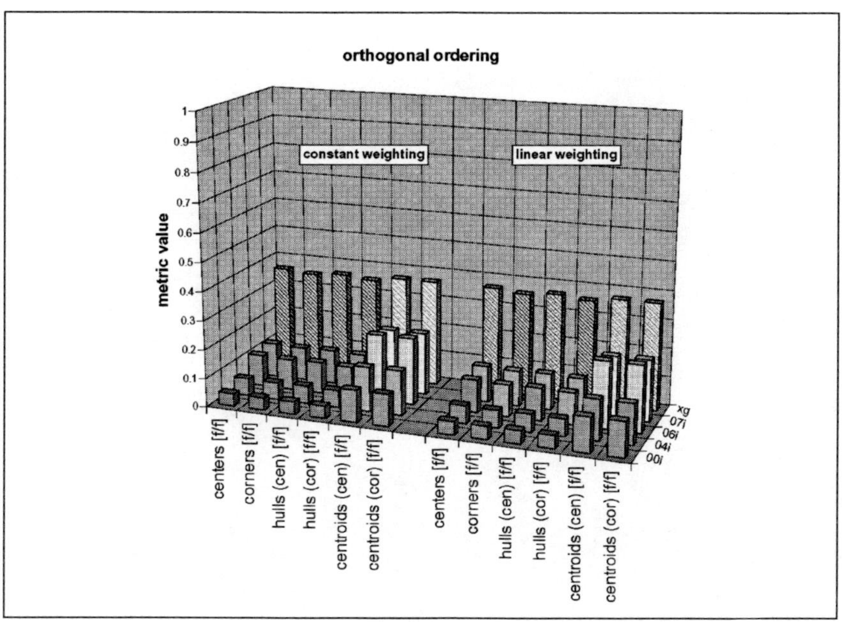

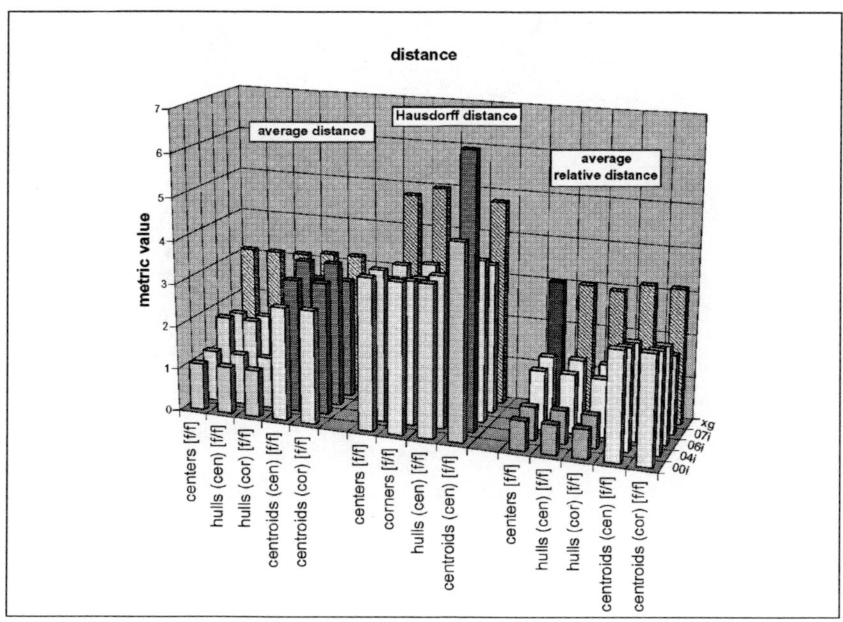

Figure 8: Selected metric values for each drawing in stage 1.

S. Bridgeman and R. Tamassia, *Difference Metrics*, JGAA, 4(3) 47–74 (2000)   64

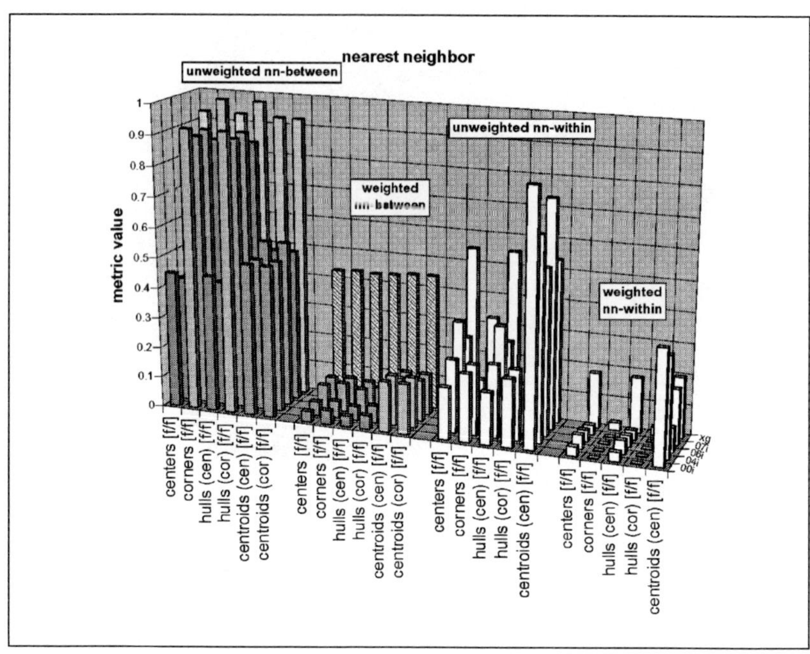

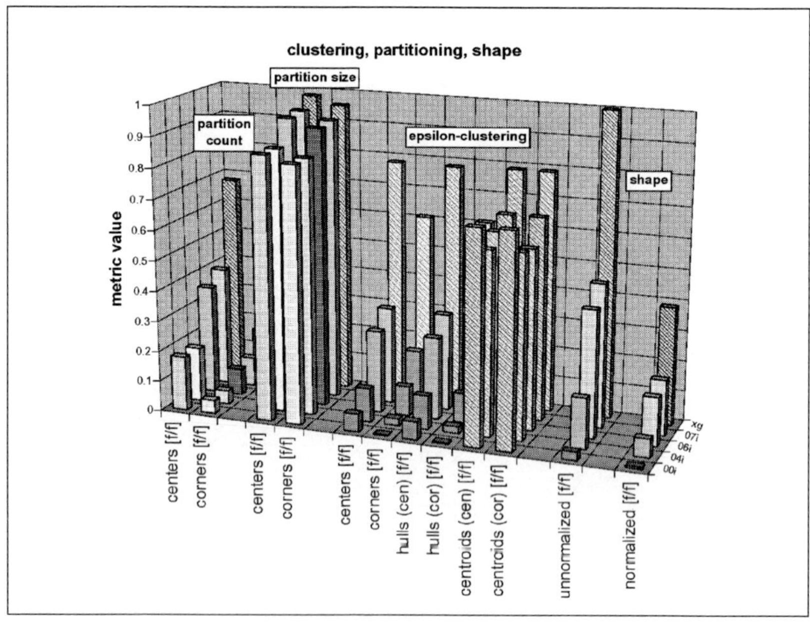

Figure 8: Selected metric values for each drawing in stage 1. (continued)

S. Bridgeman and R. Tamassia, *Difference Metrics*, JGAA, 4(3) 47–74 (2000)   65

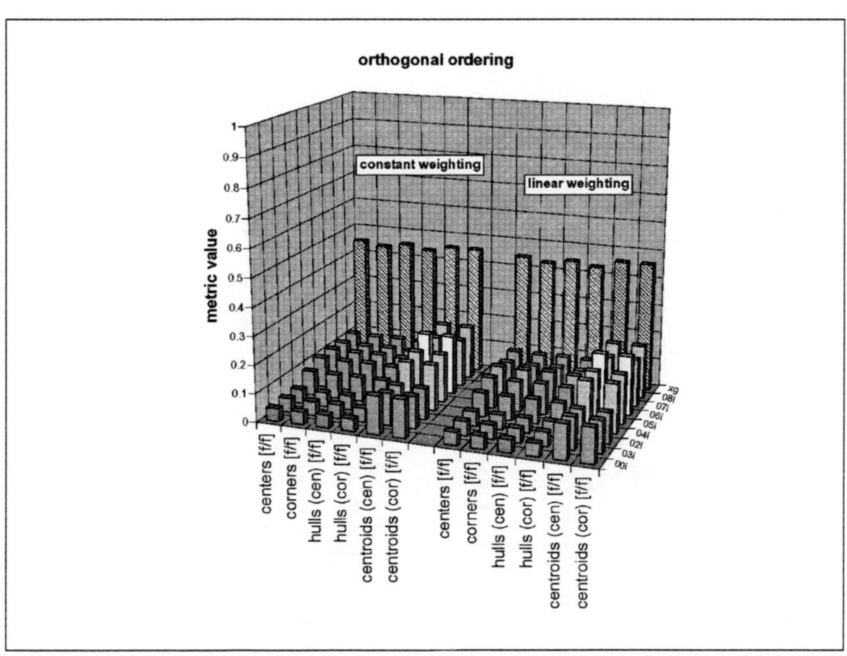

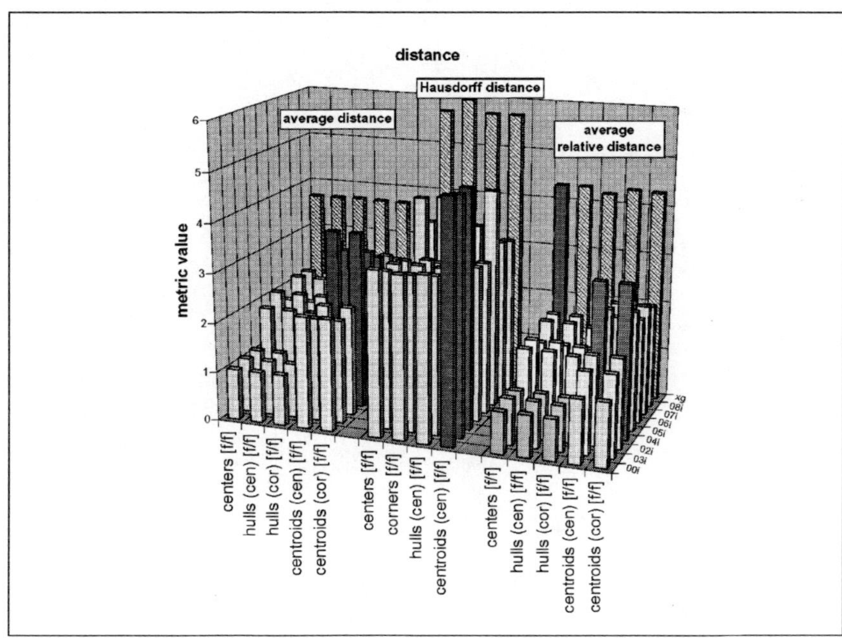

Figure 9: Selected metric values for each drawing in stage 2.

S. Bridgeman and R. Tamassia, *Difference Metrics*, JGAA, 4(3) 47–74 (2000)   66

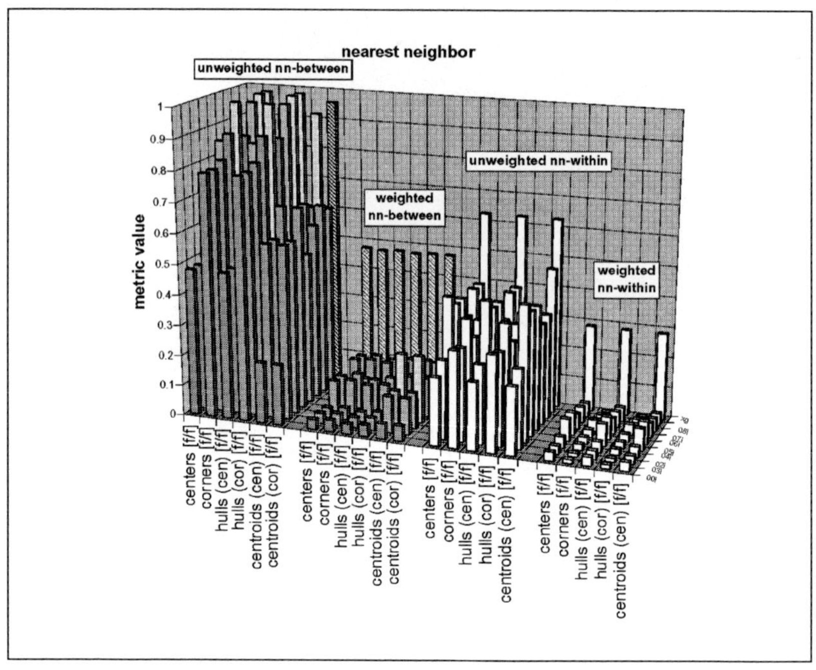

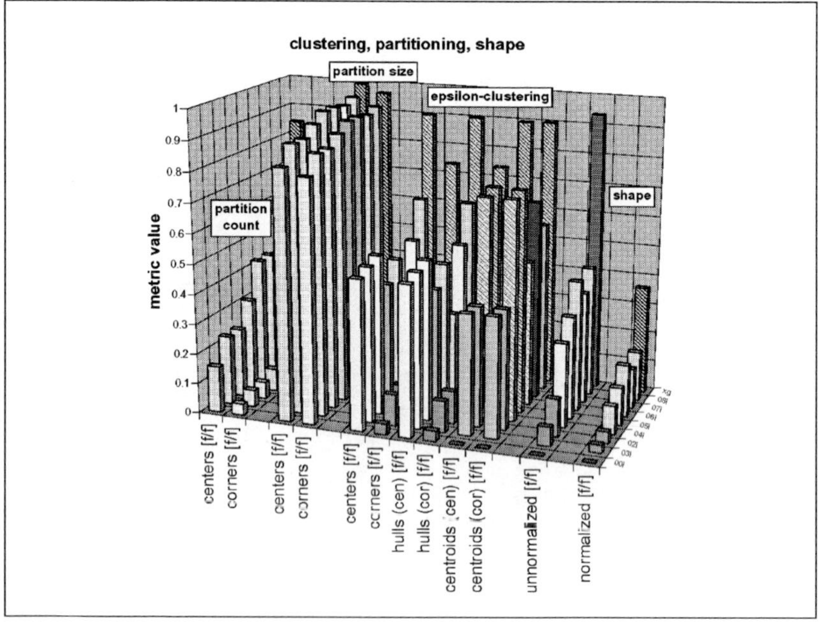

Figure 9: Selected metric values for each drawing in stage 2. (continued)

S. Bridgeman and R. Tamassia, *Difference Metrics*, JGAA, 4(3) 47–74 (2000)   67

the number of inversions: Two drawings D_i and D_j were judged to be out of order if D_i came after D_j in the metric ordering but before D_j in the human ordering, D_i and D_j were ranked equally in the metric ordering but not the human ordering, or D_i and D_j were ranked equally in the human ordering but not the metric ordering. The number of inversions was normalized by the maximum number of inversions for the sequence to adjust for different sequence lengths.

4.3   Experimental Results

Some of the metric names used in Figures 10 and 11 have modifiers:

- **nn-b** and **nn-w**: *unw* and *w* denote the unweighted and weighted versions, respectively
- **order**: *const* and *linear* denote the constant-weighted and linear-weighted versions, respectively
- **shape**: *norm* indicates that the normalized edit distance was used

4.3.1   Ordering Ability

Figure 10 shows the frequency with which different metric-and-pointset combinations did a particularly good or bad job of ordering the drawings for each graph. (Not all combinations of metrics and point sets shown in the figure were evaluted — the gray regions around the borders in Figure 10 mark combinations which were not computed.) A metric/pointset combination was flagged as doing a good job on a particular relaxation sequence if the number of inversions was noticeably lower than that for other metric/pointset combinations on the same sequence; similarly, a metric/pointset combination was flagged as doing a bad job if the number of inversions was noticeably higher than others. In a few cases no metric/pointset combination stood out as being noticeably better or worse than the others and so nothing was flagged for that sequence.

The orthogonal ordering metrics were strikingly better than most other metrics for all point sets except for those based on partition centroids. The point sets based on vertex corners fared particularly well, yielding the best ordering for over one-quarter of the sequences tested.

Shape, Euclidean distance, and the weighted nn-between metrics also stand out as being better than most of the other metrics, though they are not quite as good as orthogonal ordering.

Looking at the worst metrics, nn-within stands out as most often doing the worst job of ordering the drawings, particularly for point sets that are centroids of partitions of vertex corners. ϵ-clustering and partition size/count also do a consistently worse job of ordering.

In general, the partition centroid point sets were worse than the others, indicating that too much information is lost in these point sets to be of much use for

S. Bridgeman and R. Tamassia, *Difference Metrics*, JGAA, 4(3) 47–74 (2000)   68

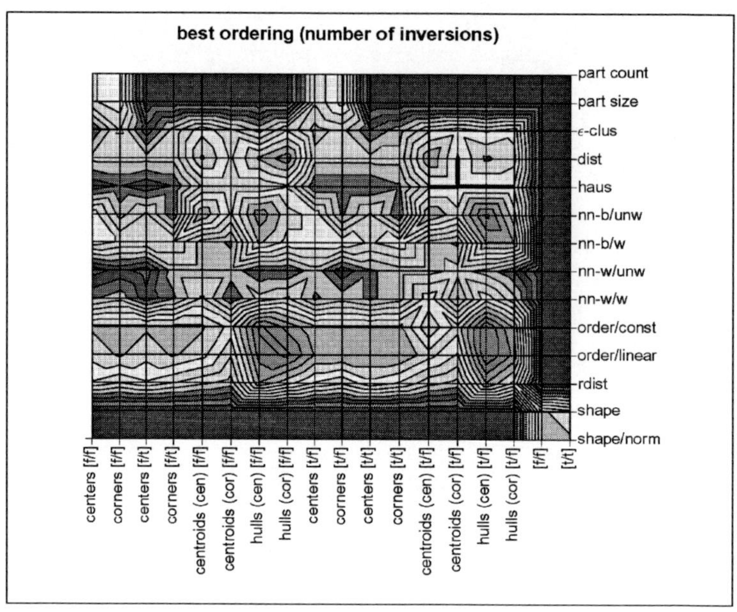

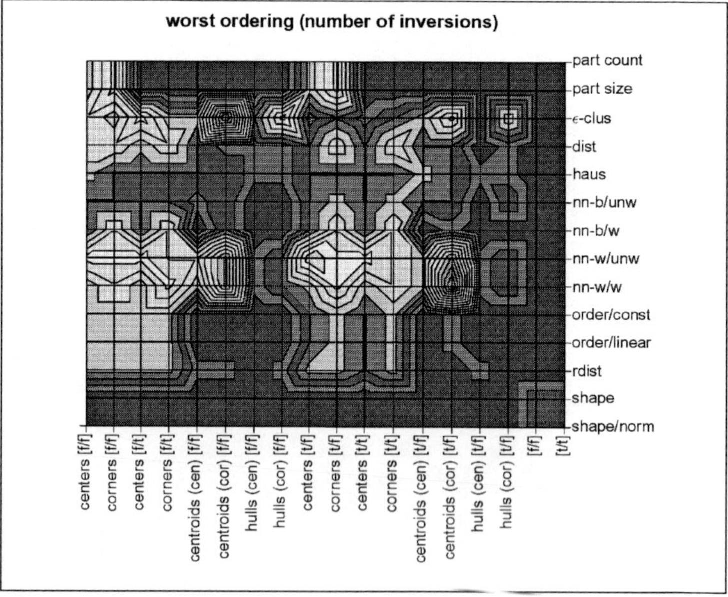

Figure 10: Purple is the lowest frequency; red is the highest. Peaks in the top picture indicate metric/pointset combinations which tend to produce orderings closest to the human-specified ordering; peaks in the bottom picture indicate combinations which tend to produce orderings most unlike the human-specified ordering.

S. Bridgeman and R. Tamassia, *Difference Metrics*, JGAA, 4(3) 47–74 (2000)   69

similarity comparisons.

It is also important to consider the variation in the relative success of a given metric/pointset combination when applied to different graphs. Many did the worst job of ordering the drawings for at least one graph, and all did the best job at least once — even the most consistently best metric (orthogonal ordering) was the best only about 30% of the time. The variation is particularly large for the partition size and count metrics, as shown by a middle-of-the-road ranking in both the best and worst plots in Figure 10. These metrics are very sensitive to the slightest movement of vertices with respect to each other, and very quickly reach their maximum values if there is much change in the drawing.

The variation in performance of the metrics for different graphs can also be seen when breaking down the results by the type of modification. Shape, average relative distance, orthogonal ordering, nn-between, and the partition size and count metrics perform better on vertex deletions. Nearly all of the metrics, except nn-within and shape, perform noticeably worse on edge insertions. The partition size and count and the Hausdorff distance metrics also perform worse on vertex insertions. This behavior is explained by noting that in InteractiveGiotto, inserting an object into the graph tends to disrupt the drawing more than a deletion.

4.3.2 Rotation Ability

The other criterion evaluated at this stage is how suitable the metrics are for solving the rotation problem. The measurement for each metric/pointset combination is the percentage of drawings in the relaxation sequence for which the correct rotation was chosen. The percentage correct for each combination, averaged over all of the experimental graphs, is shown in Figure 11. (Note that nn-within is rotation-independent and is thus not included in the chart.)

The partition size and count metrics fared the worst, getting the correct rotation only 50-54% of the time. The low results for partition size and count are largely due to many cases where the metric could not distinguish between two or more rotations. This happened most often with the "more relaxed" drawings and reflects the fact that in these drawings the partitions tend to be very small.

The ϵ-clustering metric did somewhat better, getting 60-83% of the rotations correct; the large variation is due to whether the point set was based on vertex corners (worse) or vertex centers (better). The results here are due primarily to ϵ-clustering choosing the wrong rotation, rather than being unable to choose between multiple rotations. Point sets based on vertex corners are worse than those based on vertex centers because InteractiveGiotto places vertices so that the minimum spacing between vertices is the same as the minimum vertex dimension. As a result, the ϵ-cluster for each point typically does not contain more than four points — up to two other corners of the same vertex and up to two corners of neighboring vertices. This makes ϵ-clustering using vertex corners particularly sensitive to modifications which change vertex dimensions or separate vertices which are horizontally or vertically

S. Bridgeman and R. Tamassia, *Difference Metrics*, JGAA, 4(3) 47–74 (2000)   70

near each other; however, if the vertices are spaced relatively far apart compared to the vertex size in both drawings, ϵ-clustering will report a small distance.

The unweighted nn-between metric also exhibited unsatisfactory behavior for some choices of point sets, getting 83-93% of the rotations correct. This is again the result of the metric being unable to distinguish between multiple rotations for the more relaxed drawings.

The average relative distance metric stands out as the best, averaging at least 93% of the rotations correct. This average is goes up to over 97% for the center and corner point sets.

Breaking down the results by the type of modification shows that there are again some variations between groups. The correct rotation was chosen most often when edge deletions were performed, followed by vertex insertions, edge splits, vertex deletions, and edge insertions (where the best metrics achieved only about 85% correctness). In most cases the relative performance of different metric/pointset combinations is the same as in the average of all the graphs, with a few notable exceptions:

- For graphs where an edge has been inserted, average relative distance with vertex corners and vertex centers is noticeably better and unweighted nn-between is noticeably worse than the other metrics.

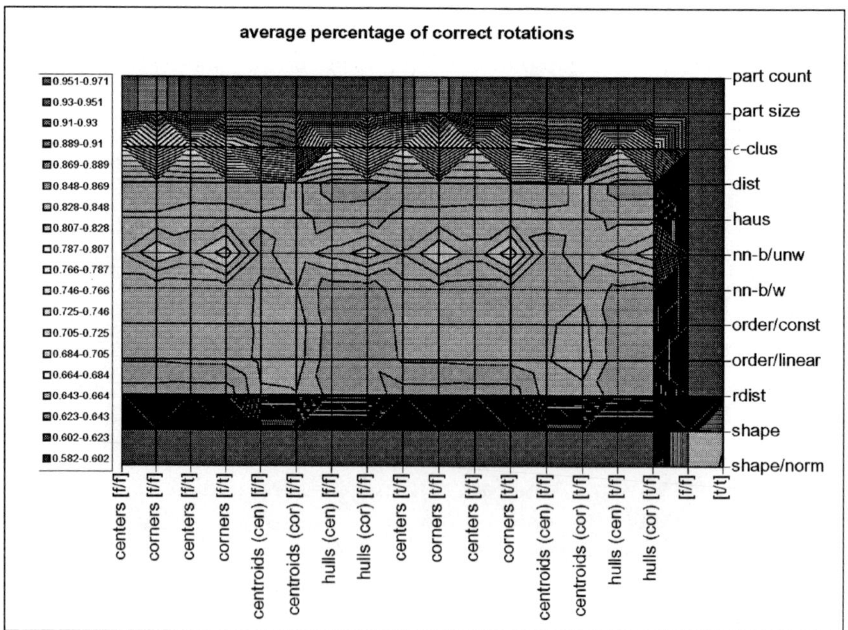

Figure 11: Purple is the lowest percentage correct (around 50%); red is the highest (94%).

S. Bridgeman and R. Tamassia, *Difference Metrics*, JGAA, 4(3) 47–74 (2000)   71

- For graphs where an edge has been split, unweighted nn-between is noticeably worse for point sets consisting of vertex corners.

4.3.3   Conclusions

If the goal is simply to solve the rotation problem, average relative distance using either corner or center point sets performs quite well. Given the success of this metric, it seems unnecessary to consider anything more complicated to solve the rotation problem.

If it is important to compare different drawings of the same graph, the orthogonal ordering metrics are the best of the metrics tested in terms of qualitative behavior. The linear-weighted version has a slight edge over the constant-weighted one. Orthogonal ordering is also nearly as good as average relative distance in solving the rotation problem. However, there is still a good deal of room for improvement — orthogonal ordering only achieved the best ordering 30% of the time, and, considering all of the metrics, the correct ordering was found for only 9 of the 63 graphs.

5   Future Work

The next step is to study in more detail those cases for which certain metric/pointset combinations perform significantly worse (or better) than the average, since there are variations in performance between individual graphs. This may also indicate combinations of the existing metrics that may work better than any single one alone.

It will also be useful to test the metrics on drawings generated by other drawing algorithms. SMILE [2], for example, provides a way of obtaining many drawings of the same graph. Since the performance of some of the metrics can be traced to characteristics of InteractiveGiotto, this may prove illuminating.

Another important task is to analyze the quantitative behavior of the metrics. This requires that each drawing be assigned (by a human) a numerical value measuring how well the user's mental map is preserved. Obtaining these values is a non-trivial task, since it is difficult for a person to judge quantitatively the difference in visual distance between two pairs of drawings, even of the same graph — Is one pair twice as different as another? Only one-and-a-half times? Five percent less? Furthermore, asking if one drawing looks more like the original than another drawing may not be exactly the same question as asking which drawing does a better job of preserving the user's mental map. The chances are that a drawing which looks more like the original will do a better job of preserving the mental map, but assuming this presupposes something about how a user's mental map is structured. A solution to this may be to design an experiment in which the user gains familiarity with the original drawing, and is then timed on her response to a question involving a new drawing of the same graph. The idea is that a faster response time means

S. Bridgeman and R. Tamassia, *Difference Metrics*, JGAA, 4(3) 47–74 (2000)   72

that the new drawing did a better job of preserving the mental map.

Other metrics can also be developed and evaluated. Metrics based on clustering and partitioning seem particularly related to how a user navigates around the drawing, and further work is needed to extend and improve these metrics. Surveying users about what they think makes two drawings more or less similar may also lead to additional metrics. Also, combinations of the current metrics may yield better results. Both current and new metrics should also be evaluated using a larger pool of graphs, including graphs of different sizes.

Finally, once suitable metrics have been identified and validated through user studies, they can be used to compare the behavior of interactive graph drawing algorithms as well as potentially providing inspiration for new drawing algorithms.

References

[1] H. Alt, O. Aichholzer, and G. Rote. Matching shapes with a reference point. *Internat. J. Comput. Geom. Appl.*, 1997. to appear.

[2] T. Biedl, J. Marks, K. Ryall, and S. Whitesides. Graph multidrawing: Finding nice drawings without defining nice. In S. Whitesides, editor, *Graph Drawing (Proc. GD '98)*, volume 1547 of *Lecture Notes Comput. Sci.*, pages 347–355. Springer-Verlag, 1998.

[3] T. C. Biedl and M. Kaufmann. Area-efficient static and incremental graph darwings. In R. Burkard and G. Woeginger, editors, *Algorithms (Proc. ESA '97)*, volume 1284 of *Lecture Notes Comput. Sci.*, pages 37–52. Springer-Verlag, 1997.

[4] U. Brandes and D. Wagner. A bayesian paradigm for dynamic graph layout. In G. Di Battista, editor, *Graph Drawing (Proc. GD '97)*, volume 1353 of *Lecture Notes Comput. Sci.*, pages 236–247. Springer-Verlag, 1998.

[5] U. Brandes and D. Wagner. Dynamic grid embedding with few bends and changes. In *Proceedings of the 9th Annual International Symposium on Algorithms and Computation (ISAAC'98)*, volume 1533 of *Lecture Notes in Computer Science*, pages 89–98. Springer-Verlag, 1998.

[6] S. S. Bridgeman, J. Fanto, A. Garg, R. Tamassia, and L. Vismara. InteractiveGiotto: An algorithm for interactive orthogonal graph drawing. In G. Di Battista, editor, *Graph Drawing (Proc. GD '97)*, volume 1353 of *Lecture Notes Comput. Sci.*, pages 303–308. Springer-Verlag, 1997.

[7] L. P. Chew, M. T. Goodrich, D. P. Huttenlocher, K. Kedem, J. M. Kleinberg, and D. Kravets. Geometric pattern matching under Euclidean motion. *Comput. Geom. Theory Appl.*, 7:113–124, 1997.

[8] R. F. Cohen, G. Di Battista, R. Tamassia, and I. G. Tollis. Dynamic graph drawings: Trees, series-parallel digraphs, and planar ST-digraphs. *SIAM J. Comput.*, 24(5):970–1001, 1995.

S. Bridgeman and R. Tamassia, *Difference Metrics*, JGAA, 4(3) 47–74 (2000)   73

[9] G. Di Battista, A. Garg, G. Liotta, R. Tamassia, E. Tassinari, and F. Vargiu. An experimental comparison of three graph drawing algorithms. In *Proc. 11th Annu. ACM Sympos. Comput. Geom.*, pages 306–315, 1995.

[10] G. Di Battista, A. Garg, G. Liotta, R. Tamassia, E. Tassinari, and F. Vargiu. An experimental comparison of four graph drawing algorithms. *Comput. Geom. Theory Appl.*, 7:303–325, 1997.

[11] P. Eades, W. Lai, K. Misue, and K. Sugiyama. Preserving the mental map of a diagram. In *Proceedings of Compugraphics 91*, pages 24–33, 1991.

[12] U. Fößmeier. Interactive orthogonal graph drawing: Algorithms and bounds. In G. Di Battista, editor, *Graph Drawing (Proc. GD '97)*, volume 1353 of *Lecture Notes Comput. Sci.*, pages 111–123. Springer-Verlag, 1997.

[13] J. E. Goodman and R. Pollack. Multidimensional sorting. *SIAM J. Comput.*, 12(3):484–507, Aug. 1983.

[14] M. T. Goodrich, J. S. B. Mitchell, and M. W. Orletsky. Practical methods for approximate geometric pattern matching under rigid motion. *IEEE Trans. Pattern Anal. Mach. Intell.* to appear.

[15] K. Imai, S. Sumino, and H. Imai. Minimax geometric fitting of two corresponding sets of points. In *Proc. 5th Annu. ACM Sympos. Comput. Geom.*, pages 266–275, 1989.

[16] K. A. Lyons, H. Meijer, and D. Rappaport. Algorithms for cluster busting in anchored graph drawing. *J. Graph Algorithms Appl.*, 2(1):1–24, 1998.

[17] A. Marzal and E. Vidal. Computation of normalized edit distance and applications. *IEEE Transactions on Pattern Analysis and Machine Intelligence*, 15(9):926–932, Sept. 1993.

[18] K. Miriyala, S. W. Hornick, and R. Tamassia. An incremental approach to aesthetic graph layout. In *Proc. Internat. Workshop on Computer-Aided Software Engineering*, 1993.

[19] K. Misue, P. Eades, W. Lai, and K. Sugiyama. Layout adjustment and the mental map. *J. Visual Lang. Comput.*, 6(2):183–210, 1995.

[20] S. Moen. Drawing dynamic trees. *IEEE Softw.*, 7:21–28, 1990.

[21] S. North. Incremental layout in DynaDAG. In *Graph Drawing (Proc. GD '95)*, volume 1027 of *Lecture Notes Comput. Sci.*, pages 409–418. Springer-Verlag, 1996.

[22] A. Papakostas, J. M. Six, and I. G. Tollis. Experimental and theoretical results in interactive graph drawing. In S. North, editor, *Graph Drawing (Proc. GD '96)*, volume 1190 of *Lecture Notes Comput. Sci.*, pages 371–386. Springer-Verlag, 1997.

[23] A. Papakostas and I. G. Tollis. Interactive orthogonal graph drawing. In *Graph Drawing (Proc. GD '95)*, volume 1027 of *Lecture Notes Comput. Sci.* Springer-

S. Bridgeman and R. Tamassia, *Difference Metrics*, JGAA, 4(3) 47–74 (2000)   74

Verlag, 1996.

[24] H. Purchase. Which aesthetic has the greatest effect on human understanding? In G. Di Battista, editor, *Graph Drawing (Proc. GD '97)*, volume 1353 of *Lecture Notes Comput. Sci.*, pages 248–261. Springer-Verlag, 1998.

[25] H. C. Purchase, R. F. Cohen, and M. James. Validating graph drawing aesthetics. In F. J. Brandenburg, editor, *Graph Drawing (Proc. GD '95)*, volume 1027 of *Lecture Notes Comput. Sci.*, pages 435–446. Springer-Verlag, 1996.

[26] K. Ryall, J. Marks, and S. Shieber. An interactive system for drawing graphs. In S. North, editor, *Graph Drawing (Proc. GD '96)*, volume 1190 of *Lecture Notes Comput. Sci.*, pages 387–393. Springer-Verlag, 1997.

[27] R. Tamassia, G. Di Battista, and C. Batini. Automatic graph drawing and readability of diagrams. *IEEE Trans. Syst. Man Cybern.*, SMC-18(1):61–79, 1988.

Journal of Graph Algorithms and Applications

http://www.cs.brown.edu/publications/jgaa/

vol. 4, no. 3, pp. 75–103 (2000)

Techniques for the Refinement of Orthogonal Graph Drawings

Janet M. Six   *Konstantinos G. Kakoulis*   *Ioannis G. Tollis*

CAD & Visualization Lab
Department of Computer Science
The University of Texas at Dallas
P.O. Box 830688, EC 31
Richardson, TX 75083-0688
{janet,kostant,tollis}@utdallas.edu

Abstract

Current orthogonal graph drawing algorithms produce drawings which are generally good. However, many times the quality of orthogonal drawings can be significantly improved with a postprocessing technique, called *refinement*, which improves aesthetic qualities of a drawing such as area, bends, crossings, and total edge length. Refinement is separate from layout and works by analyzing and then fine-tuning the existing drawing in an efficient manner. In this paper we define the problem and goals of orthogonal drawing refinement, review measures of a graph drawing's quality, and introduce a methodology which efficiently refines any orthogonal graph drawing. We have implemented our techniques in C++ and conducted experiments over a set of drawings from five well known orthogonal drawing systems. Experimental analysis shows our techniques to produce an average 37% improvement in area, 23% in bends, 25% in crossings, and 37% in total edge length.

Communicated by G. Liotta and S. H. Whitesides: submitted October 1998; revised November 1999.

Research supported in part by NIST, Advanced Technology Program grant number 70NANB5H1162 and by the Texas Advanced Research Program under Grant No. 009741-040.

J. Six et al., *Refinement of Orthogonal Drawings*, JGAA, 4(3) 75–103 (2000) 76

1   Introduction

Graphs are used in numerous applications to represent many information structures: telecommunication networks, entity-relationship diagrams, data flow charts, object-oriented software design and much more [34]. *Graph Drawing* studies the automated layout of these structures onto a two or three dimensional space where each node is represented by a polygon or circle and each edge by one or more contiguous line segments. This visualization can take form in one of many conventions: e.g., polyline, straight-line, orthogonal planar, or upward [6, 7].

Orthogonal graph drawings represent nodes with boxes and edges with polygonal chains of horizontal and vertical line segments which reside on an underlying grid. A few key applications for this style of graph drawing are project management, PERT diagramming, object-oriented analysis, database navigation, and VLSI circuit design. Much research has been conducted in this area and various algorithms exist to produce orthogonal drawings of planar [1, 12, 18, 30, 31, 33], general maximum degree four [1, 25, 28], and general higher degree graphs [2, 16, 23]. An extensive experimental study was conducted by Di Battista et. al. [8] where four general purpose orthogonal drawing algorithms were implemented and compared with respect to area, bends, crossings, edge length, and running time.

Many papers have suggested ways of evaluating the "goodness" of a graph drawing (e.g., [9, 11, 19, 22, 32]). Ding and Mateti [9] formalize the quality of data structure diagrams with nine categories: visual complexity, regularity, symmetry, consistency, modularity, size, shape, separation, and tradition. Visual complexity is the measure of ease with which a structure is communicated and includes the factors ambiguity, geometric complexity, and recognizability. Ding and Mateti also present the importance of semantic and geometric recognizability. Of course graph drawers endeavor to develop algorithms which have high semantic recognizability, but this goal is very difficult since a set of humans may have multiple interpretations of any one drawing. The authors of [9] also propose that the geometric complexity (e.g. area, crossings, bends, edge lengths) of drawings should be low, respecting the concept that the simpler a drawing appears, the simpler it will be to interpret. The last of Ding and Mateti's categories addresses a very important issue which the standard quality measures of a graph drawing do not capture: tradition. People understand a drawing better if the information is presented in a manner which is familiar. This is very difficult to qualitatively measure in a general way since people of different backgrounds may expect to see information in alternate ways.

Tamassia, Di Battista, and Batini present an extensive study of the readability of diagrams in [32]. They suggest several specific quantitative measures of quality which include: mimimization of area, balance of the drawing with respect to the horizontal or vertical axis, minimization of bends, maximization of the number of convex faces, minimization of crossings, minimization of the difference in size of the largest and smallest nodes, minimization of the total

J. Six et al., *Refinement of Orthogonal Drawings*, JGAA, 4(3) 75–103 (2000) 77

length of edges, minimization of the maximum edge length, symmetry of children in hierarchies, uniform distribution of nodes, and the upwardness of hierarchic structures. The semantics of the drawing are addressed with the following set of constraints: place a set of nodes in the center of a drawing, specify certain nodes to be a specific size, place a set of nodes on the exterior of the drawing area, cluster a set of vertices, draw a set of nodes in a specific shape and place a sequence of vertices along a straight line.

The optimization of many of the above aesthetics and constraints is known to be NP-Hard [6]. Complicating this issue is the experience that maximizing one particular quality of a drawing causes another to be significantly poor since some of these qualities work against each other. Therefore most algorithms try to layout the graph in a manner which is good for some set of aesthetics.

Current orthogonal graph drawing algorithms produce drawings which are generally good. However, many times the quality of orthogonal drawings can be significantly improved with a postprocessing step. *Refinement* is a post-processing methodology which can be applied to any orthogonal drawing. It improves the quality of a drawing by first analyzing it and then fine-tuning the drawing while keeping the majority of the layout intact. The result is a new drawing which has improved aesthetic qualities including area, bends, crossings, and edge length. Previous work includes compaction strategies [2, 31, 33] and movement of stranded nodes [13]. However, the scope of these postprocessing techniques is limited. A more comprehensive methodology is needed to further improve the aesthetic qualities of graph drawings. We presented a preliminary version of our refinement techniques in [29]. At the same conference, Fößmeier, Heß and Kaufmann discuss four techniques which reduce area, bends, and edge length, [15], yet they present no experimental results. However, most of the benefits they discuss would also be obtained by our techniques. We have focused on the development and implementation of several efficient refinement modules which work on *any* orthogonal drawing (including degree greater than four). An example of a drawing before and after refinement is shown in Figure 1.

There are two types of refinement: *interactive* and *static*. During the interactive refinement of drawings we must maintain the user's mental map [22] and are allowed to make only minimal changes. The requirements for this kind of refinement methodology are very similar to those of interactive graph drawing algorithms [3, 5, 14, 21, 22, 24, 26]. The fact that the user has already seen a drawing of a graph means that the refinement techniques must not make changes so drastic that pieces of the drawing are not recognizable. Static refinement fine tunes drawings for which we do not have to maintain the user's mental map: we are free to make any change in the embedding. Our tool performs static refinement on any orthogonal graph drawing. Using a subset of the modules, our tool can also perform interactive refinement. Certainly refinement cannot be a cure for a very poor layout because this would require the essential invocation of some other layout algorithm. Refinement fine-tunes an existing drawing by improving some layout qualities.

J. Six et al., *Refinement of Orthogonal Drawings*, JGAA, 4(3) 75–103 (2000) 78

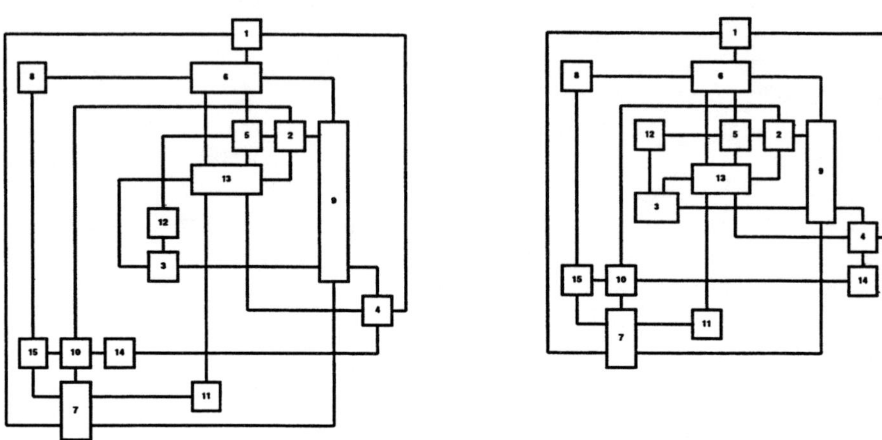

Figure 1: A graph as drawn with the GIOTTO component of Brown University's Graph Drawing Server on the left. The same drawing after refinement on the right (same scale). There is a 29% improvement in area, 13% improvement in the number of bends, 11% in the number of crossings, and 24% in the total edge length.

Our refinement techniques produced a significant 23% to 37% average improvement for each of the generally accepted quality characteristics area, bends, crossings, and total edge length in experiments over drawings from five algorithms. Since different applications require different classes of drawings and therefore need to focus on varying kinds of refinement, our system has the flexibility to vary the types and order of refinement modules called, so that a user may refine drawings in a manner specific for a particular application.

The remainder of this paper is organized as follows: we present techniques for the refinement of orthogonal graph drawings in Section 2, implementation details and the results of an experimental study in Section 3, and conclusions and future work in Section 4.

2   Refinement

During a survey of orthogonal drawings from a variety of sources, we repeatedly observed extra area, bends, edge crossings, and edge length caused by U-Turns in edges (as described in [21]), superfluous bends, poor placement of degree two nodes, two incident edges of a node crossing, nodes stranded very far from their neighbors, and unused space inside the drawing. See Figure 2 for examples.

Specifically, **U-Turns** are three contiguous edge segments which form a "U"

J. Six et al., *Refinement of Orthogonal Drawings*, JGAA, 4(3) 75–103 (2000) 79

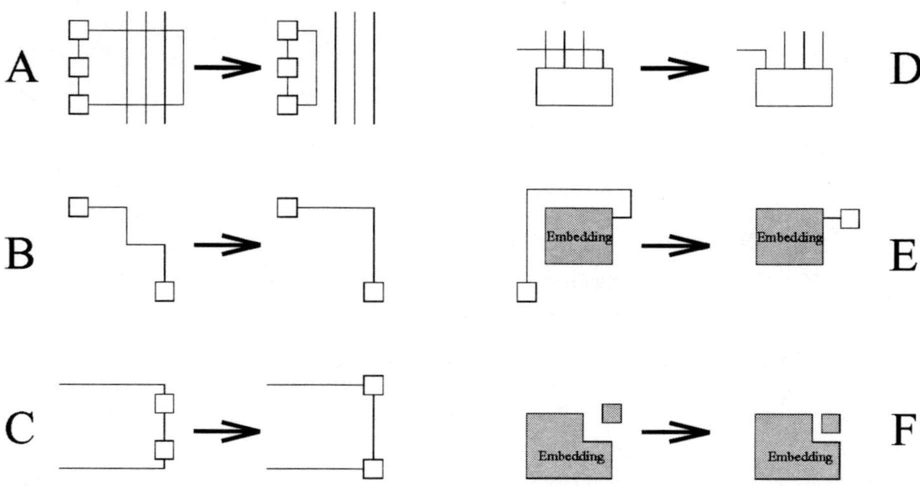

Figure 2: Examples of the problems we solve with a refinement technique: A: U-Turns, B: superfluous bends, C: poor placement of degree two nodes, D: self-crossings, E: stranded nodes, and F: extra area.

shape with the middle segment pulled far from the source and target nodes of those three segments. **Superfluous bends** are those which exist even if there is room in the drawing for an edge routing with fewer bends. Clearly U-Turns and superfluous bends can occur multiple times in edges which have four or more segments. Also, both U-Turns and superfluous bends can appear in the same edge. Poorly placed **degree two** nodes are those which are neither on a bend nor in the midst of its two incident edges. **Self crossings** are those which occur between two edges incident to the same node. Self crossings are divided into two categories: *near* and *far* self crossings. Near self crossings are those which appear north, south, east, or west of the node. Far self crossings appear northeast, northwest, southeast, or southwest of the node. A **stranded node** is a degree one node which could be placed closer to its neighbor.

Fixing a set of the above defined problems with a sequence of refinement steps will certainly reduce the visual complexity of the drawings, however we take our methodology one level further. We reduce the given graph into a simpler one. First we absorb all chains of degree two nodes into a super edge and then denote each degree one node to be a super node and determine the minimum distance needed between it and its neighbor (as is also done in [5, 23]). For example, the edge crossing between edges $(3, 13)$ and $(5, 12)$ in the left drawing of Figure 1 would not be a self crossing as defined above, but obviously it could be removed. After the reduction then the crossing between the super edge $(3, 13)$ and the edge $(3, 5)$ is discovered as a self crossing and can be removed. All refinement

J. Six et al., *Refinement of Orthogonal Drawings*, JGAA, 4(3) 75–103 (2000) 80

operations are performed on this simplified graph. The refinement techniques have been implemented to acknowledge the presence of super nodes and edges and place them in some appropriate manner. After refinement is complete the super edges are expanded in order to restore the graph to its original topology. The reduction operations allow our methodology to detect more of the above problems and therefore produce better quality drawings.

Procedure: Refine-Orthogonal-Graph-Drawing
Input: An orthogonal graph drawing, Γ, of a graph, G
Output: A new orthogonal drawing, Γ', of G with a lower visual complexity

1. Build the abstracted graph, G', of G, such that

    (a) Each chain of degree two nodes is abstracted into a super edge.

    (b) Each degree one node is denoted to be a super node and the minimum necessary distance between it and its lone neighbor is calculated. The minimum distance is directly proportional to the number of absorbed degree two nodes in the lone incident edge.

2. For each edge, e, in G'

    (a) If e contains a sequence of three edge segments which form a U-Turn edge, then pull in the middle segment of that sequence so that it is as close as possible to the tips of the U.

    (b) If e contains a sequence of three edge segments which have an extra bend and there exists room for a lower bend edge routing in the drawing then replace the current routing with the lower bend solution. This technique is very similar to the bend-stretching transformations of [33].

3. For each node, v, in G'

    (a) If v has a near self crossing, expand the node by one row or column in the appropriate direction and move the attachment point of the trouble edge to that new row or column. Place any abstracted degree two nodes so that they do not occlude any node or edge. If v has a far self crossing and is degree two, then place the node at the location of the crossing. Otherwise add one row or column and break the crossing into a knock-knee [20] edge routing. Both of these far self crossing solutions swap the attachment points of the crossing edges at v so that neither of the neighbor nodes is moved.

    (b) If v is supernode and the distance to its lone neighbor is greater than the minimum distance calculated in Step 1(b), (i.e. a stranded node) place the supernode as close to the neighbor node as space allows in Γ.

J. Six et al., *Refinement of Orthogonal Drawings*, JGAA, 4(3) 75–103 (2000) 81

4. For each superedge, *e*, expand *e* to restore the degree two nodes and verify that each degree two node is either on a bend or in the midst of its two incident edges.

5. Perform a VLSI layout compaction [10, 17, 20] to remove extra space in the drawing. □

Many improvements may be made without increasing the area of the drawing, but allowing the addition of more area may enable refinement to significantly improve other aesthetics of the drawing. For example, adding a row or column may be necessary to remove a self crossing. However this allowance should be according to user requirements and must be parametrically defined. It is possible that more area, bends, crossings, or edge length will be added with some refinement step, however we have found with our experimental study that we still gain improvements in these quality measures globally. However, we still recommend that this refinement methodology be implemented with a set of parametric options tailored to user requirements. Also, the separate refinement techniques may be performed in any order.

Refinement is an evolving methodology: we are planning to implement additional modules for improving orthogonal drawings as we discover and develop further techniques.

3 Implementation and Experimental Results

3.1 Implementation

Refine-Orthogonal-Graph-Drawing has been implemented in GNU C++ Version 2.7.2 with Tom Sawyer Software's Graph Layout Toolkit Version 2.3.1 (GLT) and a TCL/TK graphical user interface. A set of experiments has been run on a Sparc 5 running Sun OS 4.3.1.

Many interesting and challenging issues were addressed during the implementation of our refinement procedure. First we needed a mechanism to search the space within the given drawing to move pieces of the drawing without occluding uninvolved nodes and edges. We represent the space of the drawing with a dynamic orthogonal grid structure in which rows and columns may be added at any point within the space. Nodes and edges of the drawing are represented with grid segments.

As mentioned in the previous section, we recommend the use of parametric options for allowing or disallowing the addition of area, bends, crossings, or edge length. Furthermore, it is not necessary for these parametric options to be binary. For example, adding x amount of area is allowable if y crossings are removed, where x and y are some thresholds. We do not expect that any single set of parametric options would be appropriate for every application. Therefore, the definition of this set of options should be decided by the implementor. In

J. Six et al., *Refinement of Orthogonal Drawings*, JGAA, 4(3) 75–103 (2000) 82

the experimental study presented in Section 3.2 we used a set of parametric options which would allow the addition of area, bends, crossings, and edge length in order to test the strength and weaknesses of the entire set of refinement techniques. We also allow the movement of edge attachment points. As shown with the results of this experimental study, Refine-Orthogonal-Graph-Drawing overall improves these aesthetics although they may be temporarily increased during a single step.

Each of the refinement modules can be viewed as a local search technique. The module which shortens U-Turn edges searches the grid starting at the row or column next to the endpoint of the first or third edge segment (whichever is placed closer to the middle edge segment) looking for a placement which does not occlude any nodes or edges. We search the grid toward the old placement until sufficient space is found. At worst, this will be the old placement. See Figure 3. It is important to note that the edges involved in the U-Turn may

Figure 3: Fixing a U-Turn edge. The left illustration shows the nodes and edges in their original positions. The endpoints of the middle segment are shown with circles and the source and target of the U-Turn with boxes. The right illustration shows the final placement.

actually represent a chain of degree two nodes, therefore we must detect that situation and be sure to place those nodes only where there is sufficient space in the grid. This means that none of the relocated edge segments (or nodes, if this edge represents a chain of degree two nodes) can overlap any previously placed node or edge. Also, we examine each set of three contiguous segments in each edge so that we can catch more of the U-Turns. This is especially important when we are dealing with degree two chains. It is possible that we can remove bends with this refinement module. See Figure 4. For example, if there is room on the top side of node v in Figure 4 (a) for the attachment of edge (u, v) we can remove one bend. Of course, if this edge is a super edge, we must make sure that there is room for the placement of each of the degree two nodes in the grid. The illustration in Figure 4 (b) shows how multiple U-turns in a single edge can also be fixed.

The superfluous bends module examines each set of three contiguous edge segments in every edge. For each set, call one endpoint of the middle edge

J. Six et al., *Refinement of Orthogonal Drawings*, JGAA, 4(3) 75–103 (2000) 83

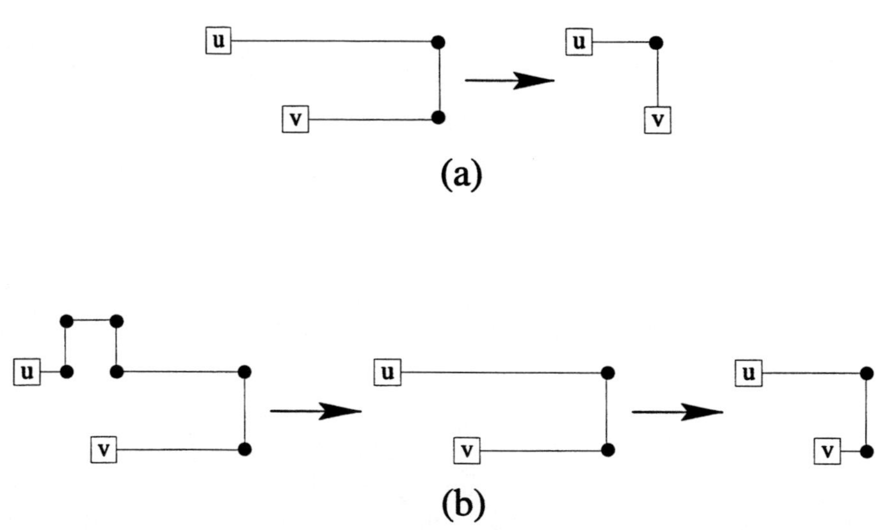

Figure 4: Removing bends with the U-Turn module. The illustration marked (a) demonstrates how one bend of a U-turn can be removed. The illustration marked (b) demonstrates how an edge with multiple U-turns is handled by Refine-Orthogonal-Graph-Drawing.

segment x and the other y. Define a and b to be the points shown in Figure 5. If space in the drawing allows, then place x at a or y at b. No occlusions are allowed. Again, it is possible that superfluous bends may appear multiple times in an edge, this is the reason for examining each set of three contiguous edge segments.

The self crossing module removes crossings as specified in Step 3 of our refinement procedure. Since the grid is dynamic, we insert the new gridlines inside the node to fix a near self crossing on the appropriate side and that automatically forces the node to grow. It is possible that a node will be grown multiple times. For far self crossings of a degree three or higher node, a row or column is inserted at that crossing (see Figure 6 for examples). If adding a row increases the size of some node, but adding a column does not, then our refinement procedure will use the solution which adds a column. The first far self crossing solution is for the case where the node is degree two while the second solution is for higher degree nodes. If the node is degree two, our procedure for refinement searches the space in the grid at the self crossing to determine if the there is enough space for the node to be placed there. If so, the node is moved and the endpoints of the incident edges are updated at the newly moved node. The positions of the neighbors are always maintained. The second solution adds two bends and therefore the superfluous bends refinement module

J. Six et al., *Refinement of Orthogonal Drawings*, JGAA, 4(3) 75–103 (2000) 84

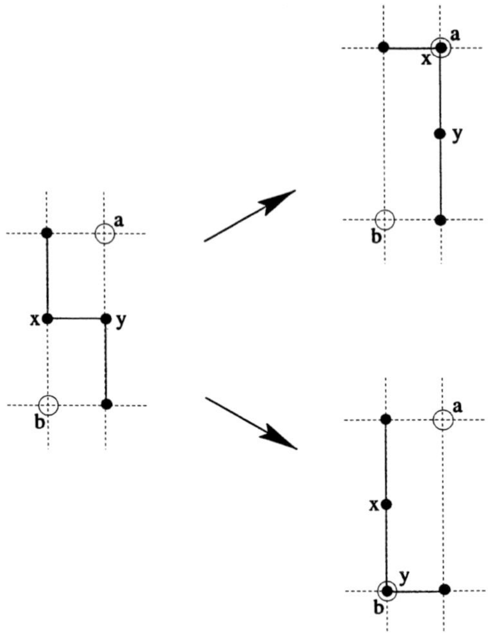

Figure 5: Superfluous bend solutions. The old edge routing is shown with a dashed line in the two illustrations on the right.

should be run on the new drawing to remove these extra bends if possible. The user has a parametric option for this module to allow the addition of rows and columns as necessary to avoid adding any crossings or to allow crossings to remain. This option gives the user the ability to decide to give priority to area or crossing reduction. We believe that reducing the number of crossings is paramount and chose the first parametric option for our experiments. In part, this decision was influenced by the study presented in [27] which showed the number of crossings to have a very significant effect on quality. This is certainly not to say that avoiding crossings will always cause the addition of area. In fact we found in our experimental study that in many drawings, the area will still be reduced while avoiding crossings. Also, note that the attachment points of the two crossing edges may be swapped so that the self crossing is removed while the positions of the neighbors are not changed. If this movement of the attachment points is not acceptable, the user can set a parametric option preventing this action.

The stranded node module searches the grid from the placement of the neighbor towards the old placement of the stranded node in order to find available space for the stranded node to be moved. At worst, this will be the old place-

J. Six et al., *Refinement of Orthogonal Drawings*, JGAA, 4(3) 75–103 (2000) 85

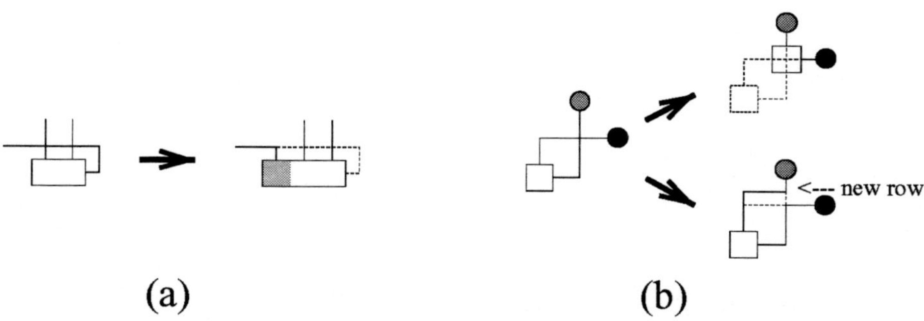

(a) (b)

Figure 6: Near(a) and far(b) self crossing solutions. Old node placements and edge routings are shown with dashed lines.

ment. If the lone incident edge of the stranded node represents a chain of degree two nodes, then this process is iteratively carried out for degree two node from the neighbor node (i.e. the node with degree higher than two) to the stranded node. See Figure 7.

Figure 7: Fixing stranded node B with an incident edge that represents a degree two chain. The refined solution appears on the right. Notice that not only are we placing related nodes closer to each other, but also unwrapping a poorly routed set of edges.

The module which fixes poor placement of degree two nodes, first expands the super edges and then visits each newly restored degree two node to verify that it lies either on a bend or in the midst of its two incident edges. See Figure 8. These refinements reduce the number of bends and produce drawings with a better distribution of node placements for the incident edges. Since the refinement modules described in this paper can be applied in any order, if there are any more refinements to be done after the completion of this module, any degree two chains should be reduced again.

The implementation of the compaction module is inspired by one dimensional VLSI layout compaction [10, 17, 20]. A one dimensional graph-based compaction is performed once each for the horizontal and vertical directions. Compaction

J. Six et al., *Refinement of Orthogonal Drawings*, JGAA, 4(3) 75–103 (2000) 86

Figure 8: Refining poor placement of degree two nodes. The illustration on the left shows the movement of degree two nodes to remove bends while the right demonstrates a node placement with better distribution.

is a NP-Hard problem [20] and therefore efficient one-dimensional compaction techniques are not guaranteed to produce optimal solutions. Similar types of compaction modules have also been successfully used in [2, 31, 33].

It is recognized that different users may want different types of drawings and need to refine a specific aesthetic quality. Also different algorithms merit the use of different types of refinement: classes of orthogonal drawing algorithms inherently have strengths and weaknesses with respect to different aesthetic criteria. Static orthogonal graph drawing algorithms either planarize and then embed with a planar algorithm or proceed in an essentially incremental fashion. While the problems described in the previous section occur with all of the orthogonal algorithms surveyed, some types of problems occur more frequently in a particular class of orthogonal algorithms. For example, planarization algorithms have a tendency to have some very long edges and place nodes far away from their neighbors. Incremental algorithms tend to have superfluous bends. Our refinement methodology automatically detects these problems and fixes them regardless of the class of algorithm used to create the drawing. Furthermore, the quality of a drawing before refinement depends heavily on a chosen algorithm, and even more heavily on the implementation of that algorithm. Our refinement methodology fixes problems caused by both the algorithms and their implementations.

Our implementation communicates with the user via a graphical user interface which allows the user to perform a desired set of refinements in any order. This flexibility adds power to refinement in that the user can refine any orthogonal drawing in a manner which is application specific. Figure 9 shows the user interface of our tool. Note the nine square buttons on the left side of the interface. These buttons invoke the refinement modules. Our design and implementation allows the user to choose the types and order of refinement to be performed. For example, if the user would like to remove only the self crossings, then that button is chosen. The refined drawing quickly appears on the can-

J. Six et al., *Refinement of Orthogonal Drawings*, JGAA, 4(3) 75–103 (2000) 87

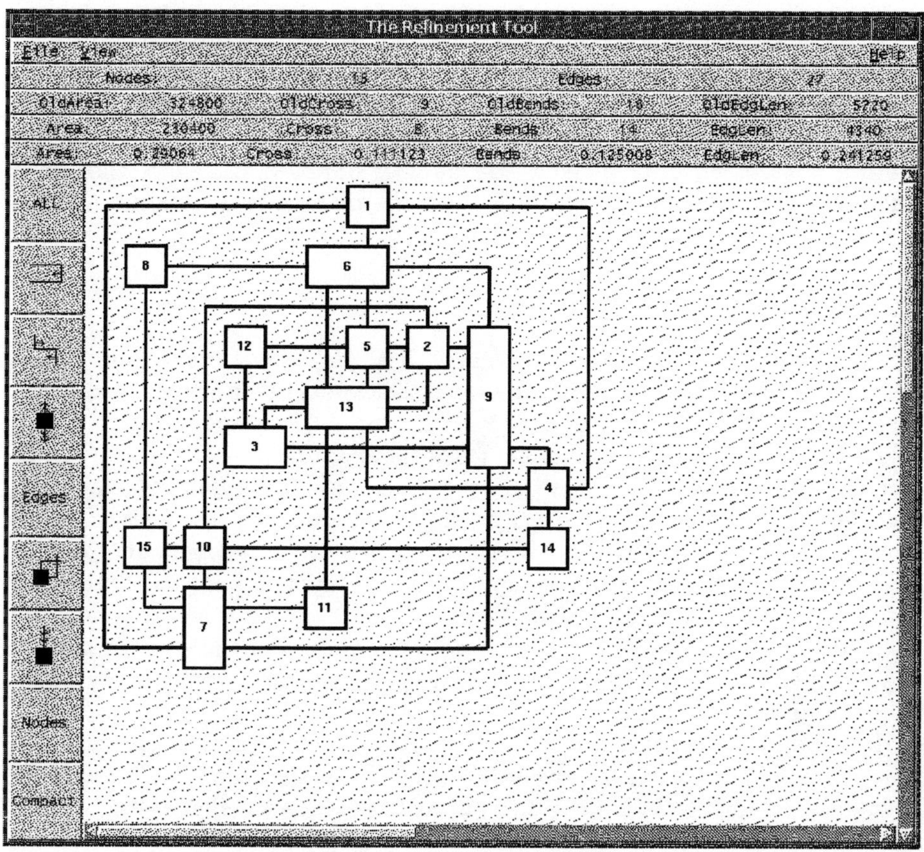

Figure 9: The Graphical User Interface for The Refinement Tool. Buttons for individual and combined refinement types appear on the left while statistics for the drawing appear at the top. The user also has options to save postscript images and view the drawings at different scales. This functionality is contained within the menu bar. In this example, refinement reduces the area by 29%, crossings by 11%, bends by 12%, and total edge length by 24%.

J. Six et al., *Refinement of Orthogonal Drawings*, JGAA, 4(3) 75–103 (2000) 88

vas and the statistical information at the top of the interface shows how many crossings were removed. The user can also choose to perform the refinement modules which deal with just the edges by selecting the "Edges" button. For the quickest application of all the refinement modules, the user chooses "ALL". The refinements are performed in the same order as their respective buttons appear on the interface. This ordering may easily be configured for each user: the individual techniques work in the same manner although the final outcome may be different.

We have designed and implemented all of our refinement techniques to take linear time with respect to the number of grid segments. Typically the number of grid segments is bounded by the number of nodes and edges and hence refinement takes $O(n + m)$ time, where n is the number of nodes and m is the number of edges.

3.2   Experimental Results

We have conducted a set of experiments with an implementation of our refinement procedure. The source drawings are 1400 layouts of the graphs used in the experimental study of [8], which we will refer to as the Rome Graphs, (available at `http://www.inf.uniroma3.it/people/gdb/wp12/ LOG.html`) which range in size from 10 to 65 nodes and are produced by the Bend-Stretch, Column, Giotto, and Pair algorithms as implemented at Brown University's Graph Drawing Server [4] (`http://loki.cs.brown.edu:8081/graphserver/`), and by GLT. Each drawing was given as input to our refinement implementation and data collected as to the improvements made in area, bends, edge crossings, and total edge length for each drawing. A table summarizing the average percent improvement for this set of aesthetics over drawings from the five layout algorithms is given in Figure 10.

In the table, the left column of percentages for each algorithm represents the average improvement over all drawings. This includes the drawings for which our techniques are unable to improve the drawing with respect to that aesthetic. The second column of percentages for each layout technique represents the average percent improvement for those drawings which our approach was able to strictly improve. The row marked δ, represents the average percent improvement made with respect to all four aesthetics for that particular algorithm. Δ represents the average percent improvement made for area, bends, crossings or edge length over all of the experimental source drawings. Note that refinement acts on the input drawing which is produced by a specific implementation of a chosen algorithm. As such, refinement improves aesthetic qualities caused by possible problems of both a chosen algorithm and the implementation of that algorithm.

Our implementation of refinement makes a 37% improvement on average in both area and total edge length. This huge improvement is due largely to the modules of our techniques which shorten long edges. As it is well known in the VLSI circuit design field, the area is usually dominated by the amount of

J. Six et al., *Refinement of Orthogonal Drawings*, JGAA, 4(3) 75–103 (2000) 89

	Bend-Stretch		Column		Giotto		Pair		GLT		Δ	
	All	Better	All	Better	All	Better	All	Better	All	Better	All	Better
Area	29	32	53	53	7	15	61	61	33	37	37	40
Bends	18	21	45	45	2	12	42	43	10	25	23	29
Crossings	10	29	49	51	2	18	39	45	23	30	25	35
Edge Length	31	32	55	55	22	24	53	53	22	26	37	38
δ	22	29	51	51	8	17	49	51	22	30		

Figure 10: Summary of results from experimental study.

J. Six et al., *Refinement of Orthogonal Drawings*, JGAA, 4(3) 75–103 (2000) 90

wiring. Likewise, the area of graph drawings is often dominated by the edge lengths. So our methods which shorten the total edge length also inherently decrease the area. Hence we see a proportional improvement in both area and total edge length. In addition the area is also reduced by the compaction phase. Although several orthogonal layout algorithms already use a compaction phase, our compaction has such a significant effect since other phases of refinement have simplified the drawing by reducing its geometric complexity. Therefore, pieces of the drawing have more freedom to move and can therefore be compacted more efficiently.

Our experiments also show about 24% improvement on average with respect to bends and crossings. This is due to the modules which particularly refine those elements.

Refinement significantly improves drawings created by each of these orthogonal algorithms. As mentioned earlier, every layout algorithm has a set of strengths and weaknesses. These strengths and weaknesses with respect to area, bends, crossings and total edge length are apparent in the numbers collected for each algorithm in our experiments. Giotto is the most evolved implementation of the Graph Drawing Server (GDS) and thus refinement has a lesser impact on these drawings. One of the main steps of Giotto finds the minimum number of bends of the embedded planarized graph [33]. Our refinement improved the number of bends by an average 2%. Improvement is possible since some of our refinement modules slightly modify the embedding. When our tool was able to improve the number of bends, the improvement was on the average 12%, with some improvements up to 33%. The planarization step of Giotto causes some nodes to be placed far from their neighbors, hence we see a more significant improvement of 22% with respect to total edge length.

Likewise we notice similar behavior with GLT. Their implementation allows each edge to have at most one bend. So the average improvement in the number of bends is 10% compared to the average 23%. Column and Pair drawings experience very significant improvements of each aesthetic. Especially notice the average 45% and 42% improvements in the number of bends and the average 55% and 53% improvements in total edge length. This is related not only to the nature of these algorithms, but also to the implementation of these algorithms in the GDS. Instead of dividing input graphs into biconnected components and performing a layout on each component, the GDS implementations of Column and Pair augment graphs to make them biconnected and then perform the layout on the augmented graph. The augmenting edges are removed during the final step and the resulting drawing shows only the input graph. This implementation decision has increased the geometric complexity of many Column and Pair drawings. Our refinement techniques provide a 10% to 31% average improvement of Bend-Stretch drawings for all aesthetics considered.

It is important to note the difference between the "All" percentages and the "Better" percentages. All five of these orthogonal algorithms produce good drawings: sometimes the number of crossings and bends is already very low, or

J. Six et al., *Refinement of Orthogonal Drawings*, JGAA, 4(3) 75–103 (2000) 91

even optimal in a few cases. Of course, the refinement tool cannot reduce the number of bends in a drawing which already has the lowest possible number of bends. Also, some layouts do not allow the embedding to be refined much.

Example drawings from Bend-Stretch, Column, Giotto, Pair, and GLT along with their refined drawings are included in Figures 11 - 19. Another such example for Giotto appears in Figure 1.

4 Conclusions and Future Work

In this paper we presented efficient postprocessing techniques to improve aesthetic qualities of any orthogonal graph drawing. Specifically, we focused on reducing the area, number of bends, number of crossings and total edge length. An experimental study conducted over a set of drawings from five well known algorithms produced very good results. An average 37% improvement was made in area, 23% in bends, 25% in crossings, and 37% in edge length.

Refinement is an evolving methodology. We plan to implement more modules of refinement as we develop further techniques to improve orthogonal graph drawings. Also, we plan to further enhance the graphical user interface of our refinement tool so that the refinement process can be even more tailored to an individual user's needs. More parametric options will be added. Further, we would like to capture the user's modal interactions with the nodes and edges so that we can further improve the quality of given drawings to some application-specific standard.

Acknowledgments

The authors would like to thank Uğur Doğrusöz and Therese Biedl for their technical advice. We are also grateful to Stina Bridgeman, Roberto Tamassia and the Graph Drawing group at Brown University for the Graph Drawing Server which is a great resource. We also wish to thank the referees for several useful suggestions.

J. Six et al., *Refinement of Orthogonal Drawings*, JGAA, 4(3) 75–103 (2000) 92

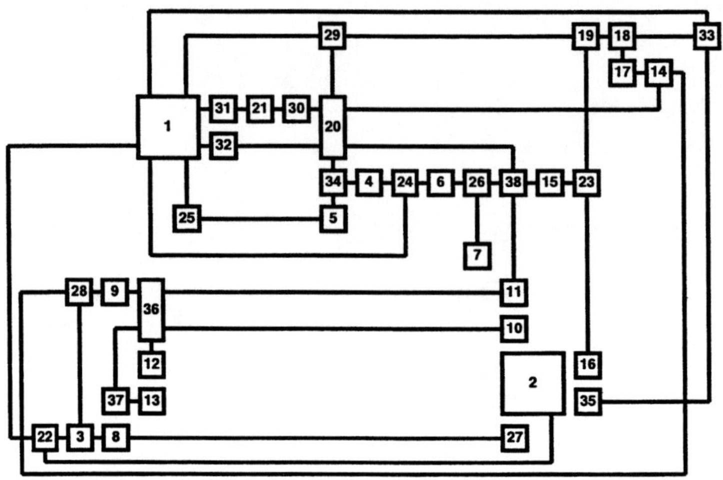

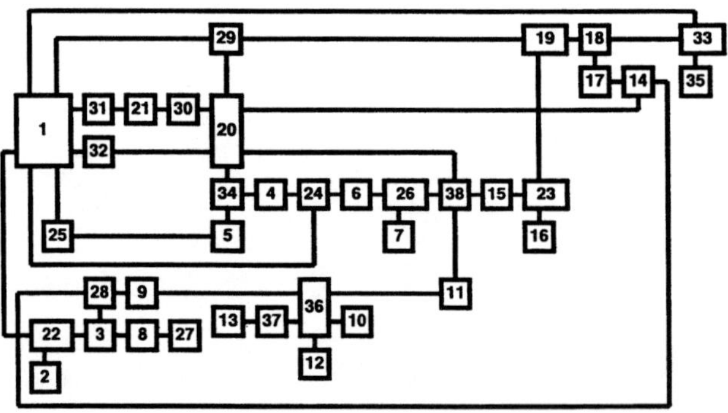

Figure 11: One of the Rome graphs as drawn with Bend-Stretch on the top. The same drawing after refinement on the bottom (same scale). There is a 34% improvement in area, 24% in the number of bends, 33% in the number of crossings, and 39% in the total edge length.

J. Six et al., *Refinement of Orthogonal Drawings*, JGAA, 4(3) 75–103 (2000) 93

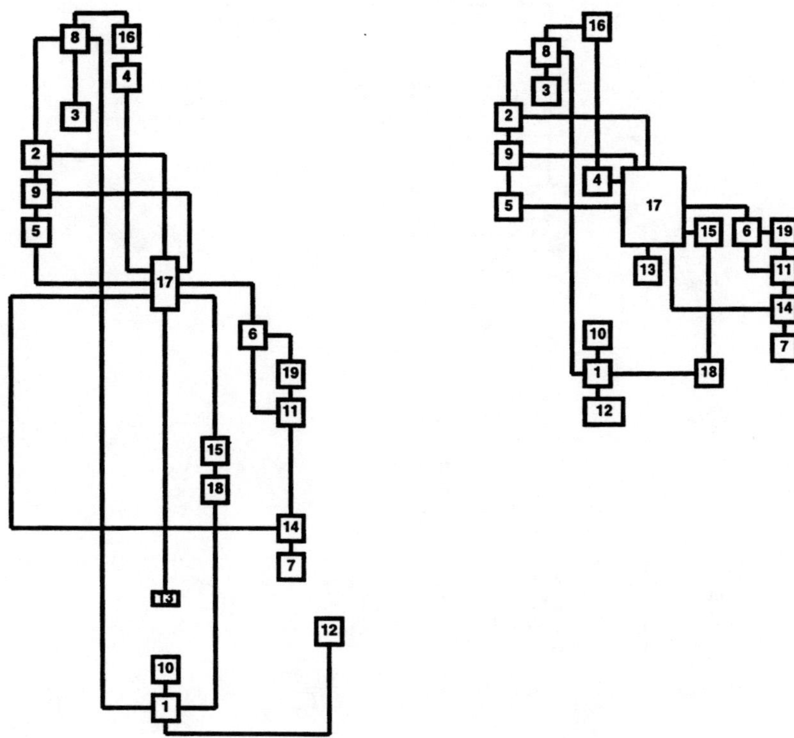

Figure 12: One of the Rome graphs as drawn with Column on the left. The same drawing after refinement on the right (same scale). There is a 47% improvement in area, 42% improvement in the number of bends, 40% in the number of crossings, and 56% in the total edge length.

J. Six et al., *Refinement of Orthogonal Drawings*, JGAA, 4(3) 75–103 (2000) 94

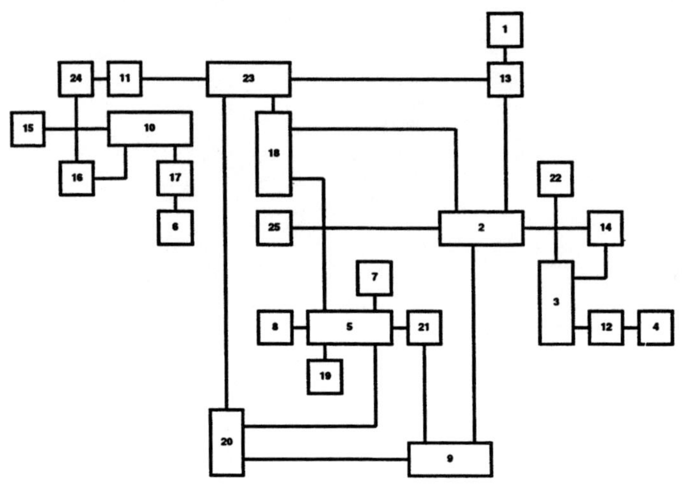

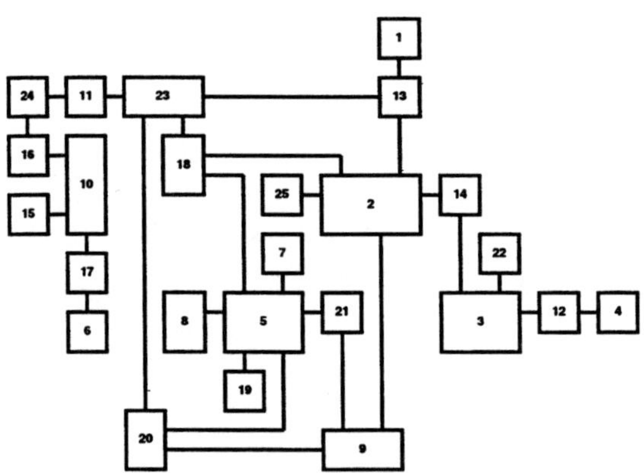

Figure 13: One of the Rome graphs as drawn with GLT on the top. The same drawing after refinement on the bottom (same scale). There is a 34% improvement in area, 40% improvement in the number of bends, 100% in the number of crossings, and 36% in the total edge length.

J. Six et al., *Refinement of Orthogonal Drawings*, JGAA, 4(3) 75–103 (2000) 95

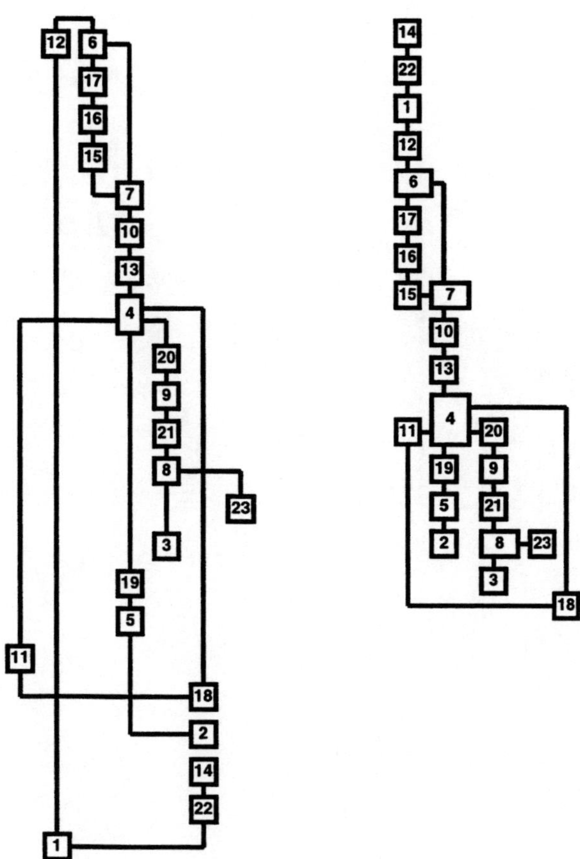

Figure 14: One of the Rome graphs as drawn with Column on the left. The same drawing after refinement on the right (same scale). There is a 46% improvement in area, 73% improvement in the number of bends, 100% in the number of crossings, and 68% in the total edge length.

J. Six et al., *Refinement of Orthogonal Drawings*, JGAA, 4(3) 75–103 (2000) 96

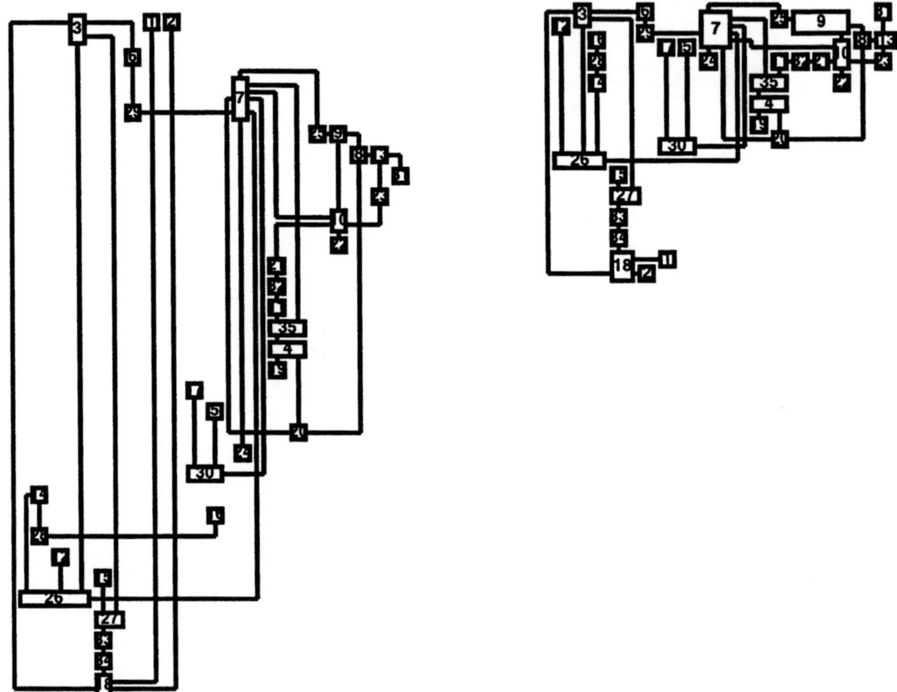

Figure 15: One of the Rome graphs as drawn with PAIR on the left. The same drawing after refinement on the right (same scale). There is a 65% improvement in area, 35% improvement in the number of bends, 56% in the number of crossings, and 66% in the total edge length.

J. Six et al., *Refinement of Orthogonal Drawings*, JGAA, 4(3) 75–103 (2000) 97

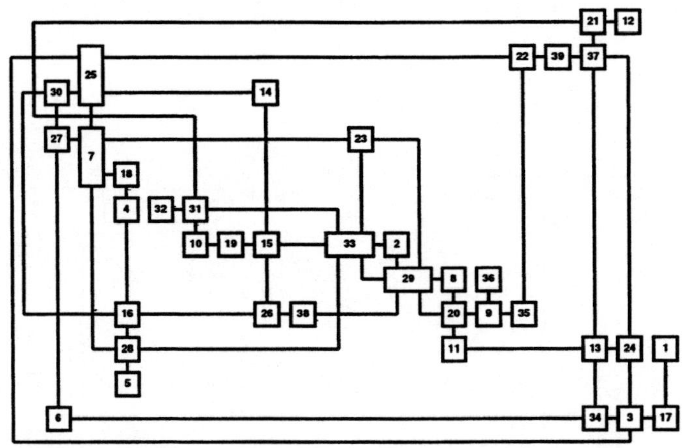

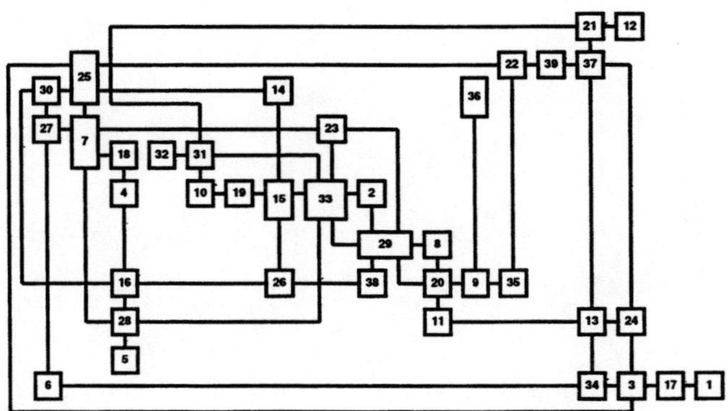

Figure 16: One of the Rome graphs as drawn with Giotto on the top. The same drawing after refinement on the bottom (same scale). There is a 19% improvement in area, 6% improvement in the number of bends, 20% in the number of crossings, and 17% in the total edge length.

J. Six et al., *Refinement of Orthogonal Drawings*, JGAA, 4(3) 75–103 (2000) 98

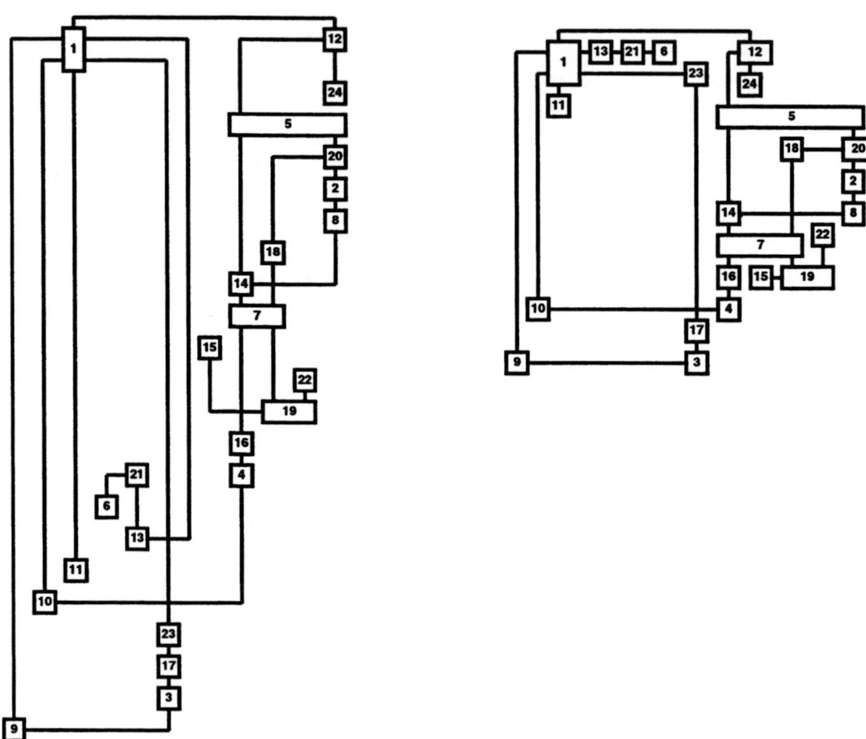

Figure 17: One of the Rome graphs as drawn with PAIR on the left. The same drawing after refinement on the right (same scale). There is a 49% improvement in area, 64% improvement in the number of bends, 50% in the number of crossings, and 60% in the total edge length.

J. Six et al., *Refinement of Orthogonal Drawings*, JGAA, 4(3) 75–103 (2000) 99

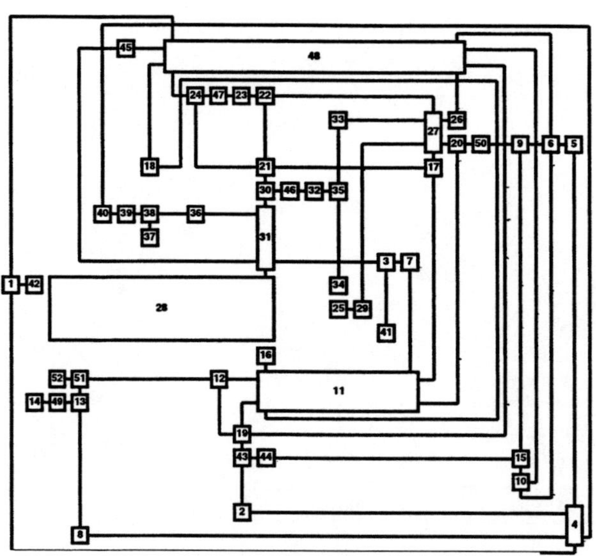

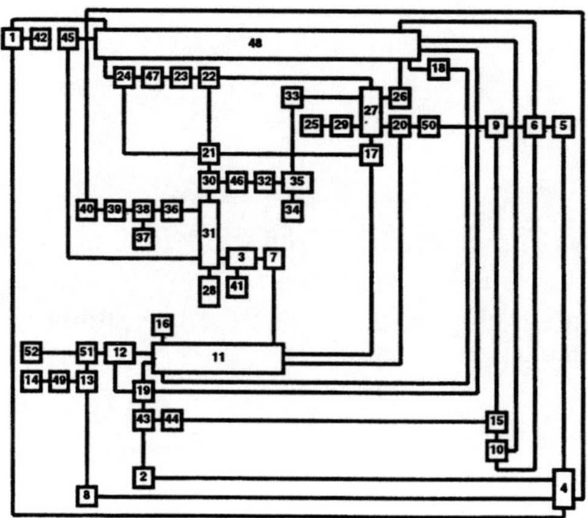

Figure 18: One of the Rome graphs as drawn with Bend-Stretch on the top. The same drawing after refinement on the bottom (same scale). There is a 35% improvement in area, 10% in the number of bends, 43% in the number of crossings, and 26% in the total edge length.

J. Six et al., *Refinement of Orthogonal Drawings*, JGAA, 4(3) 75–103 (2000)100

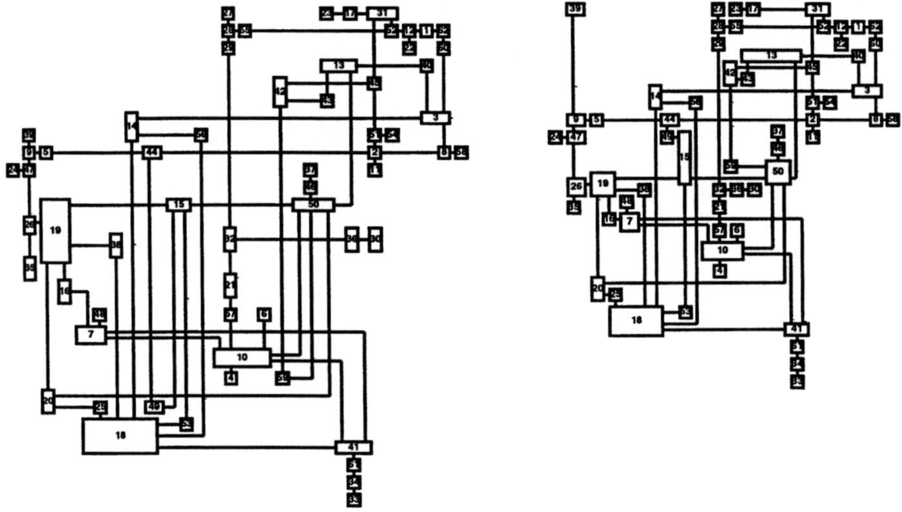

Figure 19: One of the Rome graphs as drawn with GLT on the left. The same drawing after refinement on the right (same scale). There is a 42% improvement in area, 30% improvement in the number of bends, 36% in the number of crossings, and 41% in the total edge length.

References

[1] T. Biedl and G. Kant, A Better Heuristic for Orthogonal Graph Drawings, *Computational Geometry: Theory and Applications*, 9(1998), 1998, pp. 159-180.

[2] T. C. Biedl, B. P. Madden and I. G. Tollis, The Three-Phase Method: A Unified Approach to Orthogonal Graph Drawing, *Proc. GD '97, LNCS 1353*, Springer-Verlag, 1997, pp. 391-402. Also available at http://www.utdallas.edu/~tollis.

[3] S. Bridgeman, J. Fanto, A. Garg, R. Tamassia and L. Vismara, Interactive Giotto: An Algorithm for Interactive Orthogonal Graph Drawing, *Proc. GD '97, LNCS 1353*, Springer-Verlag, 1997, pp. 303-308.

[4] S. Bridgeman, A. Garg and R. Tamassia, A Graph Drawing and Translation Service on the WWW, *Proc. GD '96, LNCS 1190*, Springer-Verlag, 1997, pp. 45-52.

[5] R. F. Cohen, G. Di Battista, R. Tamassia and I. G. Tollis, Dynamic Graph Drawings: Trees, Series-Parallel Digraphs, and Planar *ST*-Digraphs, *SIAM J. Computing*, 24(5), October 1995, pp. 970-1001.

[6] G. Di Battista, P. Eades, R. Tamassia and I. G. Tollis, Algorithms for Drawing Graphs: An Annotated Bibliography, *Computational Geometry: Theory and Applications*, 4(5), 1994, pp. 235-282. Also available at http://www.utdallas.edu/~tollis.

[7] G. Di Battista, P. Eades, R. Tamassia and I. G. Tollis, *Graph Drawing: Algorithms for the Visualization of Graphs*, Prentice-Hall, 1999.

[8] G. Di Battista, A. Garg, G. Liotta, R. Tamassia, E. Tassinari and F. Vargiu, An Experimental Comparison of Four Graph Drawing Algorithms, *Computational Geometry: Theory and Applications*, 1997, pp. 303-325. Also available at http://www.cs.brown/~rt.

[9] C. Ding and P. Mateti, A Framework for the Automated Drawing of Data Structure Diagrams, *IEEE Transactions on Software Engineering*, 16(5), 1990, pp. 543-557.

[10] J. Doenhardt and T. Lengauer, Algorithmic Aspects of One Dimensional Layout Compaction, *IEEE Transactions on Computer-Aided Design of Integrated Circuits and Systems*, 6(5), 1987, pp. 863-879.

[11] C. Esposito, Graph Graphics: Theory and Practice, *Computers and Mathematics with Applications*, 15(4), 1988, pp. 247-253.

J. Six et al., *Refinement of Orthogonal Drawings*, JGAA, 4(3) 75–103 (2000)102

[12] S. Even and G. Granot, Rectilinear Planar Drawings with Few Bends in Each Edge, Tech. Report 797, CS Dept., Technion, Israel Inst. of Tech., 1994.

[13] Jody Fanto, Postprocessing of GIOTTO drawings, Student project, Brown University, Summer 1997.

[14] U. Fößmeier, Interactive Orthogonal Graph Drawing: Algorithms and Bounds, *Proc. GD '97, LNCS 1353*, Springer-Verlag, 1997, pp. 111-123.

[15] U. Fößmeier, C. Heß and M. Kaufmann, On Improving Orthogonal Drawings: The 4M-Algorithm, *Proc. GD '98, LNCS 1547*, Springer-Verlag, 1998, pp. 125-137.

[16] U. Fößmeier and M. Kaufmann, Algorithms and Area Bounds for Non-planar Orthogonal Drawings, *Proc. GD '97, LNCS 1353*, Springer-Verlag, 1997, pp. 134-145.

[17] M. Y. Hsueh, *Symbolic Layout and Compaction of Integrated Circuits*, Ph.D. Thesis, University of California at Berkeley, Berkeley, CA, 1979.

[18] G. Kant, Drawing Planar Graphs Using the Canonical Ordering, *Algorithmica*, 16(1), 1996, pp. 4-32.

[19] C. Kosak, J. Marks and S. Shieber, Automating the Layout of Network Diagrams with Specified Visual Organization, *IEEE Transactions on Systems, Man, Cybernetics*, 24(3), 1994, pp. 440-454

[20] Thomas Lengauer, *Combinatorial Algorithms for Integrated Circuit Layout*, John Wiley and Sons, 1990.

[21] K. Miriyala, S. W. Hornick and R. Tamassia, An Incremental Approach to Aesthetic Graph Layout, *Proc. Int. Workshop on Computer-Aided Software Engineering (Case '93)*, 1993, pp. 297-308.

[22] K. Misue, P. Eades, W. Lai and K. Sugiyama, Layout Adjustment and the Mental Map, *J. of Visual Languages and Computing*, June 1995, pp. 183-210.

[23] A. Papakostas, *Information Visualization: Orthogonal Drawings of Graphs*, Ph.D. Thesis, University of Texas at Dallas, 1996.

[24] A. Papakostas, J. M. Six and I. G. Tollis, Experimental and Theoretical Results in Interactive Orthogonal Graph Drawing, *Proc. GD '96, LNCS 1190*, Springer-Verlag, 1997, pp. 371-386. Also available at `http://www.utdallas.edu/~tollis`.

J. Six et al., *Refinement of Orthogonal Drawings*, JGAA, 4(3) 75–103 (2000)103

[25] A. Papakostas and I. G. Tollis, Algorithms for Area-Efficient Orthogonal Drawings, *Computational Geometry: Theory and Applications*, 9(1998) 1998, pp. 83-110. Also available at http://www.utdallas.edu/~tollis.

[26] A. Papakostas and I. G. Tollis, Issues in Interactive Orthogonal Graph Drawing, *Proc. GD '95, LNCS 1027*, Springer-Verlag, 1995, pp. 419-430. Also available at http://www.utdallas.edu/~tollis.

[27] H. Purchase, Which Aesthetic has the Greatest Effect on Human Understanding, *Proc. of GD '97, LNCS 1353*, Springer-Verlag, 1997, pp. 248-261.

[28] M. Schäffter, Drawing Graphs on Rectangular Grids, *Discr. Appl. Math.*, 63(1995), pp. 75-89.

[29] J. M. Six, K. Kakoulis and I.G. Tollis, Refinement of Orthogonal Graph Drawings, *Proc. GD '98, LNCS 1547*, Springer-Verlag, 1998, pp. 302-315. Also available at http://www.utdallas.edu/~tollis.

[30] J. Storer, On Minimal Vertex-Cost Planar Embeddings, *Networks* 14(1984), pp. 181-212.

[31] R. Tamassia, On Embedding a Graph in the Grid with the Minimum Number of Bends, *SIAM J. Comput.*, 16(1987), pp. 421-444.

[32] R. Tamassia, G. Di Battista and C. Batini, Automatic Graph Drawing and Readability of Diagrams, *IEEE Transactions on Systems, Man, and Cybernetics*, 18(1), 1988, pp. 61-79.

[33] R. Tamassia and I. G. Tollis, Planar Grid Embeddings in Linear Time, *IEEE Trans. on Circuits and Systems CAS-36*, 1989, pp. 1230-1234.

[34] I. G. Tollis, Graph Drawing and Information Visualization, *ACM Computing Surveys*, 28A(4), December 1996. Also available at http://www.utdallas.edu/~tollis/.

Journal of Graph Algorithms and Applications

http://www.cs.brown.edu/publications/jgaa/

vol. 4, no. 3, pp. 105–133 (2000)

A Split & Push Approach to 3D Orthogonal Drawing

Giuseppe Di Battista   *Maurizio Patrignani*

Dipartimento di Informatica e Automazione, Università di Roma Tre
via della Vasca Navale 79, 00146 Roma, Italy http://www.dia.uniroma3.it/
gdb@dia.uniroma3.it   patrigna@dia.uniroma3.it

Francesco Vargiu

AIPA, via Po 14, 00198 Roma Italy
http://www.aipa.it/
vargiu@aipa.it

Abstract

We present a method for constructing orthogonal drawings of graphs of maximum degree six in three dimensions. The method is based on generating the final drawing through a sequence of steps, starting from a "degenerate" drawing. At each step the drawing "splits" into two pieces and finds a structure more similar to its final version. Also, we test the effectiveness of our approach by performing an experimental comparison with several existing algorithms.

Communicated by G. Liotta and S. H. Whitesides: submitted March 1999; revised April 2000.

Research supported in part by the ESPRIT LTR Project no. 20244 - ALCOM-IT and by the CNR Project "Geometria Computazionale Robusta con Applicazioni alla Grafica ed al CAD." An extended abstract of this paper was presented at the 6th International Symposium on Graph Drawing, GD'98, Montreal, Canada, August 1998.

G. Di Battista et al., *3D Orthogonal Drawing*, JGAA, 4(3) 105–133 (2000) 106

1 Introduction

Both for its theoretical appeal and for the high number of potential applications, research in 3D graph drawing is attracting increasing attention. In fact, low price high-performance 3D graphic workstations are becoming widely available. On the other hand the demand of visualization of large graphs increases with the popularity of the graph drawing methods and tools, and 3D graph drawing seems to offer interesting perspectives to such a demand. The interest of the 3D graph drawing community has been mainly devoted to straight-line drawings and to orthogonal drawings.

Straight-line drawings map vertices to points and edges to straight-line segments. Many different approaches to the construction of straight-line drawings can be found in the literature. For example, the method presented in [5] distributes the vertices of the graph along a "momentum curve" so that there are not crossings among the edges. The produced drawings are then "compressed" into a volume (volume of the smallest box enclosing the drawing) of $4n^3$, where n is the number of vertices of the graph to be drawn. The same paper presents another algorithm which constructs drawings without edge crossings of planar graphs with degree at most 4. It "folds" a 2-dimensional orthogonal grid drawing of area $h \times v$ into a straight-line drawing with volume $h \times v$.

Another classical approach of the graph drawing field is the force directed one [7]. It uses a physical analogy, where a graph is seen as a system of bodies with forces acting among them. These algorithms seek a configuration of the system with, possibly local, minimal energy. Force directed approaches have been exploited in 3D graph drawing to devise the algorithms presented in [4, 6, 9, 19, 26, 14, 20].

Further, the research on straight-line drawings stimulated a deep investigation on theoretical bounds. Examples of bounds on the volume of a straight-line drawing can be found in [5, 21]. Namely, in [5] it is shown that a graph can be drawn in an $n \times 2n \times 2n$ volume, which is asymptotically optimal. In [21] it is shown that, for any fixed r, any r-colorable graph has a drawing with volume $O(n^2)$, and that the order of magnitude of this bound cannot be improved.

Special types of straight-line drawings have been studied in [3, 13, 1, 16] (visibility representations) and in [18] (proximity drawings).

In an orthogonal drawing vertices are mapped to points and edges are mapped to polygonal chains composed of segments that are parallel to the axes. Also, it is quite usual to consider a drawing convention in which edges have no intersections, vertices and bends have integer coordinates, and vertices have maximum degree six.

Biedl [2] shows a linear time algorithm (in what follows we call it Slices) that draws a graph in $O(n^2)$ volume with at most 14 bends per edge. The drawing is obtained by placing all the vertices on a certain horizontal plane and by assigning a further horizontal plane to every edge, "one slice per edge".

Eades, Stirk, and Whitesides [10] propose a $O(n^{3/2})$-time algorithm, based on augmenting the graph to an Eulerian graph and on applying a variation of an algorithm by Kolmogorov and Barzdin [17]. The algorithm produces drawings

G. Di Battista et al., *3D Orthogonal Drawing*, JGAA, 4(3) 105–133 (2000) 107

that have $O(n^{3/2})$ volume and at most 16 bends per edge. We call this algorithm **Kolmogorov**.

The algorithm proposed by Eades, Symvonis, and Whitesides in [11] (we call it **Compact**) requires $O(n^{3/2})$ time and volume and introduces at most 7 bends per edge.

In the same paper [11] Eades, Symvonis, and Whitesides presented a second algorithm (we call it **Three-Bends**) whose complexity is linear, while its volume is $27n^3$, and at most 3 bends per edge are introduced. Algorithm **Three-Bends** is based on augmenting the graph to a 6-regular graph and on a coloring technique. The implementation used in the experimental comparison follows the description of the algorithm given in [11], in which the coloring phase is assumed to run in $O(n^{3/2})$ time, although in the journal version [12] of the paper [11] Eades, Symvonis, and Whitesides point out that, by using the result in [25] for the coloring phase, the actual time complexity of algorithm **Three-Bends** is $O(n)$.

Papakostas and Tollis [22] present a linear time algorithm (we call it **Interactive**) that requires at most $4.66n^3$ volume and at most 3 bends per edge. It is incremental and can be extended to draw graphs with vertices of arbitrary degree. The construction starts from a first pair of adjacent vertices, and then it adds one vertex at a time with its incident edges.

Finally, Wood [28] presents an algorithm for maximum degree 5 graphs that requires $O(n^3)$ volume and at most 2 bends per edge. Recently [27], the result has been extended to maximum degree 6 graphs using no more than 4 bends per edge. The volume is at most $2.37n^3$, the total number of bends is always less than $7m/3$, where m is the number of edges.

Although the results presented in the above papers are interesting and deep, the research in this field suffers, in our opinion, from a lack of general methodologies.

In this paper we deal with the problem of constructing orthogonal drawings in three dimensions. Namely, we experiment with several existing algorithms to test their practical applicability and propose new techniques that have a good average behavior. Our main target are graphs with vertices in the range 10–100. Such graphs are crucial in several applications [8], like conceptual modeling of databases (Entity-Relationship schemas), information system functions analysis (Data Flow Diagrams), and software engineering (modules Interaction Diagrams).

The results presented in this paper can be summarized as follows.

- We present a new method for constructing orthogonal drawings of graphs of maximum degree six in three dimensions without intersections between edges. It can be considered more as a general strategy rather than as a specific algorithm. The approach is based on generating the final drawing through a sequence of steps, starting from a "degenerate" drawing; at each step the drawing "splits" into two pieces and finds a structure more similar to its final version. The new method aims at constructing drawings without any "privileged" direction and with a routing strategy that is not decided in advance, but depends on the specific needs of the drawing.

G. Di Battista et al., *3D Orthogonal Drawing*, JGAA, 4(3) 105–133 (2000) 108

- We devise an example of an algorithm developed according to the above method. We call it `Reduce-Forks`.

- We perform an experimental comparison of algorithms `Compact`, `Interactive`, `Kolmogorov`, `Reduce-Forks`, `Slices`, and `Three-Bends` with a large test suite of graphs with at most 100 vertices. We measure the computation time and what we consider to be three important readability parameters: average volume, average edge length, and average number of bends along edges. The recent algorithms in [27] and in [12] are not included in the comparison. Our implementations try to strictly follow the descriptions given in the papers, without any further improvement.

- Our experiments show that no algorithm can be considered "the best" with respect to all the parameters. Concerning `Reduce-Forks`, we can say that it has the best behavior in terms of the readability parameters for graphs in the range 5–30, while its effectiveness decreases for larger graphs. Also, among the algorithms that have a reasonable number of bends along the edges (`Interactive`, `Reduce-Forks`, and `Three-Bends`), `Reduce-Forks` is the one that has the best behavior in terms of edge length and volume. This is obtained at the expense of an efficiency that is much worse than the other algorithms. However, the CPU time does not seem to be a critical issue for the size of graphs in the interval.

The paper is organized as follows. In Section 2 we present our approach and in Section 3 we argue about its feasibility. In Section 4 we describe Algorithm `Reduce-Forks`. In Section 5 we present the results of the experimental comparison.

The interested reader will find at our Web site a CGI program that allows the use of all the algorithms and the test suite used in the experiments (`www.dia.uniroma3.it/~patrigna/3dcube`).

2   A Strategy for Constructing 3D Orthogonal Drawings

A *drawing* of a graph represents the vertices as points in 3D space and edges as curves connecting the points corresponding to the associated vertices. An *orthogonal drawing* is such that all the curves representing edges are chains of segments parallel to the axes. A *grid* drawing is such that all vertices and bends along the chains representing the edges have integer coordinates. Further, to simplify the notation, when this does not cause ambiguities, we shall consider an orthogonal drawing as a graph with coordinate values for its vertices and bends.

We also make use of more unusual definitions that describe intermediate products of our design process. Such definitions allow us to describe "degenerate" drawings where vertices can overlap, edges can intersect, and/or can have length 0.

G. Di Battista et al., *3D Orthogonal Drawing*, JGAA, 4(3) 105–133 (2000) 109

A *01-drawing* is an orthogonal grid drawing such that each edge has either length 0 or length 1 and vertices may overlap. Observe that a 01-drawing does not have bends along the edges.

A *0-drawing* is a (very!) trivial 01-drawing such that each edge has length 0 and all vertices have the same coordinates. A *1-drawing* is a 01-drawing such that all edges have length 1 and vertices have distinct coordinates. (See Fig. 1.) Observe that while all graphs have a 01-drawing, only some admit a 1-drawing. For example the triangle graph does not admit a 1-drawing.

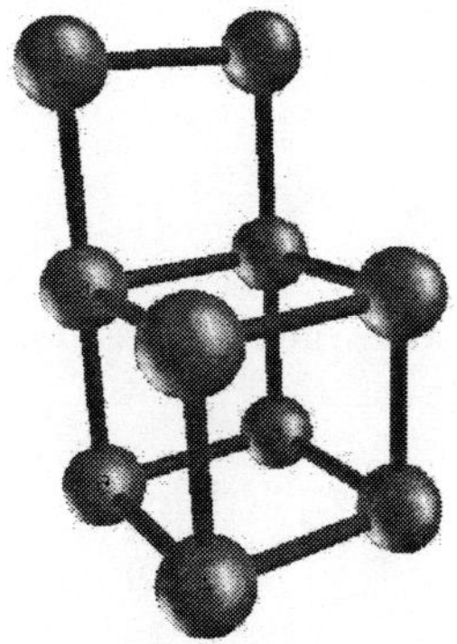

Figure 1: A 1-drawing of a graph with ten vertices.

Let G be a graph. A *subdivision* G_1 of G is a graph obtained from G by replacing some edges of G with simple paths. We partition the vertices of G_1 into vertices that belong also to G (we call them *original vertices*) and vertices that belong only to G_1 (we call them *dummy vertices*). Observe that a subdivision G_2 of G_1 is a subdivision of G. If G does not have vertices with degree greater than 6, because of the drawing algorithms mentioned in the introduction, there always exists a subdivision of G that admits a 1-drawing. From now on, unless otherwise specified, we deal only with graphs whose maximum degree is at most 6.

A *dummy path* of G_1 is a path consisting only of dummy vertices except, possibly, at the endpoints (that can be original vertices). A *planar path* of an orthogonal drawing is a maximal path whose vertices are on the same plane. A planar dummy path is *self-intersecting* if it has two distinct vertices with the same coordinates. We consider only paths with at least one edge.

Our method constructs orthogonal grid drawings with all vertices at distinct coordinates and without intersections between edges (except at the common endpoints). The drawing process consists of a sequence of steps. Each step maps a 01-drawing of a graph G into a 01-drawing of a subdivision of G. We start with a 0-drawing of G and, at the last step, we get a 1-drawing of a subdivision G_1 of G. Hence, an orthogonal grid drawing of G is obtained by replacing each path of G_1, corresponding to an edge (u, v) of G, with an orthogonal polygonal

G. Di Battista et al., *3D Orthogonal Drawing*, JGAA, 4(3) 105–133 (2000) 110

line connecting u and v.

The general strategy is as follows. Let G be a graph. We consider several subsequent subdivisions of G. We construct an orthogonal grid drawing Γ of G in four phases.

Vertex Scattering: Construct a *scattered representation* Γ_1 of G, i.e. a 01-drawing such that:

- Γ_1 is a subdivision of G,
- all the original vertices have different coordinates, and
- all the planar dummy paths are not self-intersecting.

After this phase dummy vertices may still overlap both with dummy and with original vertices.

Direction Distribution: Construct a *direction-consistent representation* Γ_2 of G, i.e. a 01-drawing such that:

- Γ_2 is a scattered representation of G,
- for each vertex v of Γ_2, v and all its adjacent vertices have different coordinates.

We call this phase Direction Distribution because after this phase the edges incident on v "leave" v with different directions. Observe that this is true both in the case v is original and in the case v is dummy.

Vertex-Edge Overlap Removal: Construct a *vertex-edge-consistent representation* Γ_3 of G, i.e. a 01-drawing such that:

- Γ_3 is a direction-consistent representation of G,
- for each original vertex v, no dummy vertex has the same coordinates of v.

After this step the original vertices do not "collide" with dummy vertices. Observe that groups of dummy vertices sharing the same coordinates may still exist.

Crossing Removal: Construct a 1-drawing Γ_4 that is a vertex-edge-consistent representation of G.

Observe that, since Γ_4 is a 1-drawing, all its vertices, both original and dummy, have different coordinates. Also, observe that an orthogonal grid drawing Γ of G is easily obtained from Γ_4.

Each of the above phases is performed by repeatedly applying the same simple primitive operation called *split*. Informally, this operation "cuts" the entire graph with a plane perpendicular to one of the axes. The vertices lying

G. Di Battista et al., *3D Orthogonal Drawing*, JGAA, 4(3) 105–133 (2000) 111

in the "cutting" plane are partitioned into two subsets that are "pushed" into two adjacent planes. A more formal definition follows.

In what follows, the term *direction* always refers to a direction that is isothetic with respect to one of the axes and the term *plane* always refers to a plane perpendicular to one of the axes. Given a direction d we denote by $-d$ its opposite direction.

Let Γ be a 01-drawing. A *split operation* has 4 parameters d, P, ϕ, ρ, where:

- d is a direction.

- P is a plane perpendicular to d.

- Function ϕ maps each vertex of Γ laying in P to a boolean.

- Function ρ maps each edge (u, v) of Γ laying in P such that $\phi(u) \neq \phi(v)$ and such that u and v have different coordinates to a boolean.

Operation $split(d, P, \phi, \rho)$, applied to Γ, performs as follows (see Fig. 2).

1. Move one unit in the d direction all vertices in the open half-space determined by P and d. Such vertices are "pushed" towards d.

2. Move one unit in the d direction each vertex u on P with $\phi(u) = true$.

3. For each edge (u, v) that after the above steps has length greater than one, replace (u, v) with the new edges (u, w) and (w, v), where w is a new dummy vertex. Vertex w is positioned as follows.

   (a) If the function $\rho(u, v)$ is not defined, then vertex w is simply put in the middle point of the segment u, v.

   (b) Else, (the function $\rho(u, v)$ is defined) suppose, without loss of generality, that $\phi(u) = true$ and $\phi(v) = false$. Two cases are possible. If $\rho(u, v) = true$, then w is put at distance 1 in the d direction from v. If $\rho(u, v) = false$, then w is put at distance 1 in the $-d$ direction from u. Roughly speaking, the function ρ is used to specify which is the orientation of the "elbow" connecting u and v.

Observe that a *split* operation applied to a 01-drawing of a graph G constructs a 01-drawing of a subdivision of G. Also, although *split* is a simple primitive, it has several degrees of freedom, expressed by the split parameters, whose usage can lead to very different drawing algorithms. Further, *split* has to be "handled with care", since by applying a "random" sequence of *split* operations there is no guarantee that the process terminates with a 1-drawing.

G. Di Battista et al., *3D Orthogonal Drawing*, JGAA, 4(3) 105–133 (2000) 112

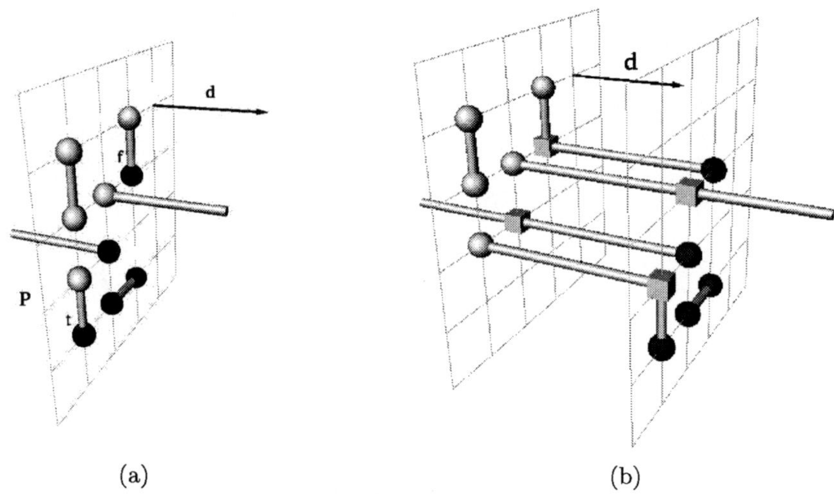

(a)   (b)

Figure 2: An example of split: (a) before the split and (b) after the split. Vertices with $\phi = true$ ($\phi = false$) are black (light grey). Edges with $\rho = true$ ($\rho = false$) are labeled t (f). The little cubes are dummy vertices inserted by the *split*.

3 Feasibility of the Approach

In this section we show how the *split* operation can be used to perform the four drawing phases described in Section 2.

Since the definition of a scattered representation requires that all the planar dummy paths are not self-intersecting and since an edge of zero length implies the existence of a self-intersecting path, we have:

Property 1 *All the edges of a scattered representation have length* 1.

We now prove that a scattered representation can always be constructed.

Theorem 1 *Let Γ_0 be a 0-drawing of a graph G. There exists a finite sequence of split operations that, starting from Γ_0, constructs a scattered representation of G.*

Proof: Consider a sequence of split operations all performed with planes perpendicular to the same axis, say the x-axis, and such that each split separates one original vertex from the others.

Namely, suppose that the n vertices of Γ_0 are labeled $v_1, \ldots, v_n$ and that all of them are positioned at the origin. For each i such that $1 \leq i \leq n - 1$ we perform $split(d, P, \phi_i, \rho)$ where:

- d is the direction of the x-axis.

G. Di Battista et al., *3D Orthogonal Drawing*, JGAA, 4(3) 105–133 (2000) 113

- P is the plane $x = 0$.

- Function ϕ_i maps vertex v_i to true and all the other vertices on P to false.

- Function ρ is not defined on any edge.

At the end of the process all vertices lie on the same line and original vertices have different coordinates. Furthermore, all the obtained dummy paths consist only of straight line segments with length 1 and with the same direction. Hence, all dummy paths are not self-intersecting. $\qquad\square$

Let u be a vertex. We call the six directions around u *access directions* of u. Consider the access direction of u determined by traversing edge (u, v) from u to v; this is the access direction of u *used* by (u, v). An access direction of u that is not used by any of its incident edges is a *free direction* of u.

Given a direction d and a vertex v, we denote by $P_{d,v}$ the plane through v and perpendicular to d. The following theorem shows that, starting from a scattered representation, we can always construct a direction-consistent representation.

Theorem 2 *Let Γ_1 be a scattered representation of a graph G. There exists a finite sequence of split operations that, starting from Γ_1 constructs a direction-consistent representation of G.*

Proof: We consider one by one each vertex u of Γ_1 with edges (u, v) and (u, w) that use the same access direction d of u. Since Γ_1 is a scattered representation of G we have that:

- u is an original vertex, and

- at least one of v and w (say v) is dummy.

Also, by Property 1 we have that all edges incident to u do not have length 0, and hence use a direction of u.

Two cases are possible. Case 1: at least one free direction d' of u is orthogonal to d; see Fig. 3.a. Case 2: direction $-d$ is the only free direction of u; see Fig. 4.a.

Case 1: We perform $split(d', P_{d',u}, \phi, \rho)$ as follows. We set $\phi(v) = true$, all the other vertices of $P_{d',u}$ have $\phi = false$. Also, $\rho(u, v) = true$, all the other edges in the domain of ρ have $\rho = false$.

After performing the *split* (see Figure 3.b), the first edge resulting from the subdivision of (u, v) uses the direction d' of u. The usage of the other access directions of u is unchanged. Also, all the other vertices still use the same access directions as before the split with the exception, possibly, of v (that is dummy).

Case 2: Let d'' be a non-free direction of u different from d. We perform the same split operation as the one of Case 1, using direction d'' instead of d'. After the *split*, (see Figure 4.b), the first edge resulting from the subdivision of (u, v) uses the direction d'' of u. At this point, since at least one direction of the free directions of u is orthogonal to d'' (direction $-d$), we can apply the same strategy of Case 1.

G. Di Battista et al., *3D Orthogonal Drawing*, JGAA, 4(3) 105–133 (2000) 114

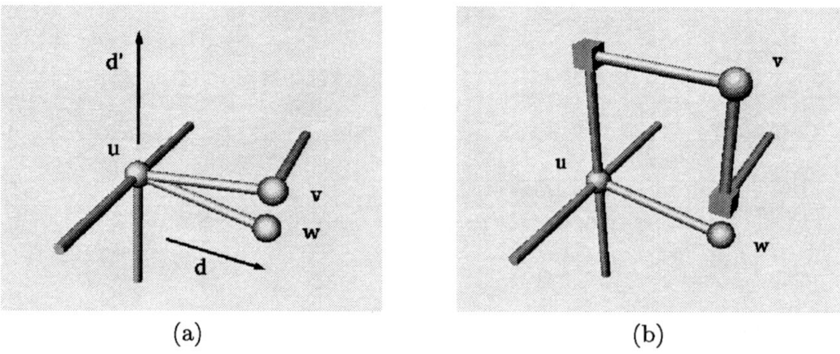

(a)   (b)

Figure 3: Case 1 in the proof of Theorem 2, before (a) and after (b) the split operation.

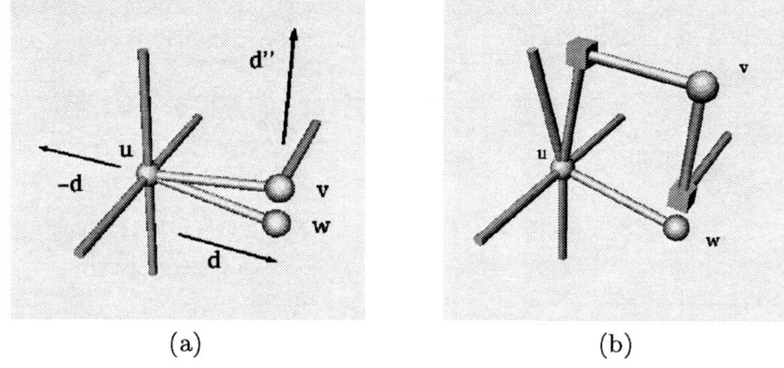

(a)   (b)

Figure 4: Case 2 in the proof of Theorem 2, before (a) and after (b) the split operation.

Finally, it is easy to observe that the above split operations preserve the properties of the scattered representations. $\square$

In the following we show that, starting from a direction-consistent representation, a vertex-edge-consistent representation can be obtained.

We define a simpler version of $split(d, P, \phi, \rho)$, called $trivialsplit(d, P)$, where ϕ is *false* for all vertices of the cutting plane, and, as a consequence, the domain of ρ is empty. Roughly speaking, $trivialsplit$ has the effect of inserting a new plane in the drawing that contains only the dummy vertices that are caused by the edge "stretchings". We use $trivialsplit$ for tackling the cases where a dummy vertex has the same coordinates of another vertex. The following property follows from the definition.

Property 2 *Operation trivialsplit does not affect the usage of the access directions of the vertices.*

G. Di Battista et al., *3D Orthogonal Drawing*, JGAA, 4(3) 105–133 (2000) 115

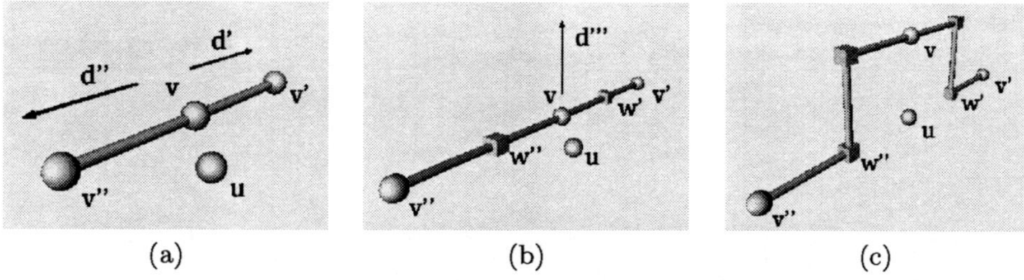

Figure 5: First case in the proof of Theorem 3.

Theorem 3 *Let Γ_2 be a direction-consistent representation of a graph G. There exists a finite sequence of split operations that, starting from Γ_2, constructs a vertex-edge-consistent representation of G.*

Proof: Consider one by one each original vertex u of Γ_2 such that there exists a dummy vertex v with the same coordinates of u. Let (v',v) and (v,v'') be the incident edges of v. By Property 1 and by the fact that Γ_2 is a scattered representation of G, it follows that v, v' and v'' have different coordinates.

Let d' and d'' be the directions of v used by (v',v) and by (v,v''), respectively (see Fig. 5.a and Fig. 6.a). We perform $trivialsplit(d',P_{d',v})$ and $trivialsplit(d'',P_{d'',v})$. After performing such operations vertex v is guaranteed to be adjacent to dummy vertices w' and w'' created by the performed splits.

Two cases are possible (see Fig. 5.b and Fig. 6.b): either $d' = -d''$ or not. In the first case we define d''' as any direction orthogonal to d'; in the second case we define d''' as any direction among d', d'', and the two directions orthogonal to d' and d''. At this point we perform a third *split*. Namely, we apply $split(d''',P_{d''',v},\phi,\rho)$ as follows (see Fig. 5.c and Fig. 6.c).

- $\phi(v) = true$, all the other vertices of $P_{d''',v}$ have $\phi = false$.

- All the edges in the domain of ρ have $\rho = true$.

Note that at this point u and v have different coordinates. Further, by Property 2 and because of the structure we have chosen for *split* we have that the entire sequence of operations preserves the properties of the direction-consistent representations, and does not generate new vertex-edge overlaps. □

In the remaining part of this section, we study how to perform the last phase of the general strategy presented in Section 2. Namely, we are going to show that, given a vertex-edge-consistent representation of a graph G, it is possible to construct a new vertex-edge-consistent representation of G that is a 1-drawing. Before introducing the corresponding theorem we need some intermediate terminology and results.

Let Γ be a 01-drawing of G. We say that two distinct vertex-disjoint planar paths p' and p'' of Γ on the same plane *intersect* if there exist two vertices one

G. Di Battista et al., *3D Orthogonal Drawing*, JGAA, 4(3) 105–133 (2000) 116

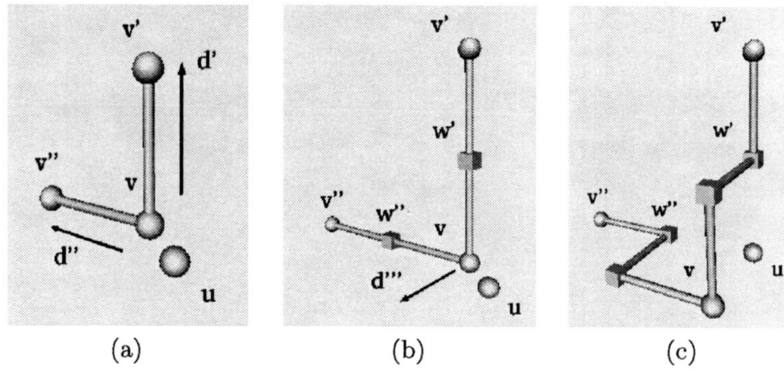

(a)   (b)   (c)

Figure 6: Second case in the proof of Theorem 3.

of p' and the other of p'' with the same coordinates. We denote by $\chi(\Gamma)$ the number of the pairs of intersecting planar dummy paths of Γ. Observe that $\chi(\Gamma)$ can be greater than one even if there are just two vertices with the same coordinates (see Fig. 7).

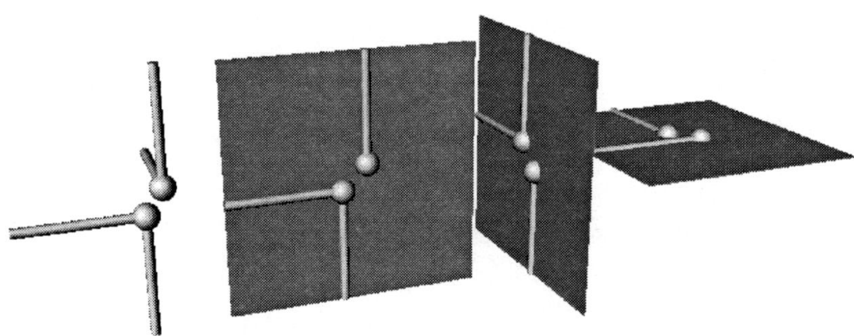

Figure 7: Two dummy vertices with the same coordinates originating 3 pairs of intersecting planar dummy paths.

Suppose we need to perform an operation $trivialsplit(d, P)$ on Γ and let Γ' be the obtained 01-drawing. Let P' be the plane of Γ parallel to P and at distance 1 from P in the d direction.

Property 3 *Plane P' does not contain any edge of Γ'.*

Of course this implies that P' does not contain any planar path. Also, because of Property 3, we have:

Property 4 *The planar dummy paths of Γ' are in one-to-one correspondence with the planar dummy paths of Γ.*

G. Di Battista et al., *3D Orthogonal Drawing*, JGAA, 4(3) 105–133 (2000) 117

Property 5 $\chi(\Gamma) = \chi(\Gamma')$.

Proof: This follows from Property 4 and from the fact that two planar dummy paths of Γ intersect if and only if the corresponding planar dummy paths of Γ' intersect. $\qquad\square$

Theorem 4 *Let Γ_3 be a vertex-edge-consistent representation of a graph G. There exists a finite sequence of split operations that, starting from Γ_3, constructs a 1-drawing of a subdivision of G.*

Proof: Since Γ_3 is vertex-edge-consistent, all original vertices have distinct coordinates, but some dummy vertices may still overlap.

If $\chi(\Gamma_3) = 0$, then Γ_3 is already a 1-drawing of G. Otherwise, we repeatedly select a pair of intersecting planar dummy paths p' and p'' (see Fig. 8.a) and "remove" their intersection, decreasing the value of χ. Such removal is performed as follows.

Let u and v be the end-vertices of p'. We have three cases:

1. exactly one of u and v (say v) is an original vertex,

2. both u and v are original vertices, or

3. both u and v are dummy vertices.

In Case 1 (see Figs. 8.a and 8.b) we perform $trivialsplit(d, P_{d,v})$ where d is the direction p' leaves v. In Case 2 we perform $trivialsplit(d', P_{d',u})$ and $trivialsplit(d'', P_{d'',v})$ where d' (d'') is the direction p' leaves u (v). In Case 3 we do not perform any $trivialsplit$.

After the above splits, by Property 5, the value of χ stays unchanged. Also, observe that the drawing is still a vertex-edge-consistent representation of G. At this point we concentrate on Case 1, the other cases are similar and are omitted for brevity. We denote by s the dummy vertex introduced along p' by $trivialsplit(d, P_{d,v})$.

Let d''' be a direction orthogonal to the plane P where p' and p'' intersect. We perform $split(d''', P, \phi, \rho)$, by setting (see Fig. 8.c):

- $\phi(x) = true$ for each vertex $x \in p'$ and $x \neq v, s$ ($false$ otherwise) and

- $\rho = true$ for all the edges in the domain of ρ.

We have that χ decreases after the split by at least one unit. It is easy to see that such a split preserves the properties of the vertex-edge-consistent representations. $\qquad\square$

In this section we have shown that *split* is a powerful tool in performing the phases of the strategy presented in Section 2. Namely, each of Vertex Scattering, Direction Distribution, Vertex-edge Overlap Removal, and Crossing Removal can be performed by a finite sequence of splits.

G. Di Battista et al., *3D Orthogonal Drawing*, JGAA, 4(3) 105–133 (2000) 118

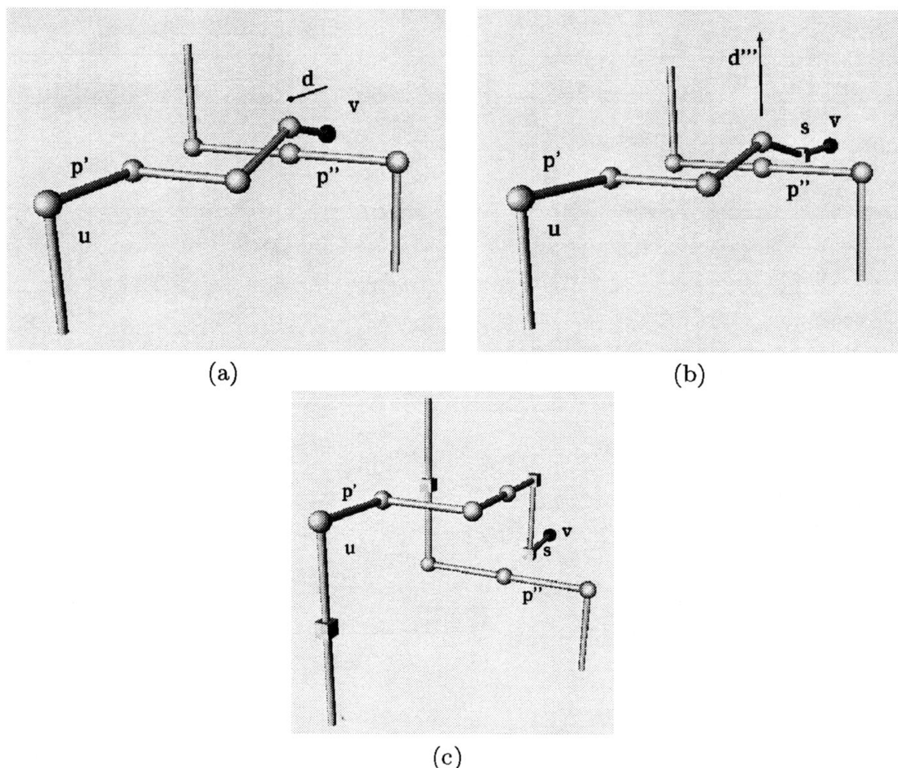

(a) (b)

(c)

Figure 8: Intersecting planar dummy paths in the proof of Theorem 4. The black vertex is original. All the other vertices are dummy. The little cubes are dummy vertices inserted by the split operations described in the proof. Slanted edges indicate the crossing.

4 The Reduce-Forks Algorithm

Algorithm Reduce-Forks is an example of an algorithm that follows the strategy described in Sections 2 and 3. Namely, the phases of the strategy are refined into several heuristics that are illustrated in the following subsections. In Section 5 Reduce-Forks will be compared with the algorithms described in Section 1.

Figs. 9 and 10 show how Reduce-Forks computes a drawing of a K_6 graph. Spheres represent original vertices while cubes represent dummy vertices. Vertices with the same coordinates are drawn inside the same box.

4.1 Vertex Scattering

An edge (u, v) is *cut* by $split(d, P, \phi, \rho)$ if u and v have different values of ϕ. Informally, they were in plane P before the split and are in different planes after the split. A pair of adjacent edges that are cut by a split is a *fork*.

G. Di Battista et al., *3D Orthogonal Drawing*, JGAA, 4(3) 105–133 (2000) 119

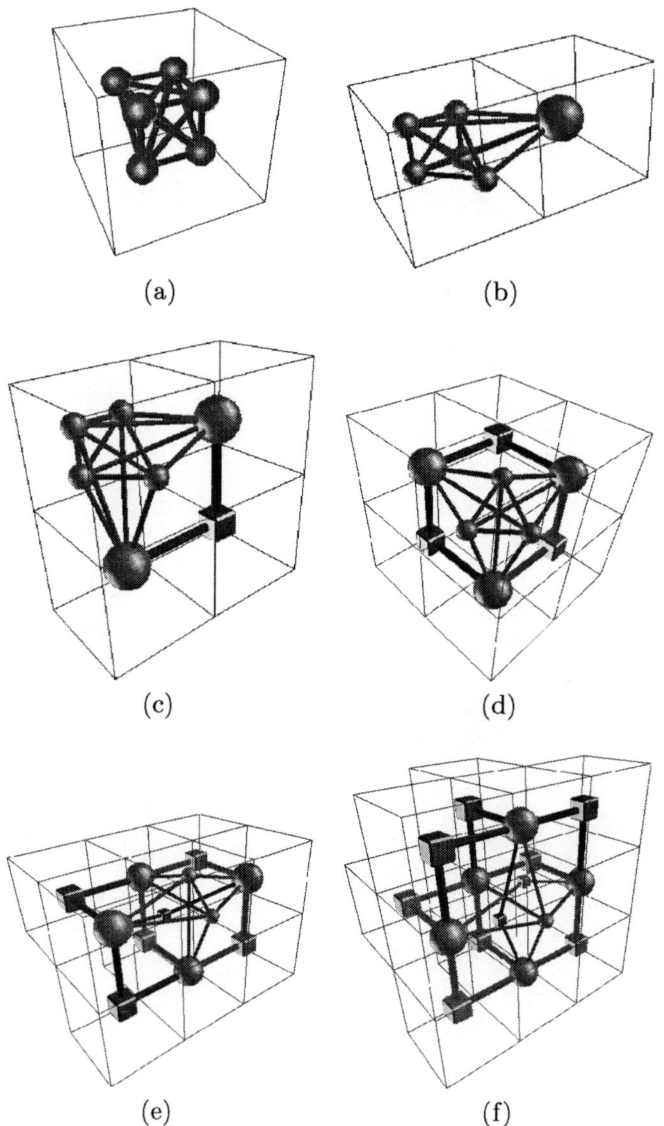

Figure 9: The Vertex Scattering phase of algorithm Reduce-Forks applied on a K_6 graph. (a) is a 0-drawing and (f) is a scattered representation.

G. Di Battista et al., *3D Orthogonal Drawing*, JGAA, 4(3) 105–133 (2000) 120

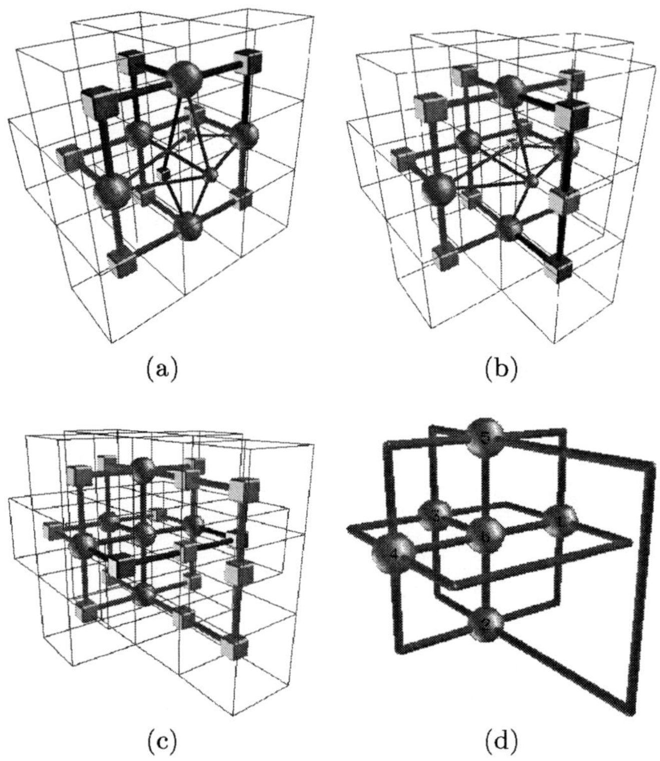

(a)   (b)

(c)   (d)

Figure 10: (a–c) the Direction Distribution phase of algorithm Reduce Forks applied on the scattered representation of the K_6 graph of Fig. 9.f (duplicated in (a) for the convenience of the reader). (d) final drawing. Observe that in this example the Vertex-edge Overlap Removal and the Crossing Removal phases are not necessary since (c) is already a 1-drawing.

G. Di Battista et al., *3D Orthogonal Drawing*, JGAA, 4(3) 105–133 (2000) 121

Roughly speaking, the heuristic of **Reduce-Forks** for Vertex Scattering works as follows. We select an arbitrary pair of original vertices u and v of G with the same coordinates. Let P', P'', and P''' be the three planes containing u and v. We consider the set of split operations with planes P', P'', and P''' and that separate u from v and perform one with "a few" forks. We choose to keep small the number of forks because each fork will require the insertion of at least one dummy vertex in the subsequent Direction Distribution phase. Such dummy vertices will become bends in the final drawing. We repeatedly apply the above strategy until a scattered representation is obtained.

More formally, observe that finding a split with no forks is equivalent to finding a *matching cut*. A matching cut in a graph is a subset of edges that are pairwise vertex disjoint (matching) and such that their removal makes the graph disconnected (cut). Unfortunately, the problem of finding a matching cut in a graph is NP-complete (see [23]). The proof in [23] is based on a reduction from the NAE3SAT problem [15] and holds for graphs of arbitrary degree.

However, a simple heuristic for finding a cut with a few forks is described below.

Consider vertices u and v. We color *black* and *red* the vertices in the two sides of the split. Each step of the heuristic colors one vertex. At a certain step a vertex can be black, red or *free* (uncolored). At the beginning u is black, v is red, and all the other vertices are free.

Colored vertices adjacent to free vertices are *active vertices*. Black (Red) vertices adjacent to red (black) vertices are *boundary vertices*. See Fig. 11. Each step works as follows.

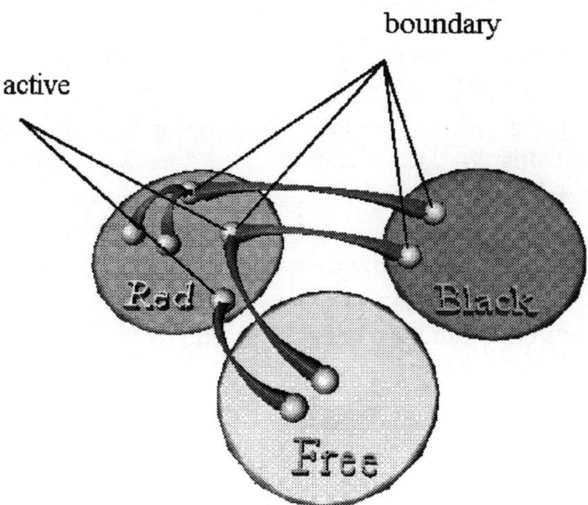

Figure 11: Red, black, and free vertices in the Vertex Scattering heuristic of Algorithm **Reduce-Forks**.

1. If a boundary active red vertex, say x, exists, then color red one free

G. Di Battista et al., *3D Orthogonal Drawing*, JGAA, 4(3) 105–133 (2000) 122

vertex y adjacent to x. The rationale for this choice is the following: since vertex x is adjacent to at least one black vertex w (from the definition of boundary vertex), by coloring y red we prevent a fork between (x, y) and (x, w). Analogously, if a boundary active black vertex exists, then color black one of its adjacent free vertices.

2. Else, if an active red vertex, say x, exists, then choose a free vertex y adjacent to x and color y red. This is done to avoid cutting edge (x, y). Analogously, if an active black vertex exists, then color black one of its adjacent free vertices.

3. Else, randomly color black or red a random free vertex.

We perform the above heuristic to each of the subgraphs induced by the vertices on P', P'', and P'''. Then we select a *split* with the plane among P', P'', and P''' where the cut with the smallest number of forks has been found.

The heuristic can be implemented to run time and space linear in the size of the current 01-drawing (a graph of maximum degree six has a linear number of edges).

Observe that, since each split gives different coordinates to at least two original vertices formerly having the same coordinates, in the worst case the heuristic is used a number times that is linear in the number of original vertices.

Fig. 9 shows the sequence of splits performed by the heuristic on the K_6 graph.

4.2 Direction Distribution

Now, for each original vertex u of G with edges (u, v) and (u, w) such that v and w have the same coordinates (at least one of v and w is dummy), we have to find a *split* that separates v from w (see Fig. 12.a). Of course there are many degrees of freedom for choosing the *split*. In Reduce-Forks a heuristic is adopted that follows the approach of the proof of Theorem 2. However, in performing the splits we try to move an entire planar dummy path rather than moving a single dummy vertex. This has the effect of both decreasing the number of bends (dummy vertices with orthogonal incident edges) introduced by the *split*, and of occasionally solving an analogous problem on the other extreme of the planar dummy path.

More formally, we apply the following algorithm.

1. Compute the (two) planar dummy paths p_v and q_v containing (u, v) (see Figs. 12.b–12.c) and the (two) planar dummy paths p_w and q_w containing (u, w) (see Figs. 12.d–12.e).

2. For each path of p_v, q_v, p_w, and q_w determine the split operations that separate the path (except for the possible original vertices) from all the other vertices that lie on its plane. For each path we have exactly two possible splits. Fig. 13 shows the effect of two possible splits on the configuration of Fig. 12.a.

G. Di Battista et al., *3D Orthogonal Drawing*, JGAA, 4(3) 105–133 (2000) 123

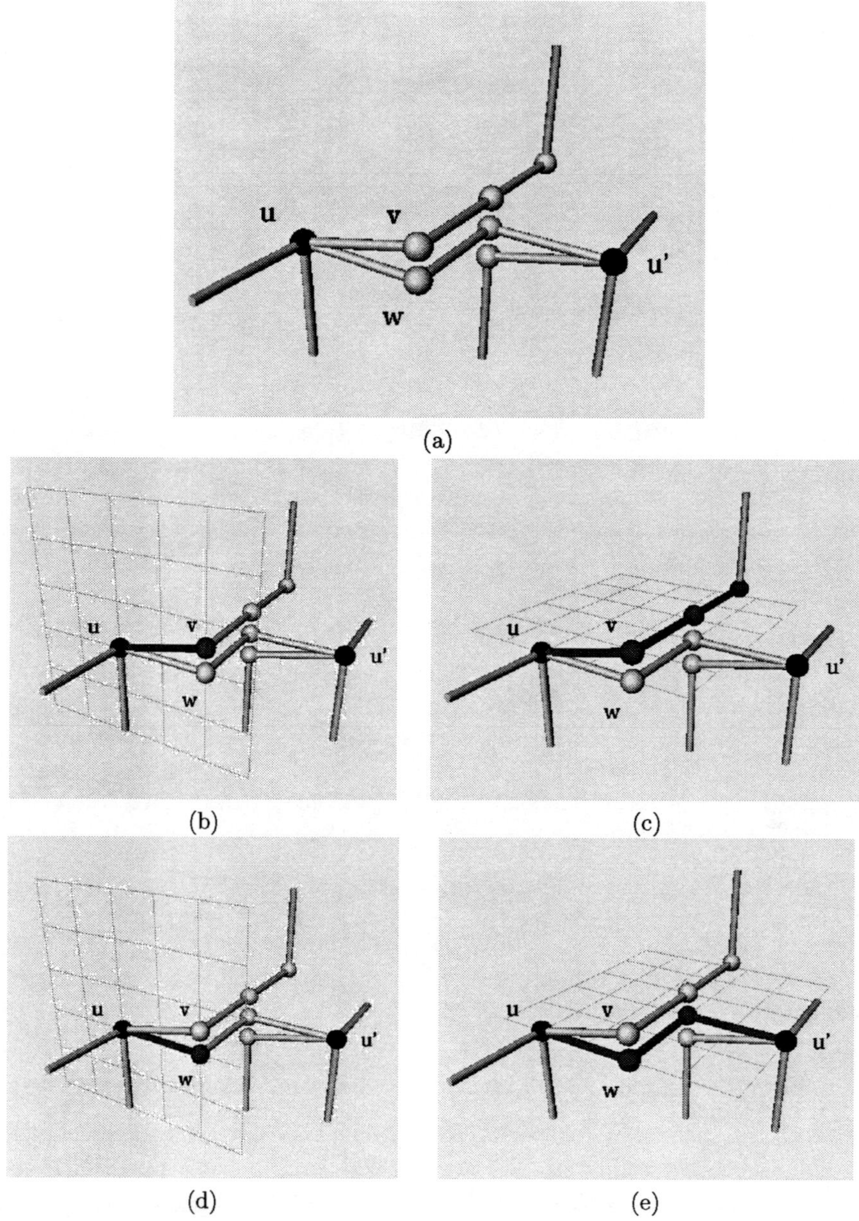

(a)

(b) (c)

(d) (e)

Figure 12: An example illustrating the heuristic adopted by the Reduce-Forks algorithm for the Direction Distribution phase. Black vertices are original. All other vertices are dummy. In (a) two vertices (v, and w) adjacent to the original vertex u, share the same coordinates. Paths p_v, q_v, p_w, and q_w are shown with dark grey in (b), (c), (d), and (e), respectively.

G. Di Battista et al., *3D Orthogonal Drawing*, JGAA, 4(3) 105–133 (2000) 124

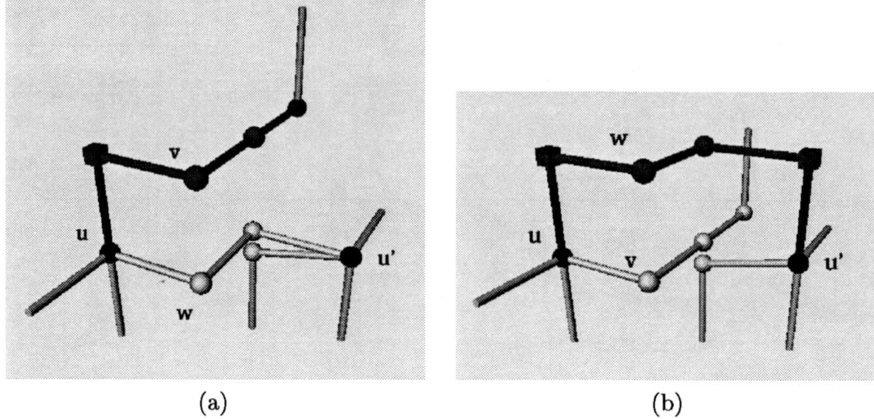

(a)　　　　　　　　　　　　　　(b)

Figure 13: The effect of applying two different split operations on the configuration shown in Fig. 12.a. In (a) and (b), the planar dummy path p_v (Fig. 12.c) and p_w (Fig. 12.e), respectively, is moved in the upward direction by the split. The split operation corresponding to the latter configuration is preferred by the **Reduce-Forks** heuristic since both u and u' become direction consistent after the split.

3. Weight the eight split operations obtained in the previous step according to the number n_d of vertices that become direction-consistent after the split and, secondarily, to the number of bends they introduce. Observe that $1 \leq n_d \leq 2$. In the example of Fig. 13 the split described by Fig. 13.b is preferred to the split described in Fig. 13.a.

4. Select and apply the split operation with minimum weight.

Observe that, since each original vertex requires at most six splits, this phase is performed with a number of splits that is, in the worst case, linear in the number of original vertices.

4.3　Vertex-Edge Overlap Removal and Crossing Removal

To perform the Vertex-Edge Overlap Removal and the Crossing Removal phases a technique is used similar to the one applied for the Direction Distribution phase. Namely, we identify a set of splits that can "do the job". We weight such splits and then apply the ones with minimum weights.

For each original vertex u of G such that v has the same coordinates as u:

1. Compute the (at most three) planar dummy paths containing v.

2. For each path computed in the previous step, determine the split operations that separate the path (except for the possible original vertices)

G. Di Battista et al., *3D Orthogonal Drawing*, JGAA, 4(3) 105–133 (2000) 125

from all the vertices that lie on its plane. For each path we have exactly two possible splits.

3. Weight the split operations obtained in the previous step according to the number of bends and/or crossings they introduce.

4. Select and apply the split operation with minimum weight.

For each pair of dummy vertices u and v having the same coordinates:

1. Compute all the planar dummy paths containing u or v.

2. Determine all the split operations that separate such paths and u from v.

3. Weight such splits according to the number of bends they introduce.

4. Select and apply the split with minimum weight.

5 Experimental Results

We have implemented and performed an experimental comparison of algorithms `Compact`, `Interactive`, `Kolmogorov`, `Reduce-Forks`, `Slices`, and `Three-Bends` with a large test suite of graphs. The experiments have been performed on a Sun SPARC station Ultra-1 by using the 3DCube [24] system. All the algorithms have been implemented in C++.

The test suite was composed of 1900 randomly generated graphs having from 6 to 100 vertices, 20 graphs for each value of vertex cardinality. All graphs were connected, with maximum vertex degree 6, without multi-edges and self-loops. The density was chosen to be in the middle of the allowed interval: the number of edges was twice the number of vertices. Note that in a connected graph of maximum degree 6, the density can range from 1 to 3. Also, as put in evidence in [8], in the practical applications of graph drawing it is unusual to have graphs with density greater than 2. The test suite is available at `http://www.dia.uniroma3.it/~patrigna/3dcube/test_suite.html`

The randomization procedure was very simple. For each graph the number of vertices and edges was set before the randomization. Edge insertions were performed on distinct randomized vertices, provided their degree was less than 6 and an edge between the two vertices did not already exist. Non connected graphs were discarded and re-randomized. The reason for choosing randomized graphs instead of real-life examples in the tests is that the 3D graph drawing field, for its novelty, still lacks well established real-life benchmark suites.

We considered two families of quality measures. For the efficiency we relied on the time performance (CPU seconds); for the readability we measured the average number of bends along the edges, the average edge length, and the average volume of the minimum enclosing box with sides isothetic to the axes. Figs. 14 and 15 illustrate the results of the comparison.

The comparison shows that no algorithm can be considered "the best". Namely, some algorithms are more effective in the average number of bends

G. Di Battista et al., *3D Orthogonal Drawing*, JGAA, 4(3) 105–133 (2000) 126

(`Interactive`, `Reduce-Forks`, and `Three-Bends`) while other algorithms perform better with respect to the average volume (`Compact` and `Slices`) or to the edge length (`Compact`, `Interactive`, and `Reduce-Forks`). More precisely:

- The average number of bends (see Fig. 14-b) is comparable for `Interactive`, `Reduce-Forks`, and `Three-Bends`, since it remains for all of them under the value of 3 bends per edge, while it is higher for `Compact` and `Slices`, and it is definitely much too high for `Kolmogorov`. Furthermore, `Reduce-Forks` performs better than the other algorithms for graphs with number of vertices in the range 5–30. `Interactive` performs better in the range 30–100. Another issue concerns the results of the experiments vs. the theoretical analysis. About `Kolmogorov` the literature shows an upper bound of 16 bends per edge [11] while our experiments obtain about 19 on average. This inconsistency might show a little "flaw" in the theoretical analysis. Further, about `Compact` the experiments show that the average case is much better than the worst case [11].

- Concerning the average edge length (see Fig. 15-a), `Reduce-Forks` performs better for graphs up to 50 vertices, whereas `Compact` is better from 50 to 100; `Interactive` is a bit worse, while the other algorithms form a separate group with a much worse level of performance.

- The values of volume occupation (see Fig. 15-b) show that `Compact` and `Slices` have the best performance for graphs bigger than 30 vertices, while `Reduce-Forks` performs better for smaller graphs.

Examples of the drawings constructed by the experimented algorithms are shown in Fig. 16.

For these considerations, we can say that `Reduce-Forks` is the most effective algorithm for graphs in the range 5–30. Also, among the algorithms that have a reasonable number of bends along the edges (`Interactive`, `Reduce-Forks`, and `Three-Bends`), `Reduce-Forks` is the one that has the best behavior in terms of edge length and volume. This is obtained at the expense of an efficiency that is much worse than the other algorithms. However, the CPU time does not seem to be a critical issue for the size of graphs in this interval. In fact, even for `Reduce-Forks`, the CPU time never exceeds 150 seconds, which is still a reasonable time for most applications.

6   Conclusions and Open Problems

We presented a new approach for constructing orthogonal drawings in three dimensions of graphs of maximum degree 6, and tested the effectiveness of our approach by performing an experimental comparison with several existing algorithms.

The presented techniques are easily extensible to obtain drawings of graphs of arbitrary degree with the following strategy.

G. Di Battista et al., *3D Orthogonal Drawing*, JGAA, 4(3) 105–133 (2000) 127

- The Vertex Scattering step remains unchanged.

- In the Direction Distribution step for vertices of degree greater than six, we first "saturate" the six available directions and then we evenly distribute the remaining edges.

- The Vertex-edge Overlap Removal step remains unchanged.

- In the Crossing Removal step we distinguish between crossings that are "needed" because of the overlay between edges that is unavoidable because of the high degree and the crossings that can be removed. For the latter type of crossings we apply the techniques presented in Section 3, while the first type of crossings are handled in a post-processing phase, where vertices are suitably expanded.

Several problems remain open.

- Devise new algorithms and heuristics (alternative to `Reduce-Forks`) within the described paradigm. Such heuristics might be based on modified versions of the 3D graph drawing algorithms listed in Section 1.

- Further explore the trade-offs among edge length, number of bends, and volume. A contribution in this direction is given by a recent paper by Eades, Symvonis and Whitesides [12].

- Measure the impact of bend-stretching (or possibly other post-processing techniques) on the performance of the different algorithms.

- Devise new quality parameters to better study the human perception of nice drawings in three dimensions.

- Set up test suites, possibly consisting of real-life graphs, specifically devoted to benchmarking 3D graph drawing algorithms.

- Although 3D graph drawing seems to offer interesting perspectives for the visualization of large graphs, the experiments presented in Section 5 show that the aesthetic quality of the drawings produced with the existing algorithms is still not sufficient to deal with large graphs (see, for example, Fig. 16). Also in this respect it would be important to improve the readability of the produced drawings, even at the expenses of a high computation time.

Acknowledgments

We are grateful to Walter Didimo, Giuseppe Liotta, and Maurizio Pizzonia for their support and friendship.

G. Di Battista et al., *3D Orthogonal Drawing*, JGAA, 4(3) 105–133 (2000) 128

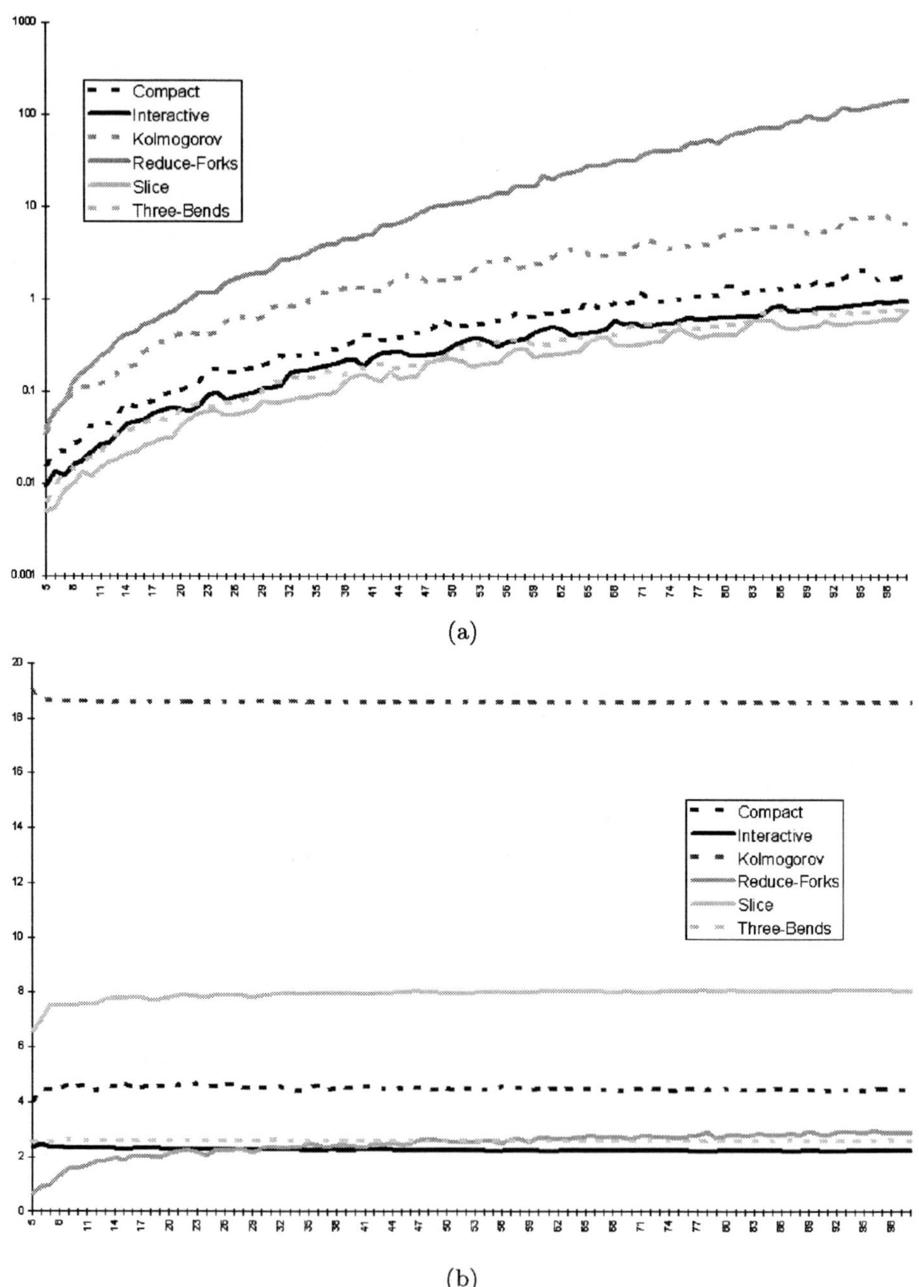

(a)

(b)

Figure 14: Comparison of Algorithms Compact, Interactive, Kolmogorov, Reduce-Forks, Slices, and Three-Bends with respect to time performance (a) and average number of bends along edges (b).

G. Di Battista et al., *3D Orthogonal Drawing*, JGAA, 4(3) 105–133 (2000) 129

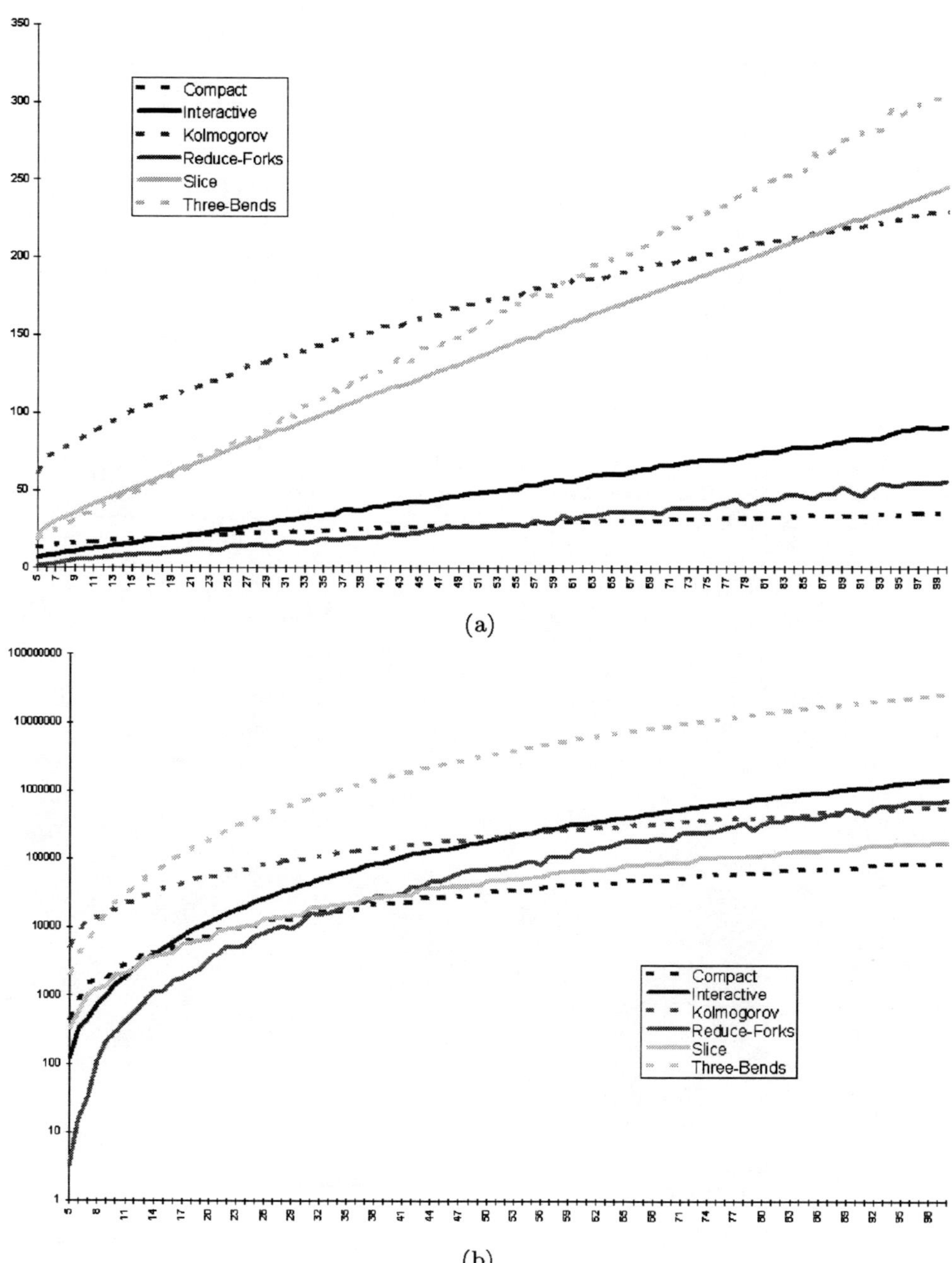

(a)

(b)

Figure 15: Comparison of Algorithms Compact, Interactive, Kolmogorov, Reduce-Forks, Slices, and Three-Bends with respect to average edge length (a) and volume occupation (b).

G. Di Battista et al., *3D Orthogonal Drawing*, JGAA, 4(3) 105–133 (2000) 130

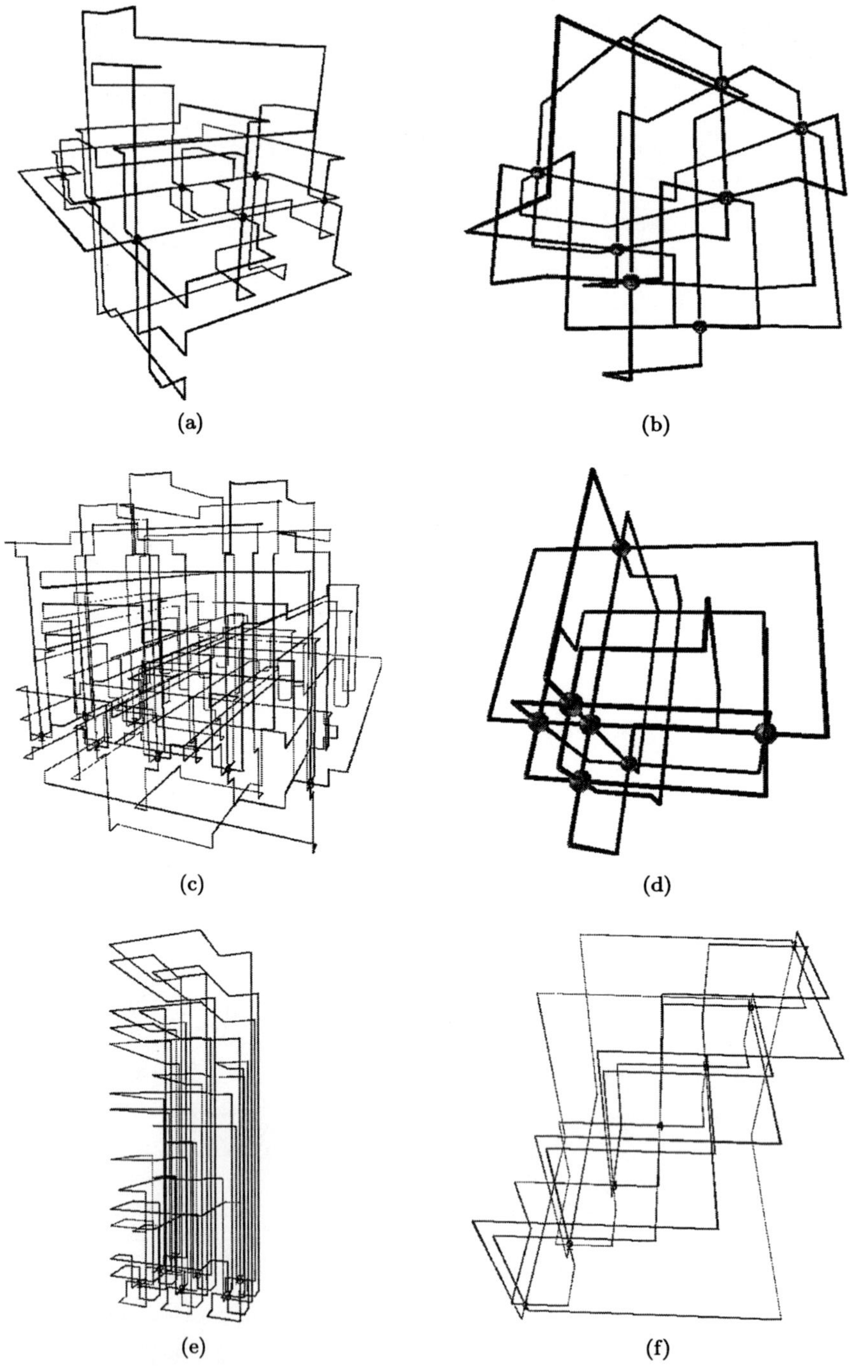

Figure 16: Three-dimensional orthogonal drawings of a K_7 as yielded by Compact (a), Interactive (b), Kolmogorov (c), Reduce-Forks (d), Slices (e), and Three-Bends (f).

G. Di Battista et al., *3D Orthogonal Drawing*, JGAA, 4(3) 105–133 (2000) 131

References

[1] H. Alt, M. Godau, and S. Whitesides. Universal 3-dimensional visibility representations for graphs. In F. J. Brandenburg, editor, *Graph Drawing (Proc. GD '95)*, volume 1027 of *Lecture Notes Comput. Sci.*, pages 8–19. Springer-Verlag, 1996.

[2] T. C. Biedl. Heuristics for *3d*-orthogonal graph drawings. In *Proc. 4th Twente Workshop on Graphs and Combinatorial Optimization*, pages 41–44, 1995.

[3] P. Bose, H. Everett, S. P. Fekete, A. Lubiw, H. Meijer, K. Romanik, T. Shermer, and S. Whitesides. On a visibility representation for graphs in three dimensions. In D. Avis and P. Bose, editors, *Snapshots in Computational and Discrete Geometry, Volume III*, pages 2–25. McGill University, July 1994. McGill technical report SOCS-94.50.

[4] I. Bruß and A. Frick. Fast interactive 3-D graph visualization. In F. J. Brandenburg, editor, *Graph Drawing (Proc. GD '95)*, volume 1027 of *Lecture Notes Comput. Sci.*, pages 99–110. Springer-Verlag, 1996.

[5] R. F. Cohen, P. Eades, T. Lin, and F. Ruskey. Three-dimensional graph drawing. *Algorithmica*, 17:199–208, 1997.

[6] I. F. Cruz and J. P. Twarog. 3d graph drawing with simulated annealing. In F. J. Brandenburg, editor, *Graph Drawing (Proc. GD '95)*, volume 1027 of *Lecture Notes Comput. Sci.*, pages 162–165. Springer-Verlag, 1996.

[7] G. Di Battista, P. Eades, R. Tamassia, and I. G. Tollis. *Graph Drawing*. Prentice Hall, Upper Saddle River, NJ, 1999.

[8] G. Di Battista, A. Garg, G. Liotta, R. Tamassia, E. Tassinari, and F. Vargiu. An experimental comparison of four graph drawing algorithms. *Comput. Geom. Theory Appl.*, 7:303–325, 1997.

[9] D. Dodson. COMAIDE: Information visualization using cooperative 3D diagram layout. In F. J. Brandenburg, editor, *Graph Drawing (Proc. GD '95)*, volume 1027 of *Lecture Notes Comput. Sci.*, pages 190–201. Springer-Verlag, 1996.

[10] P. Eades, C. Stirk, and S. Whitesides. The techniques of Kolmogorov and Bardzin for three dimensional orthogonal graph drawings. *Inform. Process. Lett.*, 60:97–103, 1996.

[11] P. Eades, A. Symvonis, and S. Whitesides. Two algorithms for three dimensional orthogonal graph drawing. In S. North, editor, *Graph Drawing (Proc. GD '96)*, volume 1190 of *Lecture Notes Comput. Sci.*, pages 139–154. Springer-Verlag, 1997.

G. Di Battista et al., *3D Orthogonal Drawing*, JGAA, 4(3) 105–133 (2000) 132

[12] P. Eades, A. Symvonis, and S. Whitesides. Three-dimensional orthogonal graph drawing algorithms. Accepted for publication by *Discrete Applied Math.*, as of February, 2000.

[13] S. P. Fekete, M. E. Houle, and S. Whitesides. New results on a visibility representation of graphs in 3-d. In F. Brandenburg, editor, *Graph Drawing (Proc. GD '95)*, volume 1027 of *Lecture Notes Comput. Sci.*, pages 234–241. Springer-Verlag, 1996.

[14] A. Frick, C. Keskin, and V. Vogelmann. Integration of declarative approaches. In S. North, editor, *Graph Drawing (Proc. GD '96)*, volume 1190 of *Lecture Notes Comput. Sci.*, pages 184–192. Springer-Verlag, 1997.

[15] M. R. Garey and D. S. Johnson. *Computers and Intractability: A Guide to the Theory of NP-Completeness.* W. H. Freeman, New York, NY, 1979.

[16] A. Garg, R. Tamassia, and P. Vocca. Drawing with colors. In *Proc. 4th Annu. European Sympos. Algorithms*, volume 1136 of *Lecture Notes Comput. Sci.*, pages 12–26. Springer-Verlag, 1996.

[17] A. N. Kolmogorov and Y. M. Barzdin. Realization of nets in 3-dimensional space. *Problems in Cybernetics*, 8:261–268, 1967.

[18] G. Liotta and G. Di Battista. Computing proximity drawings of trees in the 3-dimensional space. In *Proc. 4th Workshop Algorithms Data Struct.*, volume 955 of *Lecture Notes Comput. Sci.*, pages 239–250. Springer-Verlag, 1995.

[19] B. Monien, F. Ramme, and H. Salmen. A parallel simulated annealing algorithm for generating 3D layouts of undirected graphs. In F. J. Brandenburg, editor, *Graph Drawing (Proc. GD '95)*, volume 1027 of *Lecture Notes Comput. Sci.*, pages 396–408. Springer-Verlag, 1996.

[20] D. I. Ostry. *Some Three-Dimensional Graph Drawing Algorithms.* M.Sc. thesis, Dept. Comput. Sci. and Soft. Eng., Univ. Newcastle, Oct. 1996.

[21] J. Pach, T. Thiele, and G. Tóth. Three-dimensional grid drawings of graphs. In G. Di Battista, editor, *Graph Drawing (Proc. GD '97)*, volume 1353 of *Lecture Notes Comput. Sci.*, pages 47–51. Springer-Verlag, 1997.

[22] A. Papakostas and I. G. Tollis. Incremental orthogonal graph drawing in three dimensions. In G. Di Battista, editor, *Graph Drawing (Proc. GD '97)*, volume 1353 of *Lecture Notes Comput. Sci.*, pages 52–63. Springer-Verlag, 1997.

[23] M. Patrignani and M. Pizzonia. The complexity of the matching-cut problem. Tech. Report RT-DIA-35-1998, Dept. of Computer Sci., Univ. di Roma Tre, 1998.

G. Di Battista et al., *3D Orthogonal Drawing*, JGAA, 4(3) 105–133 (2000) 133

[24] M. Patrignani and F. Vargiu. 3DCube: A tool for three dimensional graph drawing. In G. Di Battista, editor, *Graph Drawing (Proc. GD '97)*, volume 1353 of *Lecture Notes Comput. Sci.*, pages 284–290. Springer-Verlag, 1997.

[25] A. Schrijver. Bipartite edge coloring in $o(\delta m)$ time. *SIAM J. on Computing*, 28(3):841–846, 1998.

[26] R. Webber and A. Scott. GOVE: Grammar-Oriented Visualisation Environment. In F. J. Brandenburg, editor, *Graph Drawing (Proc. GD '95)*, volume 1027 of *Lecture Notes Comput. Sci.*, pages 516–519. Springer-Verlag, 1996.

[27] D. R. Wood. An algorithm for three-dimensional orthogonal graph drawing. In S. H. Whitesides, editor, *Graph Drawing (Proc. GD '98)*, volume 1547 of *Lecture Notes Comput. Sci.*, pages 332–346. Springer-Verlag, 1998.

[28] D. R. Wood. Two-bend three-dimensional orthogonal grid drawing of maximum degree five graphs. Technical Report 98/03, School of Computer Science and Software Engineering, Monash University, 1998.

Journal of Graph Algorithms and Applications

http://www.cs.brown.edu/publications/jgaa/

vol. 4, no. 3, pp. 135–155 (2000)

Using Graph Layout to Visualize Train Interconnection Data

Ulrik Brandes   *Dorothea Wagner*

Department of Computer & Information Science
University of Konstanz
http://www.inf.uni-konstanz.de/~{brandes,wagner}
{Ulrik.Brandes,Dorothea.Wagner}@uni-konstanz.de

Abstract

We consider the problem of visualizing interconnections in railway systems. Given time tables from systems with thousands of trains, we are to visualize basic properties of the connection structure represented in a so-called train graph. It contains a vertex for each station met by any train, and one edge between every pair of vertices connected by some train running from one station to the other without halting in between.

Positions of vertices in a train graph visualization are given by the geographical location of the corresponding station. If all edges are represented by straight-lines, the result is visual clutter with many overlaps and small angles between pairs of lines. We therefore present a non-uniform approach using different representations for edges of distinct meaning in the exploration of the data. Some edges are represented by curved lines, such that the layout problem consists of placing control points for these curves. We transform it into a graph layout problem and exploit the generality of the random field layout model formulation for its solution.

Communicated by G. Liotta and S. H. Whitesides: submitted November 1998, revised October 1999.

Brandes and Wagner, *Layout of Train Graphs*, JGAA, 4(3) 135–155 (2000) 136

1   Introduction

The present layout problem arises from a cooperation with a subsidiary of the *Deutsche Bahn AG* (the central German train and railroad company), *TLC/EVA*. The aim of this cooperation is to develop data reduction and visualization techniques for the explorative analysis of large amounts of time table data from European public transport systems. These comprise mostly train schedules; however, the data may also contain bus, ferry and even some pedestrian connections. The analysis of the data with respect to completeness, consistency, changes between consecutive periods of schedule validity and so on is relevant, e.g., for quality control, (international) coordination, and pricing. Our aim is to aid visual inspection of this data, which is carried out at *TLC* to identify structural characteristics of (sub)networks and to back-up design decisions on extensions or modifications of networks. Reported future use will include evaluation support of schedules and pricing.

Figure 1 shows the kind of data that is provided. Since for even a moderately sized stop like the German part of the Konstanz main station there are about 100 trains regularly arriving or leaving, realistic input is quite large. To condense the input, a so-called *train graph* is built in the following way. For each regular stop of any train, a vertex is inserted into the graph. Two vertices are connected by exactly one edge if there is a point-to-point connection, i.e. some train runs from from one station to the other (or vice versa) without intermediate stops. Hence, the graphs considered here are simple and undirected.

An important part of the analysis is the classification of edges into two categories: *minimal* edges and *transitive* edges. Minimal edges are those corresponding to a set of continuous connections between two stations not passing through a third one. Typically, these are induced by regional trains serving minor stations. On the other hand, transitive edges correspond to connections passing through other stations without halting. These are induced by through-trains. The information contained in a train graph is therefore the existence or absence of a point-to-point connection between pairs of stations, and the classification of each connection into minimal or transitive. Graphical presentation of the train graph and an edge classification computed in the analysis is desirable.

An edge classification is easily coded using color. Figure 2(a) shows a small part of a train graph with edges colored according to a precomputed classification. Stations are positioned according to their geographical location, and all edges are drawn as straight lines. Obvious graphical problems are edge overlaps and small angles between edges.

In order to maintain geographic familiarity, we are not allowed to move vertices, and minimal edges are best depicted by straight-lines, because they usually represent actual railways and should therefore not be the cause of the problem. It seems therefore reasonable to change the representation of transitive edges to curves, as depicted in Figure 2(b). They provide the flexibility to route an edge such that overlaps and small angles are resolved. In general, representation of non-stop connections by curved lines not only helps to reduce visual clutter and ambiguity, but also directly resembles the intuition of fast

Brandes and Wagner, *Layout of Train Graphs*, JGAA, 4(3) 135–155 (2000) 137

```
*Z 05130 85    01
*G SE  8506131 8001790
*A VE 8506131 8001790 000000
*A G  8506131 8001790
8506131 Kreuzlingen              1112  (...)
8003400 Konstanz           1115 1125  (...)
8003401 Konstanz-Petersh.  1127 1128  (...)
8003416 Konstanz-Wollmat   1130 1130  (...)
8004997 Reichenau(Baden)   1132 1133  (...)
8002683 Hegne              1135 1135  (...)
8000496 Allensbach         1138 1138  (...)
8003872 Markelfingen       1143 1143  (...)
8000880 Radolfzell         1147 1149  (...)
8001059 Böhringen-Rickelsh. 1152 1152 (...)
8000073 Singen(Hohentwiel) 1158 1200  (...)
8004107 Mühlhausen(b Engen) 1206 1206 (...)
8006321 Welschingen-Neuhaus. 1209 1209 (...)
8001790 Engen              1212       (...)


(...)
8000880 Radolfzell         -58.5  -510.8  (...)
(...)
8003400 Konstanz           -43.5  -519.8  (...)
8003401 Konstanz-Petersh.  -43.5  -518.2  (...)
8003416 Konstanz-Wollmat   -45.1  -517.5  (...)
(...)
8506131 Kreuzlingen        -40.2  -524.5  (...)
(...)
```

Figure 1: Schedule of a single train and excerpts from a station list. The schedule lists all stations used by the train with arrival and departure times. Every station has a unique identification number, and coordinates are in kilometers relative to the city of Hannover (irrelevant data omitted)

Brandes and Wagner, *Layout of Train Graphs*, JGAA, 4(3) 135–155 (2000) 138

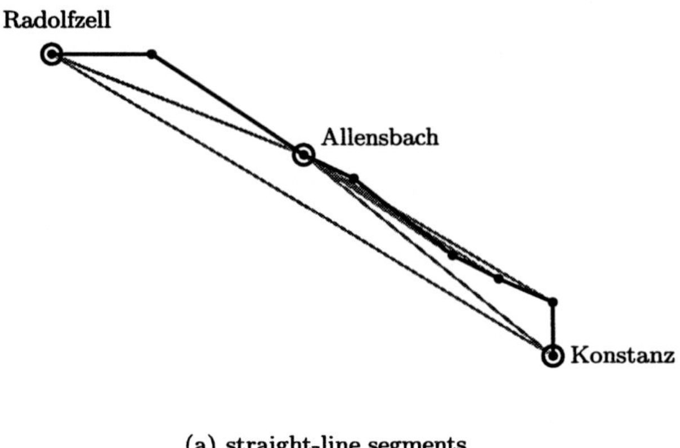

(a) straight-line segments

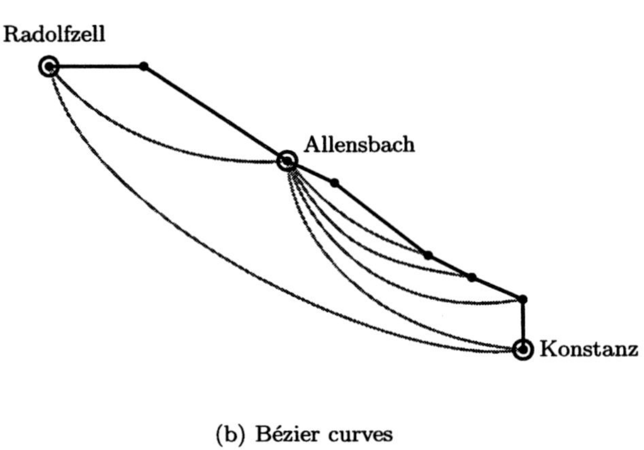

(b) Bézier curves

Figure 2: Different representations of transitive edges in a small train graph

vehicles passing by minor stops.

To render Bézier curves, control points need to be positioned. Using the framework of random field layout models introduced in [3], the problem is cast into a graph layout problem. More precisely, we consider control points to be vertices of a graph, and rules for appropriate positioning are modeled by defining edges accordingly. This way, common algorithmic approaches can be employed. Practical applicability of our approach is gained from experimental validation. In a completely different field of application, the same strategy is currently used to identify suitable layout models for social and policy networks [4, 3]. These applications are good examples of how the uniform approach of random field layout models may be used to obtain initial models for visualization problems which are not clearly defined beforehand.

The paper is organized as follows. In Section 2, we review briefly the concept of random field layout models. A specific random field model for train graph

Brandes and Wagner, *Layout of Train Graphs*, JGAA, 4(3) 135–155 (2000) 139

layout is defined in Section 3. Section 4 features a short discussion on aspects of parametrization and experiments with real-world examples.

2 Random Field Layout Models

In this section we review briefly the uniform graph layout formalism introduced in [3]. As can be seen from Section 3, model prototyping within this framework is straightforward.

Virtually every graph layout problem can be viewed as a constrained optimization problem. A layout of a graph $G = (V, E)$ is computed by assigning values to certain layout variables, subject to constraints and an objective function. Straight-line representations, for instance, are completely determined by an assignment of coordinates to each vertex. However, straight-line representations are but one special case of a layout problem. In the most general formulation, each element of a set $L = \{l_1, \ldots, l_k\}$ of arbitrary *layout elements* is assigned a value from a set of feasible values $\mathcal{X}_l$, $l \in L$. Layout elements may represent positional variables for vertices, edges, labels, and any other kind of graphical object. Therefore, L and $\mathcal{X} = \mathcal{X}^L = \mathcal{X}_{l_1} \times \cdots \times \mathcal{X}_{l_k}$ are clearly dependent on the chosen type of graphical representation. In this application, we need not constrain configurations of layout elements. Hence, all vectors $x \in \mathcal{X}$ are considered feasible *layouts*.

Objective function. In order to measure the quality of a layout, an objective function $U : \mathcal{X} \to \mathbb{R}$ is defined. Since it is difficult to judge the quality of a layout as a whole, the objective function evaluates configurations of small subsets of layout elements which mutually influence their positioning. This interaction of layout elements is modeled by an *interaction graph* $G^\eta = (L, E^\eta)$ that is obtained from a *neighborhood system* $\eta = \{\eta_l \mid l \in L\}$, where $\eta_l \subseteq L \setminus \{l\}$ is the set of layout elements for which the position assigned to l is relevant in terms of layout quality. There is an edge in E^η between two layout elements, if one is in the neighborhood of the other. The interactions are symmetric by definition, i.e. we require $l_2 \in \eta_{l_1} \Leftrightarrow l_1 \in \eta_{l_2}$ for all $l_1, l_2 \in L$, so that G^η is undirected. The set of cliques in G^η is denoted by $\mathcal{C} = \mathcal{C}(\eta)$. We define the *interaction potential* of a clique $C \in \mathcal{C}$ to be any function $U_C : \mathcal{X} \to \mathbb{R}$ for which

$$x_C = y_C \quad \Rightarrow \quad U_C(x) = U_C(y)$$

holds for all $x, y \in \mathcal{X}$, where $x_C = (x_l)_{l \in C}$. A graph layout objective function $U : \mathcal{X} \to \mathbb{R}$ is the sum of all interaction potentials, i.e. $U(x) = \sum_{C \in \mathcal{C}} U_C(x)$. By convention, the objective function is to be minimized. $U(x)$ is often called the *energy* of x, and can be interpreted as the amount of distortion in the layout.

Fundamental potentials. One advantage of separating the energy function into interaction potentials of small subsets of layout elements is that recurrent design principles can be isolated to form a toolbox of fundamental criteria. Not

Brandes and Wagner, *Layout of Train Graphs*, JGAA, 4(3) 135–155 (2000) 140

surprisingly, two central potentials are those corresponding to the forces used in the spring embedder [7]:[1]

- *Repulsion Potential:* The criterion that two layout elements k and l should not lie close to each other can be expressed by a potential

$$U_{\{k,l\}}^{(\mathrm{rep})}(x) = \mathsf{Rep}(x_k, x_l) = \frac{\varrho}{d(x_k, x_l)^2}$$

  where ϱ is a fixed constant and $d(x_k, x_l)$ is the Euclidean distance between the positions of k and l. $\mathsf{Rep}(x_k, x_l \,|\, \varrho)$ is used to indicate a specific choice of ϱ.

- *Attraction Potential:* If, in contrast, k and l should lie close to each other, a potential

$$U_{\{k,l\}}^{(\mathrm{attr})}(x) = \mathsf{Attr}(x_k, x_l) = \alpha \cdot d(x_k, x_l)^2,$$

  with α a fixed constant, is appropriate. Like above we use $\mathsf{Attr}(x_k, x_l \,|\, \alpha)$ to denote a specific choice of α.

- *Distance Potential:* Since $\mathsf{Rep}(x_k, x_l \,|\, \lambda^4) + \mathsf{Attr}(x_k, x_l \,|\, 1)$ is minimized when $d(x_k, x_l) = \lambda$, one can specify a desired distance between two layout elements (e.g. edge length) by

$$U_{\{k,l\}}^{(\mathrm{dist})}(x) = \mathsf{Dist}(x_k, x_l) = \mathsf{Rep}(x_k, x_l \,|\, \lambda^4) + \mathsf{Attr}(x_k, x_l \,|\, 1)$$

  where $\mathsf{Dist}(x_k, x_l \,|\, \lambda^4)$ is used like above.

Note that many other design rules (sufficiently large angles, vertex-edge distance, edge crossings, etc.) are easily formulated in terms of interaction potentials [3].

If layouts $x \in \mathcal{X}$ are assigned probabilities

$$P(X = x) = \frac{1}{Z} e^{-U(x)},$$

where $Z = \sum_{y \in \mathcal{X}} e^{-U(y)}$ is a normalizing constant, random variable X is a (Gibbs) random field. Both X and its distribution are called a (random field) *layout model* for G. Clearly, the above probabilities depend on the energy only, with a layout of low energy being more likely than a layout of high energy. By using a random variable, the entire layout model is described in a single object. Due to the familiar form of its distribution, a wealth of theory becomes applicable (a primer in the context of dynamic graph layout is [5]). See [13] for an overview on the theory of random fields, and some of its applications in image processing. Since random fields are used so widely, there also is a great deal of literature on algorithms for energy minimization (see e.g. [12]).

---

[1]The original spring embedder does not specify an objective function, but its gradients. The above potentials appear in [6].

Brandes and Wagner, *Layout of Train Graphs*, JGAA, 4(3) 135–155 (2000) 141

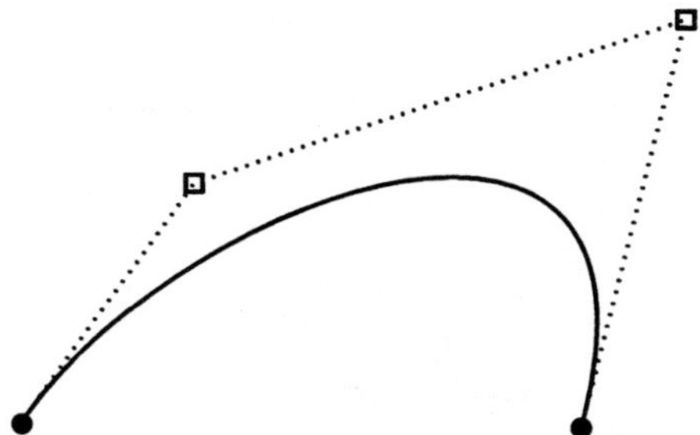

Figure 3: Bézier cubic curve [2]. Two endpoints and two control points define a smooth curve that is entirely enclosed by the convex hull of these four points

3 A Layout Model for Curved Edges

We now define a layout model for undirected train graphs $G = (V, E)$. The layout elements that need to be positioned to render Bézier curves are their control points. In fact, we may consider stations and control points to be vertices of an auxiliary graph, so that rules for favorable positioning can be modeled by auxiliary edges of appropriate desired length.

Their geographical location gives the position of all vertices corresponding to stations, and we identify these vertices with their position. Minimal edges as well as very long transitive edges are represented straight-line. For the other edges we use Bézier cubic curves (cf. Figure 3).[2] Let $\breve{E}_{\tau_1} \subseteq E$ be the set of transitive edges of length less than a threshold parameter τ_1, such that the set of layout elements consists of two control points for each edge in $\breve{E}_{\tau_1}$, $L = \left\{ b_u(e), b_v(e) \mid e = \{u, v\} \in \breve{E}_{\tau_1} \right\}$. If two Bézier points belong to the same edge, they are called *partners*. The *anchor*, $a_{b_v(e)}$, of any $b_v(e) \in L$ is v. The *default position* of all Bézier points is on the straight line through the endpoints of their edges at equal distance from their anchor and from their partner.

The position assigned to a Bézier point is influenced by its partner, its anchor, all Bézier points with the same anchor or close default positions, and all stations near the default position. Let $\{u, v\} \in \breve{E}_{\tau_1}$ be a transitive edge, and let $b \in L$ be a Bézier point of $\{u, v\}$. Given two parameters ϵ_1 and ϵ_2, consider an ellipse with major axis going through u and v. Let its radii be $\epsilon_1 \cdot \frac{d(u,v)}{2}$ and $\epsilon_2 \cdot \frac{d(u,v)}{2}$, respectively. We denote the set of all stations and Bézier points (at their default position) within this ellipse, except for b and its anchor, by $\mathcal{E}_b$. Recall that the neighborhood of some layout element consists of all those layout elements that have an influence on its positioning. Therefore, η_b equals the union of $\mathcal{E}_b \cap L$, the set of Bézier points with the same anchor as b, and (since interactions are symmetric) the set of Bézier points b' for which $b \in \mathcal{E}_{b'}$. We used $\epsilon_1 = 1.1$ and $\epsilon_2 = 0.5$ for the examples presented in Section 4.

---

[2]It will be obvious from the examples presented in Section refsec:examples why it is not useful to represent all transitive edges by Bézier curves.

Brandes and Wagner, *Layout of Train Graphs*, JGAA, 4(3) 135–155 (2000) 142

An interaction potential is defined for each design goal that a good layout of Bézier points should achieve:

- *Distance to stations.* For each Bézier point $b \in L$ of some edge $\{u, v\} \in \breve{E}_{T_1}$, there are repulsion potentials

$$\sum_{s \in \mathcal{E}_b \cap V} \mathsf{Rep}(x_b, s \,|\, (\varrho_1 \cdot \lambda_b)^4),$$

with $\lambda_b = \frac{d(u,v)}{3}$ and ϱ_1 a constant. These ensure reasonable distance from stations in the vicinity of b and can be controlled via ϱ_1. A combined repulsion and attraction potential

$$\mathsf{Dist}(x_b, a_b \,|\, (\lambda_1 \cdot \lambda_b)^4)$$

where λ is another constant, keeps b sufficiently close to its anchor a_b.

- *Distance to near Bézier points.* As is the case with near stations, a Bézier point $b_1 \in L$ should not lie too close to another Bézier point $b_2 \in \eta_{b_1}$. If b_1 is neither the partner of nor bound to b_2 (binding is defined below), we add

$$\mathsf{Rep}(x_{b_1}, x_{b_2} \,|\, \varrho_2^4 \cdot \min\{\lambda_{b_1}^4, \lambda_{b_2}^4\})$$

The desired distance between partners b_1 and b_2 is equal to the desired distance from their respective anchors,

$$\mathsf{Dist}(x_{b_1}, x_{b_2} \,|\, (\lambda_1 \cdot \lambda_{b_1})^4)$$

- *Binding.* In general, it is not desirable to have Bézier points $b_1, b_2 \in L$ with a common anchor lie on different sides of a minimal edge path through the anchor. Therefore, we *bind* them together, if λ_{b_1} does not differ much from λ_{b_2}, i.e. if $\frac{1}{\tau_2} < \frac{\lambda_{b_1}}{\lambda_{b_2}} < \tau_2$ for a threshold $\tau_2 \geq 1$, we add potentials

$$\beta \cdot \mathsf{Dist}(x_{b_1}, x_{b_2} \,|\, \lambda_2^4 \cdot (\lambda_{b_1}^4 + \lambda_{b_2}^4)/2)$$

where λ_2 is a stretch factor for the length of binding edges, and β controls the importance of binding relative to the other potentials.

In summary, the objective function is made of nothing but attraction and repulsion potentials that define an auxiliary graph layout problem in the following way: Stations correspond to vertices with fixed positions, while Bézier points correspond to vertices to be positioned. Edges of different desired lengths exist between Bézier points and their anchors, between partners, and between Bézier points bound together. Just like edge lengths, the magnitude of repulsion differs across the elements. See Figure 4 and recall that repulsion potentials are defined on local neighborhoods only. The respective influence of the different parameters is discussed in the following section.

Brandes and Wagner, *Layout of Train Graphs*, JGAA, 4(3) 135–155 (2000) 143

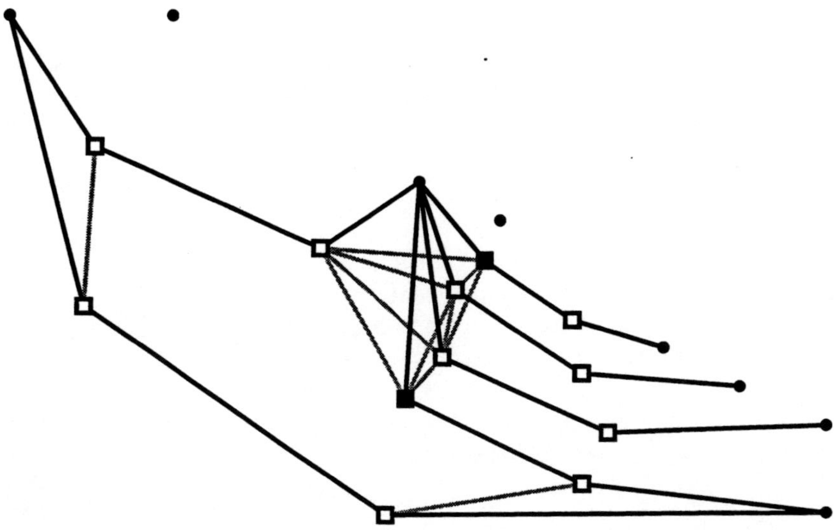

Figure 4: Auxiliary graph induced by Bézier point layout interactions for the train graph of Figure 2(b). Note that there is no binding between the two layout elements indicated by black rectangles, because their default distances from the anchor differ too much (threshold parameter τ_2)

4 Experiments

The objective function described in the previous section was obtained only after experimentation with a number of different potentials and parameters. We started with a simple combination of repulsion from stations and attraction and repulsion from partners and anchors. In fact, we then used splines to represent transitive edges. It seemed that they offered better control, since they actually pass through their control points. However, spline segments between partners tended to extend far into the layout area. After replacing splines by Bézier curves, the promising results encouraged us to try more elaborate objective functions. In particular it showed that it is useful to represent long transitive edges straight-line, which led to the introduction of threshold τ_1. A new requirement we found while discussing earlier examples with users was that incident (consecutive or nested) transitive edges should lie on one side of a path of minimal edges. Binding proved to achieve this goal, but needed to be constrained to control segments of similar desired length, because otherwise short transitive edges are deformed when bound to long ones. Threshold τ_2 therefore controls the length ratio of segments bound.

Identification of a suitable vector $\theta = (\varrho_1, \varrho_2, \lambda_1, \lambda_2, \beta, \tau_1, \tau_2)$ of parameters is a serious problem. Two nested simulated annealing computations are used in [11] to identify parameters of a spring embedder variant. In [9], a genetic algorithm is used to breed a suitable objective function. However, both meth-

Brandes and Wagner, *Layout of Train Graphs*, JGAA, 4(3) 135–155 (2000) 144

ods are heuristic in defining their objective as well as in optimizing it. Given one or more examples which are considered to be well done (e.g. by manual re-arrangement), a theoretically sound approach would be to carry out parameter estimation for random variable $X(\theta)$ describing the layout model as a function of parameter vector θ. Given a layout x, the *likelihood* of θ is

$$P(X = x \mid \theta) = \frac{1}{Z(\theta)} \exp\{-U(x \mid \theta)\}$$

where $Z(\theta) = \sum_{y \in \mathcal{X}} \exp\{-U(y \mid \theta)\}$ is the normalizing constant. A maximum likelihood estimate θ^* is obtained by maximizing the above expression with respect to θ. Unfortunately, computation of $Z(\theta)$ is practically intractable, since it sums over all possible layouts. One might hope to reduce computational demand by exploiting the locality of random fields (see e.g. [13]). Even though neighboring layout elements are clearly not independent, reasonable estimates are obtained from the *pseudo-likelihood* function [1]

$$\prod_{l \in L} \frac{1}{Z_l(\theta)} \exp\left\{-\sum_{C \in \mathcal{C} : l \in C} U_C(x \mid \theta)\right\}$$

with $Z_l(\theta) = \sum_{x_l \in \mathcal{X}_l} \exp\{-\sum_{C \in \mathcal{C} : l \in C} U_C(x \mid \theta)\}$. However, $Z_l(\theta)$ is a sum over all possible positions of layout element l, such that maximization is still intractable in this setting. So we exploit locality in a very different way, namely by experimenting with small examples in a feedback cycle. The parameters θ thus identified prove appropriate even for huge graphs, indicating that the local neighborhood definition lets the model scale well.

The rationale behind each component of $\theta = (\varrho_1, \varrho_2, \lambda_1, \lambda_2, \beta, \tau_1, \tau_2)$ is listed in Figure 5, as well as a choice of values that proved satisfactory. The effects of some parameters are demonstrated in Figure 6. It is clearly seen how increased repulsion potentials spread Bézier points (Figs. 6(a) and 6(b)). Without binding, curves tend to lie on different sides of minimal edges (Fig. 6(c)), which can even be enforced (Fig. 6(d)). This indicates why binding is a valuable refinement.

To carry out the above experiments and to generate large examples, we initially used an implementation of a fairly general random field layout module, written in C++ using LEDA [10]. It provides a set of fundamental neighborhood types and interaction potentials, to which others can be added. Since our main goals with this module are flexibility and model design, a simple simulated annealing approach is used for energy minimization. Since it turned out that the final model needed only attraction and repulsion potentials, we later replaced the module with a customized implementation of the approach of [8], which sped up energy minimization by a factor of ten. All running times given are with respect to this latter implementation executed on one 336 MHz Ultra-SPARC-II processor of a SUN Enterprise 4000/5000 running under Solaris 2.5.1 with 1024 MBytes of RAM. Note that neighborhoods are computed in a preprocessing step, and we have made no effort whatsoever to reduce its running time.

The original datasets provided by *TLC/EVA* are quite large: For a train graph of the size shown in Figure 10 (roughly 2,000 vertices and 4,000 edges),

Brandes and Wagner, *Layout of Train Graphs*, JGAA, 4(3) 135–155 (2000) 145

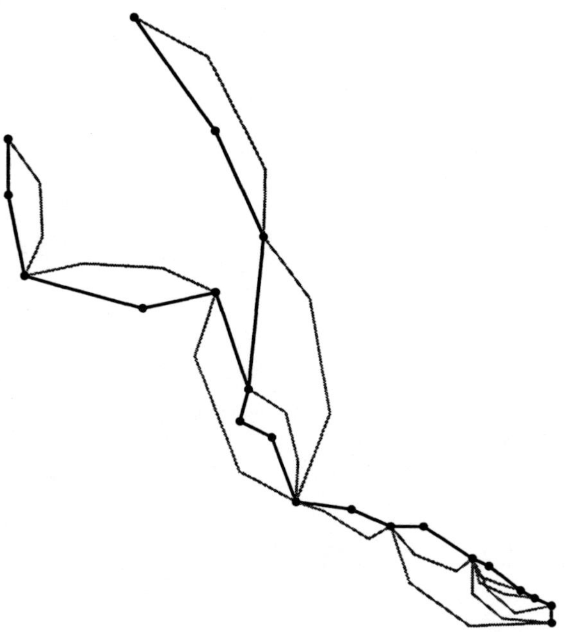

(a) Small part of a train graph with parameters $\theta =$ (0.3, 0.7, 0.7, 0.5, 0.4, 100, 2.2)

θ	controls
ϱ_1	distance of Bézier points from stations
ϱ_2	mutual distance of Bézier points
λ_1	length of control segments
λ_2	length of bands
β	importance of binding
τ_1	threshold for straight transitive edges
τ_2	threshold for binding segments of different length
ϵ_1	major axis radius of neighborhood defining ellipse
ϵ_2	minor axis radius of neighborhood defining ellipse

(b) Parameters of the train graph layout model

Figure 5: User specifiable parameters in the train graph layout model and a recommended choice applied to a small train graph. Control segments shown instead of Bézier curves (cf. Figure 6)

Brandes and Wagner, *Layout of Train Graphs*, JGAA, 4(3) 135–155 (2000) 146

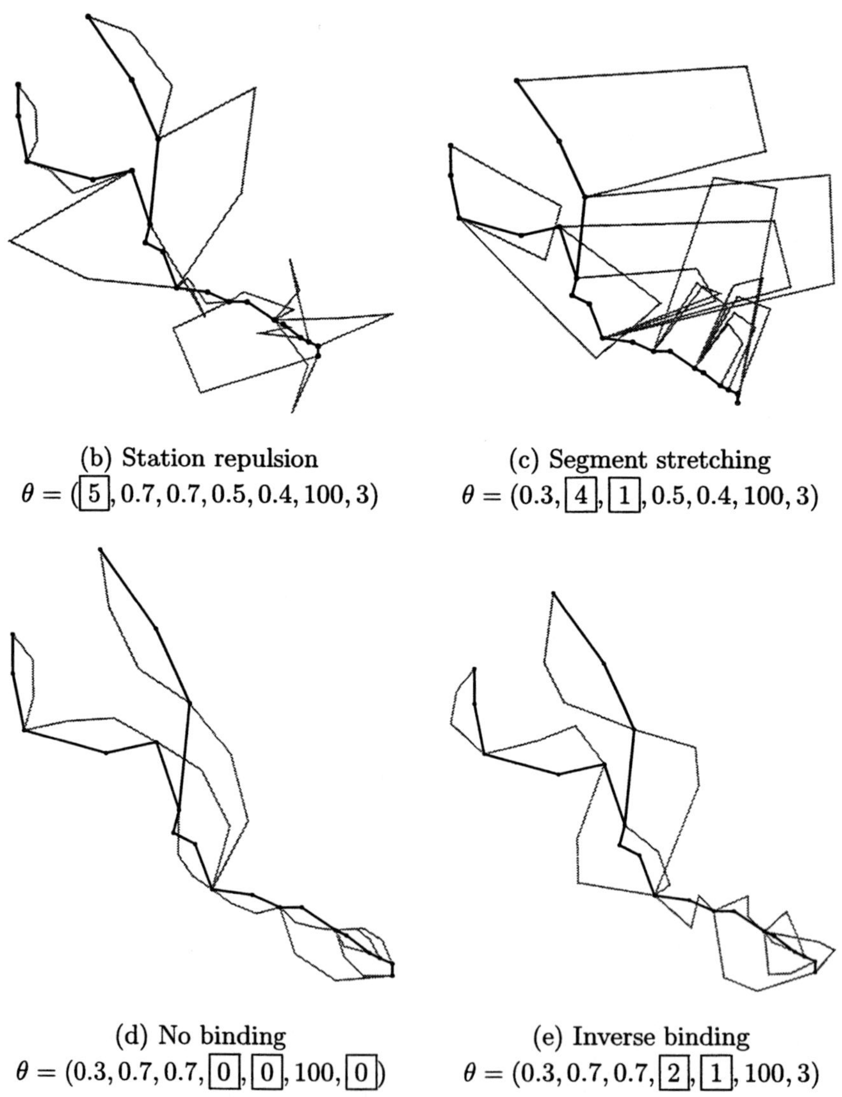

(b) Station repulsion
$\theta = (\boxed{5}, 0.7, 0.7, 0.5, 0.4, 100, 3)$

(c) Segment stretching
$\theta = (0.3, \boxed{4}, \boxed{1}, 0.5, 0.4, 100, 3)$

(d) No binding
$\theta = (0.3, 0.7, 0.7, \boxed{0}, \boxed{0}, 100, \boxed{0})$

(e) Inverse binding
$\theta = (0.3, 0.7, 0.7, \boxed{2}, \boxed{1}, 100, 3)$

Figure 6: Effects of some parameters demonstrated. For ease of comparison, control segments are shown instead of the corresponding Bézier curves. All examples have $\epsilon_1 = 1.1$ and $\epsilon_2 = 0.5$ and should be compared to Figure 5

Brandes and Wagner, *Layout of Train Graphs*, JGAA, 4(3) 135–155 (2000) 147

about 11 MBytes of time table data are evaluated. Connections are classified into minimal and transitive edges using existing code.

The first example is shown in Figure 7. The graph represents regional trains in southwest Germany. Edge classification, transformation into a layout graph, neighborhood generation, and layout computation took less than 10 seconds. The example also demonstrates how visual inspection can immediately yield some candidates for misclassified edges. Parts of the drawing are magnified in Figures 8 and 9. A few labels have been added to support geographical location of the area shown, but otherwise the drawings have not been modified. Note that connections can be told apart quite well, and that binding successfully causes incident (consecutive or nested) transitive edges to lie on the same side of minimal edges.

Larger examples are given in Figures 10 and 12. Computation times were about 5 minutes and 9 minutes, respectively, most of which was spent on determining the neighborhoods. Energy minimization took about 30 seconds and 47 seconds, respectively. One readily observes that the algorithm scales very well, i.e. increased size of the graph does not reduce layout quality on more detailed levels (Figs. 11 and 13). This is largely due to the fact that neighborhoods remain fairly local. The benefits of a length threshold for curved transitive edges is another straightforward observation, notably in Figures 12 and 13(a). Together with the ability to zoom into different regions, data exploration is well supported.

Acknowledgments

Besides our contacts at *TLC*, we would like to thank Annegret Liebers, Karsten Weihe, and Thomas Willhalm for making the train graph generation and edge classification code available. We are grateful to Frank Müller, Vaneesa Kääb, and Marco Gaertler, who carried out most of the other implementation work. We also wish to thank the referees for some helpful suggestions.

Brandes and Wagner, *Layout of Train Graphs*, JGAA, 4(3) 135–155 (2000) 148

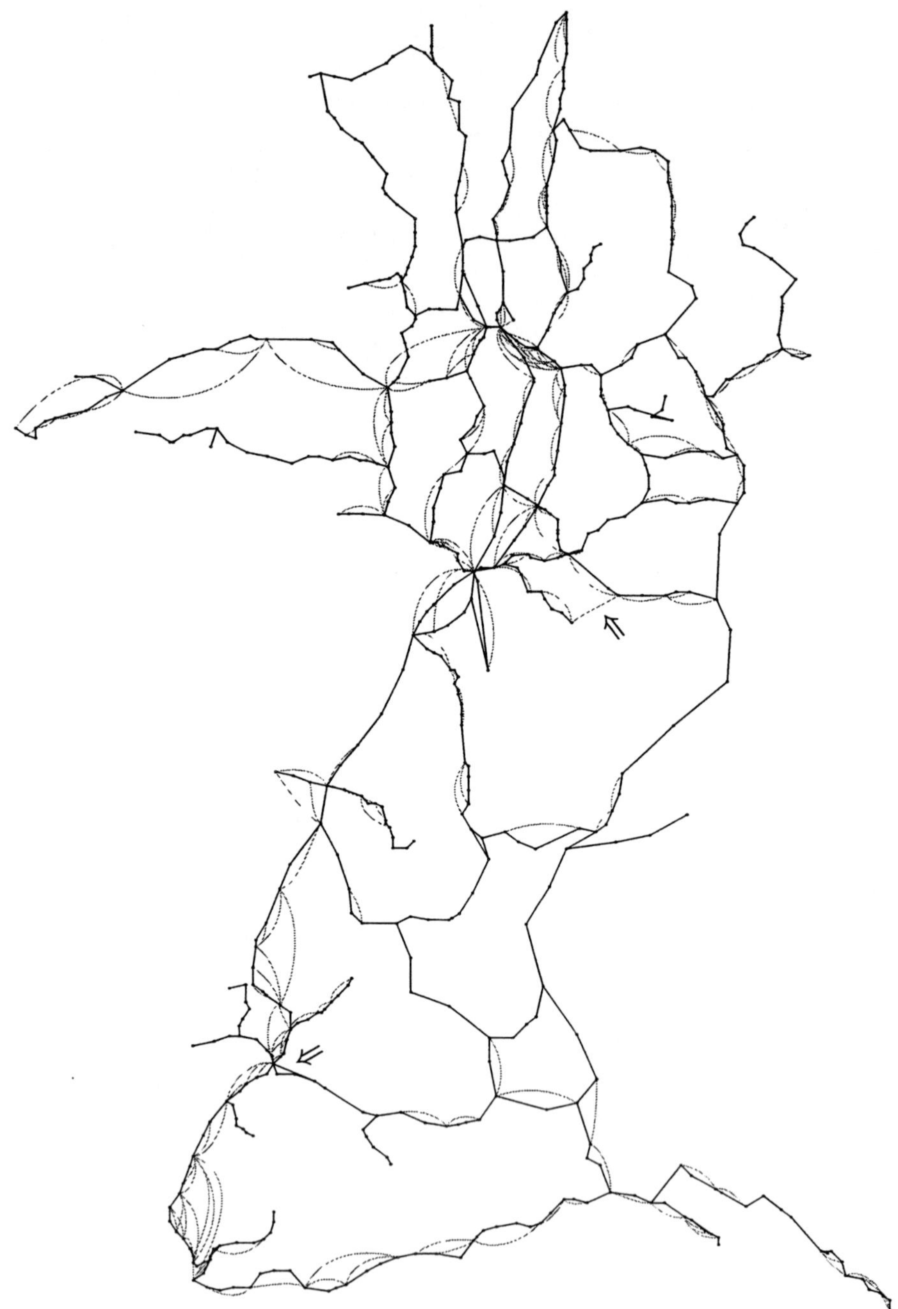

Figure 7: Regional trains in southwest Germany. 619 vertices, 876 edges (229 transitive), $\theta = (0.7, 0.3, 0.7, 0.5, 0.4, 100, 3)$. Arrows indicate two out of several edges that appear to be misclassified

Brandes and Wagner, *Layout of Train Graphs*, JGAA, 4(3) 135–155 (2000) 149

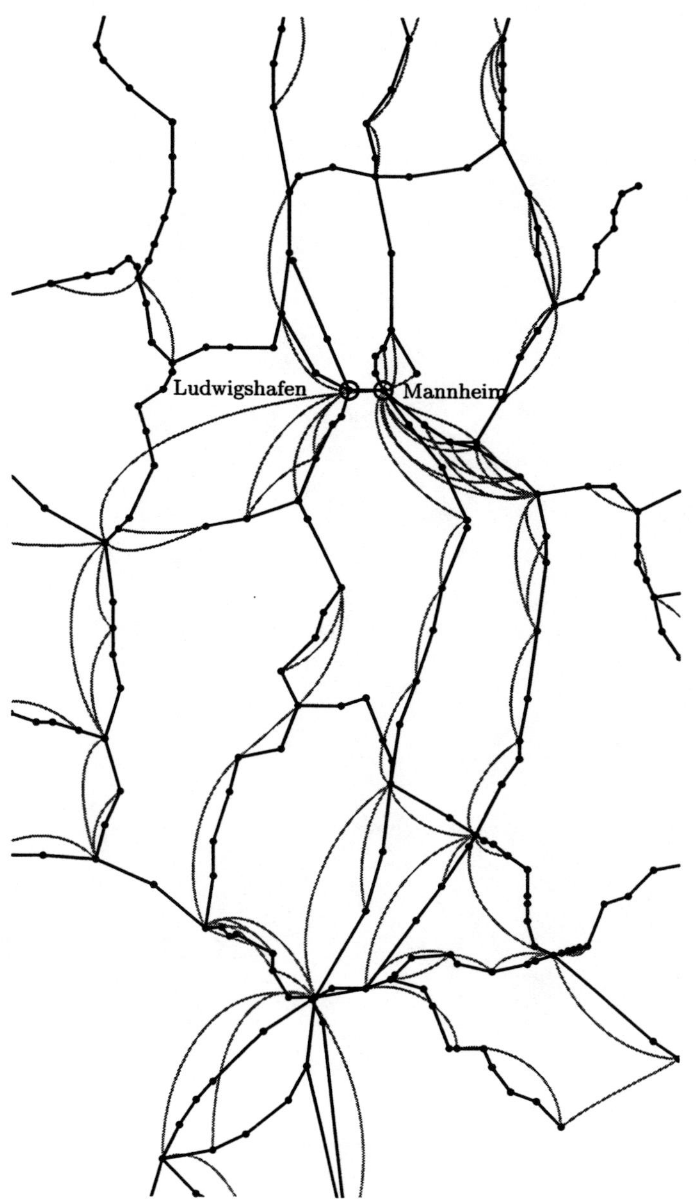

Figure 8: Magnification from Figure 7

Brandes and Wagner, *Layout of Train Graphs*, JGAA, 4(3) 135–155 (2000) 150

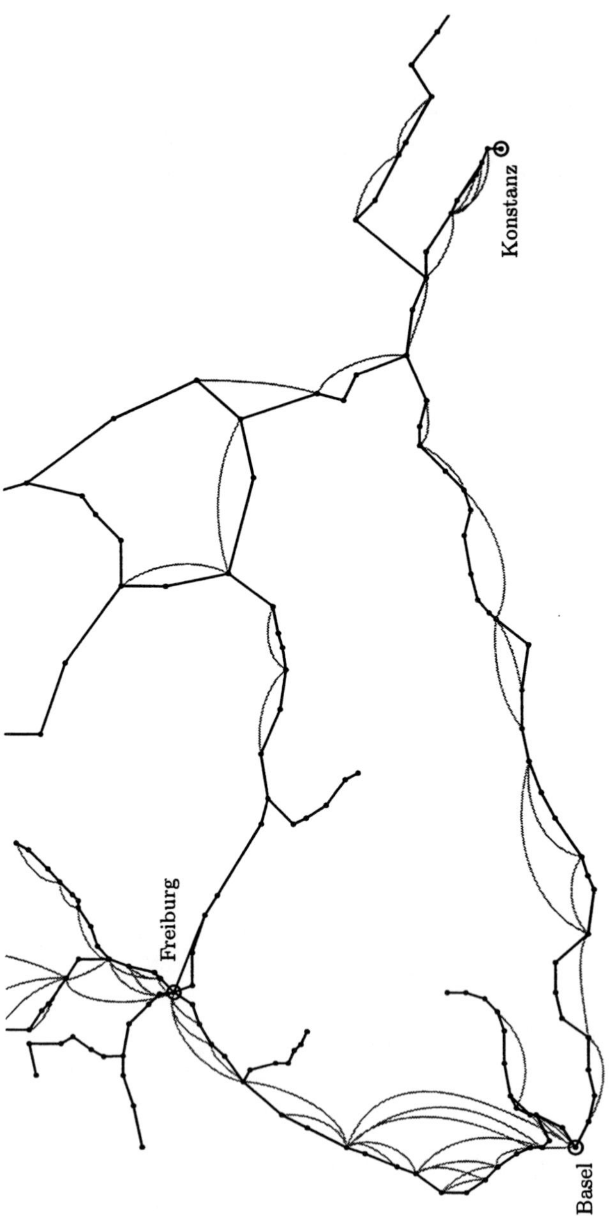

Figure 9: Magnification from Figure 7

Brandes and Wagner, *Layout of Train Graphs*, JGAA, 4(3) 135–155 (2000) 151

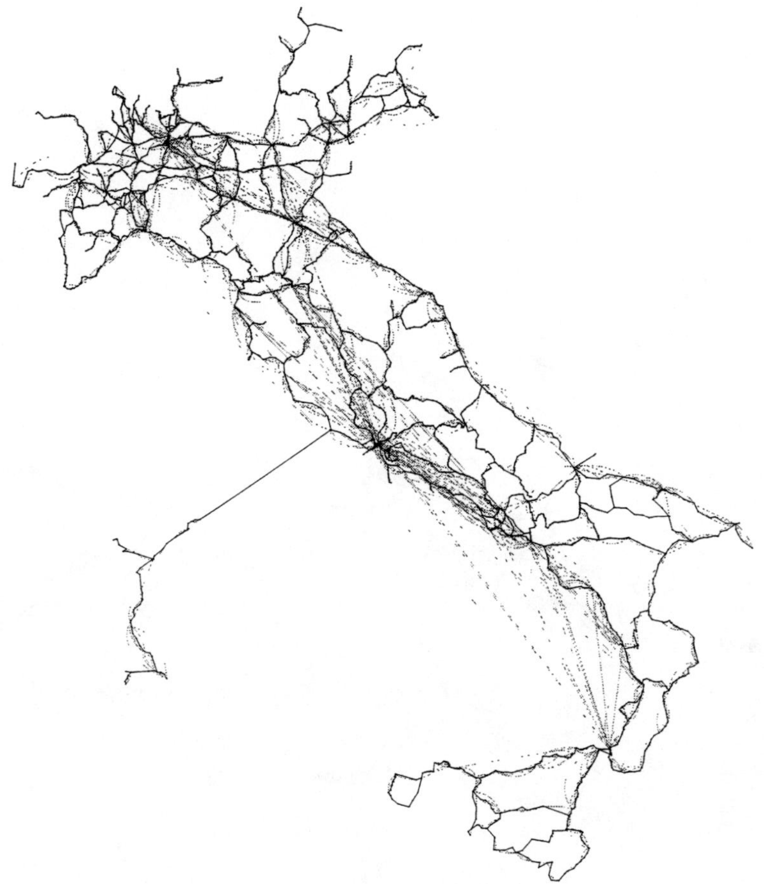

Figure 10: Italian train and ferry connections. 2,386 vertices, 4,370 edges (1,849 transitive), $\theta = (0.7, 0.3, 0.7, 0.5, 0.4, 100, 3)$

Brandes and Wagner, *Layout of Train Graphs*, JGAA, 4(3) 135–155 (2000) 152

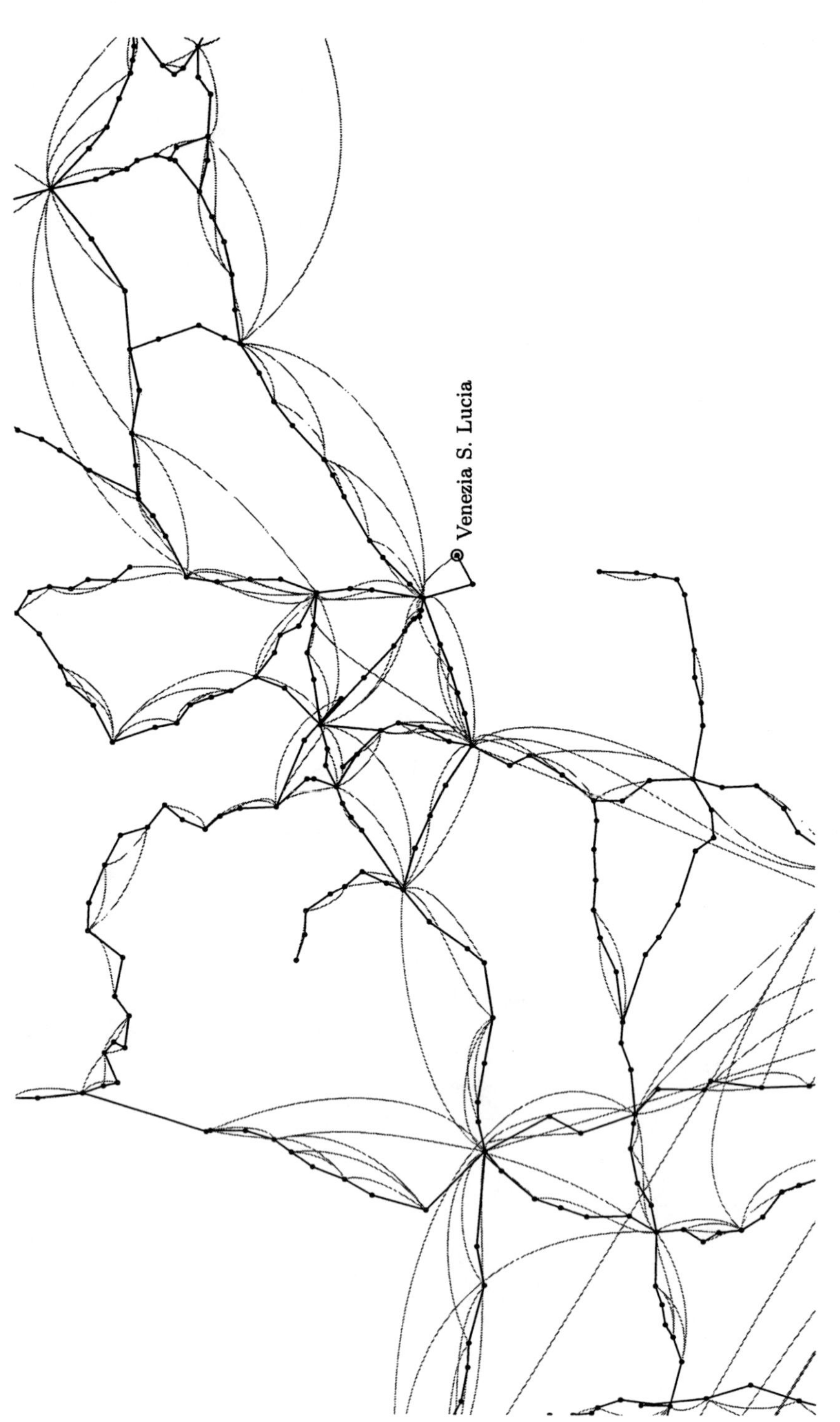

Figure 11: Magnification from Figure 10

Brandes and Wagner, *Layout of Train Graphs*, JGAA, 4(3) 135–155 (2000) 153

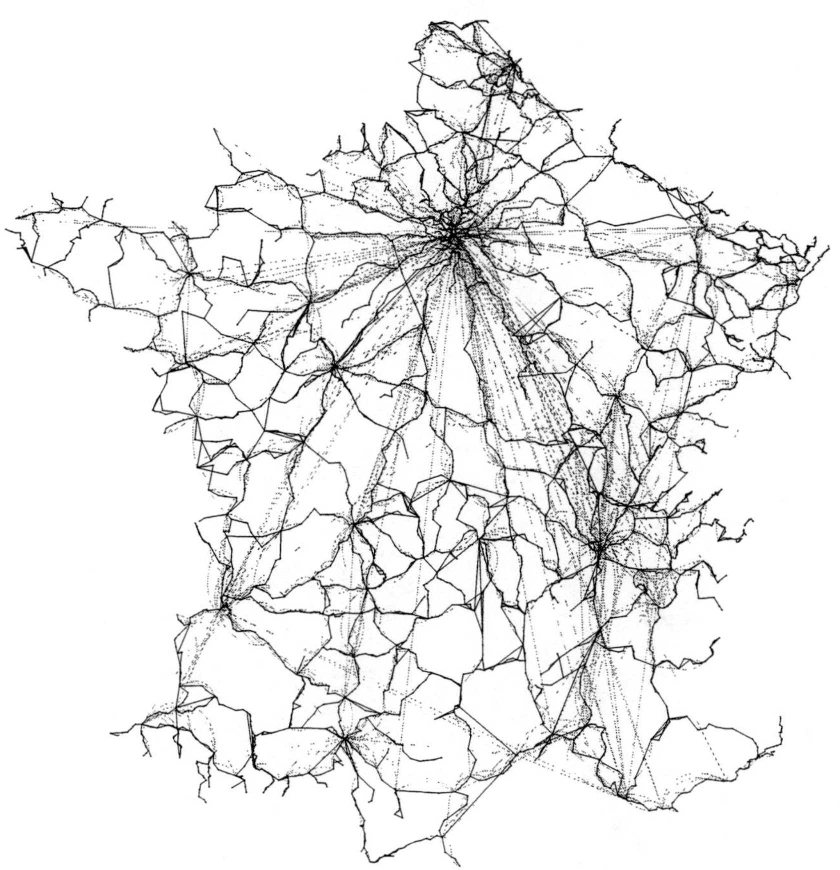

Figure 12: French connections. 4,551 vertices, 7,793 edges (2,408 transitive), $\theta = (0.7, 0.3, 0.7, 0.5, 0.4, 100, 3)$

Brandes and Wagner, *Layout of Train Graphs*, JGAA, 4(3) 135–155 (2000) 154

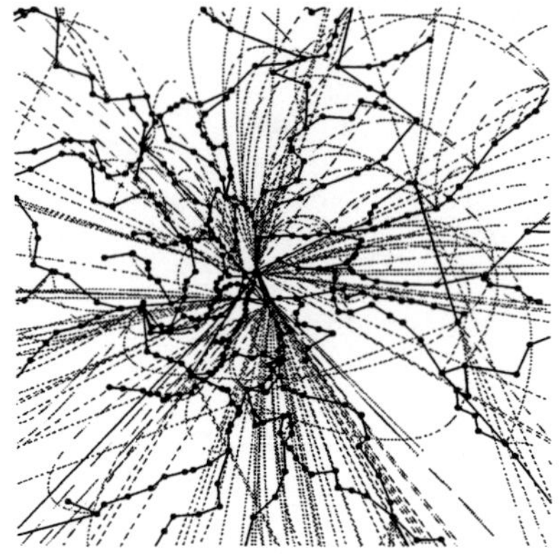

(a) Paris has six long-distance stations

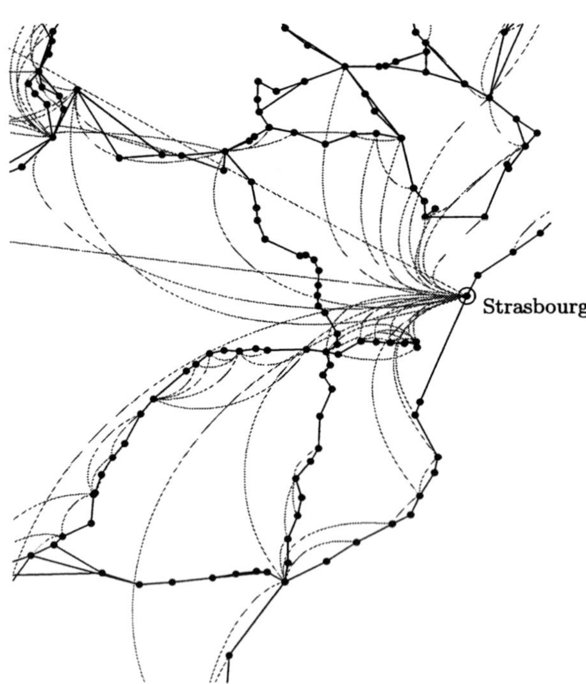

(b) Strasbourg, gateway to France

Figure 13: Magnifications from Figure 12

Brandes and Wagner, *Layout of Train Graphs*, JGAA, 4(3) 135–155 (2000) 155

References

[1] J. Besag. On the statistical analysis of dirty pictures. *Journal of the Royal Statistical Society, Series B*, 48(3):259–302, 1986.

[2] P. Bézier. *Numerical Control*. Wiley, 1972.

[3] U. Brandes. *Layout of Graph Visualizations*. PhD thesis, University of Konstanz, 1999. See `http://www.ub.uni-konstanz/kops/volltexte/1999/255/`.

[4] U. Brandes, P. Kenis, J. Raab, V. Schneider, and D. Wagner. Explorations into the visualization of policy networks. *Journal of Theoretical Politics*, 11(1):75–106, 1999.

[5] U. Brandes and D. Wagner. A Bayesian paradigm for dynamic graph layout. In G. Di Battista, editor, *Proceedings of the 5th International Symposium on Graph Drawing (GD '97)*, volume 1353 of *Lecture Notes in Computer Science*, pages 236–247. Springer, 1997.

[6] R. Davidson and D. Harel. Drawing graphs nicely using simulated annealing. *ACM Transactions on Graphics*, 15(4):301–331, 1996.

[7] P. Eades. A heuristic for graph drawing. *Congressus Numerantium*, 42:149–160, 1984.

[8] T. M. Fruchterman and E. M. Reingold. Graph-drawing by force-directed placement. *Software—Practice and Experience*, 21(11):1129–1164, 1991.

[9] T. Masui. Evolutionary learning of graph layout constraints from examples. In *Proceedings of the ACM Symposium on User Interface Software and Technology (UIST '94)*, pages 103–108. ACM, The Association for Computing Machinery, 1994.

[10] K. Mehlhorn and S. Näher. *The LEDA Platform of Combinatorial and Geometric Computing*. Cambridge University Press, 1999. Project home page at `http://www.mpi-sb.mpg.de/LEDA/`.

[11] X. Mendonça and P. Eades. Learning aesthetics for visualization. In *Anais do XX Seminário Integrado de Software e Hardware*, pages 76–88, Florianópolis, Brazil, 1993.

[12] M. Pelillo, editor. *Energy Minimization Methods in Computer Vision and Pattern Recognition (EMMCVPR '97)*, volume 1223 of *Lecture Notes in Computer Science*. Springer, 1997.

[13] G. Winkler. *Image Analysis, Random Fields and Dynamic Monte Carlo Methods*, volume 27 of *Applications of Mathematics*. Springer, 1995.

Journal of Graph Algorithms and Applications

http://www.cs.brown.edu/publications/jgaa/

vol. 4, no. 3, pp. 157–181 (2000)

Navigating Clustered Graphs Using Force-Directed Methods

Peter Eades

Basser Department of Computer Science
University of Sydney
http://www.cs.usyd.edu.au/
peter@cs.usyd.edu.au

Mao Lin Huang

Department of Computer Systems
University of Technology, Sydney
http://www.socs.uts.edu.au/
maolin@soco.uts.edu.au

Abstract

Graphs which arise in Information Visualization applications are typically very large: thousands, or perhaps millions of nodes. Current graph drawing methods successfully deal with (at best) a few hundred nodes.

This paper describes a strategy for the visualization and navigation of graphs. The strategy has three elements:

1. A layered architecture, called CGA, for handling *clustered graphs*: these are graphs with a hierarchical node clustering superimposed.

2. An online force-directed graph drawing method.

3. Animation methods.

Using this strategy, a user may view an *abridgment* of a graph, that is, a small part of the graph that is currently of interest. By changing the abridgment, the user may travel through the graph. The changes use animation to smoothly transform one view to the next.

The strategy has been implemented in a prototype system called DA-TU.

Communicated by G. Liotta and S. H. Whitesides: submitted September 1998; revised July 2000.

Eades and Huang, *Navigating Clustered Graphs*, JGAA, 4(3) 157–181 (2000)158

1 Introduction

Graphs which arise in Information Visualization applications are typically very large: thousands, or perhaps millions of nodes. Recent graph drawing competitions [5] have shown that visualization systems for classical graphs are limited to (at best) a few hundred nodes.

Attempts to overcome this problem have proceeded in two main directions:

Clustering. Groups of related nodes are "clustered" into super-nodes. The user sees a "summary" of the graph: the super-nodes and super-edges between the super-nodes. Some clusters may be shown in more detail than others. An example is in Figure 1. Note that "New South Wales" is shown in more detail than "Victoria". The clustering approach has been taken by a number of graph drawing researchers [2, 6, 13, 15], and is related to the "overview diagrams" used by some web navigation facilities [12].

Navigation. The user sees only a small subset of the nodes and edges at any one time, and facilities are provided to navigate through the graph. This approach was taken by the OFDAV system [9].

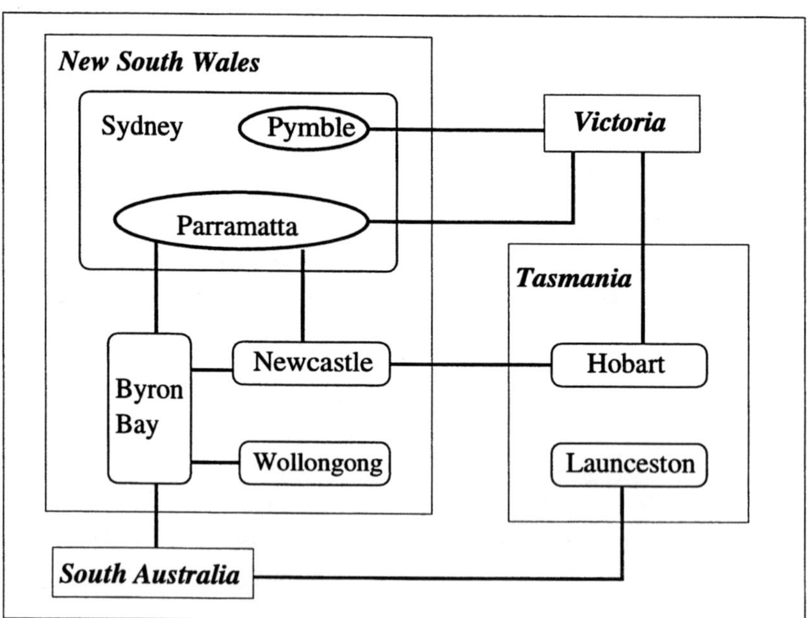

Figure 1: *A clustered graph.*

This paper introduces a strategy for combining the two approaches. The strategy has three elements:

Eades and Huang, *Navigating Clustered Graphs*, JGAA, 4(3) 157–181 (2000)159

1. **A layered architecture** for handling *clustered graphs*: these are graphs with a hierarchical node clustering superimposed (see [6]). The architecture, called CGA, is illustrated in Figure 2. This architecture supports *abridgments* of clustered graphs. These abridgments are logical views of parts of the clustered graph. Users may change their focus of interest by changing the abridgment. These changes are reflected in the *picture* of the abridgment. CGA is described in Section 2.

2. An online **force-directed graph drawing method**. This method operates at the *picture* layer of the architecture. It is a simple extension of the force-directed method from [9], described in Section 3.1.

3. **Animation methods.** Multiple animations are used to "preserve the mental map"[4], that is, to smooth the transition between pictures as the user changes focus. The animation methods are described in Section 3.2.

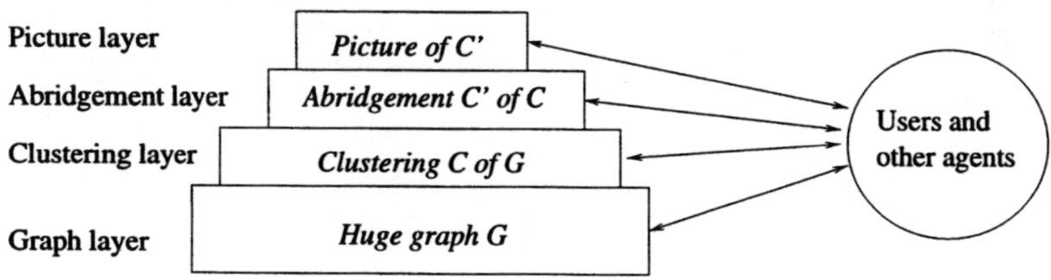

Figure 2: *The CGA architecture.*

Our strategy has been instantiated in a prototype system called DA-TU. Details of DA-TU as well as a static storyboard are in Section 4. An animated web storyboard is online at:
http://www-staff.socs.uts.edu.au/~ maolin/jgaa_demo/jgaa_demo.html.

The main purpose of this paper is to demonstrate the *feasibility* of visualizing huge graphs (with more nodes than can fit on a screen) by a combination of clustering and navigation methods. We propose that the architecture described below provides a suitable framework, and that the force-directed drawing and animation methods are suitable tools. A thorough test of this hypothesis will take many years; this paper reports on the progress that we have made to date.

2 The Architecture

The architecture CGA (*Clustered Graph Architecture*) is a design for systems in which the user manipulates data in four layers, as illustrated in Figure 2. We describe the data and methods of these four layers below.

The main aim of CGA is to separate concerns in such a way that:

Eades and Huang, *Navigating Clustered Graphs*, JGAA, 4(3) 157–181 (2000)160

- The host system need not know the whole graph. In this way, the graph can be huge (for example, it could be the whole World-Wide-Web).

- Outside agents, such as clustering algorithms and graph drawing algorithms, can be employed.

- Expertise in different areas may be confined to different layers.

2.1 The graph layer

A graph in CGA is a classical undirected graph, consisting of nodes and edges. In applications it may be a very large graph, containing many thousands of nodes. The graph may be dynamic, that is, the node and edge set may be changing; these changes may be a result of user interaction through an interface, or they may be changed by an outside agent. Further, the nodes and edges may have application-specific *attributes*, such as labels and semantics.

The changes to a graph use basic operations as follows.

G_new_node(): adds a new node to the graph, and returns an identifier for that node to the sender.

G_new_edge(u, v): adds a new edge (between existing nodes u and v) to the graph, and returns an identifier for the new edge.

G_delete_node(u): deletes node u from the graph.

G_delete_edge(e): deletes edge e from the graph.

Further, an agent can request a neighborhood of a node:

G_neighborhood(u): given a node u, this returns a list of neighbors of u.

Some more operations may be available to manage attributes of nodes and edges. For example, an elementary operation on the attributes of a node u is:

G_change_attribute(u, *attribute_id*, *attribute_value*): changes the attribute *attribute_id* to *attribute_value*.

The messages that invoke these operations may be sent and executed asynchronously, and thus, one can conceptually regard the graph as a database. If the whole graph is known, then it may be implemented by storing the graph in a database. However, in many applications the whole graph is not known (such as with web graphs), and a "graph server" implementation is appropriate.

2.2 The clustering layer

'ustered graph $C = (G, T)$ consists of an undirected graph $G = (V, E)$ and a
ed tree T such that the leaves of T are exactly the vertices of G. Each node
T represents a *cluster* of vertices of G consisting of the leaves of the subtree

Eades and Huang, *Navigating Clustered Graphs*, JGAA, 4(3) 157–181 (2000)161

rooted at ν. The tree T describes an inclusion relation between clusters; it is the *cluster tree* of C.

Figure 1 shows a clustered graph $C = (G, T)$; here the cluster tree T gives a geographical relation, and the graph G shows communication lines. Figure 3 shows the underlying graph G of C. Figure 4 shows the cluster tree T of C.

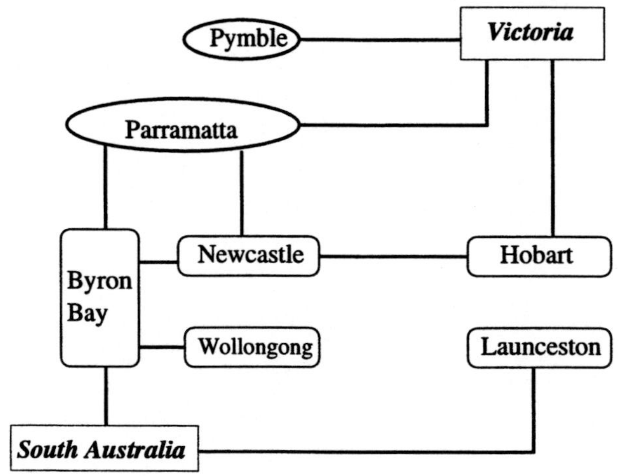

Figure 3: *Underlying graph of the clustered graph in Figure 1.*

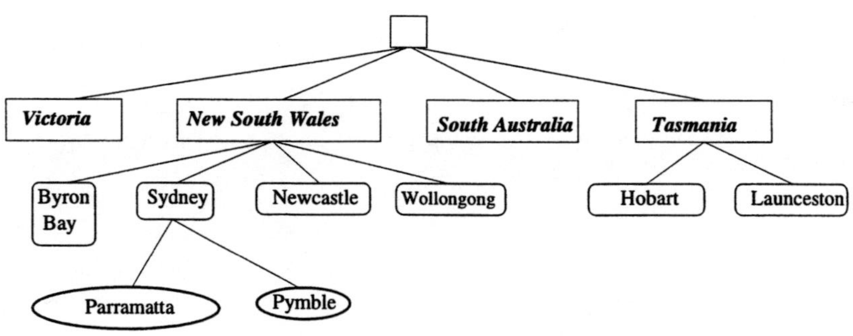

Figure 4: *Cluster tree of the clustered graph in Figure 1.*

Clustered graphs were introduced by Feng [6] as a model for relational structures with a node hierarchy. There are many closely related models; for example, the *compound graphs* of Sugiyama and Misue [16, 14] are more general; the *higraphs* of Harel [8] are far more general.

We need two elementary operations for clustered graphs.

C_create_cluster(S: set of nodes): creates a new cluster of the set S of nodes, and returns an identifier for the new cluster. As a precondition,

Eades and Huang, *Navigating Clustered Graphs*, JGAA, 4(3) 157–181 (2000)162

all nodes in S must be siblings in the cluster tree. The parent of the new cluster is the shared parent of S.

C_destroy_cluster(u: node): The children of u in the cluster tree become children of their grandparent, and u is deleted. This operation may not be applied to the root, and applying the operation of a leaf u merely deletes u.

These operations are illustrated in Figure 5.

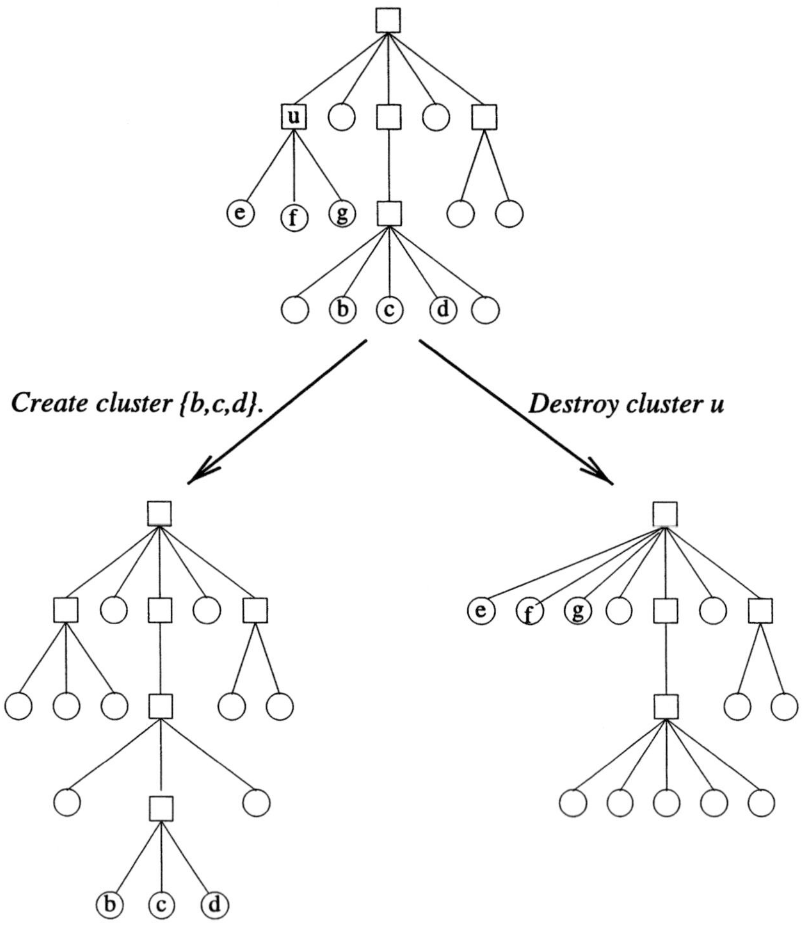

Figure 5: *Some operations at the cluster layer.*

Further, operations such as **G_new_node** at the graph layer have an effect on the cluster layer. When a new node is created at the graph layer, a message is sent to the cluster layer which makes that new node into a child of the root.

A special operation at the cluster layer is to *dismiss_clusters*; this destroys all non-leaf clusters except the root.

Eades and Huang, *Navigating Clustered Graphs*, JGAA, 4(3) 157–181 (2000)163

2.3 The abridgment layer

In applications, the whole clustered graph is too large to show on the screen; further, it is too large for the user to comprehend. This motivates the *abridgment layer*. An abridgment of the clustered graph in Figure 1 is shown in Figure 6.

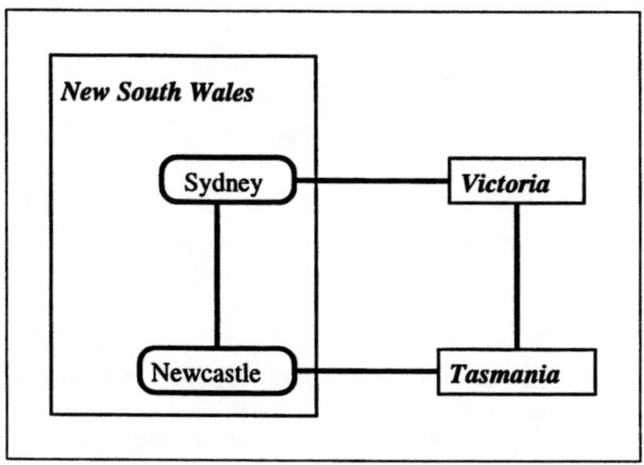

Figure 6: *An abridgment.*

We now give a formal definition of "abridgment". Suppose that U is a set of nodes of the cluster tree T. The subtree of T consisting of all nodes and edges on paths between elements of U and the root is called the *ancestor tree* of U. The set U is called the *basis* of the ancestor tree. An example of an ancestor tree is in Figure 7.

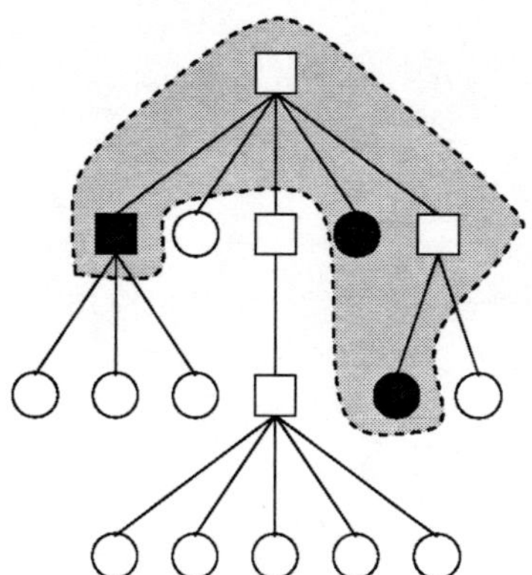

Figure 7: *The light shaded area is the ancestor tree of the dark shaded nodes.*

A clustered graph $C' = (G', T')$ is an *abridgment* of the clustered graph $C = (G, T)$ if T' is an ancestor tree of T with respect to a set U of nodes of T, and there is an edge between two distinct nodes u and v of G' if and only

Eades and Huang, *Navigating Clustered Graphs*, JGAA, 4(3) 157–181 (2000)164

if there is an edge in G between a descendent of u and a descendent of v. The vertex set of G' is the basis of the ancestor tree T'.

Figure 6 shows the abridgment of Figure 1 with basis {*Sydney, Newcastle, Tasmania, Victoria*}.

CGA has three elementary operations on abridgments; these change the basis of the abridgment.

> open_cluster(u): This is defined whenever u is in the basis U of the current abridgment. It replaces u by the children of U. The effect of this operation on the cluster tree is illustrated in Figure 8.

> close_cluster(u): All children of u in the basis of the current abridgment are deleted (from the basis), and u is added to the basis. The effect of this operation on the cluster tree is illustrated in Figure 9.

> hide(u): If u is in the basis, then it is deleted from the basis. The effect of this operation on the cluster tree is illustrated in Figure 10.

Note that the hide operation is needed so that the children of a node can be selectively hidden (otherwise, if u is in the basis then all siblings of u would need to be in the basis).

As an example, the abridgment in Figure 6 may be obtained from Figure 1 by closing "Sydney" and "Tasmania", then hiding "Byron Bay", Wollongong" and "South Australia".

The operations at the lower layers have effects at the abridgment layer.

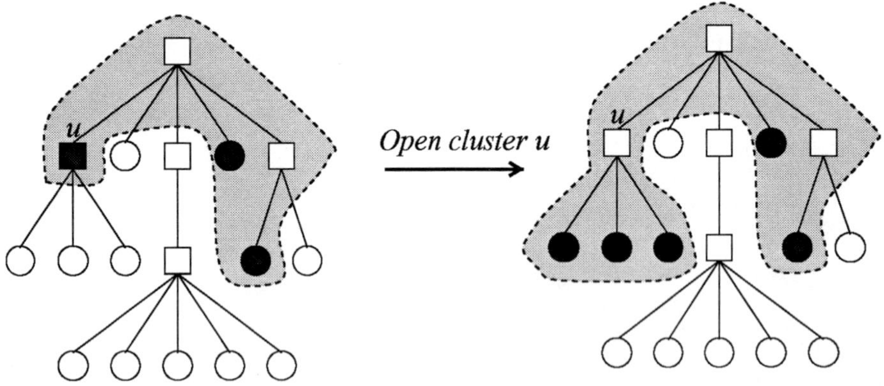

Figure 8: *The effect of the* open_cluster *operation on the cluster tree.*

2.4   The picture layer

The main purpose of CGA system is visualization; it shows animated pictures of abridgments of large clustered graphs. Note that the abridgment itself is

Eades and Huang, *Navigating Clustered Graphs*, JGAA, 4(3) 157–181 (2000)165

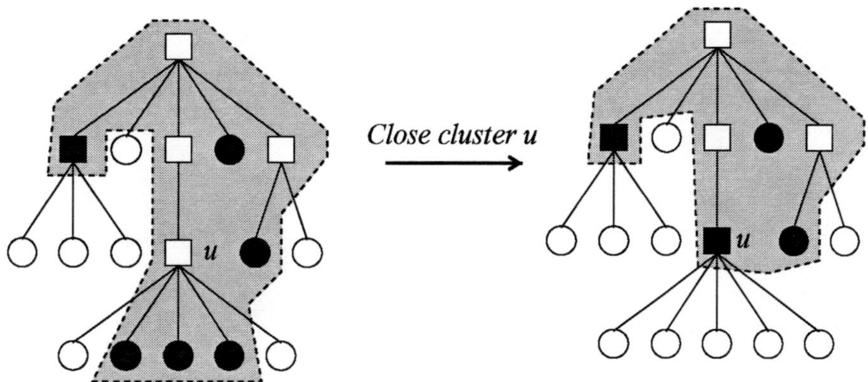

Figure 9: *The effect of the* close_cluster *operation on the cluster tree.*

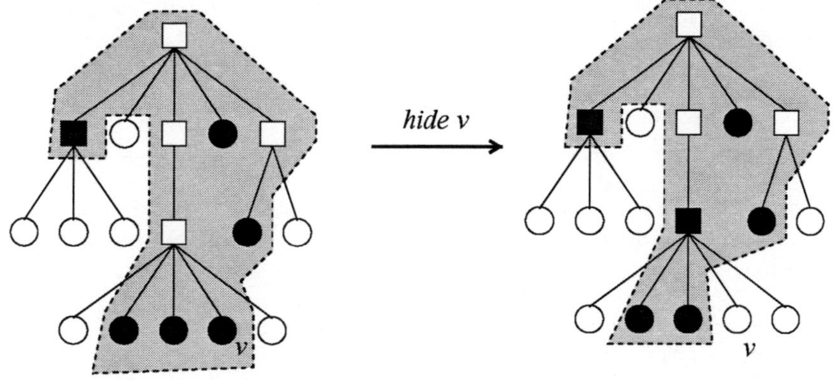

Figure 10: *The effect of the* hide *operation on the cluster tree.*

a clustered graph. Here we describe how CGA operates on static and dynamic pictures of clustered graphs.

Examples of static pictures of clustered graphs are shown in Figures 1 and 6; pictures of clustered graphs from the implementation DA-TU are in Section 4. The algorithms used by DA-TU to obtain these pictures are described in Section 3.

In CGA, a *static picture* of a clustered graph $C = (G, T)$ contains a location $p(v)$ for each vertex v of G and a route $c(u, v)$ for each edge (u, v) of G, in the same way as drawings for classical graphs. Further, a static picture has a region $b(\nu)$ of the plane for each cluster ν of T, such that if ν is a leaf of T then $b(\nu)$ is located at $p(\nu)$, and if μ is a child of ν in T then $b(\mu)$ is contained in $b(\nu)$. This definition differs slightly from Feng's definition [6] of a drawing of a clustered graph: CGA allows overlaps between regions, Feng does not. In the implementation DA-TU, methods are used to avoid overlaps, but the architecture

Eades and Huang, *Navigating Clustered Graphs*, JGAA, 4(3) 157–181 (2000)166

must allow for the existence of overlaps. The regions used by the implementation DA-TU are rectangles, and the edges are straight lines.

A *dynamic picture* of a clustered graph C consists of a sequence $D_1, D_2, \ldots, D_k$ of static pictures of C. Each D_i should differ from D_{i-1} by very little, so the sequence appears as a smooth animation. Examples of this animation are in Section 4.

Changes at the graph layer, the cluster layer, and the abridgment layer all effect the dynamic picture in CGA. Whenever such a change occurs, CGA produces a new abridgment C_{new}, which differs from the previous abridgment C_{old} by a small amount (a few nodes change). The picture layer of CGA reacts by producing a new static picture, called a *key frame*, of the new clustered graph C_{new}. Then, without user (or agent) interaction, CGA displays a dynamic picture $D_1, D_2, \ldots, D_k$ of C_{new}. This dynamic picture plays the role of "in-betweening" in animation; it continues until another change comes from a lower layer.

The picture layer has the following operations:

- `Move(u: node)`: this is the usual operation of manually moving a node in the picture.

- `Gathering`. This operation moves sibling nodes closer together. In practice, this tends to make the cluster boundaries disjoint. Examples of the gathering operation are in Section 4.

- `Scaling`. This operation can be used to increase the size of the picture to allow more detail to be seen, or reduce the size of the picture, to fit the screen.

- `Layout operations`. These are classical layout functions, which produce a picture P' of a clustered graph C' from a picture P of a clustered graph C. The layout operation is used in two circumstances:

  1. When a change from C to C' occurs at the abridgment layer, the `layout` operation responds at the picture layer by producing a picture P' of C', computed from C' and P.

  2. The layout operation is used to produce the dynamic pictures described above. That is, it produces picture D_{i+1} from D_i.

Details of the implementation of the layout operation in DA-TU are in the next section.

3 Layout and Animation

There are two requirements for algorithms to produce pictures for CGA.

- We need to produce diagrams that are easy to read and understand. A great deal of research in *graph drawing algorithms* (see, for example, [1])

Eades and Huang, *Navigating Clustered Graphs*, JGAA, 4(3) 157–181 (2000)167

has resulted in a variety of such algorithms; these take a combinatorial description of a graph and attempt to produce a drawing that is as readable as possible.

- **CGA** aims to navigate through graphs. The logical operations at the lower layers must be reflected in changes to the drawing at the picture layer. These changes must be "smooth" at the picture layer; that is, each static picture D_i should be very similar to D_{i-1}, and successive key frames should be similar.

For **DA-TU** we have used a force-directed method to achieve both requirements. Our methods are similar to those described in [9]. In Section 3.1, we describe the force model that drives this technique. Section 3.2 describes how animation is used to smooth the changes at the picture layer.

Note that other layout operations to satisfy the requirements above are possible; for an example, see [11].

3.1 The force model

In this section, we briefly outline the force model. It has been developed from several previous force models used in graph drawing [1, 3, 7, 10].

We use three types of spring forces. These operate between the vertices of the graph G in the abridgment $C = (G, T)$.

- *Internal-spring:* A spring force between a pair of nodes in G which are siblings in T.

- *External-spring:* A spring force between a pair of nodes in G which are not siblings in T.

- *Virtual-spring:* Each cluster (non-leaf node of T) ν has a *virtual node* ν' associated with it. In the picture, ν' is a point within the rectangle representing ν. Further, for each node u in G, if the parent of u in T is ν, then there is a *virtual edge* between u and ν'. A *virtual-spring* is a spring force between $u\ \nu'$ along this virtual edge.

It is best to describe these spring forces with an example; see Figure 11. The internal-spring forces on vertex c are along the edges (c, a) and (c, b); the external-spring forces on c are along the edges (c, f) and (c, g).

Virtual-springs can be described using a virtual node in each cluster. In the clusters X, Y, and Z, virtual nodes x', y', and z' are shown; each virtual node is connected to each node in its cluster by a virtual edge. Virtual-springs exert forces along these edges. Note that virtual nodes and edges are not shown in the actual picture of the clustered graph (unless the user wants to see them).

The internal-spring forces are stronger than the external-spring forces. This tends to make siblings come close together. This effect is increased by the virtual-spring forces. The *gathering* operation strengthens the virtual-spring

Eades and Huang, *Navigating Clustered Graphs*, JGAA, 4(3) 157–181 (2000)168

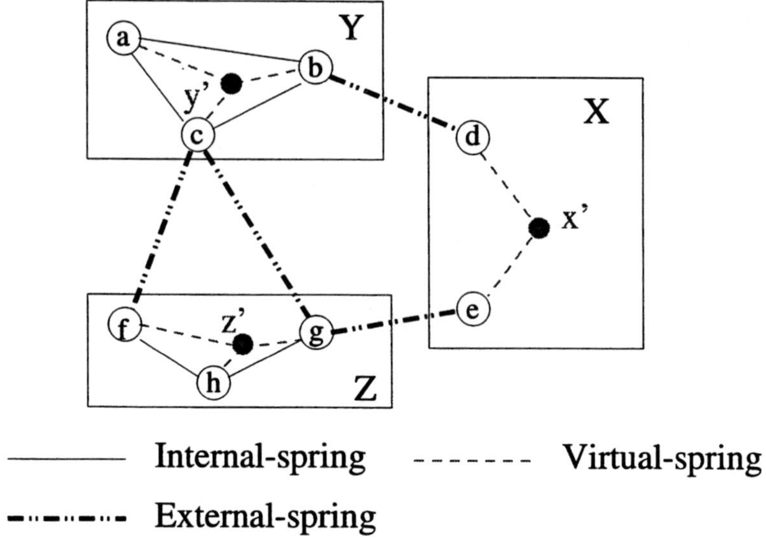

Figure 11: *Spring forces.*

forces and the internal-spring forces; this makes sibling nodes move closer together.

As well as spring forces, there are gravitational repulsion forces between all nodes.

The forces are applied additively to give an aesthetically pleasing layout of the graph. The sum of forces on each node is continually computed, and the nodes move according to the strength and direction of these forces. If the force on each node is zero, then the system has reached a locally minimal energy state, and the movement stops.

3.2 Animations

In **DA-TU**, the whole visualization is fully animated. Every change at one of the lower layers, whether triggered by the user or by another agent, is animated at the picture layer.

This animation may reduce the cognitive effort of the user in recognizing change; we aim to preserve the user's "mental map" [4].

There are several types of animation that are implemented in our system. These are detailed below.

- *Force animation.* The force model above is used in the animation of all changes to picture. The forces move the nodes toward a minimum energy state, and this movement gives the sequence $D_1, D_2, \ldots, D_k$ of drawings to smoothly change the layout from abridgment to the next. Overall, the animation driven by the forces is the most important mechanism for the

Eades and Huang, *Navigating Clustered Graphs*, JGAA, 4(3) 157–181 (2000)169

smooth transition between changes at the layers lower than the picture layer.

- *Animated addition and deletion* of nodes and edges. When a node is deleted or added to the abridgment, we use animated *shrinking* or *growing* to help the user identify nodes that are disappearing or appearing. Disappearing nodes lose their connecting edges, then shrink until they disappear. Appearing nodes are displayed in a reduced size with their edges, then grow to be full-sized nodes. This avoids the sudden disappearing/appearing of node images on the screen, which can disturb the user's mental map.

- *Animated cluster regions.* Nodes move according to the force model, and the rectangular regions representing clusters follow the nodes. At all times, each cluster is the minimum enclosing rectangle of its children. These rectangles move smoothly as the positions of the children change. This is especially important for the gathering operation; the clusters separate smoothly.

- *Animated scaling.* Animated shrinking scales down a picture if it gets too large for the screen. Further, animated enlarging is used to increase the size of the picture; this enables the user to see details and makes direct manipulation operations of the picture easier. In fact, animated scaling is implemented by simply changing the forces in the springs; for example, to shrink the whole picture, every spring is increased in strength.

- *Animated cluster closing/opening:* When closing a cluster, we firstly use animation to reduce the size of closed region of this cluster. As soon as the size reaches a certain threshold layer we smoothly replace the representation of the cluster from its opened form (a red line bounded rectangle) to its closed form (a small black line bounded rectangle).

  When opening a cluster, we firstly smoothly replace the representation of the cluster from its closed form to its opened form. The children and relevant links of the cluster are smoothly added into the layout. Then we smoothly enlarge the size of this cluster to its normal size.

- *Camera animation.* Camera animation moves the whole drawing. It is optional. It can be used, for example, to move specific nodes of interest to the center of the screen.

All these animations operate in parallel.

4 Implementation

The strategy described in Sections 2 and 3 has been implemented in a prototype system called DA-TU[1]. A screenshot from DA-TU is in Figure 12; from DA-TU, the

---

[1] "Big map" in Mandarin.

Eades and Huang, *Navigating Clustered Graphs*, JGAA, 4(3) 157–181 (2000)170

user can call many of the operations described in Section 2. Note that operations at the graph layer are not implemented.

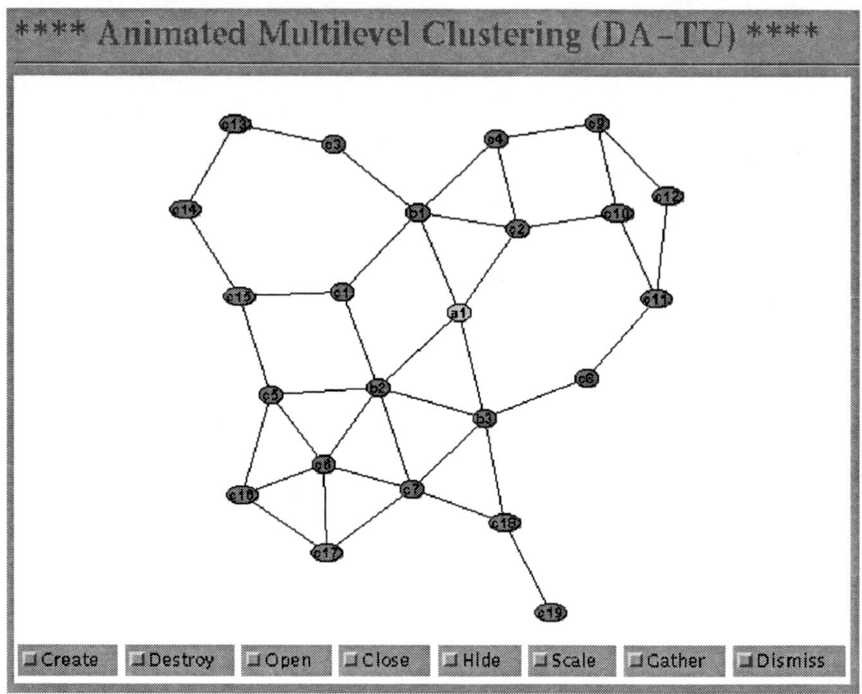

Figure 12: A screen from DA-TU.

This section records interactive sessions with DA-TU. They illustrate the basic operations of the system and how it works to achieve a better quality of the layout of clustered graphs.

4.1   Session one

The first session is a storyboard applet at:
http://www-staff.socs.uts.edu.au/~ maolin/jgaa_demo/jgaa_demo.html.
using this applet, the reader may step through several types of operation. This session has nine steps:

1. A graph in DA-TU. The animated layout is computed by a spring algorithm. Note node $a1$ is fixed in the middle of the screen; it is the current focus.

2. The user creates five clusters. However, there are some overlaps among the clusters.

Eades and Huang, *Navigating Clustered Graphs*, JGAA, 4(3) 157–181 (2000)171

3. The user applies multiple forces in the 'gathering' operation to eliminate the overlaps. This may improve the readability of the layout. Note the 'virtual edges' and 'virtual node' in the cluster to the far left. This holds the nodes within the cluster together.

4. The same layout as shown in last one; however, the virtual edges and node are invisible. However, the 'virtual spring' applied between non-adjacent vertices $c8$ and $c19$ still operates.

5. Two clusters, *Cluster_1* and *Cluster_4* are 'closed' by clicking on regions occupied by these two clusters, in the 'close' mode. The user also applies a little 'gathering' to keep the clusters disjoint.

6. Open the cluster *Cluster_4* by clicking on the cluster in the 'open' mode. Some more 'gathering' is applied to make the rectangles disjoint.

7. Close the cluster $< b2, c5, c6, c7, c16, 17 >$ by clicking on its region, in the 'close' mode.

8. Open the clusters *Cluster_1* and then *Cluster_3*. This is done by clicking on each of the clusters, one by one, in the 'open' mode. Again, some 'gathering' is applied.

9. Dismissing the whole cluster tree from the graph. This is done by clicking on the 'dismiss' button in control panel.

4.2 Session two

The second session consists of Figures 13 to 20. This session shows operations on the graph in Figure 12.

The first few figures show the results of operations *create_cluster* and *gathering*, followed by an illustration of the virtual springs, and the *close_cluster* operation.

4.3 Session three

The next session begins with Figure 16. The *gathering* operation, some *close_cluster* operations, and an *open_cluster* operation are shown. Finally, a special operation *dismiss_clusters* is shown.

5 Conclusions

CGA provides an architecture for handling graphs visually. The architecture is aimed at handling huge graphs, which may be too large to fit onto the screen. The user begins with a picture of an abridgment of the graph. This abridgment may be changed by user interaction or by an outside agent. Navigation through the graph is basically moving between abridgments.

Eades and Huang, *Navigating Clustered Graphs*, JGAA, 4(3) 157–181 (2000)172

The architecture has been implemented in a prototype system DA-TU. This provides a simple interface to the functions of CGA. It also uses a force-directed animated layout algorithm to ensure that the picture is aesthetically pleasing, and that transitions between pictures do not destroy the "mental map" [4].

Although DA-TU provides a proof of the main concepts of CGA, it is far from complete. Our future plans include the implementation of a graph drawing system within an application domain (such as web graphs) in which huge graphs commonly occur.

Eades and Huang, *Navigating Clustered Graphs*, JGAA, 4(3) 157–181 (2000)173

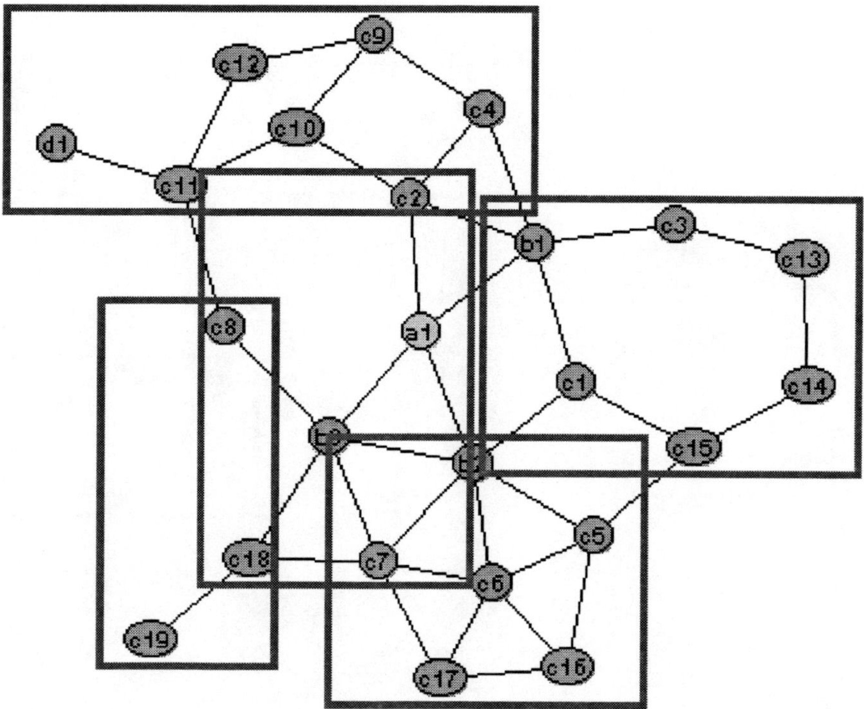

Figure 13: The user creates five clusters on the graph in Figure 12. However, this layout has five overlaps among the clusters.

Eades and Huang, *Navigating Clustered Graphs*, JGAA, 4(3) 157–181 (2000)174

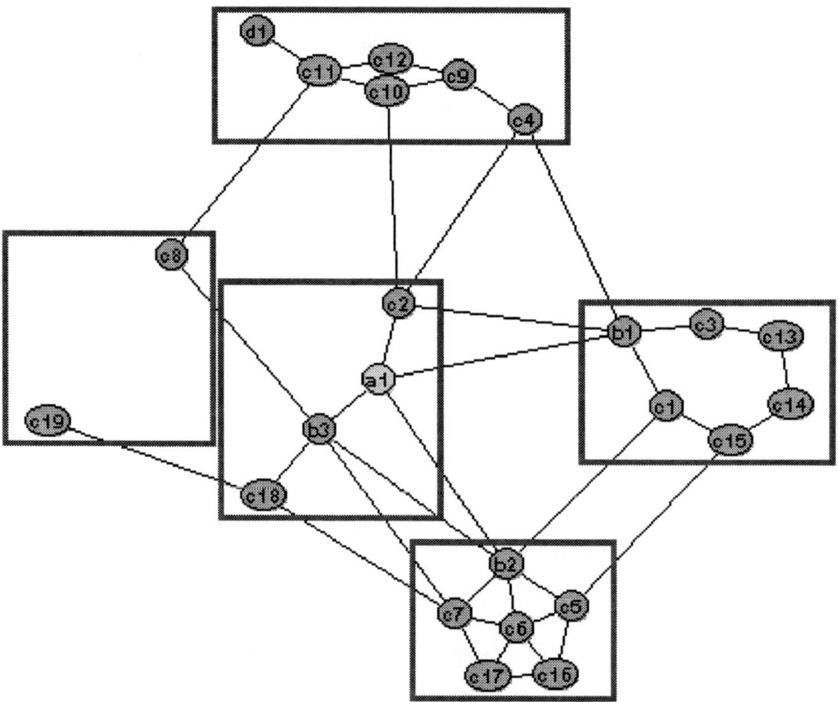

Figure 14: The user applies the spring forces in the "gathering" operation to eliminate the overlaps in Figure 13. This greatly improves the readability of the layout for the user.

Eades and Huang, *Navigating Clustered Graphs*, JGAA, 4(3) 157–181 (2000)175

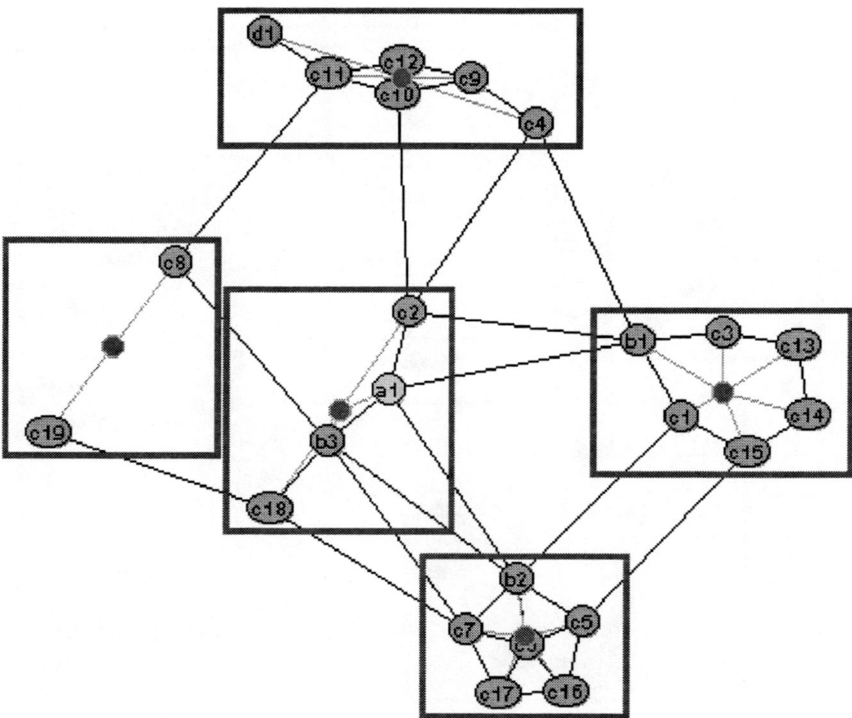

Figure 15: The same layout as shown in Figure 14; however, the virtual nodes and edges are shown. Note the "virtual springs", that is, an attractive force applied between non-adjacent vertices $c8$ and $c19$.

Eades and Huang, *Navigating Clustered Graphs*, JGAA, 4(3) 157–181 (2000)176

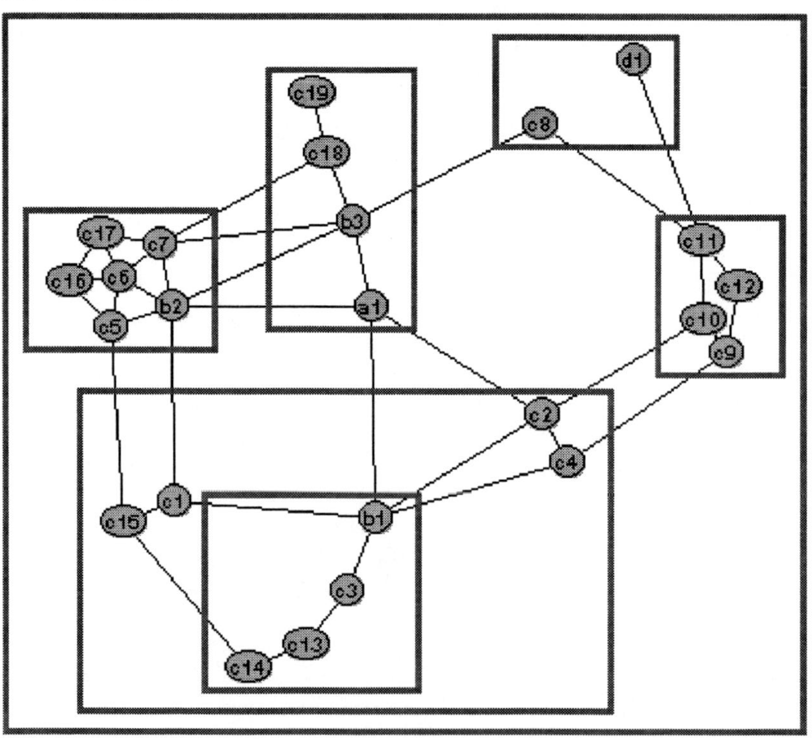

Figure 16: A clustered graph C with 4 layers, after the application of spring forces in the "gathering" mode.

Eades and Huang, *Navigating Clustered Graphs*, JGAA, 4(3) 157–181 (2000)177

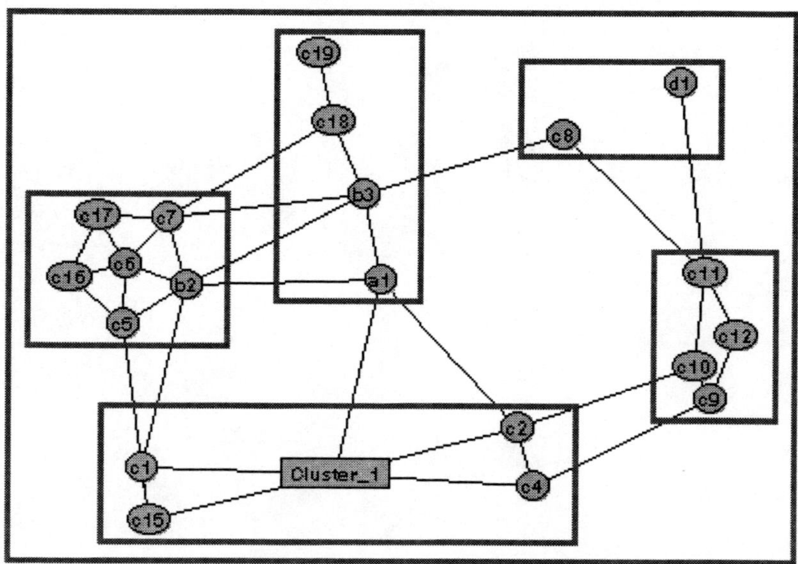

Figure 17: Closing one cluster, $\nu = \{b1, c3, c13, c14\}$, in Figure 16 by clicking on ν in the "close" mode. The representation of the closed cluster ν is a small rectangle.

Eades and Huang, *Navigating Clustered Graphs*, JGAA, 4(3) 157–181 (2000)178

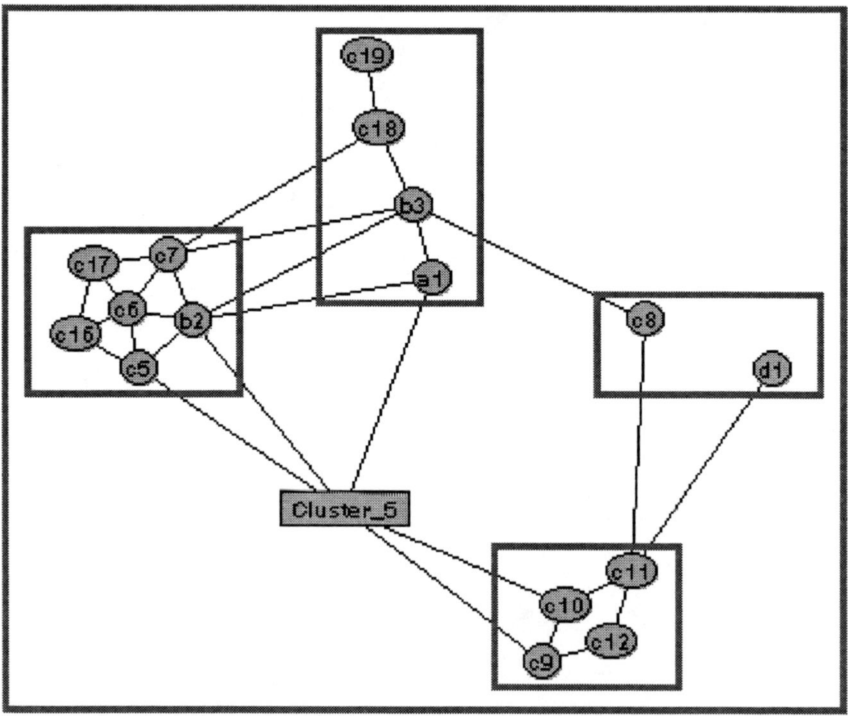

Figure 18: Continuing, closing the cluster $\{c1, c2, c4, c15, Cluster\_1\}$ in Figure 17.

Eades and Huang, *Navigating Clustered Graphs*, JGAA, 4(3) 157–181 (2000)179

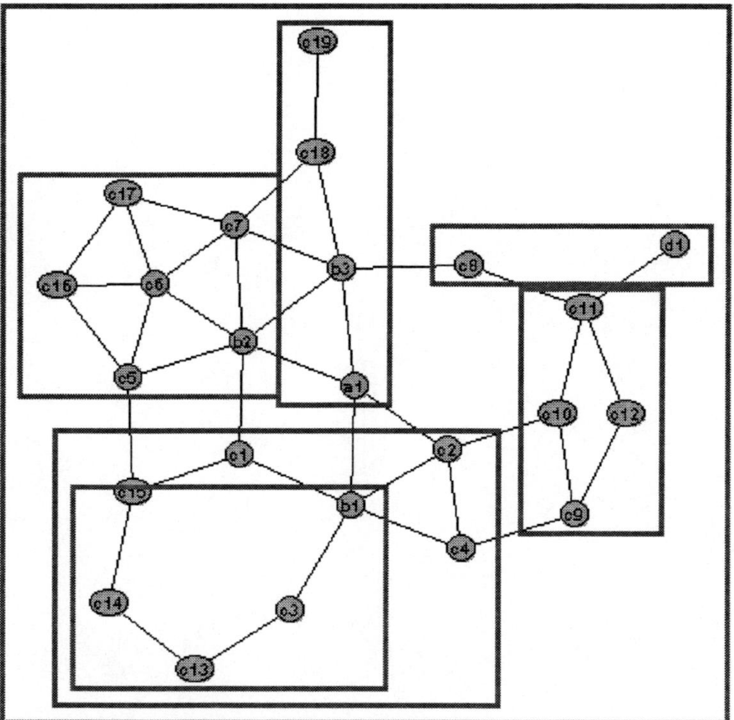

Figure 19: Opening the cluster *Cluster_5* in Figure 18, and then opening the cluster *Cluster_1*. This is done by clicking on both visual rectangles of two clusters one by one under in the "open" mode.

Eades and Huang, *Navigating Clustered Graphs*, JGAA, 4(3) 157–181 (2000)180

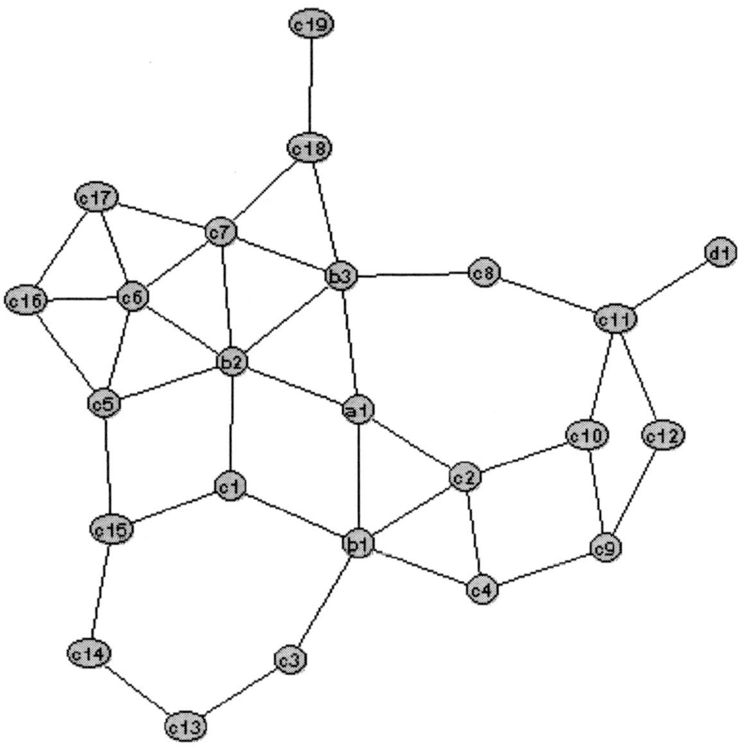

Figure 20: Dismissing the whole cluster tree from the layout as shown in Figure 19, and returning to the graph with no clusters. This could be done by destroying the clusters one by one; however, a "dismiss" button is provide to destroy all clusters at once.

Eades and Huang, *Navigating Clustered Graphs*, JGAA, 4(3) 157–181 (2000)181

References

[1] G. D. Battista, P. Eades, R. Tamassia, and I. G. Tollis. Graph drawing - algorithms for the visualization of graphs. *Prentice-Hall, Englewood cliffs, NJ*, 1999.

[2] F. Bertault and M. Miller. Graph drawing 99. In *An algorithm for drawing compound graphs*, 1999. (to appear).

[3] P. Eades. A heuristic for graph drawing. *Congr. Numer.*, 42:149–160, 1984.

[4] P. Eades, W. Lai, K. Misue, and K. Sugiyama. Preserving the mental map of a diagram. In *Proceedings of Compugraphics 91*, pages 24–33, 1991.

[5] P. Eades and J. Marks. Graph drawing contest report. In R. Tamassia and I. G. Tollis, editors, *Graph Drawing (Proc. GD '94)*, volume 894 of *Lecture Notes Comput. Sci.*, pages 143–146. Springer-Verlag, 1995.

[6] Q. Feng. Algorithms for drawing clustered graphs. In *PhD thesis, Department of Computer Science and Software Engineering, The University of Newcastle, Australia.*, 1997.

[7] T. Fruchterman and E. Reingold. Graph drawing by force-directed placement. *Softw. – Pract. Exp.*, 21(11):1129–1164, 1991.

[8] D. Harel. On visual formalisms. *Commun. ACM*, 31(5):514–530, 1988.

[9] M. L. Huang, P. Eades, and J. Wang. On-line animated visualization of huge graphs using a modified spring algorithm. *Journal of Visual Languages and Computing*, 9(4), No. vl980094:623–645, 1998.

[10] T. Kamada. Symmetric graph drawing by a spring algorithm and its applications to radial drawing, 1989.

[11] S. Moen. Drawing dynamic trees. *IEEE Software*, 7:21–8, 1990.

[12] S. Mukherjea and J. Foley. Navigational view builder: A tool for building navigational views of information spaces. In *ACM SIGCHI '94 Conference Companion, pages 289-290*, 1994.

[13] S. C. North. Drawing ranked digraphs with recursive clusters. *preprint, Software Systems and Research Center, AT&T Laboratories.*, 1993.

[14] G. Sander. Layout of compound directed graphs. In *Technical Report A/03/96, University of the Saarlands*, 1996.

[15] K. Sugiyama. A cognitive approach for graph drawing. *Cybernetics and Systems: An International Journal*, 18:447–488, 1987.

[16] K. Sugiyama and K. Misue. Visualization of structural information: Automatic drawing of compound digraphs. *IEEE Transactions on Systems, Man and Cybernetics*, 21(4):876–896, 1991.

Journal of Graph Algorithms and Applications

http://www.cs.brown.edu/publications/jgaa/

vol. 4, no. 3, pp. 183–191 (2000)

Dynamic WWW Structures in 3D

Ulrik Brandes   *Vanessa Kääb*   *Andres Löh*   *Dorothea Wagner*
Thomas Willhalm

Department of Computer & Information Science
University of Konstanz
78457 Konstanz, Germany
http://www.inf.uni-konstanz.de/
~{brandes,kaeaeb,loeh,wagner,willhalm}
{Ulrik.Brandes, Vanessa.Kaeaeb, Andres.Loeh, Dorothea.Wagner,
Thomas.Willhalm}@uni-konstanz.de

Abstract

We describe a method for three-dimensional straight-line representation of dynamic directed graphs (such as parts of the World Wide Web). It has been developed on the occasion of the 1998 Graph Drawing Contest and constitutes a customized blend of techniques known from the Graph Drawing literature. Since we feel that they may be of interest to others facing similar graph drawing problems, some technical details are mixed in.

The animation of the contest graph and accompanying material are available from the *Journal of Graph Algorithms and Applications*'s Web site.

Communicated by G. Liotta and S. H. Whitesides: submitted November 1998, revised October 1999.

Brandes et al., *Dynamic WWW Structures in 3D*, JGAA, 4(3) 183–191 (2000)184

1 Introduction

Since 1993, a contest is organized along with the annual *Symposium on Graph Drawing* [7, 8, 10, 11, 9]. It usually features graphs from different applications posing distinct problems with respect to graphical presentation. In the present case [9], a dynamic graph of links between World Wide Web (WWW) pages was given as a list of link additions and deletions, shown in Figure 1. The contestants should

"...depict the content and structure of the graph as it evolves."

Our primary interest was to study the implied dynamic graph layout problem. Hence, we largely ignored content and focussed on evolving structure.

```
add www.att.com/catalog/consumer -> www.att.com
add www.att.com/att -> www.att.com/catalog/consumer
add www.att.com -> www.att.com/catalog/consumer
add www.att.com/catalog/consumer -> www.att.com/cmd/custcare
add www.att.com/catalog/consumer -> www.att.com/write
add www.att.com -> www.att.com/worldnet
add www.att.com/catalog/consumer -> www.att.com/terms.html
add www.att.com/att -> www.att.com/learningnetwork
add www.att.com/catalog/consumer -> www.att.com/cgi-bin/ppps.cgi
add www.att.com -> www.att.com/catalog/small_business
add www.att.com/whatsnew -> www.att.com/catalog/consumer
add www.att.com/catalog/small_business -> www.att.com/cgi-bin/bmd_cart.cgi
add www.att.com/catalog/consumer -> www.att.com/cmd/jump
add www.att.com/att -> www.att.com/catalog
add www.att.com/textindex.html -> www.att.com/catalog/small_business
add www.att.com/catalog/small_business -> www.att.com/bmd/tollfree
add www.att.com/whatsnew -> www.att.com/worldnet/wmis
add www.att.com/catalog/consumer -> search.att.com
add www.att.com/catalog/small_business -> www.att.com/bmd/jump
add www.att.com/features -> www.att.com/rock
add www.att.com/catalog/small_business -> www.att.com
add www.att.com/home -> www.att.com/catalog/consumer
delete www.att.com -> www.att.com/catalog/consumer
add www.att.com/catalog/small_business -> www.att.com/services
delete www.att.com/att -> www.att.com/learningnetwork
delete www.att.com/catalog/consumer -> www.att.com/cmd/custcare
add www.att.com/catalog/small_business -> www.att.com/write
add www.att.com/net -> www.att.com/worldnet/wis/sky/signup.html
add www.att.com/catalog/small_business -> www.att.com/bmd/products
add www.att.com/catalog/small_business -> search.att.com
add www.att.com/net -> www.att.com/worldnet/wmis
add www.att.com/catalog/small_business -> www.att.com/terms.html
delete www.att.com/catalog/small_business -> www.att.com/bmd/tollfree
add www.att.com/att -> www.att.com/catalog/small_business
add www.att.com/news -> www.att.com/catalog/consumer
add www.att.com/whatsnew -> www.att.com/catalog/small_business
add www.att.com/net -> www.att.com/worldnet/intranet
delete www.att.com/whatsnew -> www.att.com/catalog/consumer
add www.att.com/catalog/small_business -> www.att.com/bmd/custcare
add www.att.com/catalog/consumer -> www.att.com/cgi-bin/cart.cgi
delete www.att.com/home -> www.att.com/catalog/consumer
add www.att.com/catalog/small_business -> www.att.com/cmd
add www.att.com/catalog/consumer -> www.att.com/services
add www.att.com/worldnet -> www.att.com/worldnet/intranet
add www.att.com/services -> www.att.com/catalog
delete www.att.com/catalog/consumer -> www.att.com
add www.att.com/services -> www.att.com/catalog/consumer
delete www.att.com/catalog/consumer -> www.att.com/services
add www.att.com/services -> www.att.com/catalog/small_business
add www.att.com/whatsnew -> www.att.com/news
delete www.att.com/whatsnew -> www.att.com/worldnet/wmis
add www.att.com/catalog/consumer -> www.att.com/cmd/products
delete www.att.com/net -> www.att.com/worldnet/wis/sky/signup.html
delete www.att.com/catalog/small_business -> www.att.com/services
add www.att.com/speeches -> www.att.com/speeches/index96.html
add www.att.com/worldnet -> www.att.com/worldnet/wmis
delete www.att.com/att -> www.att.com/catalog
add www.att.com/textindex.html -> www.att.com/catalog/consumer
delete www.att.com/services -> www.att.com/catalog
add www.att.com/news -> www.att.com/catalog/small_business
add www.att.com/international -> www.att.co.uk
delete www.att.com/catalog/small_business -> search.att.com
add www.att.com/business -> www.att.com/catalog/small_business
delete www.att.com/textindex.html -> www.att.com/catalog/consumer
add www.att.com/write -> www.catalog.att.com/cmd/jump
```

Figure 1: Data given for Graph A of the 1998 Graph Drawing Contest

By definition the graph starts out empty, and directed edges, sometimes introducing new vertices, are added and deleted. To get a first impression of

Brandes et al., *Dynamic WWW Structures in 3D*, JGAA, 4(3) 183–191 (2000)185

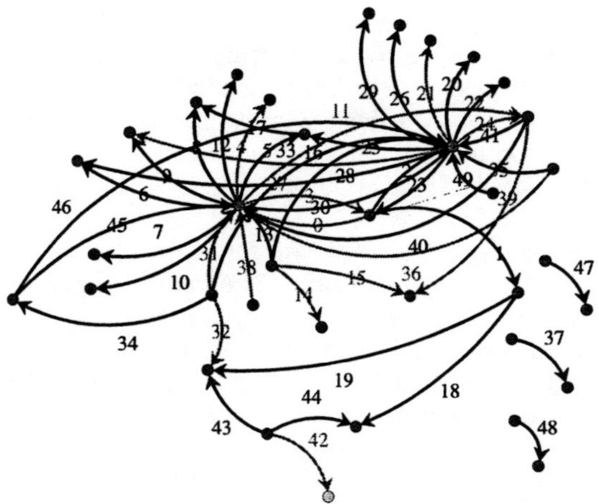

Figure 2: 2D representation with insertion (black edges) or deletion (gray edges) times. Edge curvature indicates the order of creation of incident vertices

the graph's structure, in particular of its size and density, it was input in a graph editor (LEDA's [12] GraphWin), with edges colored differently, if they are inserted and later deleted. See Figure 2.

We found that the graph was sufficiently small and sparse to try animating a straight-line representation in space. For such an animation we essentially had to cope with three aspects, each of which is addressed in a subsequent paragraph. First of all, a dynamic layout model had to be devised, specifying vertex trajectories. Secondly, unlike in two-dimensional representations, a point from which to view the graph had to be selected carefully, because otherwise substantial parts of the structure might be occluded. And finally, a rendering had to be specified. In Section 5 we conclude with lessons learned from this project.

2 Dynamic Graph Layout

Since no dynamic three-dimensional straight-line layout algorithm was known to us, we did our own experiments using random field layout models, a flexible modeling scheme introduced in [1]. A prototypical implementation of a corresponding layout module was available to us because of its use in other projects.[1]

Our dynamic layout model consists of a sequence of static layouts and is hence described in two steps. First, structures are mapped into space by static layout models, which are then extended into a sequence of dependent layouts, in sync with graph modifications.

---

[1]One of these projects is described in this issue [3], together with a brief overview of the modeling framework.

Brandes et al., *Dynamic WWW Structures in 3D*, JGAA, 4(3) 183–191 (2000)186

Static layout model. Suppose we are given a single directed graph $G = (V, A)$, and suppose further that vertices ought to be represented by points in space, while edges ought to be represent by straight lines. It is then sufficient to compute a vector $p = (p_v)_{v \in V}$ of vertex positions. Any such vector is called a *layout*. An objective function defined on p is used to formally capture the quality of a layout, defined by local criteria such as uniform vertex distribution, uniform edge lengths, and sufficient distance between vertices and those edges they are not incident to.

The corresponding terms of our objective function (see Table 1), which is very similar to [5], are called *potentials*, since the objective function itself can be interpreted as an *energy* measuring layout distortion with respect to the above criteria. The additional potentials account for the contest graph being directed and, most of the time, disconnected. To avoid that disconnected components drift arbitrarily far apart, one representative vertex of each component is connected to the layout space origin by an invisible edge.[2] Even though WWW links form a general directed graph, they tend to contain hierarchical, tree-like substructures (if purely navigational links are omitted, as is the case in the contest data).[3] To encode the direction of edges graphically, we hence favor them pointing downward, thus displaying edge directions in context rather than locally by using cone-shaped arrows at the end of an edge. Note that this is similar to layered [14] and upward [6] layouts, however in 3D. Our rotation potentials are an adaptation of rotative forces used in [13] to align edges with an imaginary magnetic field. Note that squaring a normalized angle renders angles larger than $\frac{\pi}{2}$ (upward pointing edges) more severe a distortion than small deviations (downward pointing edges not exactly parallel to the the z-axis).

Dynamic layout model. The contest data implies a sequence $G^{(0)}, \ldots, G^{(65)}$ of directed graphs, where $G^{(0)}$ is the empty graph. To produce a smooth animation, a sequence of static layouts was computed. Vertex positions in these layouts were then used as break points of splines, i.e. smooth curves interpolating these points and thus defining vertex trajectories for the animation.

For every $G^{(t)}$ with $0 < t \leq 65$ we computed three layouts, $p^{(t,0)}, p^{(t,1)}, p^{(t,2)}$, in slightly different ways described below. For graceful animation it is necessary that consecutive layouts depend on each other. For simplicity, we let each layout depend only on its predecessor, so that the sequence of layouts has the form

$$
\begin{aligned}
\downarrow & G^{(1)} \\
p^{(1,0)} & \to p^{(1,1)} \to p^{(1,2)} \\
& \downarrow G^{(2)} \\
& p^{(2,0)} \to p^{(2,1)} \to p^{(2,2)} \\
& \quad \downarrow G^{(3)} \\
& \quad p^{(3,0)} \to p^{(3,1)} \to p^{(3,2)} \\
& \qquad \vdots
\end{aligned}
$$

---

[2] For the contest graph, these vertices could be selected manually, since in each component there is at least one vertex that is present the entire life of the component. In general, one could use the vertex closest to the origin whenever the previous representative is deleted.

[3] Presumably because of hierarchical content organization and storage in likewise filesystems. An alternative considered briefly was therefore to use horizontal layers defined by subdirectories.

Brandes et al., *Dynamic WWW Structures in 3D*, JGAA, 4(3) 183–191 (2000)187

potential	definition	effect
repulsion	for each pair of non-adjacent vertices $u, v \in V$ evaluate $$\frac{c_\delta^4}{d(p_u, p_v)^2}$$	spreads vertices evenly in space
distance	for each pair of adjacent vertices $u, v \in V$ evaluate $$\frac{c_\delta^4}{d(p_u, p_v)^2} + d(p_u, p_v)^2$$	causes edges to have length roughly equal to c_δ
repulsion	for each pair $v \in V$ and $(u, w) \in A$ evaluate $$\frac{c_\delta}{d(p_v; p_u, p_w)^2}$$	prevents edges going through vertices
attraction	for a representative $v \in V$ of each component evaluate $$d(p_v, (0, 0, 0))^2$$	prevents components from drifting apart by tying them to the origin
rotation	for each $(u, v) \in A$ evaluate $$c_\varrho \cdot c_\delta^3 \cdot \left(\frac{\arccos \frac{z_u - z_v}{d(p_u, p_v)}}{\frac{\pi}{2}} \right)^2$$	favors edges pointing downward

Table 1: Building blocks of a static layout objective function for a directed graph $G = V, A)$, where $p_v = (x_v, y_v, z_v)$ is a point in space, $d(p_u, p_v)$ denotes the Euclidean distance between positions p_u and p_v, and $d(p_v; p_u, p_w)$ denotes the smallest Euclidean distance between p_v and any point on the straight line connecting p_u and p_w. We used $c_\delta = 60$ and $c_\varrho = 0.25$

The first layout of each graph $G^{(t)}$, $p^{(t,0)}$, was computed by minimizing the static layout objective function, subject to $p_v^{(t,0)} = p_v^{(t-1,0)}$ for those vertices v that are present in both $G^{(t)}$ and $G^{(t-1)}$. That is, the at most two newly introduced vertices are placed optimally in the unchanged previous layout.

It is argued in [2] that dependencies between layouts should be modeled by adding a stability term to the objective function. Good choices for stability terms are difference metrics [4] capturing the notion of change between two layouts. A natural candidate for straight-line embeddings is the sum of Euclidean distances between positions of corresponding vertices. To compute $p^{(t,i)}$, $i = 1, 2$, we hence augmented the static layout model with stability potentials

$$\sigma_i \cdot d \left(p_v^{(t,i)}, p_v^{(t,i-1)} \right)^2$$

for all vertices v, attracting vertices to their previous position. Constant σ_i controls the strength of the attraction and thus relative importance of stability. We used $\sigma_1 = 1.5$ and $\sigma_2 = 0.75$. For obvious reasons, this notion of stability is called *anchoring,* and it has an interesting probabilistic interpretation [2].

Brandes et al., *Dynamic WWW Structures in 3D*, JGAA, 4(3) 183–191 (2000)188

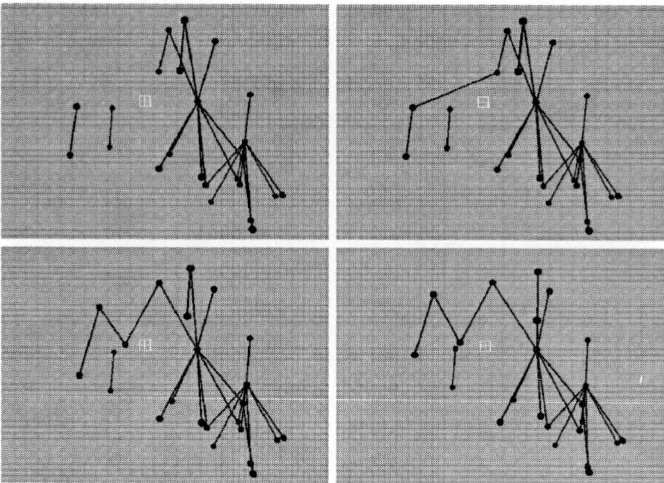

Figure 3: Insertion of an edge (directed left to right)

The resulting minimization problems were solved approximately by simulated annealing, an iterative method for combinatorial optimization, again because we already had the implementation available. Though simulated annealing is notoriously slow, its running time was negligible in this context (see below). Quality of the 195 layouts generated was controlled using simple VRML[4] output (ball-and-stick representations, see Figure 3).

3 Viewpoints

There were no truly three-dimensional media available to display the animation. We therefore had to select two-dimensional projections of the graph, i.e. positions of an observer in space and his or her directions of view. For an overview of viewpoint selection we refer to [15]. Here, we used a simple heuristic to find a decent viewpoint under perspective projection once after each modification of the graph. In the course of the animation we let the observer follow a spline through these points while focusing on the origin.

Good viewpoints are those that are not bad. A viewpoint is said to be bad, if it yields a projection in which an item hides another one or false incidences are suggested. A projection is said to yield a vertex-vertex, a vertex-edge, or an edge-edge *occlusion*, if two vertices, a vertex and an edge, or a two edges coincide in the projection, but not in the three-dimensional layout. The random viewpoint of Figure 3 is bad.

For a given distance from the origin, we can imagine the observer to sit on a sphere centered at the origin. Recall that we let the observer focus on the origin at all times. A point on the sphere yields an occlusion, if it is collinear with either two vertices, a vertex and an edge, or two edges. Because the points on

---

[4]Virtual Reality Modeling Language, see `http://www.web3d.org/vrml/`. Since we had to use VRML 1.0, the effectiveness of the dynamic model was checked using a script that remotely loaded the output files into the VRML browser, one at a time.

Brandes et al., *Dynamic WWW Structures in 3D*, JGAA, 4(3) 183–191 (2000)189

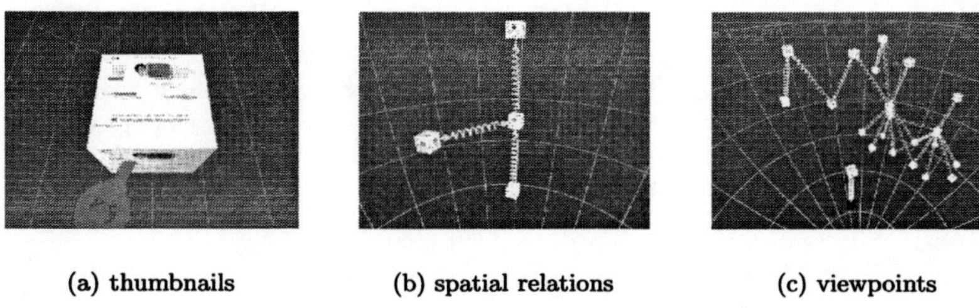

| (a) thumbnails | (b) spatial relations | (c) viewpoints |

Figure 4: Three out of 3250 frames

the sphere that are collinear with any two points on a pair of non-coplanar edges cover large areas on the sphere, we ignore edge-edge occlusions altogether. In general we cannot expect to find any point on the sphere not yielding an edge-edge occlusion, anyway.

The *goodness* of a viewpoint is defined to be its minimum great-circle distance from any occlusion. This measure of goodness is called *rotational separation* in [15]. For the animation we approximated the best viewpoint under rotational separation within great-circle distance at most $\pi/2$ from the previous one to suppress large rotations, by placing an spherical grid on the observer's semi-sphere and exhaustively searching for the best grid point.

Finally, we fit the graph onto the screen by adjusting the radius of the observer sphere such that all vertices are just inside the finite projection area.

4 Rendering

As a small hint on the actual content of the graph, vertices were represented by boxes with corresponding WWW pages texture-mapped onto their sides (thumbnail representations).[5] Figure 4(a) shows the initial frame in which positions have been modified manually to yield a more attractive get-go. Springs are used to render edges, just because they are reminiscent of the physical analogy underlying the layout. It is a by-product that their coils also visualize the stress on an edge. It can be seen from Figure 4(b) that referred pages are hooked to their referring pages. In the case of bidirectional edges, just one spring is shown but both ends are hooked. Edges added or deleted in a modification are highlighted during that event. A spherical grid surrounding the graph helps to distinguish between vertex movements and movements of the viewpoint.

Spline break points and rendering information are specified in a POV-Ray[6] scene description file. Using the ray tracer, we generated 50 still frames per modification, which were subsequently converted into an animation with 25 frames per second (among others, in MPEG-1[7] format).

---

[5]One page could not be retrieved and is therefore represented all white. Towards the end of the animation, two boxes seem to not have thumbnails, but in fact the corresponding WWW pages have a black background.

[6]Persistence of Vison Ray Tracer, see http://www.povray.org/.

[7]Moving Picture Experts Group, see http://drogo.cselt.stet.it/mpeg/

Brandes et al., *Dynamic WWW Structures in 3D*, JGAA, 4(3) 183–191 (2000)190

5 Conclusion

The final animation can be retrieved from the accompanying WWW page.[8] Compared to the time it took to render all 3250 frames (one and a half nights on half a dozen workstations), layout and viewpoint selection times were negligible (35 minutes on a SUN Ultra-1 workstation, 80% for viewpoints). We believe that both layout and viewpoint selection can be improved to interactive speed. Since the layout objective function has no discrete terms, standard force-directed methods may be applied. Moreover, very few iterations are needed, since the initial layouts are by definition close to a locally optimal solution. Using newer features not present in VRML 1.0 and a sufficiently fast browser, we therefore do think that our approach can be customized for interactive applications. Given the convincing result of our off-line experiments, we definitely encourage research in this direction. The main open problem will be to deal with directed cycles, since, depending on the rest of the graph, they either cause several layers to be lifted onto one, or destroy the otherwise emergent layering altogether.

In working on this project, we found that the contest is a unique opportunity to experiment with graph drawing techniques and to spread knowledge of them. Three of the authors were students at that time, but certainly all of us have learned more about graph drawing and short term project management. The task nicely split into almost independent subtasks (layout, viewpoints, rendering) and most of the time pragmatic solutions would be preferred over quality or running time improvements. The restricted time span, gradual (visually comprehensible!) improvement of intermediate results, and the competitive nature of a contest combined to keep motivation high. It goes without saying that the party we were able to throw with the prize money (featuring a continuously running animation on a large canvas) was considered a rewarding finale.

Acknowledgments

We wish to thank the referees for greatly improving the presentation. Special thanks to the editors of this issue for giving us the opportunity to write this non-standard kind of paper.

References

[1] U. Brandes. *Layout of Graph Visualizations*. PhD thesis, University of Konstanz, 1999. See `http://www.ub.uni-konstanz/kops/volltexte/1999/255/`.

[2] U. Brandes and D. Wagner. A Bayesian paradigm for dynamic graph layout. G. Di Battista, editor. *Proc. 5th Intl. Symp. Graph Drawing (GD '97)*, *Lecture Notes in Computer Science* vol. 1353, pages 236–247. Springer, 1997.

[3] U. Brandes and D. Wagner. Using graph layout to visualize train interconnection data. In this issue.

---

[8]The animation submitted to the Graph Drawing Contest was rendered at 640 × 480 pixels. Due to its immense file size, a reproduction with 320 × 240 pixels is provided instead.

[4] S. Bridgeman and R. Tamassia. Difference metrics for interactive orthogonal graph drawing algorithms. In this issue.

[5] I. F. Cruz and J. P. Twarog. 3D graph drawing with simulated annealing. F. J. Brandenburg, editor. *Proc. 3rd Intl. Symp. Graph Drawing (GD '95)*, *Lecture Notes in Computer Science* vol. 1027, pages 162–165. Springer, 1996.

[6] G. Di Battista and R. Tamassia. Algorithms for plane representations of acyclic digraphs. *Theoretical Computer Science*, 61:175–198, 1988.

[7] P. Eades and J. Marks. Graph drawing contest report. R. Tamassia and I. G. Tollis, editors. *Proc. 2nd Intl. Symp. Graph Drawing (GD '94)*, *Lecture Notes in Computer Science* vol. 894, pages 143–146. Springer, 1995.

[8] P. Eades and J. Marks. Graph-drawing contest report. F. J. Brandenburg, editor. *Proc. 3rd Intl. Symp. Graph Drawing (GD '95)*, *Lecture Notes in Computer Science* vol. 1027, pages 224–233. Springer, 1996.

[9] P. Eades, J. Marks, P. Mutzel, and S. C. North. Graph drawing contest report. S. H. Whitesides, editor. *Proc. 6th Intl. Symp. Graph Drawing (GD '98)*, *Lecture Notes in Computer Science* vol. 1547, pages 423–435. Springer, 1998.

[10] P. Eades, J. Marks, and S. C. North. Graph-drawing contest report. In S. C. North, editor, *Proc. 4th Intl. Symp. on Graph Drawing (GD '96)*, *Lecture Notes in Computer Science* vol. 1190, pages 129–138. Springer, 1996.

[11] P. Eades, J. Marks, and S. C. North. Graph-drawing contest report. G. Di Battista, editor. *Proc. 5th Intl. Symp. Graph Drawing (GD '97)*, *Lecture Notes in Computer Science* vol. 1353, pages 438–445. Springer, 1997.

[12] K. Mehlhorn and S. Näher. *The* LEDA *Platform of Combinatorial and Geometric Computing*. Cambridge University Press, 1999. Project home page http://www.mpi-sb.mpg.de/LEDA/.

[13] K. Sugiyama and K. Misue. Graph drawing by the magnetic spring model. *Journal on Visual Languages and Computing*, 6(3):217–231, 1995.

[14] K. Sugiyama, S. Tagawa, and M. Toda. Methods for visual understanding of hierarchical system structures. *IEEE Transactions on Systems, Man and Cybernetics*, 11(2):109–125, 1981.

[15] R. Webber. *Finding the Best Viewpoint for Three-Dimensional Graph Drawings*. PhD thesis, University of Newcastle, 1998. See http://www.cs.mu.oz.au/~rwebber/research/thesis/.

Volume 4:4 (2000)

Journal of Graph Algorithms and Applications

http://www.cs.brown.edu/publications/jgaa/

vol. 4, no. 4, pp. 1–26 (2000)

Clustering in Trees: Optimizing Cluster Sizes and Number of Subtrees

Susanne E. Hambrusch   *Chuan-Ming Liu*

Department of Computer Sciences
Purdue University
West Lafayette, IN 47907, USA
http://www.cs.purdue.edu
seh@cs.purdue.edu   liucm@cs.purdue.edu

Hyeong-Seok Lim

Chonnam National University
Kwangju, 500-757, Korea
hslim@chonnam.chonnam.ac.kr

Abstract

This paper considers partitioning the vertices of an n-vertex tree into p disjoint sets $C_1, C_2, \ldots, C_p$, called clusters so that the number of vertices in a cluster and the number of subtrees in a cluster are minimized. For this NP-hard problem we present greedy heuristics which differ in (i) how subtrees are identified (using either a best-fit, good-fit, or first-fit selection criteria), (ii) whether clusters are filled one at a time or simultaneously, and (iii) how much cluster sizes can differ from the ideal size of c vertices per cluster, $n = cp$. The last criteria is controlled by a constant α, $0 \leq \alpha < 1$, such that cluster C_i satisfies $(1 - \frac{\alpha}{2})c \leq |C_i| \leq c(1 + \alpha)$, $1 \leq i \leq p$. For algorithms resulting from combinations of these criteria we develop worst-case bounds on the number of subtrees in a cluster in terms of c, α, and the maximum degree of a vertex. We present experimental results which give insight into how parameters c, α, and the maximum degree of a vertex impact the number of subtrees and the cluster sizes.

Communicated by G. Liotta: submitted November 1999, revised August 2000.

---

1. Hambrusch's research supported in part by the National Science Foundation under Grant 9988339-CCR.

2. Lim's research supported in part by Korea Science and Engineering Foundation under Contract No. 98-0102-07-01-3.

S. E. Hambrusch et al., *Clustering in Trees*, JGAA, 4(4) 1–26 (2000)   2

1 Introduction

Tree clustering partitions the vertices of a given tree into disjoint sets, called clusters, subject to optimizing one or more objective functions. Tree clustering arises in parallel and distributed computing environments and external memory systems. For a tree representing an external search structure, the created clusters correspond to the blocks. Clusters should minimize the number of blocks as well as the access to external storage devices [1, 4, 7, 12]. For a tree representing data flow and communication requirements in a parallel and distributed environment, partitioning the vertices corresponds to assigning tasks to processors. The goal is to balance processor loads and to minimize communication between processors [6, 10, 11]. Not surprisingly, the combinatorial nature of clustering problems makes finding optimal solutions computationally intractable for most realistic situations [4, 5, 7, 14].

Let T be a tree with $n = cp$ vertices, $c \geq 2$. We assume that edges and vertices have no associated weights. A clustering of T partitions the vertices into p sets, $C_1, C_2, \ldots, C_p$. We consider generating clusters when the number of vertices assigned to different clusters should be as equal as possible and the number of subtrees assigned to every cluster should be minimized. While minimizing these two cost measures simultaneously captures desirable features for the above applications, it is an NP-hard problem.

An ideal load is achieved when every cluster contains c vertices. This corresponds to every block containing c data items and every processor assigned c tasks, respectively. Achieving an ideal load is straightforward in the absence of weights[1]. Our second cost measure is the number of subtrees in a cluster. For parallel and distributed applications, minimizing the number of subtrees enhances locality and decreases communication. When generating blocks for external tree structures, load and blocknumber are often optimized [4, 8, 12, 13]. The blocknumber measures the number of blocks needed during a search from the root to a leaf in the tree. Minimizing the blocknumber and achieving ideal load is NP-hard [7]. Existing heuristics first assign to every block a single subtree and then achieve a better load by partitioning selected subtrees [7, 8, 13]. This approach can assign many subtrees to a block and result in high I/O. Our approach is to minimize the number of subtrees and the load simultaneously. We refer to [9] for a more detailed discussion on the relationship between the blocknumber and the number of subtrees.

Achieving an ideal load and minimizing the maximum number of subtrees in the clusters is NP-hard [9]. We note that deciding whether there exists a clustering having an ideal load and every cluster containing one subtree can be done in linear time. However, deciding whether there exist clusters of size c with every cluster containing at most 3 subtrees is already NP-complete. An ideal load is desirable, but generating clusters of size of c is not always necessary.

In this paper we introduce the concept of α-clustering to capture such a tolerated slackness in cluster sizes. Given a tree T with $n = cp$ vertices and

---

[1]The existence of weights on the vertices results in an NP-hard problem, as clustering becomes a bin-packing like problem.

S. E. Hambrusch et al., *Clustering in Trees*, JGAA, 4(4) 1–26 (2000)   3

a parameter α, $0 \leq \alpha < 1$, an α-clustering generates p clusters so that every cluster C_i satisfies $(1 - \frac{\alpha}{2})c \leq |C_i| \leq c(1+\alpha)$, $1 \leq i \leq p$. For $\alpha = 0$, we generate an exact clustering; i.e., $|C_i| = c$. The clustering algorithms presented are greedy heuristics. They differ in (i) the identification of subtrees (i.e., whether a best-fit, good-fit, and first-fit selection criteria is used), (ii) the order in which clusters are filled (i.e., whether clusters are filled one at a time or simultaneously), and (iii) different values of α which control how much cluster sizes are allowed to differ from the ideal size of c vertices per cluster. Our work provides insight into how cluster sizes and number of subtrees in a cluster are impacted by the value of α, the maximum degree d in the tree, the relationship between c and d, the subtree selection method, as well as the order in which clusters are filled. We develop worst-case upper bounds on the number of subtrees and the cluster sizes and provide experimental results supporting our claims.

The paper is organized as follows. In Section 2 we describe the ingredients of our clustering algorithms and prove that the cluster forming approaches generate cluster sizes in the required range. Section 3 presents the two single fill clustering algorithms along with asymptotic bounds on the number of subtrees in a cluster. Section 4 discusses the simultaneous fill algorithms. The experimental performance of the algorithms is discussed in Section 5.

2   Overview of the Clustering Algorithms

In this section we discuss the framework underlying our α-clustering algorithms. Figure 1 gives time and number of subtrees bounds for four α-clustering algorithms presented in this paper. Throughout, d is the maximum degree of a vertex in T.

The quantities $\log_{\frac{d-2}{d-1}} \frac{\alpha}{2}$ and $\log_{\frac{d-1}{d}} \frac{\alpha}{4}$ should be read as $\min\{c, \log_{\frac{d-2}{d-1}} \frac{\alpha}{2}\}$ and $\min\{c, \log_{\frac{d-1}{d}} \frac{\alpha}{4}\}$, respectively. Note that when $\alpha = 0$, the stated minima generate c. Figure 2 shows these two quantities (independent of c) for the range of degrees considered in this paper. Observe that the upper bounds can exceed the trivial bound of at most c vertices in a cluster.

2.1   Single versus Simultaneous Cluster Forming

Our algorithms assign subtrees to clusters in either a single fill or a simultaneous fill mode. Algorithms based on the *single fill mode* determine the subtrees for cluster C_i before generating cluster C_{i+1}. Algorithms based on a *simultaneous fill mode* assign subtrees to clusters without this restriction. Symultaneous fill algorithms may assign one subtree to each cluster in one iteration or use current cluster sizes to decide which cluster receives the next subtree. When $\alpha > 0$, single fill as well as simultaneous fill need to ensure that cluster sizes are within the required bounds. For example, if too many clusters are underfull (i.e., have $|C_i| < c$), the remaining vertices of T may force a cluster to exceed the upper bound. Figure 3 gives the outline of a generic single fill algorithm. The quantity $remain_i$ represents the total number of vertices to be made up due to underfull

S. E. Hambrusch et al., *Clustering in Trees*, JGAA, 4(4) 1–26 (2000) 4

Algorithm	Time	Maximum number of subtrees
SingFill-BF	$\Theta(np)$	$\lceil \log_{\frac{d-2}{d-1}} \frac{\alpha}{2} \rceil$
SingFill-FF	$\Theta(n)$	$\min\{c, d * \lceil \frac{\log c}{\log d} \rceil\}$
SimulFill-BF	$O(np \log_{\frac{d-1}{d}} \frac{\alpha}{4})$	$\lceil \log_{\frac{d-1}{d}} \frac{\alpha}{4} \rceil$
SimulFill-GF	$O(n \log_{\frac{d-1}{d}} \frac{\alpha}{4})$	$\lceil \log_{\frac{d-1}{d}} \frac{\alpha}{4} \rceil$

Figure 1: Bounds achieved by our clustering algorithms.

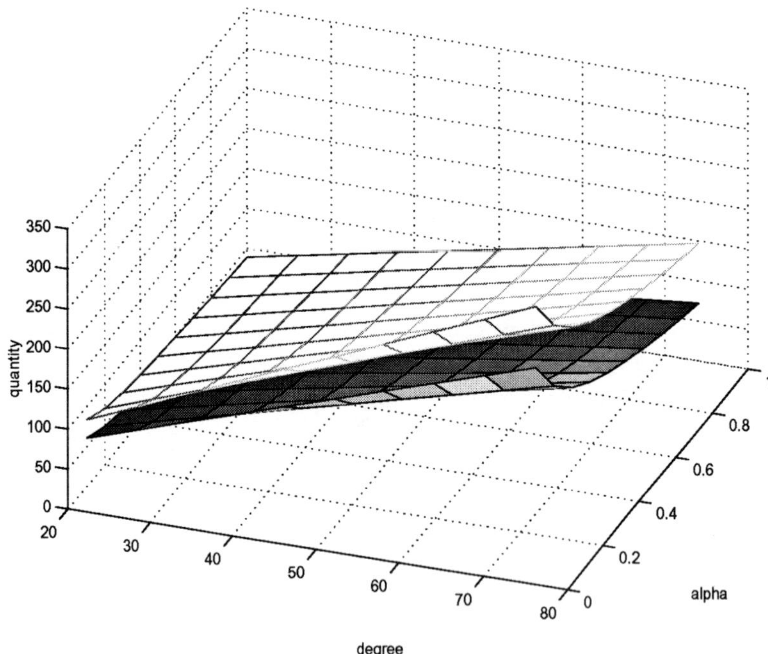

Figure 2: Comparing the quantities of $\log_{\frac{d-2}{d-1}} \frac{\alpha}{2}$ (filled grid) and $\log_{\frac{d-1}{d}} \frac{\alpha}{4}$ (non-filled grid) for different degrees.

S. E. Hambrusch et al., *Clustering in Trees*, JGAA, 4(4) 1–26 (2000)   5

clusters. Lemma 1 shows that $c + remain_i$ never exceeds the upper bound on the cluster size.

Algorithm Generic-SingFill
Input: tree $T = (V, E)$, $n = cp$, and parameter α
Output: $C_1, C_2, \ldots, C_p$ representing the p clusters of an α-clustering

1. Initialize each cluster as an empty set.

2. $remain_0 = 0$

3. **for** $i = 1$ **to** $p - 1$ **do**
   $target_i = c + remain_{i-1}$
   $remain_i = target_i$
   while $(|C_i| < (1 - \frac{\alpha}{2}) \times target_i)$ **do**

   (a) Determine a subtree $T' = (V', E')$ with $|V'| \leq remain_i$ using one of the subtree finding methods

   (b) Update: $T = T - T'$; $C_i = C_i \cup V'$
       $remain_i = remain_i - |V'|$

   endwhile
   endfor

4. $C_p = V$

Figure 3: Description of Algorithm Generic-SingFill.

The different ways of determining subtrees are described in Section 2.2. The following lemma shows that Algorithm Generic-SingFill generates cluster sizes which fall within the range needed for the α-clustering. The number of subtrees in a cluster depends on how subtrees are selected and bounds will be given when individual algorithms are described.

Lemma 1 *Cluster C_i generated by Algorithm Generic-SingFill satisfies* $(1 - \frac{\alpha}{2})c \leq |C_i| \leq c(1 + \alpha)$, $1 \leq i \leq p$.

Proof: Consider first the $p - 1$ clusters generated within the while-loop. Since $target_i \geq c$ and the algorithm terminates with $|C_i| \geq (1 - \frac{\alpha}{2}) \times target_i$, $1 \leq i \leq p - 1$, the lower bound on the cluster size is satisfied for the first $p - 1$ clusters. The upper bound of $|C_i| \leq c(1 + \alpha)$ is shown as follows. At the end of the first iteration we have $remain_1 \leq \frac{\alpha}{2}c$. Hence, $target_2 \leq c + \frac{\alpha}{2}c$ and $remain_2 \leq \frac{\alpha}{2}c + (\frac{\alpha}{2})^2 c$ at the end of the second iteration. In general,

$$target_i \leq c + remain_{i-1} \text{ and}$$

$$remain_i \leq \frac{\alpha}{2} \times target_i.$$

S. E. Hambrusch et al., *Clustering in Trees*, JGAA, 4(4) 1–26 (2000)   6

Hence, $target_i \leq c + \frac{\alpha}{2} \times target_{i-1}$ and

$$target_i \leq c \sum_{k=0}^{i-1} (\frac{\alpha}{2})^k < \frac{2}{2-\alpha} \times c.$$

For $0 < \alpha < 1$, we have $\frac{2}{2-\alpha} < 1 + \alpha$. Thus, $target_i < c(1 + \alpha)$ and the upper bound on the cluster size holds for the first $p - 1$ clusters.

Cluster C_p is assigned the remaining vertices of tree T. Since $\sum_{i=1}^{p-1} |C_i| + remain_{p-1} = (p - 1)c$, we have $|C_p| = c + remain_{p-1}$. Since $remain_{p-1} \leq \frac{\alpha}{2} \times target_{p-1}$ and $target_{p-1} < \frac{2c}{2-\alpha}$, we have $remain_{p-1} \leq \frac{\alpha}{2-\alpha} \times c$. Hence, $c \leq |C_p| \leq c + \frac{\alpha}{2-\alpha} \times c \leq c(1 + \alpha)$. $\square$

Algorithm Generic SimulFill
Input: tree $T = (V, E)$, $n = cp$, and parameter α
Output: $C_1, C_2, \ldots, C_p$ representing the p clusters of an α-clustering

Initialize $C_i = \emptyset$ and $remain_i = c$, $1 \leq i \leq p$.
PHASE 1: Generate p safe clusters.
while there exists a cluster which is not safe **do**
for $i = 1$ **to** p **do**
if cluster C_i is not safe **then**

1. Determine the next subtree $T' = (V', E')$ with $|V'| \leq remain_i$ using one of the subtree finding methods.

2. Update: $T = T - T'$; $C_i = C_i \cup V'$
   $remain_i = remain_i - |V'|$

endfor
endwhile

PHASE 2: Assign the remaining vertices of T.
Update $remain$-entries: $remain_i = \alpha c + remain_i$, $1 \leq i \leq p$.
while tree T is not empty **do**
for $i = 1$ **to** p **do**
if tree T not empty and cluster C_i not full **then**

1. Determine the next subtree $T' = (V', E')$ with $|V'| \leq remain_i$ using one of the subtree finding methods.

2. Update: $T = T - T'$; $C_i = C_i \cup V'$
   $remain_i = remain_i - |V'|$

endfor
endwhile

Figure 4: Description of Algorithm Generic-SimulFill.

S. E. Hambrusch et al., *Clustering in Trees*, JGAA, 4(4) 1–26 (2000) 7

We now turn to the simultaneous filling of clusters. As for single fill, we need to ensure that deficits in cluster sizes can be made up by other clusters without exceeding the upper bound of $(1+\alpha)c$. Our clustering algorithms based on the simultaneous fill mode create the clusters in two phases, as evident from the outline given in Figure 4. We say cluster C_i is *safe* if $(1 - \frac{\alpha}{2})c \leq |C_i| \leq c$. In Phase 1, we generate p safe clusters. The number of iterations executed in Phase 1 equals the maximum number of subtrees assigned to a safe cluster. After Phase 1, every cluster size lies within the required range. However, not all vertices of the tree may have been assigned to clusters yet.

Phase 2 assigns the remaining vertices of tree T to the safe clusters. We say cluster C_i is *full* if $|C_i| \geq (1 + \frac{\alpha}{2})c$. Once a cluster becomes full, no more assignments are made to it. The while-loop is executed until all vertices of T have been assigned to a cluster. A cluster may thus not receive any additional vertices in Phase 2. In particular, when $\alpha = 0$, all vertices of T are assigned to clusters in Phase 1.

From the way Algorithm Generic-SimulFill forms clusters it is clear that the number of vertices assigned to a cluster lies in the required range determined by α. The number of subtrees assigned to a cluster depends on how subtrees are identified and bounds on the number of subtrees are developed in Section 4.

We conclude this section with a brief comparison of the two cluster filling modes. The advantage of the single-fill mode is that at the time cluster C_i is filled, the final sizes of the first $i - 1$ clusters are known. A single-fill algorithm fills cluster C_i using α and information on how underfull previous clusters are. A single-fill algorithm tries to make up an earlier created deficit as soon as possible. The advantage of the simultaneous-fill mode is that during its first few iterations, every cluster has a chance to find subtrees in a large tree. This can lead to Phase 1 generating safe clusters consisting of few trees in each cluster. As will be discussed in Section 5.2, these characteristics show up in the experimental results. At the same time, corresponding disadvantages show up as well. For example, the final clusters created by a single-fill algorithm select subtrees from a relatively small tree. Since the number of subtree choices is now limited, these final clusters can end up being assigned a large number of subtrees.

2.2 Identifying Subtrees

In this section we sketch the three methods used by the clustering algorithms for identifying subtrees. Assume we are to determine the next subtree for cluster C_i. Let $remain_i$ be the maximum number of vertices that can still be assigned to C_i (without exceeding the upper bound on the cluster size of C_i).

Suppose we remove an edge $e = (u, v)$ in T. Then, T is divided into two subtrees. Let $T_{e,u} = (V_{e,u}, E_{e,u})$ (resp. $T_{e,v} = (V_{e,v}, E_{e,v})$) be the subtree containing vertex u (resp. v), but not edge e. Recall that d is the maximum degree of a vertex. The subtree $T' = (V', E')$ of T is found using one of the following:

S. E. Hambrusch et al., *Clustering in Trees*, JGAA, 4(4) 1–26 (2000)   8

- *Best-Fit:* Determine an edge $e = (u, v)$ and vertex u such that $|V_{e,u}| \leq remain_i$ and $|V_{e,u}|$ is a maximum. Set $T' = T_{e,u}$.

- *Good-Fit:* Choose the first tree T' encountered in the traversal of T with $\lceil remain_i/d \rceil \leq |V'| \leq remain_i$.

- *First-Fit:* Choose the first tree T' encountered in the traversal of T with $|V'| \leq remain_i$.

The different tree selection methods result in algorithms with different running times. Clustering algorithms using best-fit selection traverse, in the worst case, the entire tree T to find one subtree T'. For clustering algorithms based on good-fit and best-fit the running time depends on whether single-fill or simultaneous-fill is used. For single-fill, our implementations perform one tree traversal when forming one cluster. For simultaneous-fill, one traversal of the tree identifies p subtrees, one for every cluster. We refer to Figure 1 for running times and upper bounds on the number of subtrees in a cluster. A major focus of our experimental work is whether the use of the best-fit subtree selection results in significantly better clusters and thus justifies the increase in time.

3   Single Fill Clustering

We now present two single clustering algorithms, Algorithm SingFill-BF based on best-fit and Algorithm SingFill-FF based on first-fit subtree selection. Algorithm SingFill-BF creates one cluster by performing one traversal of the tree, and thus achieves a $\Theta(np)$ running time. Algorithm SingFill-FF determines all clusters during a single traversal of the tree, and thus has an $\Theta(n)$ running time. We do not consider good-fit subtree selection for single fill clusterings. Good-fit subtree selection can be implemented to achieve $O(np)$ time, as does best-fit (which determines better fitting subtrees). The good-fit strategy is used in the simultaneous fill algorithms described in Section 4.

3.1   Algorithm SingFill-BF

Algorithm SingFill-BF corresponds to the generic single fill algorithm described in Figure 3 with the best-fit subtree selection. We describe an $O(np)$ time implementation and then show that the number of subtrees in a cluster is bounded by $\min\{c, \lceil \log_{\frac{d-2}{d-1}} \frac{\alpha}{2} \rceil\}$.

A straightforward $O(np \log_{\frac{d-2}{d-1}} \frac{\alpha}{2})$ time bound is obtained by searching the current tree for the next subtree giving the best fit. The implementation described below determines the subtrees for one cluster in $O(n)$ time by using a queue to efficiently locate the subtrees giving the best fit.

Consider the beginning of the i-th iteration. Tree T now corresponds to the original tree from which the vertices assigned to clusters $C_1, \ldots, C_{i-1}$ have been removed. Before entering the while-loop of iteration i, we determine for all edges $e = (u, v)$ in tree T the quantities $|V_{e,u}|$ and $|V_{e,v}|$. A priority queue

S. E. Hambrusch et al., *Clustering in Trees*, JGAA, 4(4) 1–26 (2000)   9

Q in the form of an array of size $target_i$ is used to represent selected subtree entries. Subtree $T_{e,u} = (V_{e,u}, E_{e,u})$ is an entry in queue Q at index $|V_{e,u}|$ if the following two conditions hold:

1. $|V_{e,u}| \leq remain_i$ and

2. for every edge $e' = (u', v)$ with $u' \neq u$ we have $|V_{e',v}| > remain_i$.

Condition (1) selects for queue Q only those subtrees that "fit" (i.e., they do not exceed the remaining capacity). Condition (2) selects, among all subtrees that fit, the ones that are as large as possible. Using standard tree computations and traversals, queue Q can be set up in $O(n)$ time.

Step 3(a) of SingFill-BF determines the next best fitting subtree by scanning array Q starting at position $remain_i$. The subtree is found by scanning left, looking for the first non-empty entry in Q. Let $T' = T_{e,u}$ be the subtree chosen. Before $remain_i$ is decreased in Step 3(b), we update array Q. The entry representing subtree $T_{e,u}$ is deleted. Before the next subtree is selected, we "break up" subtrees which are now too large while satisfying conditions (1) and (2). Entries corresponding to subtrees larger than $remain_i - |V_{e,u}|$ are no longer needed. To record appropriate subtrees of these trees, we proceed as follows. Scan array Q from the position which contained $T_{e,u}$ to the left to position $remain_i - |V_{e,u}|$. Let $T_{b,x}$ be a subtree encountered during this scan, $b = (x, y)$. The entry corresponding to $T_{b,x}$ is deleted and every vertex adjacent to x (excluding y) is considered. Let w be such an adjacent neighbor. If $|V_{(w,x),w}| \leq remain_i - |V_{e,u}|$, condition (1) is satisfied. Observe that we do not need to check whether condition 2 is satisfied: since it was satisfied for tree $T_{e,u}$, it is also satisfied for $T_{(w,x),w}$. We thus insert $T_{(w,x),w}$ into Q. On the other hand, if condition (1) does not hold for subtree $T_{(w,x),w}$ (i.e., $|V_{(w,x),w}| > remain_i - |V_{e,u}|$), the vertices adjacent to w (excluding x) are considered for insertion. This process continues until subtrees of small enough size are found. During the entire while-loop of Step 3, an edge is considered at most a constant number of times. Thus the maintenance of array Q costs $O(n)$ time. The $O(np)$ overall time follows.

The correctness of the above approach relies on the subtrees represented in queue Q being disjoint. The existence of disjoint subtrees when creating clusters $C_1, \ldots, C_{p-2}$ is guaranteed since we have $n - |V_{e,u}| > 2c$ for every subtree in Q. For iteration $p - 1$, subtrees represented in Q may not be disjoint. In our implementation, iteration $p - 1$ does thus not use the queue, but it explicitly traverses the remaining tree for finding best fitting, disjoint subtrees. This does not impact the $O(np)$ overall time.

We now turn to bounding the number of subtrees in a cluster. The first lemma relates the size of subtree T' to $remain_i$.

Lemma 2 *Assume edge $e = (u, v)$ and vertex u are selected in Step 3(a) of the i-th iteration of Algorithm SingFill. Then, $|V_{e,u}| \geq \frac{remain_i}{d-1}$.*

Proof: Assume this is not true and let $T_{e,u}$ be a best fitting subtree satisfying $|V_{e,u}| < \frac{remain_i}{d-1}$. For any edge $e' = (u', v)$ incident to vertex v, we either have

S. E. Hambrusch et al., *Clustering in Trees*, JGAA, 4(4) 1–26 (2000)   10

- $|V_{e',u'}| \leq |V_{e,u}| < \frac{remain_i}{d-1}$ (i.e., subtree $T_{e',u'}$ could be chosen, but does not give a better fit), or

- $|V_{e',u'}| > remain_i$ (i.e., subtree $T_{e',u'}$ is too large).

There must exist at least one vertex u' with $|V_{e',u'}| > remain_i$. (To be precise, there must exist at least two such vertices.) Otherwise $|V_{e',u'}| < \frac{remain_i}{d-1}$ would hold for every vertex u' adjacent to v and thus for subtree $T_{e',v}$ we would have $|V_{e',v}| < \frac{remain_i}{d-1} \times (d-1) < remain_i$. This would contradict that $T_{e,u}$ is a best fitting subtree.

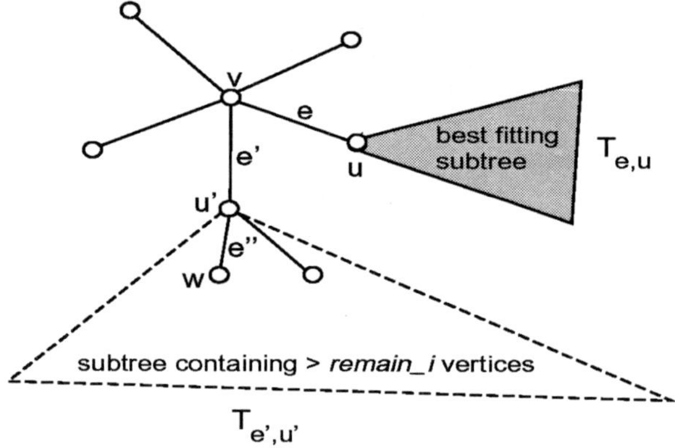

Figure 5: Illustrating the position of edges e, e', and e''.

We arrive at a contradiction for the assumption $|V_{e,u}| < \frac{remain_i}{d-1}$ by considering a subtree in $T_{e',u'}$ with $|V_{e',u'}| > remain_i$. Vertex u' is incident to at least one edge $e'' = (u', w)$ with $|V_{e'',w}| \geq \frac{remain_i}{d-1}$. This situation is illustrated in Figure 5. The case $|V_{e'',w}| \leq remain_i$ would imply that the subtree rooted at w is a better fit than $T_{e,u}$ and give a contradiction. If $|V_{e'',w}| \geq remain_i$, we apply the same argument using edge e'' in the role of e'. A subsequent step leads to a contradiction. Hence, $|V_{e,u}| \geq \frac{remain_i}{d-1}$. $\qquad\square$

Lemma 3 *The number of subtrees assigned to a cluster by Algorithm SingFill-BF is at most* $\min\{c, \lceil \log_{\frac{d-2}{d-1}} \frac{\alpha}{2} \rceil\})$.

Proof: Let $t(i, j)$ be the minimum size of the subtree selected at the j-th step of the i-th iteration of the while-loop. We set $t(i, 0) = target_i$. From Lemma 2 it follows that $t(i, 1) = \frac{t(i,0)}{d-1}$ and $t(i, 2) = \frac{t(i,0) - t(i,1)}{d-1} = t(i, 0)\frac{d-2}{(d-1)^2}$. The j-th step of the while loop removes a subtree of size $t(i, j) = t(i, 0)\frac{(d-2)^{j-1}}{(d-1)^j}$. The total number of vertices in cluster C_i after m steps of the while loop is thus

$$t(i, 0) \sum_{j=1}^{m} \frac{(d-2)^{(j-1)}}{(d-1)^j} = \left(1 - \left(\frac{d-2}{d-1}\right)^{m-1}\right) * target_i.$$

The while loop terminates when $(1 - (\frac{d-2}{d-1})^m) \times target_i > (1 - \frac{\alpha}{2}) \times target_i$. This implies that the number of subtrees assigned to cluster C_i is bounded by $\log_{\frac{d-2}{d-1}} \frac{\alpha}{2} = \frac{1 - \log \alpha}{\log(d-1) - \log(d-2)}$. Since no cluster contains more than c vertices, the claimed bound follows. $\square$

The following theorem summarizes our discussion:

Theorem 4 *Algorithm SingFill-BF determines an α-clustering for an n-vertex tree T in time $\Theta(np)$, $n = cp$. The number of subtrees assigned to a cluster is bounded by* $\min\{c, \lceil \log_{\frac{d-2}{d-1}} \frac{\alpha}{2} \rceil\})$.

3.2 Algorithm SingFill-FF

In this section we describe Algorithm SingFill-FF, a single fill clustering algorithm using first-fit subtree selection. We describe the algorithm for the case $\alpha = 0$. Its generalization to arbitrary values of α's uses *target* and *remain*-entries as described in Algorithm Generic-SingFill in Figure 3.

Algorithm SingFill-FF uses the results of a weighted postorder numbering on a rooted version of tree T to form the clusters. Let r be an arbitrary vertex of T chosen as the root. With T rooted towards r, we determine the *weighted postorder number* of every vertex as follows. Let u be a vertex with children $v_1, v_2, \ldots, v_k$. The children are arranged by non-increasing sizes of subtrees; i.e., $|V_{(v_i,u),v_i}| \geq |V_{(v_{i+1},u),v_{i+1}}|$ for every i, $1 \leq i < k$. With the children ordered this way, perform a postorder traversal of T. Let $post(u)$ be the postorder number assigned to vertex u. Then, vertex u belongs to cluster $C_{\lceil post(u)/c \rceil}$. Figure 6 shows clusters C_1 and C_2 for the sketched tree. Ordering the children of all vertices by size can be done in $O(n)$ time. One implementation uses the fact that subtree sizes are bounded by n and thus all sizes can be indexed into an array of size n, allowing an $O(n)$ time rearranging. The assignment of vertices to clusters based on the weighted postorder traversal number can thus be done in $O(n)$ time. In the remainder of this section we show that the number of subtrees in a cluster is bounded by $\min\{c, d * \lceil \frac{\log c}{\log d} \rceil\}$.

W.l.o.g. assume the formation of cluster C_i starts at vertex u and only vertices in the subtree rooted at u are in cluster C_i. If this is not the case, the vertices in C_i having smaller postorder numbers form one subtree. For illustration, consider vertex a in Figure 6. Cluster C_2 contains vertices in the subtree rooted at a and the vertices not in this subtree form one tree as indicated. We ignore this one subtree when counting subtrees. Let $v_1, v_2, \ldots, v_k$, $k \leq d$, be the children of u. Assume cluster C_i receives the subtrees rooted at $v_1, \ldots, v_{l_1-1}$ and some of the vertices in the subtree rooted at v_{l_1}, $l_1 \geq 2$. The number of vertices needed from the subtree rooted at v_{l_1} is at most c/l_1. If more vertices were needed, the use of the weighted postorder numbering (i.e., $|V_{(u,v_j),v_j}| \geq |V_{(u,v_{j+1}),v_{j+1}}|$ and $|V_{(u,v_j),v_j}| > c/l_1$, $1 \leq j \leq l_1 - 1$) would imply that C_i contains more than c vertices.

To show the claimed bound on the number of subtrees in C_i we first show that after the inclusion of $d - 1$ subtrees into cluster C_i, the cluster misses at most c/d vertices. In other words, the first $c - c/d$ vertices selected by

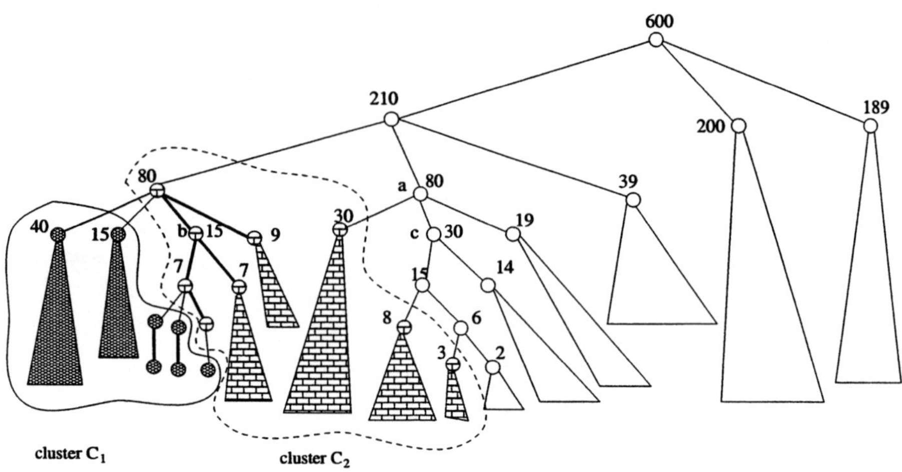

Figure 6: Forming exact clusters using weighted postorder numbers. The tree has $n = 600$, $c = 60$, $d = 10$; integers next to vertices represent the number of vertices in the subtree.

the algorithm induce at most $d - 1$ subtrees. Observe that "the first $c - c/d$ vertices" refers to the $c - c/d$ vertices in C_i and in the subtree rooted at u with the smallest postorder numbers. We then apply the same argument to the at most c/d remaining vertices. This results in at most $\min\{c, \lceil \frac{\log c}{\log d} \rceil\}$ iterations, each iteration contributing at most $d - 1$ subtrees.

The subtrees rooted at $v_1, \ldots, v_{l_1-1}$ represent $l_1 - 1$ subtrees in C_i. To avoid conflict in notation, rename $v_{l_1} = u_{l_1}$. The algorithm then continues including vertices from the subtree rooted at u_{l_1}. At vertex $u_{l_{j-1}}$, we include subtrees rooted at children of $u_{l_{j-1}}$ and identify at most one subtree rooted at child u_{l_j} which contains more vertices than needed. More specifically,

- u_{l_j}'s left siblings are roots of subtrees included into C_i and

- not all vertices in the subtree rooted u_{l_j} are needed for C_i.

Assume the process of including subtrees and identifying subtrees of size larger than needed considers vertices $u_{l_1}, u_{l_2}, \ldots, u_{l_t}$. See Figure 7 for an illustration. Observe that we assume $l_j \geq 2$. If for a vertex $u_{l_{j-1}}$ the subtree rooted at its leftmost child contains more vertices than needed, vertex $u_{l_{j-1}}$ does not appear in this enumeration. For example, for the tree shown in Figure 6, vertex a would appear in the enumeration, but vertex c would not.

As already stated, the maximum number of vertices needed for cluster C_i from the subtree rooted at u_{l_1} is $\frac{c}{l_1}$. Using the same argument, the number of vertices needed for cluster C_i from the subtree rooted at u_{l_j} is at most $\frac{c}{l_1 l_2 \ldots l_j}$. We stop the process of including subtrees into cluster C_i at vertex u_{l_j} when the actual number of vertices needed from the subtree rooted at u_{l_j} is smaller than

S. E. Hambrusch et al., *Clustering in Trees*, JGAA, 4(4) 1–26 (2000)   13

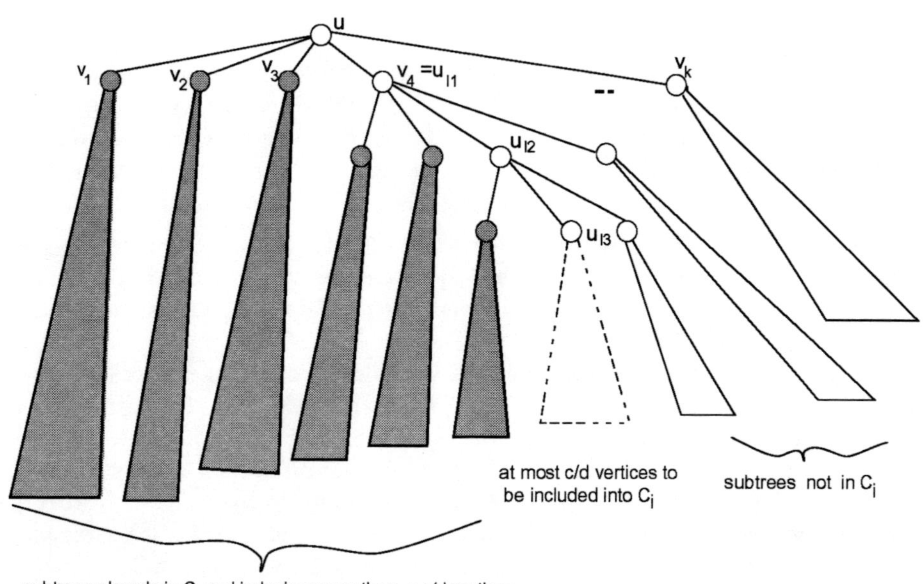

Figure 7: Illustrating vertices $u_{l_1}, u_{l_2}, \ldots, u_{l_t}$ and the subtrees in cluster C_i for $l_1 = 4$, $l_2 = 3$, $l_3 = 2$, and $t = 3$.

c/d for the first time. For cluster C_1 in the tree shown in Figure 6, the first iteration of this process stops at vertex b when C_1 already contains 55 vertices. Only 5 more vertices are needed and $5 < 6 = c/d$. It follows that

$$\frac{c}{l_1 l_2 \ldots l_t} \geq \frac{c}{d}$$

and $l_1 l_2 \ldots l_t \leq d$. Cluster C_i contains already $l_1 + l_2 + \ldots + l_t - t$ subtrees and we have $l_j \geq 2$, $1 \leq j \leq t$. The number of subtrees already in C_i (i.e., $\sum_{j=1}^{t}(l_j - 1)$) is maximized and $l_1 l_2 \ldots l_t \leq d$ is satisfied for $t = 1$ and $l_1 = d$. Hence, the first $c - c/d$ vertices in cluster C_i induce at most $d - 1$ subtrees.

This above argument is repeated for the subtree with root u_{l_t}. The goal is to include the remaining (i.e., at most c/d) vertices into cluster C_i. The next $c/d - c/d^2$ vertices assigned to cluster C_i induce at most $d - 1$ subtrees. After δ applications of the argument, $\frac{c}{d^\delta}$ vertices remain to be assigned to cluster C_i. This implies that $c \geq d^\delta$ and $\delta \leq \frac{\log c}{\log d}$.

The total number of subtrees assigned to cluster C_i is thus at most $\min\{c, d * \lceil \frac{\log c}{\log d} \rceil\}$. This bound on the number of subtrees also holds for $\alpha > 0$. We conclude this section with the following theorem.

Theorem 5 *Algorithm SingFill-FF determines an α-clustering for a given n-vertex tree T in time $\Theta(n)$. The number of subtrees assigned to a cluster is bounded by $\min\{c, d * \lceil \frac{\log c}{\log d} \rceil\}$.*

4   Simultaneous Fill Clustering

In this section we describe our clustering algorithms based on simultaneous cluster filling. To turn Algorithm Generic-SimulFill described in Figure 4 into a complete algorithm, we need to specify the subtree selection and the order in which clusters are considered. Algorithm SimulFill-GF uses the good-fit subtree selection and has $O(n \log_{\frac{d-1}{d}} \frac{\alpha}{4})$ running time. Algorithm SimulFill-BF uses best-fit subtree selection and achieves $O(np \log_{\frac{d-1}{d}} \frac{\alpha}{4})$ time. First-fit subtree selection can be implemnetd to achieve the same performance as SimulFill-GF. Since good-fit determines better fitting subtrees, we do not consider first-fit for simultaneous fill algorithms.

We first present Algorithm SimulFill-GF which considers clusters by non-decreasing *remain*-entries. This order is crucial for achieving the claimed time bound. Since the *remain* entries are between 0 and c, sorting the *remain*-entries costs $O(n)$ time per iteration. Recall that SimulFill-GF selects subtrees T' which satisfy $\lceil remain_i/d \rceil \leq |V'| \leq remain_i$. Let r be an arbitrary vertex of T chosen as the root. The algorithm first roots tree T at r. Next, it determines for every vertex v the number of vertices in the subtree rooted at v. Let $s(v)$ be this quantity. Rooting the tree and the computation of the $s(v)$-entries can be done in $O(n)$ time [3].

One **for**-loop in Phases 1 or 2 makes one traversal of the current tree and assigns one subtree to every cluster (if the cluster still qualifies for receiving vertices). Phase 1 executes iterations until every cluster is safe and the number of iterations equals the maximum number of subtrees assigned to a safe cluster.

Assume the last step determined a subtree for cluster C_i. Assume vertex v has children $u_1, u_2, \ldots, u_k$ and cluster C_i is assigned the subtree rooted at vertex u_l, $1 \leq l \leq k$. Hence, $remain_i/d \leq s(u_l) \leq remain_i$ and for $1 \leq j < l$ we have $s(u_j) < remain_i/d$ (i.e., the subtree rooted at u_j is too small for C_i). After the subtree rooted at u_l has been assigned to cluster C_i, a subtree for cluster C_{i+1} is determined. Note that we have $remain_i \leq remain_{i+1}$. The next paragraph sketches how the next subtree for C_{i+1} is found. The $O(n)$ time bound for one iteration follows from the way the tree traversal identifies subtrees.

In order to determine the subtree to be assigned to cluster C_{i+1}, the traversal first considers the remaining children of vertex v, namely vertices $u_{l+1}, \ldots, u_k$. Observe that since the subtrees rooted at $u_1, u_2, \ldots, u_{l-1}$ were not large enough for cluster C_i, they are also not large enough for C_{i+1} (since clusters are considered by non-decreasing *remain*-entries). If there exists a vertex u_j with $s(u_j) \geq remain_{i+1}/d$, $l+1 \leq j \leq k$, a subtree for C_{i+1} is found in a tree rooted at one of the siblings of u_l. The traversal considers thus vertices not yet traversed in the current iteration. Assume that for all vertices u_j, $l+1 \leq j \leq k$, we have $s(u_j) < remain_{i+1}/d$. The traversal now backs up the tree and considers vertex v next. For vertex v we maintain the number of vertices in its subtree which have already been assigned to clusters in the current iteration. If the current subtree rooted at vertex v does satisfy the size requirements for C_{i+1}, an assignment is made. Observe that the subtree rooted at vertex v cannot be

S. E. Hambrusch et al., *Clustering in Trees*, JGAA, 4(4) 1–26 (2000)   15

too large (since all children have subtrees which are too small). If the subtree rooted at v is too small, we continue with the parent of v, say v'. We repeat the same process of first considering the children of v' not previously considered and, if no suitable subtree is found, we consider the subtree rooted at v'. Again, we know that the children of v' considered earlier are not the roots of a big enough subtree and thus do not need to be checked. It follows that each cluster receives a subtree while executing one traversal of the tree and thus one iteration takes $O(n)$ time.

Phase 2 proceeds with the subtree selection and the ordering of the clusters as Phase 1. The while-loop is executed until all vertices of T have been assigned to a cluster. A cluster may thus not receive any additional vertices in Phase 2. The total time spent in Phases 1 and 2 is $O(n)$ times the maximum number of subtrees assigned to a cluster. The bound on the number of subtrees is given in the proof of the following theorem.

Theorem 6 *Algorithm SimulFill-GF determines an α-clustering for a tree T of n vertices in time $O(n \log_{\frac{d-1}{d}} \frac{\alpha}{4})$. The number of subtrees assigned to a cluster is bounded by $\min\{c, \lceil \log_{\frac{d-1}{d}} \frac{\alpha}{4} \rceil\}$.*

Proof: From the conditions Phase 1 and 2 impose on the cluster sizes it follows that $(1 - \frac{\alpha}{2})c \leq |C_i| \leq (1+\alpha)c$, $1 \leq i \leq p$. The number of iterations within each phase gives an upper bound on the number of subtrees assigned to a cluster. Using an argument similar to that used in Lemma 2, it follows that the algorithm can always find a subtree T' such that $|V_{T'}| \geq remain_i/d$. (Since the tree is rooted and not all subtrees are considered in a rooted tree, the bound is $|V_{T'}| \geq remain_i/d$ instead of $|V_{T'}| \geq remain_i/(d-1)$.) Assume cluster C_i is safe after m iterations of Phase 1. We have $target_i = c$ and, using the argument in the proof of Lemma 3, we have

$$(1 - (\frac{d-1}{d})^m) \times c > (1 - \frac{\alpha}{2}) \times c.$$

This implies that the number of iterations in Phase 1 bounded by $\lceil \log_{\frac{d-1}{d}} \frac{\alpha}{2} \rceil$. In Phase 2, we have $target_i = (1+\alpha)c - |C_i|$ with $\alpha c \leq target_i \leq \frac{3}{2}\alpha c$. Assume cluster C_i is full after m iterations of Phase 2. Then,

$$(1 - (\frac{d-1}{d})^m) \times \alpha c > \frac{\alpha}{2} \times c.$$

Hence, the number of iterations of Phase 2 is bounded by $\lceil \log_{\frac{d-1}{d}} \frac{1}{2} \rceil$ and the claimed bound on the total number of iterations follows. $\square$

Algorithm SimulFill-BF uses best-fit selection for determining the subtrees. Determining a subtree may result in a complete traversal of the current tree. Our implementation considers the clusters by non-increasing *remain*-entries. Even though this ordering does not impact the worst-case bounds, the approach of looking for large subtrees in large trees tends to produce better experimental results.

S. E. Hambrusch et al., *Clustering in Trees*, JGAA, 4(4) 1–26 (2000)   16

Theorem 7 *Algorithm SimulFill-BF determines an α-clustering for a given n-vertex tree T in time $O(np \log_{\frac{d-1}{d}} \frac{\alpha}{4})$. The number of subtrees is a cluster in bounded by $\min\{c, \lceil \log_{\frac{d-1}{d}} \frac{\alpha}{4} \rceil\}$.*

Proof: Using best-fit for the subtree selection results in a new traversal whenever a subtree is assigned to a cluster. This increases time to the stated bound. The bound on the number of subtrees is as in SimulFill-GF. $\square$

5   Experimental Results

In this section we discuss the performance of the different clustering algorithms and show how parameters α, c, and d impact cluster sizes and number of subtrees in the clusters. We considered synthetically generated trees with n ranging from $1,000$ to $6,000$. Ideal cluster sizes considered varied from $c = 10$ to $c = 500$ and the maximum degree varied from $d = 20$ to $d = 74$. We used four classes of synthetic trees. All trees were created level by level and the classes differ on how the degree of a vertex is determined. Class 1 assumes that every degree between 1 and d is equally likely for every vertex. Class 2 assumes that the probability of a vertex being a leaf is significantly higher (we used 0.5 instead $1/d$) and that, once a vertex is identified as a non-leaf, every degree is equally likely. Class 3 generates degrees using a normal distribution. Trees in classes 1 to 3 are generated level-by-level starting with the root. Class 4 generates B-trees [2]. Trees in class 4 are created by specifying number of leaves and the value of B (which corresponds to the maximum degree). Trees are then generated from the leaves towards the root and for a non-root, interior vertex, every degree between $B/2$ and B is equally likely.

For all trees, we report the mean, median, and the maximum of number of subtrees in a cluster and the cluster sizes. When we report, for example, the median number of subtrees in a cluster for a tree, we report the mean of the medians of the p clusters over 10 different trees within the same class. The different classes of trees exhibit the same performance trend for trees with the same n, c, d, and α values. As is discussed in the next two sections, we observed that the choice of α and the relationship between c and d has significant impact on the performance. The plots shown in this paper are for trees in classes 1 and 4.

Given the NP-completeness of the problem and the considered tree sizes, we did not generate optimal results. Comparing the algorithms gives interesting and relevant insight into the different strategies as well as the parameter choices. The implementation of the clustering algorithms was done in Java. The implementations have no hidden constants and are based on the same data structures. We do not report actual running times and expect the running times to follow the asymptotic worst-case bounds established.

S. E. Hambrusch et al., *Clustering in Trees*, JGAA, 4(4) 1–26 (2000)   17

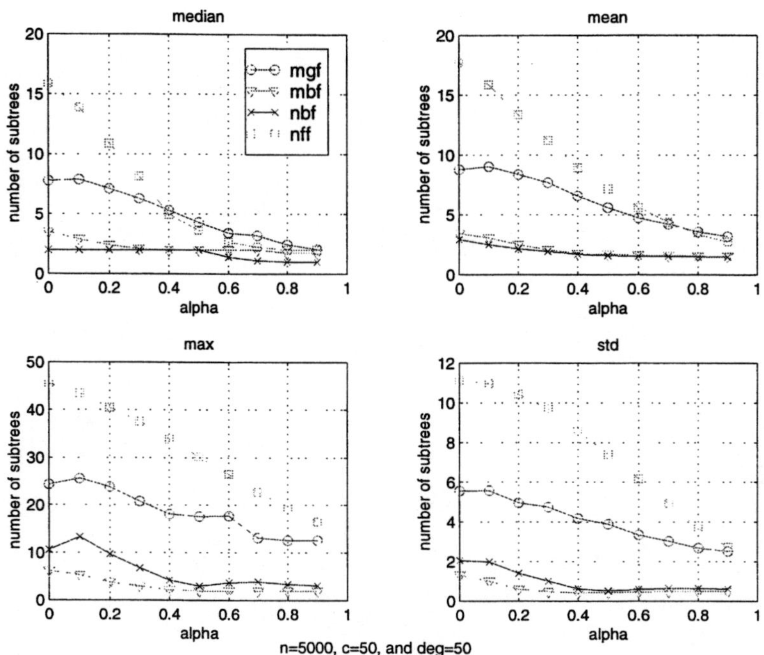

Figure 8: Comparing the number of subtrees for SimulFill-GF ○, SingFill-FF □, SimulFill-BF ◁, SingFill-BF × for trees in class 1.

5.1 Comparing Clustering Algorithms

In this section we discuss the performance of Algorithms SingFill-FF, SimulFill-GF, SingFill-BF, and SimulFill-BF with respect to the number of subtrees and cluster sizes for synthetically generated trees belonging to classes 1 and 4. The graphs show results for ten α-values, $\alpha = j/10$, $j \in \{0, 1, 2, \ldots, 9\}$. Graphs for trees in class 1 (i.e., trees in which every degree is equally likely) have $n = 5,000$, $c = 50$, and a maximum degree of 50. Graphs for trees in class 4 (i.e., B-trees) have $5,000$ leaves.

Figures 8 and 9 show results for the number of subtrees. The graphs show trends which were observed for all trees classes and cluster sizes considered. Algorithm SingFill-FF generates clusters containing the largest number of subtrees. This holds when we consider the median, mean, and maximum number of subtrees (over all clusters and over 10 trees of the same type). This is not surprising since SingFill-FF simply arranges subtrees by size and proceeds greedily without further optimizations. Algorithms SimulFill-BF and SingFill-BF consistently outperform the two algorithms based on first-fit and good-fit with respect to minimizing the number of subtrees in the clusters. The relationship between the two best-fit approaches is examined in detail in the next section.

For all four clustering algorithms Figures 8 and 9 show a "leveling off" in the number of subtrees as α increases. Overall, our experimental work suggests that

S. E. Hambrusch et al., *Clustering in Trees*, JGAA, 4(4) 1–26 (2000)   18

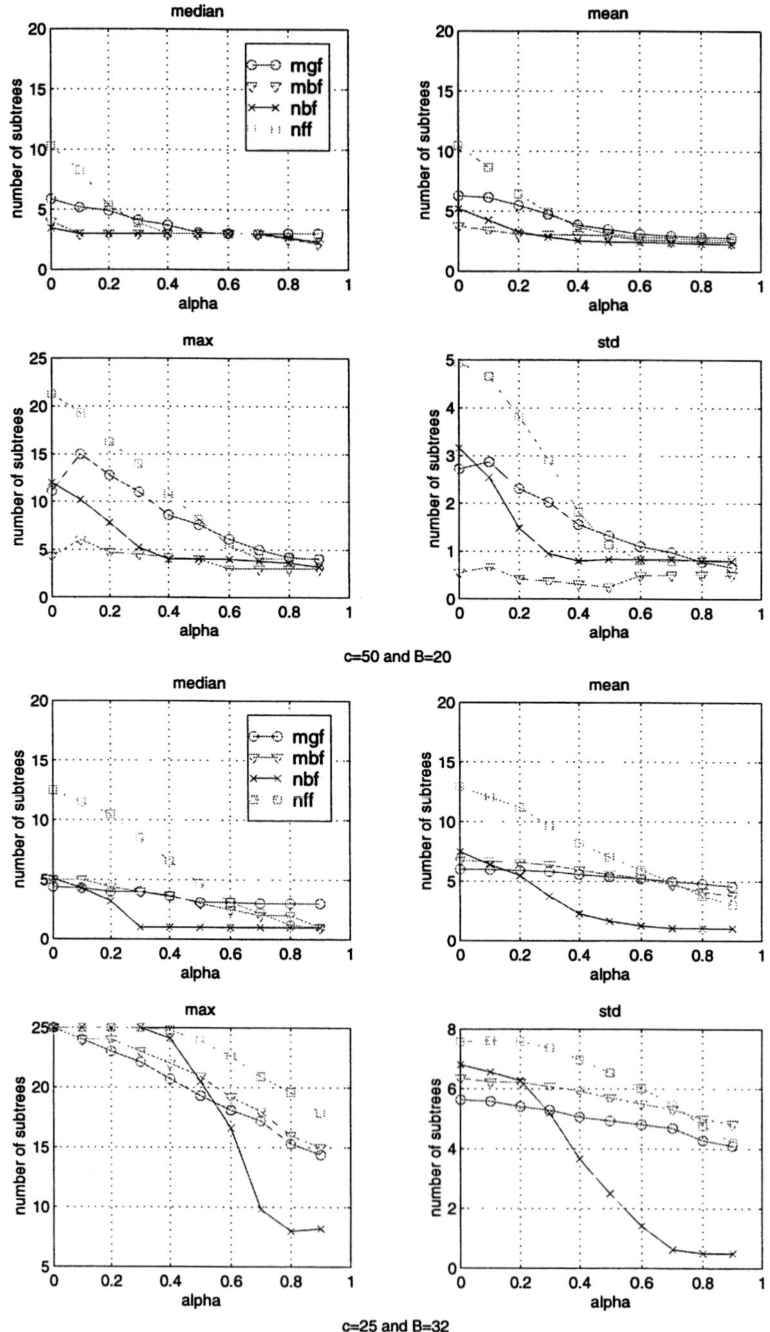

Figure 9: Comparing the number of subtrees for SimulFill-GF ∘, SingFill-FF □, SimulFill-BF ◁, SingFill-BF × for B-trees with 5,000 leaves; upper four graphs have $B = 20$ and $c = 50$; lower four have $B = 32$ and $c = 25$.

S. E. Hambrusch et al., *Clustering in Trees*, JGAA, 4(4) 1–26 (2000)   19

using α greater than 0.4 has little impact on reducing the number of subtrees. We next comment on a relationship between c and d we observed throughout. As the maximum degree exceeds c and a tree contains a significant number of nodes with high degree, it becomes harder - and at times impossible - to keep the number of subtrees in a cluster small. For trees in class 1 with $c = 50$ and values of d smaller than 50, we observed a decrease in the number of subtrees for all algorithms. For degrees larger than 50, we observed an increase. However, the relationship between the four algorithms remains stable. Figure 9 shows this for two types of B-trees. The upper four graphs show a typical behavior for the case when the maximum degree is smaller than c. In the lower four graphs the maximum degree exceeds c (32 versus 25). All algorithms produce clusters with more subtrees. In particular, the maximum number of subtrees is equal (or close to) to the possible worst case of c vertices in a cluster for small values of α. Plots in Section 5.2 which vary c or d will also reflect this characteristic.

We use trees in class 1 to illustrate observed behavior on the cluster sizes as α increases. Figure 10 shows the median, mean, and maximum difference between achieved and ideal cluster size (i.e., the quantities $|C_i - c|$). Clearly, as α increases, the differences in the cluster sizes continue to increase. Algorithm SimulFill-GF fills the clusters closer to the limits set by α than any other algorithm. In Figure 10, the maximum cluster sizes generated by the two simultaneous fill algorithms are consistently higher compared to those of the single fill algorithms. This is a characteristic of the simultaneous filling of clusters. Recall that the simultaneous filling of clusters proceeds in two phases: the first phase generates safe clusters and it does not allow a cluster size to exceed c. Clusters exceeding size c are generated in the second phase. This approach ensures correct cluster sizes, but it also makes it more likely that there exist clusters which are close to the extremes of the required range.

5.2 Comparing SingFill-BF and SimulFill-BF

We now turn to comparing the two best-fit clustering algorithms, SimulFill-BF and SingFill-BF. All graphs shown in this section were obtained using trees in class 1, but are representative for all types of trees we considered. Figure 11 shows typical results for the number of subtrees when α ranges from 0 to 0.9 and c ranges from 10 to 500. In the trees used, the maximum degree is 44. For small c values (up to around $c = d$), we observe a large number of subtrees for both best-fit algorithms. Note the significant increase in the maximum number of subtrees (and the different scale used). For larger c values, we see the number of subtrees decrease as α increases and we observe a leveling off around $\alpha = 0.4$ with respect the mean, median, and maximum number of subtrees in the clusters. Figure 12 also illustrates the observed leveling off for the mean number of subtrees and Algorithm SingFill-BF.

From our experimental results we can conclude that SingFill-BF outperforms SimulFill-BF with respect to the mean and the median number of subtrees. When considering the maximum number of subtrees, we see that SimulFill-BF outperforms SingFill-BF. This behavior showed up in all trees we considered

S. E. Hambrusch et al., *Clustering in Trees*, JGAA, 4(4) 1–26 (2000)   20

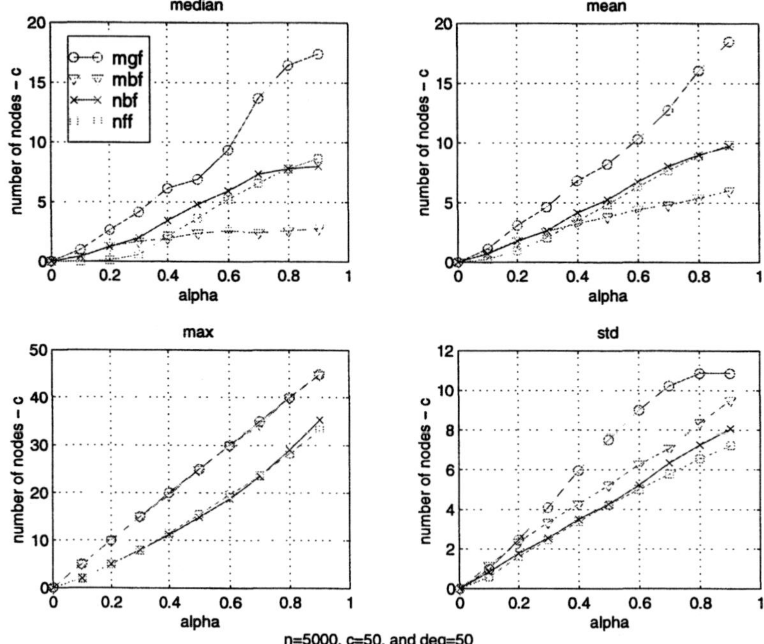

Figure 10: Comparing cluster sizes for SimulFill-GF ∘, SingFill-FF □, SimulFill-BF ◁, SingFill-BF × for trees in class 1.

S. E. Hambrusch et al., *Clustering in Trees*, JGAA, 4(4) 1–26 (2000)   21

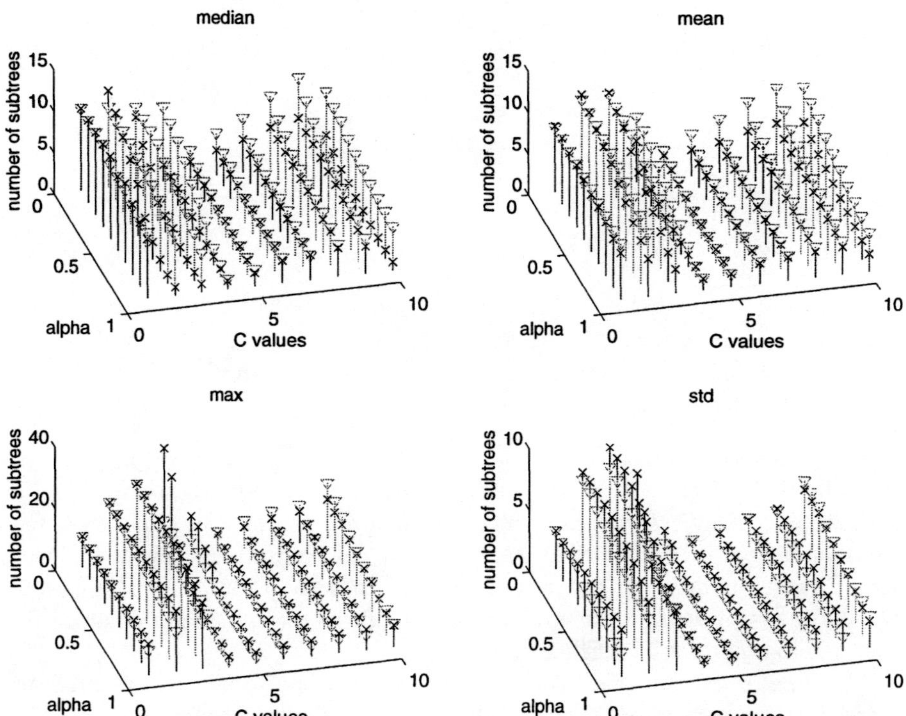

Figure 11: Comparing the number of subtrees for SimulFill-BF $\triangledown$ and SingFill-BF $\times$ when c and α change; $n = 5,000$ and $d = 44$; c values are [10, 20, 25, 40, 50, 100, 125, 200, 250, 500].

S. E. Hambrusch et al., *Clustering in Trees*, JGAA, 4(4) 1–26 (2000)   22

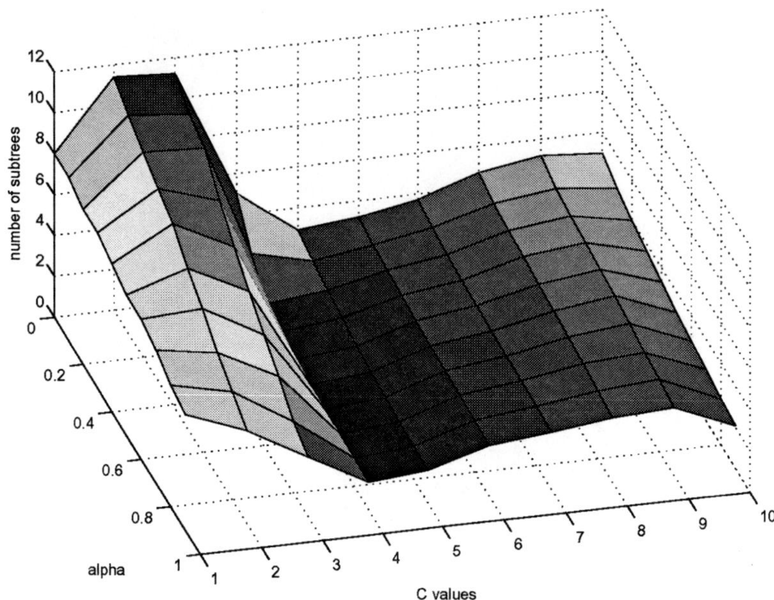

Figure 12: Mean number of subtrees in clusters for Algorithm SingFill-BF as α and c change for $n = 5,000$ and $d = 44$; c values are [10, 20, 25, 40, 50, 100, 125, 200, 250, 500].

and reflects a characteristic between the two approaches. Algorithm SingFill-BF is able to delay creating clusters containing a large number of subtrees until the remaining tree is small. The final iterations of SingFill-BF generate clusters with a larger number of subtrees compared to what SimulFill-BF generates. This happens since the small tree remaining allows fewer choices, creating thus a large maximum for SingFill-BF.

The final set of experimental results examines the impact of the maximum degree on the performance of the two best-fit algorithms. We show data obtained for $n = 5,000$, $c = 50$, and d in the range from 20 to 74. In Figure 13 we again see that Algorithm SimulFill-BF outperforms SingFill-BF with respect to the maximum number of subtrees placed in a cluster, but that SingFill-BF gives better results for the mean and median values. For the large degrees ($d = 62, 68$, and 74 in Figure 13), we observed a significantly larger number of subtrees for the mean, median, as well as the maximum. This confirms the relationship of c and d discussed earlier.

Figure 14 illustrates cluster sizes for $c = 50$ as maximum degrees and α increase. As to be expected, increasing α generates for both algorithms clusters whose sizes vary more and more from the ideal size of c. SimulFill-BF generates clusters much closer to their upper and lower limits, as was already mentioned in the discussion in single versus simultaneous fill for Figure 10. Using Figures 13 and 14 and $d = 26$ for SimulFill-BF, we see a maximum of 5 subtrees in the

S. E. Hambrusch et al., *Clustering in Trees*, JGAA, 4(4) 1–26 (2000)   23

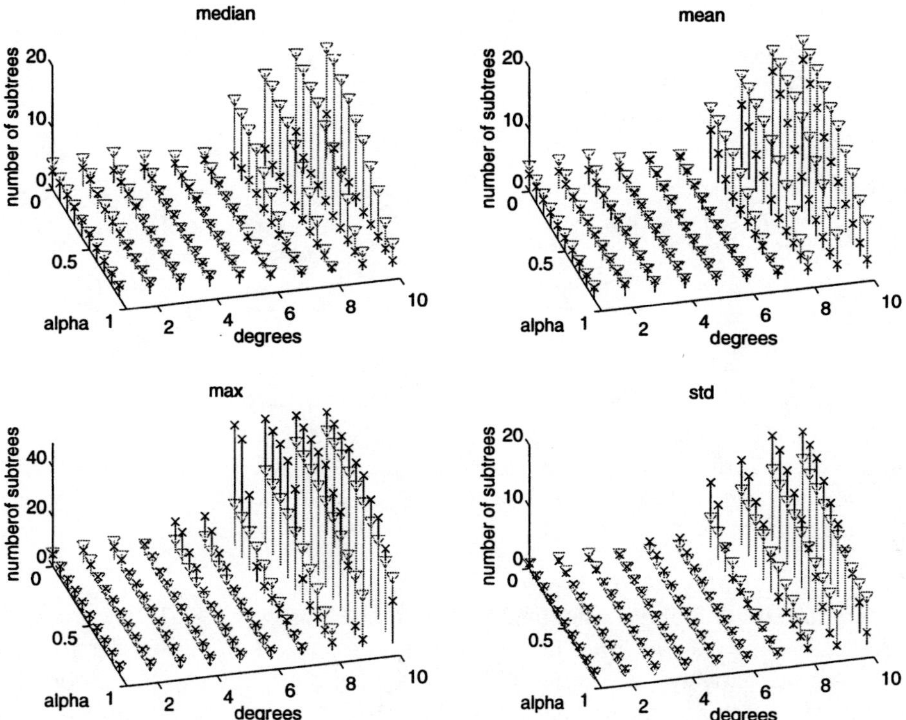

Figure 13: Comparing the number of subtrees for SimulFill-BF $\triangledown$ and SingFill-BF $\times$ when the maximum degree d and α change; $n = 5,000$ and $c = 50$; d values are $[20, 26, 32, 38, 44, 50, 56, 62, 68, 74]$.

S. E. Hambrusch et al., *Clustering in Trees*, JGAA, 4(4) 1–26 (2000)   24

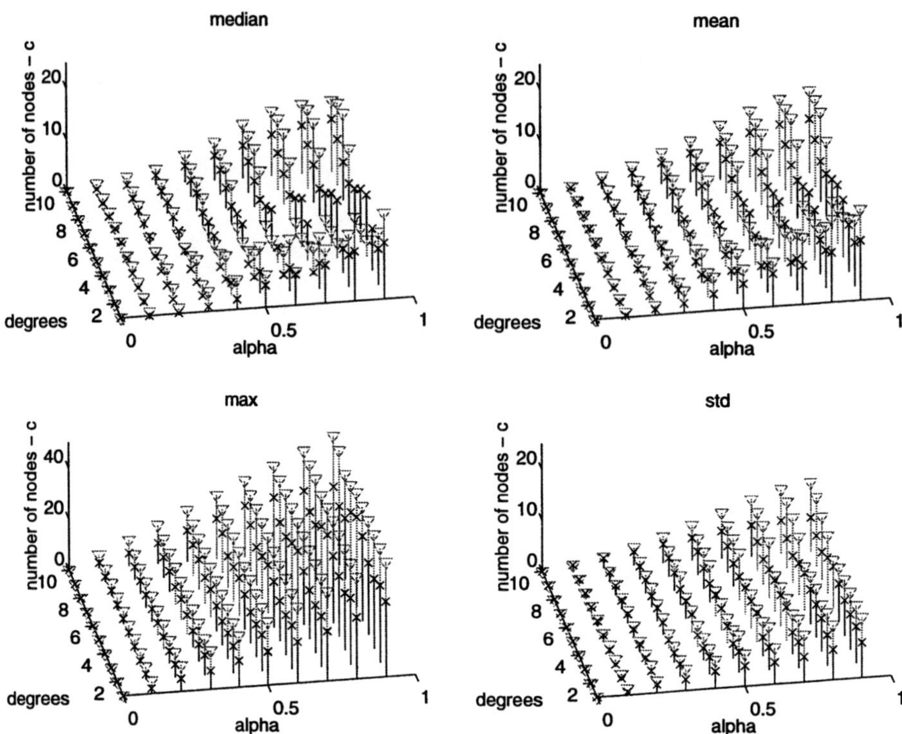

Figure 14: Comparing cluster sizes for SimulFill-BF $\triangledown$ and SingFill-BF $\times$ when d and α change; $n = 5,000$ and $c = 50$; d values are $[20, 26, 32, 38, 44, 50, 56, 62, 68, 74]$.

clusters for $\alpha = 0.5$ and a maximum of 5 subtrees for $\alpha = 0.8$. In both cases, there exists clusters which are filled to the upper limit of 75 and 90 vertices in a cluster, respectively. While increasing α beyond 0.4 tends not to reduce the number of subtrees, it does generate clusters sizes lying in larger ranges.

6   Conclusion

We presented algorithms for α-clustering the vertices of a tree when cluster sizes need to lie in a range defined by α and the number of subtrees assigned to a cluster should be minimized. In addition to input parameter α, the algorithms differ in the identification of subtrees and the order in which clusters are filled. We described efficient implementation of the clustering algorithms and established upper bounds on the number of subtrees in a cluster. Our experimental results provided insight into how the maximum degree d, the relationship between c and d, the value of α, the subtree selection method, and the order in which clusters are filled impact the number of subtrees and the cluster sizes. In particular, our experimental result show that as α increases, the reduction in the

S. E. Hambrusch et al., *Clustering in Trees*, JGAA, 4(4) 1–26 (2000)   25

number of subtrees slows down considerably, but the differences between cluster sizes continues to increase. Overall, we observed that the best-fit clustering algorithm filling one cluster at a time generates consistently good results.

7   Acknowledgments

We thank the referees for their helpful and constructive comments.

References

[1] J. Banerjee, W. Kim, S.-J. Kim, and J. F. Garza. Clustering a DAG for CAD databases. *IEEE Transactions on Software Engineering (SE)*, 14(11), Nov. 1988.

[2] R. Bayer and E. McCreight. Organization and maintenance of large ordered indices. *Acta Informatica*, 1:173–189, 1972.

[3] T. Cormen, C. Leiserson, and R. Rivest. *Introduction to algorithms*. The MIT Press, 1990.

[4] A. A. Diwan, S. Rane, S. Seshadri, and S. Sudarshan. Clustering techniques for minimizing external path length. In *Proceedings of the 22-nd International Conference on Very Large Data Bases*, pages 342–353, 1996.

[5] A. Farley, S. Hedetniemi, and A. Proskurowski. Partitioning trees: matching, domination, and maximum diameter. *International Journal of Computer and Information Sciences*, 10(1):55–61, Feb. 1981.

[6] A. Gerasoulis and T. Yang. On the granularity and clustering of directed acyclic task graphs. *IEEE Transactions on Parallel and Distributed Systems*, 4(6):686–701, June 1993.

[7] J. Gil and A. Itai. How to pack trees. *Journal of Algorithms*, 32, 1999.

[8] S. Hambrusch and C.-M. Liu. Data replication for external searching in static tree structures. In *Proceedings of the Ninth ACM International Conference on Information and Knowledge Management*, Nov. 2000.

[9] C.-M. Liu. Searching in static, external memory data structures. Ph.d. thesis, Purdue University, Department of Computer Sciences, in progress.

[10] P. Maheshwari and H. Shen. An efficient clustering algorithm for partitioning parallel program. *Parallel Computing*, 24(5-6):893–909, June 1998.

[11] D. M. Nicol and D. R. O'Hallaron. Improved algorithms for mapping pipelined and parallel computations. *IEEE Transactions on Computers*, 40(3):295–306, Mar. 1991.

[12] M. H. Nodine, M. T. Goodrich, and J. S. Vitter. Blocking for external graph searching. *Algorithmica*, 16(2):181–214, Aug. 1996.

[13] S. Ramaswamy and S. Subramanian. Path caching: A technique for optimal external searching. In *Proceedings of the Thirteenth ACM Symposium on Principles of Database Systems*, volume 13, pages 25–35, 1994.

[14] R. Schrader. Approximations to clustering and subgraph problems on trees. *Discrete Applied Mathematics*, 6:301–309, 1983.

Volume 5:1 (2001)

Journal of Graph Algorithms and Applications

http://www.cs.brown.edu/publications/jgaa/

vol. 5, no. 1, pp. 1–74 (2001)

Planarizing Graphs — A Survey and Annotated Bibliography

Annegret Liebers

Department of Computer and Information Science
University of Konstanz, Germany
http://www.inf.uni-konstanz.de/~liebers/
Annegret.Liebers@uni-konstanz.de

Abstract

Given a finite, undirected, simple graph G, we are concerned with operations on G that transform it into a planar graph. We give a survey of results about such operations and related graph parameters. While there are many algorithmic results about planarization through edge deletion, the results about vertex splitting, thickness, and crossing number are mostly of a structural nature. We also include a brief section on vertex deletion.

We do not consider parallel algorithms, nor do we deal with on-line algorithms.

Communicated by A. Gibbons: submitted June 1996; revised December 1998 and January 2001.

Research supported in part by DFG grant Wa 654/10-2.

Contents

1   Introduction

Many problems in discrete mathematics and combinatorial optimization can be viewed as graph problems. Graphs immediately come to mind for modeling networks of all kinds, but also seemingly unrelated problems from areas like transportation or warehousing can turn out to be, e.g., network flow problems, and their solution involves algorithms on graphs [AMO93].

Graphs that can be drawn without edge crossings (i.e. *planar* graphs) have a natural advantage for visualization, but also other graph problems can be easier to solve when restricted to this special class of graphs. "Easier" might mean that a special algorithm for planar graphs may have a better asymptotic time complexity than the best known algorithm for general graphs, or even that an intractable problem may become tractable if restricted to planar graphs. The former case applies for example to the Vertex- and Edge-Disjoint Menger Problems [RLWW97, Wei97].

The latter case, however, seems to be relatively rare [Joh85, p. 440]: There is a polynomial time algorithm for Max Cut restricted to planar graphs [GJ79, Problem ND16], and Vertex Coloring is NP-complete for general graphs, even for a fixed number $k \geq 3$ of colors [GJ79, Problem GT4], but is trivially solvable for a fixed number $k \geq 4$ for planar graphs by virtue of the Four Color Theorem. See [JT95, Section 2.1.] for a discussion of the original proof by Appel and Haken, and of algorithms for actually finding a coloring of a planar graph, also in light of the new proof [RSST96] of the Four Color Theorem.

When visualizing nonplanar graphs, a natural approach is to draw the graph in a way as close to planarity as possible (for example with as few edge crossings as possible). This is one of the problems of graph drawing, a field that has grown tremendously within the last decade [DETT94, DETT99].

In any case there is great interest in the question of how far from being planar a given graph is. We survey ways of transforming a nonplanar graph into a planar graph and discuss measures for the nonplanarity of a graph. We concentrate on sequential algorithms for the off-line case, i.e. we do not consider parallel or on-line algorithms.

One approach is to look for the largest induced planar subgraph of a nonplanar graph. Finding an induced subgraph is equivalent to deleting vertices from a graph and will be discussed in Section 2. It does not seem to be a very common approach, and there is relatively little literature about it.

Another approach is to look for the largest planar subgraph (without the restriction to induced subgraphs). Since deleting an edge from a graph is a less "drastic" operation than deleting a vertex together with all its incident edges, it is not surprising that finding a planar subgraph of a nonplanar graph (i.e. deleting edges) has been studied much more intensively. There is a large amount of literature about finding a planar subgraph, with an emphasis on algorithmic results. They are the subject of Section 3.

Another technique for planarizing a graph is vertex splitting. There are relatively few algorithmic results about vertex splitting, but it turns out that there are many different structural results involving this operation. Section 4 describes the vertex splitting operation as it relates to graph planarization.

Vertex deletion, edge deletion, and vertex splitting are operations performed on single vertices or edges of the graph in question, i.e. they are local operations. Section 5 discusses partitioning the whole graph into several planar layers, hence following a global approach. The greater the number of layers needed, the further away from planarity the graph is. There seem to be few algorithmic results about finding this *thickness* of a graph, but there are many structural results about thickness within topological graph theory.

Section 6 discusses the problem of drawing a graph so that there are as few edge crossings as possible in the drawing. Again, most results about the *crossing number* of a graph are of a structural nature. Finally, Section 7 mentions the concept of *coarseness*.

We do not study hierarchical graph models such as presented in [Len89, FCE95], nor do we discuss hypergraphs [Ber73, Ber89] or infinite graphs [Kön90].

The remainder of the introduction gives definitions and terminology concerning graphs in Section 1.1, and then gives a brief introduction to planar graphs in Section 1.2. Section 1.3 lists some generalizations of planarity. For an introduction to algorithms and the definition and use of $O(\cdots)$ and $\Omega(\cdots)$ for asymptotic bounds, the reader is referred to textbooks on algorithms, for example [CLR94]. The complexity classes P and NP and the concept of NP-completeness are also discussed in [CLR94], but a more thorough treatment can be found in [GJ79] and [Pap94].

1.1   Graphs

There are many textbooks on graph theory.[1] Some of the standard ones are [Har69, BM76, Tut84, CL96]. For a focus on algorithmic graph theory, see for example [Eve79, Gol80, GM84, Gib85, Lee90, TS92], and for topological graph theory, see [GT87, BL95]. Another recent text is also [Wes96, Wes01].

We will now give some definitions and notation concerning graphs that are used throughout the text.

A *finite, undirected, simple graph* G, denoted $G = (V, E)$, consists of a finite vertex set V and a set of undirected edges $E \subseteq \{\{u, v\} \mid u \in V, v \in V, u \neq v\}$. The *end vertices* of an edge $e = \{u, v\} \in E$, u and v, are said to be *adjacent*. u is said to be a *neighbor* of v and vice versa. Furthermore, u and v are said to be *incident* to e (and vice versa). For brevity we often write uv instead of $\{u, v\}$. From now on, when we speak of a graph, we always mean a finite, undirected, simple graph.

The number of edges incident to a vertex u is called the *vertex degree* (or simply *degree*) of u. The *minimum (maximum) degree* of a graph G is the minimum (maximum) degree of all vertices of G. The minimum and maximum degrees of a graph are denoted by δ and Δ, respectively. If all vertices of a

---

[1] The first textbook devoted solely to graph theory was [Kön36] by König. [Kön90] is the first English translation. The history of graph theory is presented in [BLW76], [Wil86], [Fou92, Section 1.1], for instance.

graph have the same degree d, the graph is called *d-regular* (or just *regular*). A 3-regular graph is also called *cubic*.

A graph is usually visualized by representing each vertex through a point in the plane, and by representing each edge through a curve in the plane, connecting the points corresponding to the end vertices of the edge. We usually do not distinguish between a vertex and the point representing it, or between an edge and the curve representing it. Such a representation is called a *drawing* of the graph if no two vertices are represented by the same point, if the curve representing an edge does not include any point representing a vertex (except that the endpoints of the curve are the points representing the end vertices of the edge), and if two distinct edges have at most one point in common.

Given a graph $G = (V, E)$, a graph $G' = (V', E')$ is called a *subgraph* of G if $V' \subseteq V$ and $E' \subseteq \{uv \mid u \in V', v \in V', \text{ and } uv \in E\}$. If furthermore $V' = V$ then G' is said to be a *spanning* subgraph of G. If $V' \subset V$ or $E' \subset E$ (or both) then G' is said to be a *proper* subgraph of G. A graph $G'' = (V'', E'')$ is called a *vertex induced* (or simply *induced*) subgraph of G if $V'' \subseteq V$ and $E'' = \{uv \mid u \in V'' \text{ and } v \in V'' \text{ and } uv \in E\}$. In that case we call G'' the *subgraph of G induced by V''*.

If $G_1 = (V_1, E_1)$ and $G_2 = (V_2, E_2)$ are two (not necessarily distinct) subgraphs of a graph $G = (V, E)$, then the subgraph $G' = (V_1 \cup V_2, E_1 \cup E_2)$ of G is called the *union* of G_1 and G_2.[2]

·Given a graph $G = (V, E)$, a sequence $v_0 e_1 v_1 e_2 v_2 \ldots e_k v_k$ is called a *path* in G if the $k + 1$ vertices $v_0 \ldots v_k$ are elements of V, if they are pairwise distinct except possibly v_0 and v_k, and if v_{i-1} and v_i are the end vertices of e_i for $1 \leq i \leq k$. k is called the *length* of the path. We also say that the path *connects* the vertices v_0 and v_k. If additionally $v_0 = v_k$, the path is called a *cycle*. The length of a shortest cycle in G is called the *girth* of G. If G has no cycles, it is said to be *acyclic* and the girth is undefined (but note that an acyclic graph is always planar).

We denote with P_n the graph consisting only of a path of length $n - 1$, where the end vertices of the path are not identical. P_n has n vertices and $n - 1$ edges. C_n denotes a graph consisting of a cycle of length n, having n vertices and n edges. If a path in a graph G includes all vertices of G it is called a *Hamilton path*. If additionally this path is a cycle, it is called a *Hamilton cycle*. Observe that in Figure 6 on page 24, graph 13 contains a Hamilton path, but no Hamilton cycle, whereas graph 14 contains both.

If for every pair of vertices u and v of a graph $G = (V, E)$ there is a path in G connecting u and v then G is said to be *connected*. Otherwise G is said to be *disconnected*. If $V' \subseteq V$ is a vertex set such that the subgraph G' of G induced by V' is connected and such that for every set V'' with $V' \subset V'' \subseteq V$ the subgraph of G induced by V'' is disconnected, then G' is said to be a *connected component* (or simply *component*) of G.

Given a graph $G = (V, E)$ and a vertex $v \in V$ we say that the subgraph G' of G induced by $V \setminus \{v\}$ is obtained by *deleting v from G*. If G' has more

---

[2]Note that the term *union* is sometimes defined differently (see for example [Har69, p. 21]).

A. Liebers, *Planarizing Graphs*, JGAA, 5(1) 1–74 (2001)

connected components than G then v is said to be a *cut vertex* of G. If at least k vertices have to be deleted from G before the resulting graph is disconnected, or before the resulting graph consists of a single vertex, then G is said to be *k-connected*. Observe that if a graph is 1-connected, then it is connected, and that a connected graph with at least 3 vertices and without cut vertices is 2-connected. In Figure 6, graph 7 has two cut vertices. Graph 16 is 2-connected, but it is not 3-connected.

Analogous definitions exist for edges: Given a graph $G = (V, E)$ and an edge $e \in E$ we say that the subgraph $G' = (V, E \setminus \{e\})$ of G is obtained by *deleting* e from G. If G' has more connected components than G then e is said to be a *cut edge* of G. If at least k edges have to be deleted from G before the resulting graph is disconnected, then G is said to be *k-edge-connected*. The graph consisting of a single vertex is defined to be 0-edge-connected.

If for a graph $G = (V, E)$, $V' \subseteq V$ is a vertex set such that the subgraph of G induced by V' is 2-connected and such that for every set V'' with $V' \subset V'' \subseteq V$ the subgraph of G induced by V'' is not 2-connected, then we call the subgraph of G induced by V' a *2-connected block* (or simply a *block*) of G.

If an edge $e = uv$ of a graph $G = (V, E)$ is replaced by a path $ue'v_e e''v$ introducing a new vertex $v_e \notin V$, then we say that the graph $G' = (V \cup \{v_e\}$, $(E \setminus \{e\}) \cup \{e', e''\})$ is obtained from G by *subdividing* the edge e. If a graph G'' is obtained from G by any number of (possibly zero) subdivisions of edges then G'' is called a *subdivision* of G. It will be clear from the context whether the term subdivision refers to the operation of subdividing an edge or to the resulting graph. For an illustration of subdivisions, see Figure 7 on page 28.

For a graph $G = (V, E)$ and an edge $e = uv \in E$, the graph G' obtained from G by deleting e, identifying u and v and by removing all edges $f \in \{ux \mid x \in V$, $x \neq u$, $x \neq v$, $ux \in E$, and $vx \in E\}$, is said to have been obtained from G by *contracting* the edge e. In other words, contracting an edge means identifying its two end vertices and making the resulting graph simple by deleting loops and multiple edges. A graph obtained from a subgraph of G by any number (including zero) of edge contractions is said to be a *minor* of G. A subgraph of G is always a minor of G, but not vice versa. In Figure 6 on page 24, the graph G is a minor of graphs 1 through 6 and 9 through 18, but it is not a minor of graphs 7 and 8. For another illustration of graph minors, see Figure 13 on page 38.

Besides the paths P_n and the cycles C_n, the following special graphs appear throughout the text:

For $n \geq 2$, the *complete graph*, denoted K_n, consists of n vertices together with all possible $\binom{n}{2}$ edges. So in K_n every vertex is adjacent to every other vertex. We define K_1 to be the graph consisting of a single vertex. K_2 is a single edge with its two end vertices, and K_3 is a triangle.

The *complete bipartite graph*, denoted K_{n_1,n_2}, consists of two disjoint vertex sets $V = \{v_1, \ldots v_{n_1}\}$ and $W = \{w_1, \ldots w_{n_2}\}$ and the edge set $E = \{v_i w_j \mid 1 \leq i \leq n_1$ and $1 \leq j \leq n_2\}$ of all edges between vertices in V and vertices in W. Note that $K_{n_1,n_2} = K_{n_2,n_1}$.

The *hypercube of dimension n*, denoted Q_n, is the graph with 2^n vertices

A. Liebers, *Planarizing Graphs*, JGAA, 5(1) 1–74 (2001)   7

where each vertex has a label consisting of an n-digit binary number between $0\ldots0$ and $1\ldots1$ and with an edge connecting two vertices if and only if the labels of the vertices differ in a single digit. Observe that Q_n has $n \cdot 2^{n-1}$ edges, that $Q_1 = K_2$ and that $Q_2 = C_4$. For further properties of hypercube graphs see [HHW88].

A connected, acyclic graph is called a *tree*. A tree with n vertices has $n - 1$ edges.

1.2   Planar Graphs

The class of planar graphs has been widely studied, and many of the textbooks mentioned above contain chapters about planar graphs [Har69, BM76, Tut84, Gib85, GT87, TS92, CL96, Wes96, Wes01]. A wealth of literature studies properties of planar graphs, algorithms for solving problems on planar graphs, and how close other graphs are to planarity. The latter topic results in algorithms that transform a given graph into a planar graph. These results are briefly summarized in Section 4.2 of the annotated graph drawing bibliography by Di Battista et al. [DETT94].

The book by Nishizeki and Chiba [NC88] is a thorough treatment of planar graphs, with an emphasis on algorithms. [Nis90] can be seen as an update of [NC88]. Johnson [Joh85] surveys the algorithmic complexity of problems on graphs, including problems on planar graphs.

A graph G is said to be *planar* if it admits a drawing such that no two edges contain a common point except possibly a common end vertex. Such a drawing of a planar graph is called a *planar embedding* (or simply an *embedding*) of G. Wagner [Wag36], Fáry [Fár48], and Stein [Ste51] independently showed that every planar graph has an embedding in which the edges are straight line segments. This result also follows from Schnyder's characterization of planarity [Sch89].

Given a planar graph G together with an embedding, each connected subset of the plane that is delimited by a closed curve consisting of vertices and edges of G is called a *face* of the embedding. A face is said to be *incident* to the vertices and edges it is delimited by (and vice versa). All faces except one are bounded subsets of the plane. The unbounded face is called the *outer* face.

Figure 1 on page 12 shows the nonplanar graph G as well as two planar graphs G_1 and G_2. The drawing for G_1 is not an embedding, but the drawing for G_2 is. In Figure 2 on page 14, the graphs G_1, G_2, and G_3 are planar, and the drawing given for each of them is an embedding. The embedding for G_1 contains three faces, one incident to four vertices, another incident to five vertices, and a third one (the outer face) incident to seven vertices.

A planar graph together with an embedding is also called a *plane* graph. For a connected plane graph G with n vertices, m edges and f faces, Euler found the following formula:

$$n - m + f = 2 \qquad \text{(Euler 1750)} \qquad (1)$$

This can be shown by an induction over m (see for example [NC88]). Note that if a planar graph with $n \geq 3$ vertices has as many edges as possible, then each face is incident to exactly three vertices (for otherwise an additional edge could be added, dividing a face that is incident to more than three vertices into two faces, without violating planarity). Euler's formula together with this observation yields the following well known corollary:

$$m \leq 3n - 6 \qquad \text{(for } n \geq 3\text{)} \tag{2}$$

We now turn our attention to the question of deciding whether a given graph is planar. We first note that we can restrict ourselves to 2-connected graphs as stated by Kelmans [Kel93]: Clearly a graph is planar if and only if each of its connected components is planar. Furthermore, a connected graph is planar if and only if each of its 2-connected blocks is planar. [Kel93] goes on to show that we may even restrict our attention to 3-connected graphs.

First we will give some of the known characterizations of planar graphs. We start with Steinitz's Theorem, relating planar graphs to 3-dimensional polytopes. Given a 3-dimensional polytope P, its *edge graph* $G_P = (V_P, E_P)$ is formed as follows. Let V_P be the set of 0-dimensional faces[3] of P (i.e. the so-called vertices of P) and let E_P be the set of 1-dimensional faces of P (the so-called edges of P). Recalling that a polytope is convex by definition and that all graphs considered here are simple, Steinitz's Theorem [SR34] can be stated as follows [Whi84, p. 53],[RZ95]:

Theorem 3 (Steinitz 1922) *A graph G is the edge graph of a 3-dimensional polytope if and only if G is planar and 3-connected.*

For a proof, see [Grü67, Chapter 13]. As an example, observe that K_4 is the edge graph of a tetrahedron.

The most well known characterization of planar graphs is probably the one by Kuratowski [Kur30, KJ83]:

Theorem 4 (Kuratowski [Kur30]) *A graph G is planar if and only if it does not contain a subdivision of K_5 or $K_{3,3}$ as a subgraph.*

The graphs K_5 and $K_{3,3}$ are the complete graph on 5 vertices and the complete bipartite graph on two times three vertices as defined above. A subdivision of K_5 or $K_{3,3}$ that is contained as a subgraph in some graph G is called a *Kuratowski subgraph* of G. A proof of Kuratowski's Theorem can be found in [NC88] or [GT87], for example. The theorem was strengthened by Wagner [Wag37b], and, independently, by Hall [Hal43]. Kelmans [Kel93] states the stronger version as follows:

Theorem 5 (Wagner [Wag37b], Hall [Hal43]) *A 3-connected graph G distinct from K_5 is planar if and only if it does not contain a subdivision of $K_{3,3}$ as a subgraph.*

Wagner [Wag37a], and, independently, Harary and Tutte [HT65] give another characterization that can be stated in the following way:

---

[3]Note the difference between the face of a plane graph and the face of a polytope.

A. Liebers, *Planarizing Graphs*, JGAA, 5(1) 1–74 (2001) 9

Theorem 6 (Wagner [Wag37a], Harary and Tutte [HT65]) *A graph G is planar if and only if it does not contain K_5 or $K_{3,3}$ as a minor.*

For further characterizations of planar graphs see for example [Whi33, Mac37], [Sch89, dFdM96], [NC88, BS93, Kel93, ABL95], [dV90, dV93, Sch97], and [TT97].

An algorithm for determining whether a given graph is planar was first developed by Auslander and Parter [AP61] and Goldstein [Gol63]. Hopcroft and Tarjan [HT74] improved it to run in linear time. [Wil80] and [Mut94, p. 39] discuss the development of this result and give additional references. The algorithm tests the planarity of a given graph for each of its 2-connected blocks using the following idea recursively: Let $G = (V, E)$ be 2-connected. Let now $T = (V, E')$ be a depth first search tree[4] of G with root v, and let C be a cycle containing v and consisting of edges from E' plus one edge from $E \setminus E'$. For each edge e of G that is not part of C but that has at least one end vertex in C, consider a certain subgraph G_e of G and test (recursively) whether it can be embedded in the plane with certain edges bordering the outer face. After this has been done for each edge e emanating from C, test whether the embeddings of the different subgraphs G_e can be merged to embed G in the plane. [DETT99, Section 3.3] describes this algorithm in detail, and [Meh84, Section IV.10] additionally shows that it can be implemented in linear time.

This algorithm by Hopcroft and Tarjan tests whether a given graph is planar, but it is not obvious how to extract an embedding for the graph from it, if the graph is planar. Mutzel et al. [MMN93, MM96] modified the planarity testing algorithm to then also yield a *combinatorial embedding* of the graph in linear time, i.e. for each vertex a cyclic list of the incident edges so that the graph can be embedded in the plane obeying these edge sequences. Given a combinatorial embedding of a planar graph G with n vertices, de Fraysseix et al. construct a straight line embedding of G on a grid of size $2n - 4$ by $n - 2$ in time $\mathcal{O}(n \log n)$ [dFPP90]. This result was improved to a linear time algorithm finding a straight line embedding on a grid of size $n - 2$ by $n - 2$ by Schnyder [Sch90a]. See [DETT94, Section 5][DETT99, Chapter 4] for further discussions on drawing planar graphs.

Another linear time planarity testing algorithm was developed by Lempel, Even, and Cederbaum [LEC67]. They define an *st-numbering* as follows: Let $G = (V, E)$ be a 2-connected graph, and let $\{s, t\} \in E$ be an edge of G. An *st*-numbering is a bijection $f : V \rightarrow \{1, 2, \ldots, |V|\}$ such that $f(s) = 1$, $f(t) = |V|$, and such that for every $v \in V \setminus \{s, t\}$ there are vertices u and w in V with $\{u, v\} \in E$, $\{v, w\} \in E$, and $f(u) < f(v) < f(w)$.

[LEC67] shows that an *st*-numbering always exists. The idea of the planarity testing algorithm is this: For a 2-connected graph G, compute an *st*-numbering, and then try to build up a planar graph by starting with the vertex with *st*-number 1 and by adding the vertices of G together with their incident edges one by one according to their ascending *st*-numbers.

---

[4]For a description of depth first search, see for example [Meh84, Sections IV.4 and IV.5] or [TS92, Chapter 11.7].

A. Liebers, *Planarizing Graphs*, JGAA, 5(1) 1–74 (2001) 10

Even and Tarjan [ET76] showed that an *st*-numbering can be computed in linear time using depth first search. Using this result, and introducing a data structure called *PQ*-trees, Booth and Lueker [BL76] improved Lempel, Even, and Cederbaum's planarity testing algorithm to run in linear time.

The algorithm was modified to also yield a combinatorial embedding for the graph if it is planar by Chiba et al. [CNAO85]. [Eve79, Section 8.4] and [TS92, Section 11.11] describe the original algorithm [LEC67], and [Kan93, Section 2.2.2] describes the implementation [BL76] using *PQ*-trees.

Recently, two different, new, planarity testing and embedding algorithms have been proposed [SH99, BM99].

1.3 Generalizations of Planarity

Just as planar graphs are graphs embeddable in the 2-dimensional plane, we can consider graphs embeddable in other surfaces. By *surface* we mean a topological space that is a compact 2-manifold. A surface is characterized by its property of being either *orientable* or *nonorientable*, and by its *genus*. The *sphere* is the most simple orientable surface. It has genus 0. Informally speaking, the orientable surface S_g of genus $g \geq 0$ is the sphere with g handles attached to it. So S_0 denotes the sphere itself, whereas S_1 is also known as the *torus*. For the orientable surface S_g, the *Euler characteristic* of S_g is defined to be $E(S_g) = 2 - 2g$. See [WB78] and [Whi84, Chapters 5 and 6] for precise definitions and further explanations, in particular for the nonorientable case.

Note that the 2-dimensional plane is not compact, so it is not a surface in the above sense. But embedding a graph in the plane is equivalent to embedding it in the sphere (see [Whi84, Chapter 5] or [NC88, Section 1.3], for example).

The orientable (nonorientable) genus g of a graph G is defined to be the smallest g so that G can be embedded in an orientable (nonorientable) surface of genus g. It is NP-hard to determine the genus of a given graph [Tho89]. [DR91] provides an algorithm to determine the orientable genus of a graph. The running time of the algorithm is superexponential in the genus. Given an arbitrary but fixed surface S, [Moh96] presents a linear time algorithm that, for a given graph G, either finds an embedding of G in S, or finds a minimal forbidden subgraph H of G that cannot be embedded in S.

Besides considering different surfaces in which to embed a graph, further generalizations of planarity result when weaker forms of embedding a graph in a surface are considered. Graphs that can be drawn in a surface S so that each edge is involved in at most k edge crossings are called *k-embeddable in S*. So planar graphs are precisely the 0-embeddable graphs in the plane. [Sch90b] and [PT97] study 2-embeddable and k-embeddable graphs in the plane, respectively.

Considering graphs that can be drawn in the plane so that there are no k pairwise crossing edges, we get the planar graphs for $k = 2$. [AAP$^+$96] shows that for graphs with no three pairwise crossing edges and n vertices, the number of edges is in $O(n)$, and calls such graphs *quasi-planar*. For general k, see also [PSS94, PSS96] and [Val97, Val98] for recent work and further references.

A. Liebers, *Planarizing Graphs*, JGAA, 5(1) 1–74 (2001)   11

Chen et al. [CGP98] study intersection graphs of planar regions with disjoint interiors and call them *planar map graphs*. This generalizes planar graphs since planar graphs may be defined as the intersection graphs of planar regions with disjoint interiors such that no four regions meet at a point.

Yet another way of generalizing the concept of planarity is to weaken the characterizations of planarity that involve the Kuratowski graphs, (subdivisions of) K_5 and $K_{3,3}$, as subgraphs or minors of a graph. The result are four classes of graphs: Graphs that do not contain K_5 as a minor (or that do not contain a subdivision of K_5 as a subgraph) have been studied, and similarly for $K_{3,3}$ (see for example [Bar83, Khu90, KM92, NP94, MP95, Che96, Che98, JMOS98]).

2 Vertex Deletion

Given a graph $G = (V, E)$, we can transform it into a planar graph $G' = (V', E')$ in a trivial way by deleting all but four vertices of V from G together with all their incident edges. G' is then a tetrahedron (K_4) or a subgraph thereof, and hence planar. But we would hope to retain more than four vertices of the original graph and still obtain a planar subgraph. This section investigates the question of deleting as few vertices as possible (together with their incident edges) from a given graph G to make it planar. It seems that deleting vertices is too drastic an operation on a given graph to be useful in practice. The author is only aware of few results investigating vertex deletion for planarization.

Definition 7 (maximum induced planar subgraph)   *If a graph* $G' = (V', E')$ *is an induced planar subgraph of a graph* $G = (V, E)$ *such that there is no induced planar subgraph* $G'' = (V'', E'')$ *of* G *with* $|V''| > |V'|$, *then* G' *is called a* maximum induced planar subgraph *of* G.

So the problem of deleting as few vertices as possible from a graph so that the resulting graph is planar means to find, for a given graph G, a maximum induced planar subgraph of G.

Problem 8 (Maximum Induced Planar Subgraph [GJ79, Pr. GT21])
Given a graph $G = (V, E)$ *and a positive integer* $K \leq |V|$, *is there a subset* $V' \subseteq V$ *with* $|V'| \geq K$ *such that the subgraph of* G *induced by* V' *is planar?*

Lewis and Yannakakis [LY80] showed that this problem is NP-complete. [LY80] is based on independent work by the two authors and actually shows a far more general result:

Theorem 9 [LY80] *If* Π *is a graph property satisfying the following conditions*

1. *There are infinitely many graphs for which* Π *holds.*

2. *There are infinitely many graphs for which* Π *does not hold.*

3. *If* Π *holds for a graph* G *and if* G' *is an induced subgraph of* G, *then* Π *holds for* G'.

A. Liebers, *Planarizing Graphs*, JGAA, 5(1) 1–74 (2001)   12

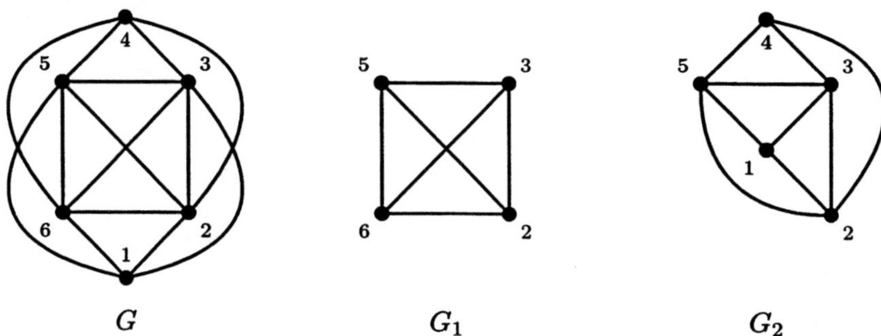

$$G \qquad\qquad G_1 \qquad\qquad G_2$$

Figure 1: G is a nonplanar graph (note that G contains K_5 as a subgraph). G_1 is a maximal induced planar subgraph of G. G_2 is another maximal induced planar subgraph of G, and G_2 is also a maximum induced planar subgraph of G.

then the following problem is NP-complete: Given a graph $G = (V, E)$ and a positive integer $K \leq |V|$, is there a subset $V' \subseteq V$ with $|V'| \geq K$ such that Π holds for the subgraph of G induced by V'?

Note that the graph property of being planar satisfies the three conditions of Theorem 9. Independently from Lewis and Yannakakis, Krishnamoorthy and Deo [KD79] also showed the NP-completeness of a whole range of vertex deletion problems including the maximum induced planar subgraph problem.

Djidjev and Venkatesan [DV95] show that for a graph G with n vertices and with orientable genus g, there exists a set of $4\sqrt{gn}$ vertices whose removal planarizes G, and that the size of this planarizing vertex set is optimal up to a constant factor. The proof is constructive and can be transformed into an $\mathcal{O}(n + g)$ time algorithm to find such a planarizing vertex set if the graph G is given together with an embedding on an orientable surface of genus g. But recall that it is NP-hard to determine the genus of a given graph [Tho89]. [DV95] goes on to show that if no such embedding of the graph is given, a planarizing vertex set of size $\mathcal{O}(\sqrt{gn\log(2g)})$ can be found in time $\mathcal{O}(n\log(2g))$. This algorithm recursively uses a graph partitioning algorithm also by Djidjev [Dji85]. However, no indications of computational studies or existing implementations are given. [DV95] improves results of [Dji84] and [HM87, Hut89], and also considers the nonorientable case.

Since Maximum Induced Planar Subgraph is an NP-complete problem, we also consider an easier problem:

Definition 10 (maximal induced planar subgraph)   *If a graph $G' = (V', E')$ is an induced planar subgraph of a graph $G = (V, E)$ such that every subgraph of G induced by a vertex set $V'' = V' \cup \{v\}$ with $v \in V \setminus V'$ is nonplanar, then G' is called a* maximal induced planar subgraph *of G.*

For a given graph G we want to find a maximal induced planar subgraph. Note that every maximum induced planar subgraph is also a maximal induced planar

A. Liebers, *Planarizing Graphs*, JGAA, 5(1) 1–74 (2001)   13

subgraph, but not vice versa. Observe that a maximal induced planar subgraph is maximal with respect to inclusion of its vertex set, whereas a maximum induced planar subgraph is maximal with respect to the cardinality of its vertex set. Analogous definitions concerning the edge set will be used in Section 3. Figure 1 illustrates maximal and maximum induced planar subgraphs.

A straightforward way of finding, for a given graph G with n vertices and m edges, a maximal induced planar subgraph is the Greedy Algorithm: The input is a graph $G = (V, E)$ with n vertices and m edges. The output is a maximal induced planar subgraph $G' = (V', E')$ of G. We start with G' as the empty graph (so $V' = \emptyset$ and $E' = \emptyset$). One vertex of V after the other is taken and either added to V' (if the subgraph of G induced by V' remains planar) or discarded, until every vertex of V has been considered. The order in which the vertices of V are considered is arbitrary. Considering the worst case time complexity of this algorithm, we have to perform a planarity test and to update V' and E' in each of the n iterations. Each planarity test takes $O(n + m)$ time in the worst case. Each update of V' takes $O(1)$ time. All updating operations for E' together take $O(m)$ time. Thus the overall time complexity of the Greedy Algorithm is in $O(n \cdot m)$ (assuming that G is connected, so that $m \in \Omega(n)$). The resulting vertex set V' usually depends on the order in which the vertices of V are considered. However, the author is not aware of work investigating the impact of different vertex orderings on the resulting maximal induced planar subgraph.

3 Edge Deletion and Skewness

If a graph $G = (V, E)$ with an edge $e \in E$ is transformed into a graph $G' = (V, E \setminus \{e\})$ then we say that G' was obtained from G by *edge deletion*. By repeatedly deleting edges from a given nonplanar graph G, G can be transformed into a planar graph G'. In this section, we are interested in planarizing G by deleting as few edges as possible.

Deleting edges from a given graph G in order to transform G into a graph G' with a particular property is a common approach (see for example [SC89, Sen90]). We will only discuss edge deletion with the purpose of planarization, a topic that has been studied intensively, and that has applications in graph drawing [DETT99], for example.

Definition 11 (maximum planar subgraph, skewness)   *If a graph $G' = (V, E')$ is a planar subgraph of a graph $G = (V, E)$ such that there is no planar subgraph $G'' = (V, E'')$ of G with $|E''| > |E'|$, then G' is called a* maximum planar subgraph *of G, and the number of deleted edges, $|E| - |E'|$, is called the* skewness *of G.*

So the skewness of a graph G is 0 if and only if G is planar. The problem of finding, for a given graph G, a maximum planar subgraph is NP-hard [LG79]. It will be discussed in Section 3.1. For some graph classes, the skewness is known: The complete graph K_n has $n(n-1)/2$ edges. For $n \geq 3$, it has a

A. Liebers, *Planarizing Graphs*, JGAA, 5(1) 1–74 (2001)   14

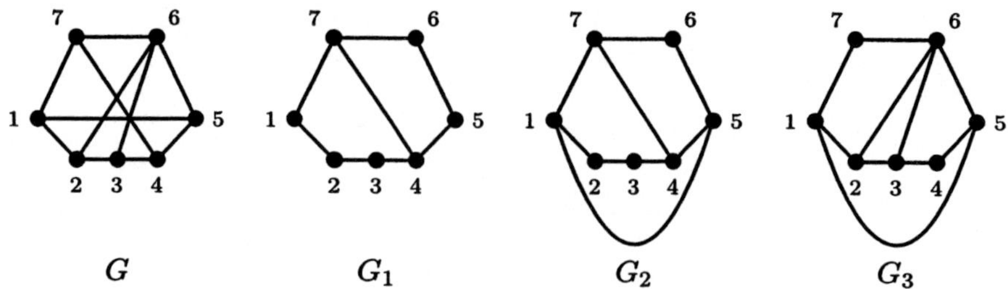

Figure 2: G is a nonplanar graph. Note that G contains $K_{3,3}$ as a minor (contract edge $\{2,3\}$). G_1 is a planar subgraph of G, but it is not a maximal planar subgraph: Edge $\{1,5\}$ can be added to G_1 without destroying planarity. The result is G_2. Another maximal planar subgraph of G is G_3. G_3 is also a maximum planar subgraph.

planar subgraph with $3n - 6$ edges. Since a planar graph with $n \geq 3$ vertices cannot have more than $3n - 6$ edges (Equation 2), the skewness of the complete graph K_n is $n(n - 1)/2 - (3n - 6) = (n - 3)(n - 4)/2$ for $n \geq 3$. A similar argument shows that the skewness of the complete bipartite graph K_{n_1,n_2} is $n_1 \cdot n_2 - 2(n_1 + n_2) + 4$ for $n_1 \geq 2$ and $n_2 \geq 2$. The skewness of the hypercube of dimension n, Q_n, is $2^n(n - 2) - n \cdot 2^{n-1} + 4$ [Cim92].

Definition 12 (maximal planar subgraph) *If a graph $G' = (V, E')$ is a planar subgraph of a graph $G = (V, E)$ such that every graph $G'' \in \{(V, E' \cup \{e\}) \mid e \in E \setminus E'\}$ is nonplanar, then G' is called a* maximal planar subgraph *of G.*

In other words a maximal planar subgraph is maximal with respect to inclusion of its edge set, whereas a maximum planar subgraph is maximal with respect to the cardinality of its edge set. Observe that every maximum planar subgraph is also a maximal planar subgraph, but not vice versa. Also note the analogy with Definitions 7 and 10 concerning the vertex set of a graph. Figure 2 illustrates maximal and maximum planar subgraphs.

Finding a maximum planar subgraph is an NP-hard problem, and Section 3.1 discusses this result. But a maximal planar subgraph can be found in polynomial time, as will be seen in Section 3.2. Finally, Section 3.3 discusses approximative and heuristic approaches for finding a large planar subgraph. It also considers the weighted version, where edges are assigned nonnegative edge weights, and where the goal is to find a planar subgraph with total edge weight as large as possible.

For another survey of algorithms for planarization through edge deletion, see Mutzel [Mut94, Chapter 5].

3.1 Finding a Maximum Planar Subgraph

In this section, we study the following problem:

A. Liebers, *Planarizing Graphs*, JGAA, 5(1) 1–74 (2001)   15

Problem 13 (Maximum Planar Subgraph [GJ79, Problem GT27])
Given a graph $G = (V, E)$ and a positive integer $K \leq |E|$, is there a subset $E' \subseteq E$ with $|E'| \geq K$ such that the graph $G' = (V, E')$ is planar?

Liu and Geldmacher [LG79], and, independently, Yannakakis [Yan78][5], and, also independently, Watanabe et al. [WAN83],[6] showed that this problem is NP-complete. The proof of Liu and Geldmacher is a two step reduction using the following problems:

Problem 14 (Vertex Cover [GJ79, Problem GT1])[7]   *Given a graph $G = (V, E)$ and a positive integer $K \leq |V|$, is there a vertex cover of size K or less for G, i.e. is there a subset $V' \subseteq V$ of vertices with $|V'| \leq K$ such that for each edge $uv \in E$ at least one of its end vertices u and v belongs to V'?*

Problem 15 (Hamilton Path in Graphs Without Triangles) *Given   a graph $G = (V, E)$ that does not contain a cycle of length 3, and given two vertices $u \in V$ and $v \in V$, does G contain a Hamilton path from u to v?*

Karp [Kar72] shows Vertex Cover to be NP-complete. [LG79] first reduces Vertex Cover to Hamilton Path in Graphs Without Triangles, and then reduces this problem to Maximum Planar Subgraph. Recently, Faria, Figueiredo, and Mendonça [FFM98a, FFM01] have shown that Maximum Planar Subgraph is even NP-complete for cubic graphs (see Section 4.2).

Djidjev and Venkatesan [DV95] show that for a graph G with m edges, maximum vertex degree Δ, and orientable genus g, there exists a set of $4\sqrt{\Delta gm}$ edges whose removal planarizes G, and that the size of this planarizing edge set is optimal up to a constant factor. If G is connected and has n vertices, then there exists a set of $4\sqrt{2\Delta g(n + 2g - 2)}$ edges whose removal planarizes G. The proofs are constructive and can be transformed into $\mathcal{O}(n + g)$ time algorithms to find such planarizing edge sets if the graph G is given together with an embedding on an orientable surface of genus g. But recall that it is NP-hard to determine the genus of a given graph [Tho89]. [DV95] states that if no such embedding of the graph is given, a planarizing edge set of size $\mathcal{O}(\sqrt{\Delta gm} \log(2g))$ can be found in time $\mathcal{O}(m \log(2g))$.

[DV95] improves results in [Dji84]. No indications of computational studies or existing implementations for these algorithmic results are given though. For the corresponding results concerning planarizing vertex sets see Section 2.

3.2 Finding a Maximal Planar Subgraph

The problem of finding a maximal planar subgraph for a given graph G with n vertices and m edges is solvable in polynomial time. A straightforward way

---

[5][Yan78] shows the NP-completeness of several edge deletion problems. The journal version [Yan81] does not contain the NP-completeness proof for the maximum planar subgraph problem anymore but refers to [LG79], which uses similar ideas as presented in [Yan78, Yan81].

[6]Watanabe et al. [WAN83] show the NP-completeness for a whole class of edge deletion problems.

[7]Liu and Geldmacher call this problem Vertex Edge-Cover.

of finding a maximal planar subgraph is the Greedy Algorithm: The input is a graph $G = (V, E)$ with n vertices and m edges. The output is a maximal planar subgraph $G' = (V, E')$ of G. We start with $G' = (V, \emptyset)$ and build up E' by considering one edge e of E after the other. For each $e \in E$, e is added to E' if $G' = (V, E')$ remains planar, and discarded otherwise. We stop either after all edges of E have been considered, or when $|E'|$ becomes equal to $3n - 6$ (since a planar graph cannot have more than $3n - 6$ edges). For each edge of E that is considered we need to perform a planarity test for a graph with n vertices and at most $3n - 6$ edges. Each planarity test takes linear time, i.e. $O(n)$ in the worst case. The remaining operations like updating E' take $O(1)$ time per edge. Thus the worst case time complexity is in $O(n \cdot m)$.

The standard algorithms for planarity testing [HT74, BL76] are rather complicated to implement. Therefore, algorithms for finding a maximal planar subgraph are sought that not only have a better worst case time complexity than the algorithm described above, but that are also less involved.

T. Chiba, Nishioka, and Shirakawa [CNS79] propose an algorithm based on the planarity testing algorithm [HT74]. They achieve a worst case time complexity of $O(n \cdot m)$, the same as that of the Greedy Algorithm.

A whole series of results about better polynomial time algorithms for finding a maximal planar subgraph starts with [OT81], which also claims to give an $O(n \cdot m)$ algorithm. In contrast to [CNS79], [OT81] is based on the planarity testing algorithm [LEC67, BL76] using PQ-trees. The algorithm starts with one vertex as the initial planar subgraph and then adds one vertex (together with as many of its incident edges as possible) at a time. But [TJS86] points out that the subgraph generated by this algorithm is not always a maximal planar subgraph, and that it is not even always a spanning subgraph. [JST89, JTS89] claim to amend the problem and give two $O(n^2)$ algorithms, one to find a spanning planar subgraph of a 2-connected graph G, and one to find a maximal planar subgraph by augmenting the previously found spanning planar subgraph. The latter algorithm is shown to be incorrect by [Kan92, Kan93], claiming to show how to correct the algorithm. [Lei94, JLM97] in turn point out that the result in [Kan92] is not correct either and discuss the difficulties of using PQ-trees.

Di Battista and Tamassia [DT89, DT96b][DT90, DT96a] define and use $SPQR$-trees to describe the recursive decomposition of a 2-connected graph into its 3-connected components. [DT89] obtains an $O(m \log n)$ time algorithm for finding a maximal planar subgraph as a byproduct of an algorithm for *incremental planarity testing*. An incremental (or *dynamic*) planarity testing algorithm maintains a data structure representing a planar graph $G = (V, E)$ and can handle requests of the following types: *a)* For two vertices v_1 and v_2 in G with $v_1 v_2 \notin E$, determine whether G stays planar if the edge $v_1 v_2$ is added to G. *b)* If $v_1 \in V$, $v_2 \in V$, $v_1 v_2 \notin E$, add the edge $v_1 v_2$ to G (assuming the corresponding request of type a yields a positive answer). *c)* Add a new vertex to G.

Independently, Cai, Han, and Tarjan [CHT93] developed an $O(m \log n)$ algorithm to find a maximal planar subgraph of a graph G with n vertices and m edges. Their algorithm is based on a new version of the Hopcroft and Tarjan

A. Liebers, *Planarizing Graphs*, JGAA, 5(1) 1–74 (2001)

planarity testing algorithm [HT74].

La Poutré [La 94] presents algorithms for incremental planarity testing that yield an $O(n + m \cdot \alpha(m, n))$ time algorithm for the maximal planar subgraph problem (where $\alpha(m, n)$ is the functional inverse of the Ackermann function). This result was improved to linear time complexity by Djidjev [Dji95], and, independently, by Hsu [Hsu95].

Given a graph $G = (V, E)$, Djidjev [Dji95] first computes a depth first search tree of G. This spanning tree of G is the initial planar subgraph $G' = (V, E')$ of G. Then for each edge $e \in E \setminus E'$ it is determined whether the graph $(V, E' \cup \{e\})$ is still planar. If so, e is added to E'. The order in which the edges in $E \setminus E'$ are considered is chosen in a sophisticated way so that, with $O(1)$ amortized time per test and insert operation for each edge $e \in E \setminus E'$, the overall time complexity is linear. Many intricate data structures are needed to achieve the $O(1)$ amortized time per test and insert operation. Two of them are BC-trees to describe the decomposition of a connected planar graph into its 2-connected components[8] and $SPQR$-trees to describe the decomposition of a 2-connected graph into its 3-connected components [DT96a]. Djidjev's algorithm is linear and therefore asymptotically best possible. However, it is so involved that a linear implementation seems difficult to achieve.

[Hsu95] also starts with a depth first search tree of the given graph $G = (V, E)$, and then determines a postordering of the vertices of G. The postordering is a labeling $l : V \rightarrow \{1, \ldots, n\}$ so that if u is an ancestor of v in the depth first search tree, then $l(u) > l(v)$. The initial planar subgraph G' of G is empty, and the vertices are added in ascending order of their labels. So in step i of the algorithm, the vertex with label i (and the edges incident to it) are added to G'. Note that G' is not necessarily connected at all times. According to [Hsu95], the way in which the vertices are added and in which for each edge it is decided whether the edge can be added to G' without destroying planarity ensures the construction of a maximal planar subgraph in linear time. So the algorithm also achieves the asymptotically best possible time complexity, and it appears to be less complicated than that of Djidjev. However, the conference version [Hsu95] does not contain the details of the algorithm and of the proof of its correctness.

3.3 Approximations and Heuristics

First consider a trivial approximation for finding a maximum planar subgraph by observing that for a given graph G with n vertices, any spanning tree of G is a planar subgraph with $n - 1$ edges (assume that G is connected), and that a spanning tree can be found in linear time. Furthermore, a planar subgraph of G cannot have more than $3n - 6$ edges (see Equation 2). So if E' is the edge set of a spanning tree for a given graph G, and if E^* is the edge set of a maximum

---

[8]In [Har69] these trees are called block-cutpoint trees.

A. Liebers, *Planarizing Graphs*, JGAA, 5(1) 1–74 (2001) 18

planar subgraph of G, then the ratio $\frac{|E'|}{|E^*|}$ is bounded (see also [Cim92]):

$$\frac{|E'|}{|E^*|} = \frac{n-1}{|E^*|} \geq \frac{n-1}{3n-6} > \frac{1}{3} \tag{16}$$

This bound was improved for the first time by Călinescu et al. to $\frac{4}{9}$ [CFFK96, CF96, CFFK98].

Let's turn our attention to the weighted version of the Maximum Planar Subgraph Problem:

Problem 17 (Weighted Maximum Planar Subgraph)   *Given a graph $G = (V, E)$ with a nonnegative edge weight $w(e)$ for each edge e, and a positive number K, is there a subset $E' \subseteq E$ with $\sum_{e \in E'} w(e) \geq K$ such that the graph $G' = (V, E')$ is planar?*

Being a generalization of the Maximum Planar Subgraph Problem, Weighted Maximum Planar Subgraph is NP-complete as well.

The Greedy Algorithm of Section 3.2 finds a maximal planar subgraph, which will be at least as good as just taking a spanning tree. [KH78, DFF85, FGG85] use this greedy approach for a heuristic for the Weighted Maximum Planar Subgraph problem: Instead of considering the edges in arbitrary order, they consider them in an order of nonincreasing weight. This Greedy Heuristic does involve repeated planarity testing, and even though planarity testing can be done in linear time, the algorithms are rather complicated. The following heuristics avoid planarity testing.

The Deltahedron Heuristic [FR78, FGG85] starts with a tetrahedron (K_4) as the initial planar subgraph and then adds one vertex at a time, placing each new vertex in one of the faces of the current planar subgraph (see the left part of Figure 5 on page 21 for an illustration). The sequence in which the vertices are added is determined by a vertex weight W that can be defined in various ways, as discussed below. Figure 3 shows the Deltahedron Heuristic in detail. Note that in contrast to the Greedy Heuristic, the Deltahedron Heuristic does not necessarily yield a maximal planar subgraph of the input graph.

[FGG85] assigns the vertex weights as the sum of the weights of incident edges: $W(v) = W_{sum}(v) = \sum_{u \in V} w(uv)$. [DFF85] suggests to use the maximum of the vertex weights instead of the sum: $W(v) = W_{max}(v) = \max_{u \in V}\{w(uv)\}$, and also provides a worst case analysis for the performance of the Greedy Heuristic and the two versions of the Deltahedron Heuristic.

Definition 18 (worst case ratio) *Let P denote an instance of the Weighted Maximum Planar Subgraph Problem with a graph $G = (V, E)$ and positive edge weights $w(e)$ for $e \in E$. If A is an algorithm that finds a planar subgraph $G' = (V, E')$ of G, and if $E^* \subseteq E$ is an optimal edge set, i.e. if $G^* = (V, E^*)$ is planar, and if $w(E^*) = \sum_{e \in E^*} w(e)$ is as large as possible, then the* worst case ratio, *denoted ρ_A, is defined as*

$$\rho_A = \inf_P \frac{w(E')}{w(E^*)} \; .$$

A. Liebers, *Planarizing Graphs*, JGAA, 5(1) 1–74 (2001)   19

Input: A graph $G = (V, E)$ and real edge weights $w(e) \geq 0$ for $e \in E$
Output: A planar subgraph $G' = (V, E')$ of G
1   Build a complete graph $G_K = (V, E_K)$ from G by adding an edge
     $e = uv$ with $w(e) = 0$ for each pair uv of nonadjacent vertices in V.
     Let $EE = E_K \setminus E$ be this set of "extra" edges.
2   Assign a vertex weight $W(v)$ to each $v \in V$.
3   Sort the vertices by vertex weight in nonincreasing order and
     let a, b, c, and d be the first four vertices in that order.
4   Let G' be the tetrahedron on a, b, c, and d,
     i.e. let G' have edge set $E' = \{ab, ac, ad, bc, bd, cd\}$.
     Let $T = \{abc, acd, abd, bcd\}$ be the set of faces of G'.
     By construction, each face of G' is a triangle.
5   As long as there is a vertex in V that has not yet been added to G':
6   Let v be a vertex with largest weight among those not yet in G'.
7   Choose a face $xyz \in T$ such that
         $w(xv) + w(yv) + w(zv)$ is as large as possible.
8   Add v to the face xyz, i.e. set $E' = E' \cup \{xv, yv, zv\}$,
         and set $T = (T \setminus \{xyz\}) \cup \{xyv, yzv, zxv\}$.
(G' is now a triangulated graph with n vertices and $3n - 6$ edges.)
9   If E' contains "extra" edges from EE, eliminate them from E'.

Figure 3: The Deltahedron Heuristic for finding a planar subgraph with large edge weights.

Clearly $\rho_A \leq 1$ for any algorithm A. The closer ρ_A is to 1, the better A performs (in the worst case).

[DFF85] shows that the Deltahedron Heuristic with vertex weights W_{sum} can be arbitrarily bad in the general case but has a performance guarantee in the unweighted case (i.e. if the edge weights are restricted to 0 and 1). But for the unweighted case, the worst case ratio of the trivial approximation using a spanning tree (Equation 16) is higher anyway.

The Deltahedron Heuristic with vertex weights W_{max} and the Greedy Heuristic both have performance guarantees in the general case. Figure 4 lists the results presented in [DFF85]. They show that the Greedy Heuristic is the best algorithm as far as worst case analysis is concerned. Figure 4 also lists the time complexities of the above algorithms as given in [FGG85].

[FGG85] suggests improving the result of the Deltahedron Heuristic by edge replacement or vertex relocation operations in a postprocessing phase and additionally discusses the wheel expansion approach of [EFG82].

To compare the performance of these heuristics, computational results are reported in detail in [FGG85]. Complete graphs with 10, 20, 30, and 40 vertices and with a normal distribution on the edge weights with mean value 100 and standard deviations in the range from 5 to 30 are generated, and planar sub-

A. Liebers, *Planarizing Graphs*, JGAA, 5(1) 1–74 (2001)   20

algorithm A	worst case ratio ρ_A	worst case time complexity
Greedy Heuristic	$\frac{1}{3}$	$O(n^3)$
Deltahedron Heuristic with vertex weights W_{sum}	0	$O(n^2)$
Deltahedron Heuristic with vertex weights W_{sum} in the unweighted case	$\frac{1}{6} \leq \rho_A \leq \frac{2}{9}$	
Deltahedron Heuristic with vertex weights W_{max}	$\frac{1}{6}$	

Figure 4: The results of [DFF85] show the worst case performance of three algorithms for finding a planar subgraph with a large sum of edge weights. The worst case time complexity of the algorithms for an input graph with n vertices is given in [FGG85].

graphs of these are constructed with the Deltahedron Heuristic, the improved Deltahedron Heuristic, the wheel expansion heuristic (each with vertex weights W_{sum} and each in two versions depending on the choice of the initial K_4), and the Greedy Heuristic. For each one of the altogether 102 planar subgraphs constructed, the performance ratio of the respective algorithm was never below 0.87, where the improved Deltahedron Heuristic with vertex weights W_{sum}[9] and the Greedy Heuristic outperformed the other heuristics, and the Greedy Heuristic was better than the improved Deltahedron Heuristic. The performance ratio for the Greedy Heuristic was never below 0.91. The Greedy Heuristic, however, required 5 to 10 times the run time of any of the other heuristics.

Besides the worst case analysis mentioned above, [DFF85] also analyzes a simplification of the Deltahedron Heuristic: The vertices are considered in arbitrary order instead of in order of nonincreasing vertex weights. In the worst case, this heuristic can be arbitrarily bad, even in the unweighted case. But [DFF85] shows that under the assumption that the edge weights are independent and that they are chosen from a probability density restricted to a bounded interval of the nonnegative reals, as the number n of vertices tends to infinity, the probability that the performance ratio $\frac{w(E')}{w(E^*)}$ is below $1 - n^{-0.1}$ tends to zero.

Leung [Leu92] generalizes the Deltahedron Heuristic. Starting with a tetrahedron (K_4), a planar subgraph is built such that in each step, the current planar subgraph has only triangular faces. In each step, a single vertex and three incident edges (as in the Deltahedron Heuristic) or a set of three vertices

---

[9]There are no computational results for the Deltahedron Heuristic with vertex weights W_{max}.

A. Liebers, *Planarizing Graphs*, JGAA, 5(1) 1–74 (2001)   21

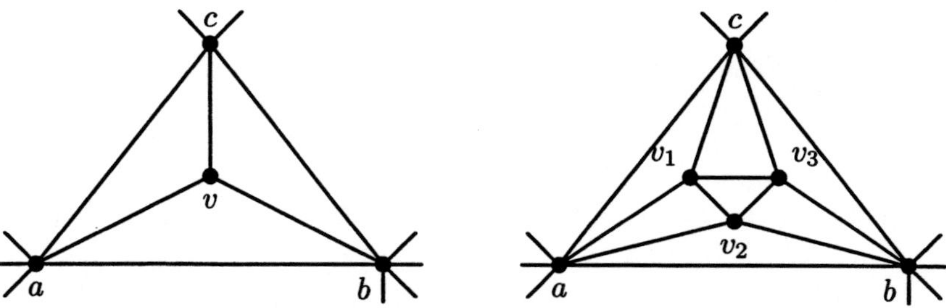

Figure 5: A step in the Deltahedron Heuristic for finding a planar subgraph with large edge weights [FR78, FGG85, DFF85] (left; see also Figure 3), or in its generalization [Leu92](left or right). In the operation on the left, vertex v and 3 incident edges are inserted into face abc. In the operation on the right, vertices v_1, v_2, v_3 and 9 incident edges are inserted into face abc.

and nine incident edges are placed in one of the faces of the current planar subgraph as illustrated in Figure 5. Unlike in the Deltahedron Heuristic, the vertices to be inserted are not chosen in any predetermined ordering, but in each step the vertex or the set of three vertices, and the face into which to insert them, is determined so that the gain in edge weights per inserted vertex in this step is best possible. The worst case time complexity of this approach is $O(n^4 \log n)$. Computational results are carried out, generating the test base in much the same way as [FGG85]. They suggest that the results of the generalized Deltahedron Heuristic are better than the ones achieved by the original Deltahedron approach discussed in [FR78, FGG85], but there is no comparison with the improved Deltahedron Heuristic of [FGG85] or with the Greedy Heuristic.

A completely different approach is taken by Jünger and Mutzel, who use a heuristic based on polyhedral combinatorics within a branch and cut framework [Mut94, JM96].

[JM96] reports computational results where the branch and cut heuristic was applied to various graphs known from the literature with 10 to 100 vertices. In many cases, a provably optimal solution, or at least a solution that is better than the previously known one, could be found. But the running time needed is usually significantly larger than the running time of other algorithms. In fact, Jünger and Mutzel interrupt their algorithm when a time limit of 1000 CPU seconds is reached. They find that the easiest problem instances are sparse graphs and very dense graphs, and that for weighted graphs the performance of their branch and cut heuristic is much worse than for unweighted graphs.

For the unweighted case (i.e. all edge weights are 1) there are still other approaches. Cimikowski [Cim92] suggests a heuristic based on spanning trees. Suppose a graph G with n vertices and m edges is 2-connected and has two edge disjoint spanning trees whose union is planar. Then this union forms a planar subgraph and has $2n - 2$ edges. If the graph does not have two such spanning trees, some heuristic edge manipulations are performed, so that the output is still a spanning planar subgraph, but without a guaranteed number of edges. If two spanning trees exist, they can be found in $O(m^2)$ [RT85].

Takefuji and Lee [TL89, TLC91] and Goldschmidt and Takvorian [GT94] each propose a two-phase heuristic for finding a planar subgraph with as many edges as possible. In the first phase, a linear ordering of the vertices is determined. The vertices are placed on a line according to that ordering. In the second phase edges are placed above or below the line. The resulting planar subgraph is thus embedded in a book with two pages. The techniques used for each phase are very different in [TL89] and [GT94]. [TL89] places the vertices in a random order in the first phase and uses a neural network technique for the second phase.

[GT94] argues that it is useful to attempt to order the vertices of the input graph $G = (V, E)$ according to a Hamiltonian cycle in the first phase. Given an ordering of the vertices on a line, in the second phase a partition of E into three sets A, B, and C must be determined so that $|A| + |B|$ is as large as possible, and so that no two edges of A (B) intersect if all edges of A (B) are placed above (below) the line of vertices. The edges in C are not part of the planar subgraph. If we imagine the vertices of G to lie on the real line, then each edge $e \in E$ can be regarded as an interval defined by its two end vertices. Let $H = (E, F)$ be a graph such that each edge of G is a vertex of H. Let e_1, e_2 be two edges of G and thus two vertices of H, and let i_1 and i_2 be the intervals corresponding to the edges e_1 and e_2 in G. e_1 and e_2 are connected by an edge in H if and only if the intervals i_1 and i_2 intersect but none is contained in the other. Thus H is an overlap graph (also called circle graph). Finding the sets A, B and C as described above is now equivalent to finding a maximum induced bipartite subgraph of the overlap graph H. Finding a maximum induced bipartite subgraph of an overlap graph is NP-complete [SL89].

[GT94] now uses the following greedy algorithm to construct a maximal induced bipartite subgraph of an overlap graph: Find a maximum independent vertex set in H (the vertices of this set are then the edges in A), delete it from H, and find a maximum independent set in the remaining graph (the vertices of this set are then the edges in B). Since a maximum independent set of an overlap graph can be found in polynomial time [Gav73], this algorithm runs in polynomial time also. [GT94] shows that the number of vertices in the maximal induced bipartite subgraph is at least 0.75 times the number of vertices of a maximum bipartite subgraph.

Computational results reported by Goldschmidt and Takvorian [GT94] compare their implementation of their heuristic with their implementation of [TL89] on a set of 19 graphs with 10 to 150 vertices and two larger graphs with 300 and 1000 vertices, respectively. For each instance, their heuristic finds at least as good a solution as [TL89]. For the graphs with 50 or more vertices, the solution of [GT94] is even dramatically better than that of [TL89]. But note that the test base is small, that it is unclear how representative it is, and that even the results of [GT94] might still be very far away from an optimal solution.

A. Liebers, *Planarizing Graphs*, JGAA, 5(1) 1–74 (2001)   23

The approach of [GT94] is further refined by Resende and Ribeiro [RR97]. They apply a greedy randomized adaptive search procedure (GRASP), a meta-heuristic for combinatorial optimization [FR95], to the problem of planarizing a graph through edge deletion. Experimental results using most graphs from the test base in [GT94] as well as graphs with up to 300 vertices collected by Cimikowski are discussed in [RR97, RR98]. They indicate that the GRASP compares favorably with the results of [GT94]. In comparison with the branch and cut heuristic [Mut94, JM96], however, the situation is not so clear: On some instances the branch and cut heuristic is clearly better, on others the GRASP outperforms the branch and cut heuristic. The latter happens in particular when the time limit set for the branch and cut heuristic is reached so that the computation is halted and the best solution found until then is reported.

Still further results presenting and comparing different heuristics are given in [Cim94, Cim95a, Cim97][Com92].

For algorithmic results, and in particular for approximations and heuristics, computational results are an important performance measure, both regarding the quality of the result of the algorithm and the running time needed. But a fair comparison of algorithms with each other on the basis of computational results is usually difficult, if not impossible, since the implementation of an algorithm and the graphs used for the test strongly influence the computational results. With this in mind, the comparisons of algorithms made in this section have to be considered with caution.

4   Vertex Splitting and Splitting Number

The vertex splitting operation on a graph is the reversal of identifying two vertices:

Definition 19 (vertex splitting) *If* $G' = (V', E')$ *is a graph with two vertices* $v_1 \in V'$ *and* $v_2 \in V'$, *and if* $G = (V, E)$ *is the graph obtained from* G' *with*

$$
\begin{aligned}
V &= (V' \setminus \{v_1, v_2\}) \cup \{v\} \quad \text{and} \\
E &= (E' \setminus \{uv_i \mid u \in V' \text{ and } i \in \{1,2\} \text{ and } uv_i \in E'\}) \\
&\quad \cup \{uv \mid u \in V \setminus \{v\} \text{ and } (uv_1 \in E' \text{ or } uv_2 \in E')\}
\end{aligned}
$$

then we say that G' *was obtained from* G *by* splitting *the vertex* v.

If a graph G' has been obtained from a graph G by a (possibly empty) sequence of vertex splitting operations, we call G' a *splitting* of G. Note that even if there is a vertex $x \in V'$ such that $xv_1 \in E'$ and $xv_2 \in E'$, no multiple edges are formed in G by the vertex identification operation. Likewise, no loop vv is formed in G, even if $v_1v_2 \in E'$.

The vertex identification of given vertices v_1 and v_2 in a given graph G' yields a unique graph G. But the opposite is not true: Given a graph G and one of its vertices v, there are many ways to split this vertex. Given the graph $G = K_3$, for example, and one of its vertices v, there are six ways to perform a

A. Liebers, *Planarizing Graphs*, JGAA, 5(1) 1–74 (2001)   24

vertex splitting at v such that the resulting graphs are pairwise non-isomorphic (see Figure 6).

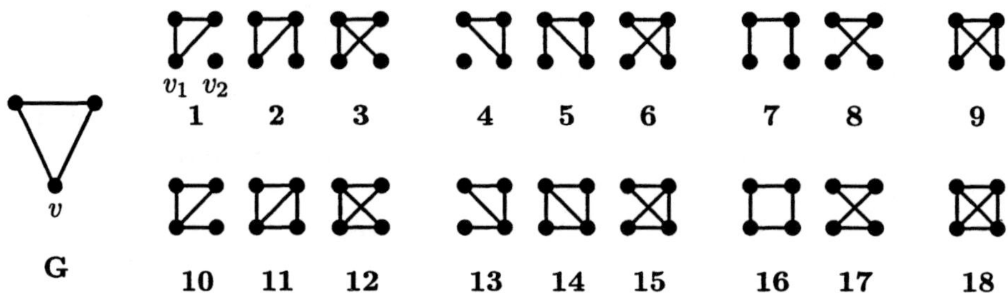

Figure 6: Eighteen possible ways to split $G = K_3$ at v. Essentially, there are only six ways to split v in G: The graphs numbered 1 and 4 are isomorphic. The graphs numbered 2, 3, 5, 6, 10, and 13 are isomorphic. The graphs numbered 7 and 8 are isomorphic. The graphs 9, 11, 12, 14, and 15 are isomorphic. The graphs 16 and 17 are isomorphic.

One might want to define a vertex splitting in a more general way as the reversal of identifying k vertices of a graph at once, where $k \geq 2$. So a splitting of a vertex v would result in vertices $v_1 \ldots v_k$ so that the adjacencies of $v_1 \ldots v_k$ cover the adjacencies of v in the original graph. But since splitting a vertex k ways can always be regarded as $(k - 1)$ successive vertex splitting operations where each vertex splitting is only a 2-way-splitting, we restrict our definition of vertex splitting to splitting a vertex v into exactly two vertices v_1 and v_2.

The vertex splitting operation has appeared in very different contexts. Note, for example, that decomposing a graph into its 2-connected blocks means performing a vertex splitting at every cut vertex. Already Steinitz and Rademacher [SR34] observed (as restated in [Sch91]) that every triangulation of the plane can be generated from a planar embedding of K_4 by vertex splitting operations (note that a planar embedding of K_4 is in itself a triangulation of the plane). Similar results hold for other surfaces [Bar82, Bar87, BE88, BE89, Bar90, Bar91, Sch91, MM92, FMN94]. The vertex splitting operation is central to Tutte's [Tut61, Tut66], Slater's [Sla74], Chen and Kanevsky's [CK93], and Gubser's [Gub93] characterizations of various classes of graphs.

Given a graph $G = (V, E)$ and a function $f : V \to I\!N$, Nash-Williams [NW79, NW85a, NW85b, NW87] answers questions of the following type: Can G be transformed into a graph H by a sequence of vertex splitting operations such that H has a certain property and such that each $v \in V$ results in $f(v)$ vertices $v_1, v_2, \ldots, v_{f(v)}$ in H? The work in [Arc84], [Yap81, Yap83], and [Sel88] about graph coloring also uses vertex splitting, and Mayer and Ercal [ME93a, ME93b] attack the following NP-hard problem with genetic algorithms: Given a directed, acyclic graph G and a positive number δ, determine a set X of vertices in G with minimum cardinality so that performing a vertex splitting operation on

A. Liebers, *Planarizing Graphs*, JGAA, 5(1) 1–74 (2001)   25

each vertex in X transforms G into a graph where the length of the longest directed path is at most δ.

Eades and Mendonça [Men94, EM96] address the problem of finding a planar embedding for a graph G with edge weights such that for each edge uv, the Euclidean distance between u and v in the layout is proportional to the weight of the edge uv. In general, finding such a layout for a given graph G with given weights is impossible, but by applying proper vertex splitting operations to G, G can be transformed into a graph H that admits a layout with the desired property. Determining the least number of vertex splitting operations required to achieve this is NP-complete [Men94, Section 4.4.1],[ELMM95]. Heuristics are given to solve the problem.

For the remainder of this section, we are concerned with vertex splitting operations as a means to planarize a given graph: Given a graph G, we want to know the smallest number k, so that G can be planarized by k vertex splitting operations. In other words:

Definition 20 (splitting number) *Given a graph G, the* splitting number *of G, denoted $\sigma(G)$, is the smallest number k, so that G can be obtained from a planar graph G' by k vertex identifications (of 2 vertices each).*

Clearly $\sigma(G) = 0$ if and only if G is planar. If a planar graph G' was obtained from a graph G by vertex splitting operations, we call G' a *planar* splitting of G. If additionally, G' was obtained by $\sigma(G)$ vertex splitting operations, we call G' an *optimal* planar splitting of G. For a general surface S, $\sigma(G, S)$ denotes the smallest number k, so that G can be obtained from a graph G' by k vertex identifications, where G' is embeddable in S.

Investigation of the splitting number seems to have started with the work of Hartsfield, Jackson and Ringel in the 1980s about lower bounds for the splitting number and about splitting vertices of complete graphs and of complete bipartite graphs so that the resulting graph is embeddable in a given surface as described in Sections 4.1 and 4.3 [JR85, JR84a, JR84b, HJR85, Har86, Har87]. Section 4.2 describes the work of Eades, Faria, Figueiredo and Mendonça on establishing the NP-hardness of finding the splitting number for a given graph [EM93, Men94, FFM98a, FFM01].

4.1 Lower Bounds for the Splitting Number

First consider the different ways of vertex splitting as illustrated in Figure 6. The graphs numbered 1 and 7 (and all graphs isomorphic to them) have the same number of edges as the original graph K_3. The other graphs have more edges than K_3: In the graphs numbered 10 through 18, v_1v_2 is an additional edge, and in graphs such as the ones numbered 2 or 18, some vertex u that was adjacent to v in K_3 is now adjacent to both v_1 and v_2. For each u that was adjacent to v and is now adjacent to both v_1 and v_2, we call one of the edges uv_1 and uv_2 *superfluous*. Likewise, we call an edge v_1v_2 superfluous. We say a vertex splitting is *proper* if it does not create superfluous edges, and if none of the resulting vertices v_1 and v_2 is isolated. Otherwise we call it *improper*.

A. Liebers, *Planarizing Graphs*, JGAA, 5(1) 1–74 (2001)

Now observe that when splitting vertices of a graph G with the goal of planarizing it, we can restrict our attention to proper vertex splittings. For assume we obtain a planar graph G' from G by using improper vertex splittings. Now perform the same sequence of vertex splittings on G again, but in each vertex splitting, leave out all superfluous edges. Also, skip all the vertex splittings that create an isolated vertex. This yields a graph G''. Since G'' is a subgraph of G' and since G' is planar, G'' is also planar.

The upper bound for the number of edges for planar graphs from Equation 2 immediately yields a lower bound for the splitting number: Let G be a graph with n vertices and m edges, and let $\sigma(G)$ be the splitting number of G. Let G' be a graph obtained from G by $\sigma(G)$ vertex splitting operations so that G' is planar. Then G' has $n' = n + \sigma(G)$ vertices, and by the above argument about superfluous edges, we can construct G' in such a way that $m' = m$. Since G' is planar, Equation 2 says that it has at most $m' \leq 3n' - 6$ edges (for $n' \geq 3$). Since $m = m'$, this implies $n' \geq \frac{1}{3}m + 2$ for $n' \geq 3$. Every graph on $n \leq 4$ vertices is planar, so for $n \leq 4$ we have $n' = n$. For $n \geq 5$, we have $n' \geq n$. Therefore, the condition $n' \geq 3$ is equivalent to the condition $n \geq 3$, and with $\sigma(G) = n' - n$ we obtain the lower bound

$$\sigma(G) \geq \left\lceil \frac{1}{3} \cdot m - n + 2 \right\rceil \quad \text{for } n \geq 3 \tag{21}$$

If we know the girth g of a graph G with n vertices and m edges, a better bound for $\sigma(G)$ can be derived [JR85]: Let again G' be a graph obtained from G by $\sigma(G)$ vertex splitting operations so that G' is planar. Let n' and m' be the number of vertices and edges of G', respectively. Let f' be the number of faces of G' in a given planar embedding. Euler's formula for planar graphs (Equation 1) says $n' - m' + f' = 2$. If g' is the girth of G', then every face of G' is incident to at least g' edges. Furthermore, if m'' edges are incident to exactly two faces, then $f' \cdot g' \leq 2 \cdot m'' \leq 2 \cdot m'$. Combining this inequality with Euler's formula, we have

$$2 + m' - n' = f' \leq \frac{2m'}{g'} \ .$$

Since $n' = n + \sigma(G)$, $m' = m$, and $g' \geq g$, we have

$$2 + m - n - \sigma(G) \leq \frac{2m}{g'} \leq \frac{2m}{g} \ .$$

Since $\sigma(G)$ is an integer, we can conclude

$$\sigma(G) \geq \left\lceil m - \frac{2m}{g} - n + 2 \right\rceil \tag{22}$$

Note that if a graph G has cycles, but its girth is not known, combining $g \geq 3$ with Equation 22 yields Equation 21. This is not surprising, since the formula $m' \leq 3n' - 6$ follows from Euler's formula $2 + m' - n' = f'$ with the observation that each of the f' faces is incident to at least 3 edges.

A. Liebers, *Planarizing Graphs*, JGAA, 5(1) 1–74 (2001)   27

[JR85] provides this lower bound for a general surface S of Euler characteristic $E(S)$:

$$\sigma(G) \geq \left\lceil m - \frac{2m}{g} - n + E(S) \right\rceil \tag{23}$$

[JR85] also shows that this bound is achieved for all complete bipartite graphs on any surface S.

4.2   Finding the Splitting Number of a Graph

It has only recently been shown that determining the splitting number of a given graph G is an NP-hard problem [FFM98a, FFM01]. The investigation of the complexity status of the splitting number problem begins with Mendonça's definition of the following two problems and his proof that the first one is NP-complete [Men94, Section 4.3.1]:

Problem 24 (Eligible Set Split Planar Graph [Men94, Section 4.3.1]) *Given a graph $G = (V, E)$, a subset of vertices $S \subseteq V$, and a positive integer $K \leq |S|$, can G be transformed into a planar graph G' by K or less vertex splitting operations that involve only vertices in S? The vertices in S are called eligible vertices*[10].

Problem 25 (Split Planar Graph [Men94, Section 4.3.1])   *Given a graph $G = (V, E)$ and a positive integer $K < |E|$, can G be transformed into a planar graph G' by K or less vertex splitting operations?*

A reduction from the Maximum Planar Subgraph Problem (see also Section 3.1) shows that Eligible Set Split Planar Graph is NP-complete.

Problem 26 (Maximum Planar Subgraph [GJ79, Problem GT27]) *Given a graph $G = (V, E)$ and a positive integer $K \leq |E|$, is there a subset $E' \subseteq E$ with $|E'| \geq K$ such that the graph $G' = (V, E')$ is planar?*

Theorem 27 [Men94, Section 4.3.1] *Eligible Set Split Planar Graph is NP-complete.*

For the proof, let the graph $G = (V, E)$ and the positive integer $K \leq |E|$ be an arbitrary instance of Maximum Planar Subgraph. Construct an instance of Eligible Set Split Planar Graph as follows: Replace each edge $e = uv \in E$ with a path $ue'v_e e''v$, i.e. subdivide each edge e once. Call the resulting graph H. Let $K' = |E| - K$ (note that $K' \leq |E| = |S|$), and let $S = \{v_e \mid e \in E\}$ be the set of vertices created through the subdivisions. H, S, and K' define an instance of Eligible Set Split Planar Graph. G has a planar subgraph with K or more edges if and only if H can be planarized by K' vertex splitting operations on S. For if G has a planar subgraph $G' = (V, E')$ with $|E'| \geq K$ edges, then for each edge $e \in E \setminus E'$, split the vertex v_e in H so that one of the copies of v_e, v_{e1}, is

---

[10]In [Men94], we actually have "$K < |S|$"

A. Liebers, *Planarizing Graphs*, JGAA, 5(1) 1–74 (2001)   28

incident to e', and the other one, v_{e2}, is incident to e'', and v_{e1} and v_{e2} are not adjacent. The resulting graph H' is planar and the number of vertex splitting operations was $k' = |E| - |E'| = K' + K - |E'| \leq K'$. On the other hand, if there are $k' \leq K'$ vertex splitting operations on vertices in S that transform H into a planar graph H', then for each vertex $v_e \in S$ that was involved in a vertex splitting, delete the corresponding edge e from G. The resulting graph G' is planar since H' is planar, and the number of deleted edges is $l \leq k' \leq K'$, so G' has $|E'| = |E| - l \geq |E| - K' = K$ edges. Figure 7 shows the steps of this reduction for $G = K_{3,3}$.

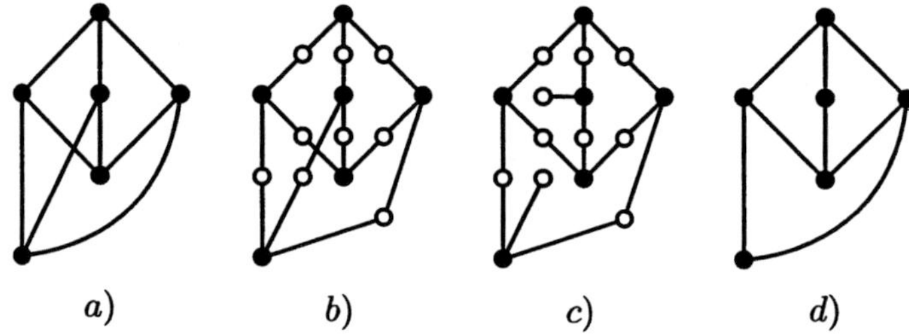

a)   b)   c)   d)

Figure 7: Illustration of the reduction for Eligible Set Split Planar Graph from Maximum Planar Subgraph with $K_{3,3}$. a) $K_{3,3}$. b) Every edge is subdivided by a white vertex. c) One of the white vertices needs to be split to planarize the graph in b). d) Alternatively, the deletion of the edge that was subdivided by the vertex split in c) yields a planar subgraph.

A similar transformation does not seem to work for Split Planar Graph, but Mendonça points out that for the class of graphs with vertex degree not greater than 3, Split Planar Graph and Maximum Planar Subgraph are equivalent [Men94, Section 4.3.1]. If Maximum Planar Subgraph restricted to graphs with vertex degree not greater than 3 were known to be NP-complete, then the following reduction would yield the NP-completeness of Split Planar Graph: A graph $G = (V, E)$ with vertex degrees not greater than 3 has a planar subgraph $G' = (V, E')$ with $|E'| \geq K$ edges if and only if G can be transformed into a planar graph G by less than or equal to $|E| - K$ vertex splitting operations. For assume $E' \subseteq E$ with $|E'| \geq K$ exists so that $G' = (V, E')$ is planar. Then for each edge $e = uv \in E \setminus E'$, perform a proper splitting operation on either u or v in G so that one of the resulting two vertices is only incident to e. The resulting graph is planar, and the number of vertex splitting operations was $|E| - |E'| \leq |E| - K$. On the other hand, assume that G can be planarized by K' vertex splitting operations. Then each (proper) splitting operation yields at least one vertex v with degree 1. Let E'' be the set of edges incident to those vertices. $|E''| \leq K'$. Then $G' = (V, E \setminus E'')$ is a planar graph with $|E| - |E''| \geq |E| - K'$ edges.

Faria, Figueiredo, and Mendonça have now settled the complexity status of Split Planar Graph:

A. Liebers, *Planarizing Graphs*, JGAA, 5(1) 1–74 (2001)   29

Theorem 28 [FFM98a, FFM01] *Split Planar Graph is NP-complete, even when restricted to cubic graphs.*

The proof uses a rather involved reduction from 3-SAT [GJ79, Problem LO2], where for an instance of 3-SAT with n variables and m clauses, a graph of maximum degree 3 with more than $1200 \cdot n^3 \cdot m^2$ vertices is constructed. A variation of the reduction where every vertex has degree exactly 3 is also given, completing the proof of the above theorem. [FFM98a, FFM01] then observes that the NP-completeness of Split Planar Graph for cubic graphs implies the NP-completeness of Maximum Planar Subgraph when restricted to cubic graphs.

Eades and Mendonça, in their work towards a layout system for diagrams, have developed and implemented a heuristic for planarizing a graph through vertex splitting [EM93] and [Men94, Sections 4.3.2 and 4.3.3]. It is based on Lempel, Even and Cederbaum's planarity testing algorithm and its implementation using PQ-tree algorithms by Booth and Lueker [LEC67, BL76] mentioned in Section 1.2 and uses ideas of [JST89, JTS89]. The vertices of the original graph are considered one at a time. A vertex is added to the graph being constructed if the resulting graph remains planar. Otherwise, the vertex is split and both copies of the vertex are added so that the resulting graph is planar. The running time of the heuristic is in $\mathcal{O}(n^2)$ for graphs with n vertices. There seem to be no computational studies on the performance of this heuristic.

4.3   Results for Particular Classes of Graphs

This section discusses the results about the splitting number of complete bipartite graphs and complete graphs. The splitting number of the hypercube of dimension 4, Q_4, is 4 [FFM98b], and the splitting number is also known for the Cartesian product of an m-cycle C_m and an n-cycle C_n. The latter result allows the construction of a graph with genus g and splitting number σ, for any integers $\sigma \geq g \geq 1$ [Sch86].

The first class of graphs for which the splitting number was determined was the class of complete bipartite graphs. Note that the complete bipartite graph K_{n_1,n_2} is planar if and only if $n_1 \in \{1,2\}$ or $n_2 \in \{1,2\}$. The girth of K_{n_1,n_2} is 4 (for $n_1, n_2 \geq 2$), so the lower bound 22 yields

$$
\begin{aligned}
\sigma(K_{n_1,n_2}) &\geq \left\lceil n_1 \cdot n_2 - \frac{2n_1 n_2}{4} - n_1 - n_2 + 2 \right\rceil \\
&= \left\lceil \frac{(n_1 - 2)(n_2 - 2)}{2} \right\rceil
\end{aligned}
\tag{29}
$$

In [JR85, JR84b], Jackson and Ringel show that this lower bound is also an upper bound. Again, they consider the general case of transforming $G = K_{n_1,n_2}$ into a graph G' that is embeddable in a surface S with Euler characteristic $E(S)$. They show that if S is a closed orientable or nonorientable surface, then $\sigma(K_{n_1,n_2}, S) = \max\left(\left\lceil \frac{(n_1-2)(n_2-2)}{2} \right\rceil - 2 + E(S), 0 \right)$. Recall that embedding a graph in the plane is equivalent to embedding it in the sphere. The sphere is commonly referred to as S_0, and $E(S_0) = 2$, so we have

A. Liebers, *Planarizing Graphs*, JGAA, 5(1) 1–74 (2001)   30

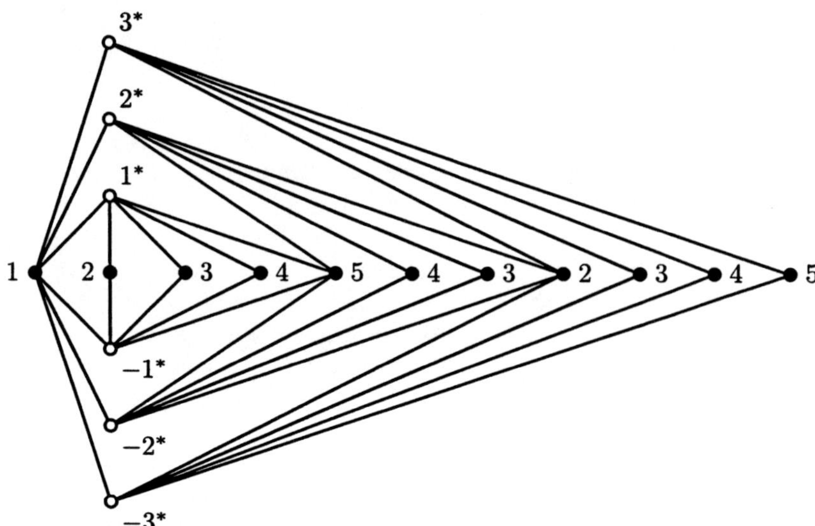

Figure 8: An optimal planar splitting of $K_{6,5}$, as constructed in the proof of Theorem 30. Observe that if we identify, for each j, $2 \leq j \leq 5$, all vertices labeled j, we obtain the original graph $G = K_{6,5}$. Counting the number of such vertex identifications shows that $K_{6,5}$ can be constructed from this graph by $\sigma = \left\lceil \frac{(6-2)(5-2)}{2} \right\rceil$ vertex identifications.

Theorem 30 (Jackson, Ringel [JR85, JR84b]) *The splitting number of the complete bipartite graph* K_{n_1,n_2} *is*

$$\sigma(K_{n_1,n_2}) = \left\lceil \frac{(n_1 - 2)(n_2 - 2)}{2} \right\rceil \qquad \text{for } n_1, n_2 \geq 2 \ .$$

Figure 8 illustrates the idea of the constructive proof given in [JR84b] for the case that n_1 or n_2 is an even number.

The second class of graphs for which the splitting number was found is the class of complete graphs. This result is much more involved than the one for complete bipartite graphs. First recall that for $n \geq 5$, the complete graph K_n is nonplanar. If less than $(n - 4)$ vertex splitting operations are performed on a graph K_n with $n > 5$, the resulting graph G' contains (at least) 5 vertices that were not involved in splitting operations. These 5 vertices induce the nonplanar graph K_5, so G' cannot be planar. This yields the trivial lower bound

$$\sigma(K_n) \geq n - 4 \tag{31}$$

With the girth $g = 3$ the lower bounds (21) and (22) both yield

$$\sigma(K_n) \geq \left\lceil \frac{1}{3} \binom{n}{2} - n + 2 \right\rceil = \left\lceil \frac{(n-3)(n-4)}{6} \right\rceil \qquad \text{for } n \geq 3 \tag{32}$$

A. Liebers, *Planarizing Graphs*, JGAA, 5(1) 1–74 (2001)   31

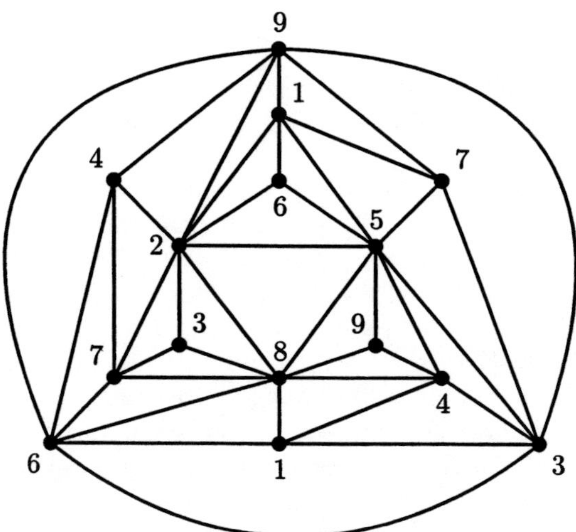

Figure 9: An optimal planar splitting of K_9 as exhibited in [HJR85]. $\sigma(K_9) = 6$. Note that there are superfluous edges connecting vertices labeled 1 and 6, 3 and 7, and 9 and 4, respectively.

The lower bound (31) is only interesting for $n = 6$ and $n = 7$. For $n \geq 8$, the bound (32) is greater than or equal to the bound (31).

Hartsfield, Jackson and Ringel show that except for $n = 6, 7$ or 9 the lower bound (32) is also an upper bound. Unlike the result of Theorem 30 for K_{n_1,n_2}, this result does not extend to general surfaces. In the conference presentations [JR85] and [JR84a], partial results towards finding the splitting number of K_n are announced. [HJR85] then presents a proof for the following theorem:

Theorem 33 (Hartsfield, Jackson, Ringel [HJR85]) *The splitting number of the complete graph K_n is*

$$\sigma(K_n) = \begin{cases} \left\lceil \frac{(n-3)(n-4)}{6} \right\rceil & \text{for } n \geq 3 \text{ and } n \notin \{6,7,9\} \\ \left\lceil \frac{(n-3)(n-4)}{6} \right\rceil + 1 & \text{for } n \in \{6,7,9\} \end{cases}.$$

For each n, $3 \leq n \leq 8$, Theorem 33 yields the higher one of the lower bounds (31) and (32). But for $n = 9$, (31) and (32) both yield 5 splitting operations as a lower bound, whereas Theorem 33 yields $\sigma(K_9) = 6$. Figure 9 shows a planar splitting with 6 splitting operations for K_9. [HJR85] explains that the proof for $\sigma(K_9) = 6$ involves checking many cases and that Mark Jungerman has verified the proof using a computer.

The proof for large n is a meticulous case analysis for the congruence classes of n modulo 12. It is actually carried out in a dual formulation of the problem: A planar splitting of K_n is represented as a map where the countries represent the vertices. Countries that correspond to vertices with the same label belong to a common *empire*. Two empires e_i, e_j are *adjacent* if there exist countries c_i and c_j belonging to the empires e_i and e_j, respectively, that share a common border. Countries whose corresponding vertices are adjacent in the planar splitting share a common border in the map.

Finding an optimal planar splitting of K_n is then equivalent to finding a map with n mutually adjacent empires where the overall number of countries is minimum. Figure 10 shows an optimal planar splitting of K_{10} and the corresponding map. This map was actually found by Jungerman's program mentioned above [JR84a, HJR85]. Finding maps of mutually adjacent empires is an old problem: According to [JR84a], Heawood found a map of 12 mutually adjacent empires of 2 countries each in 1890 [Hea90]. Note that indeed $\sigma(K_{12}) = 12$.

The splitting number of the complete graph on the torus is given in [Har86], and [HJR85, Har87] give results about the splitting number of the complete graph on two nonorientable surfaces.

5   Thickness

In Sections 2, 3, and 4, we have performed the operations vertex deletion, edge deletion, and vertex splitting on a graph G with the goal of obtaining a new planar graph G'. We now ask for a collection of planar subgraphs of a given graph G, the union of which is G:

Definition 34 (thickness) *The* thickness *of a graph G, denoted $\theta(G)$, is the minimum number of planar subgraphs of G whose union is G.*

Clearly the thickness of a graph is 1 if and only if the graph is planar.

As an example, consider the two planar subgraphs of $K_{3,3}$ whose union is $K_{3,3}$ in Figure 11, and the three planar subgraphs of K_9 whose union is K_9 in Figure 12. Since $K_{3,3}$ is nonplanar, the exhibition of two planar subgraphs of $K_{3,3}$ whose union forms $K_{3,3}$ shows that $\theta(K_{3,3}) = 2$. The thickness of K_9 is not so easily determined: Figure 12 only shows that $\theta(K_9) \leq 3$. [BHK62] shows that indeed $\theta(K_9) = 3$. (Alternative proofs are provided in [Tut63a, Wes86].) See Section 5.3 for further results about the thickness of complete and complete bipartite graphs.

Since each planar subgraph of a given graph G with $n \geq 3$ vertices and m edges can have at most $3n - 6$ edges (Equation 2), we obtain an immediate lower bound for the thickness of G:

$$\theta(G) \geq \left\lceil \frac{m}{3n - 6} \right\rceil \quad \text{for } n \geq 3 \tag{35}$$

For upper bounds, see page 38.

Observe that if the graphs in Figure 11 were printed onto slides, the two planar subgraphs given could actually be placed on top of each other so that each vertex labeled i in the first subgraph lies exactly on top of the vertex

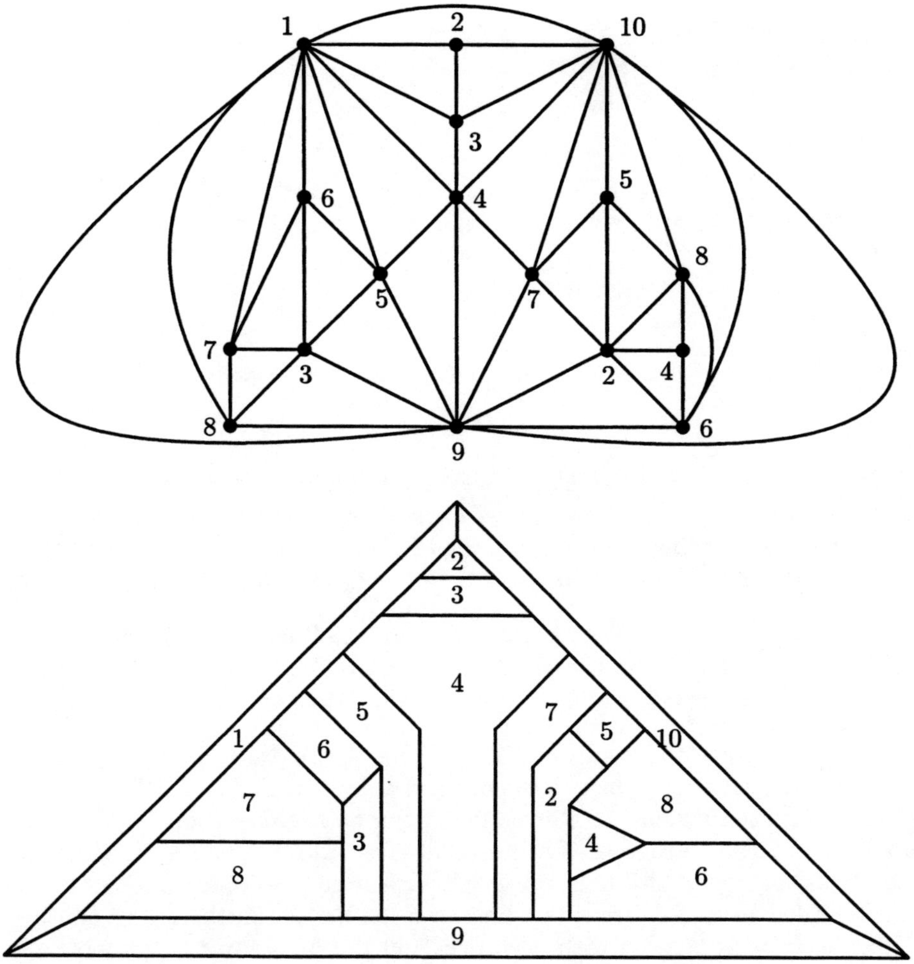

A. Liebers, *Planarizing Graphs*, JGAA, 5(1) 1–74 (2001)   33

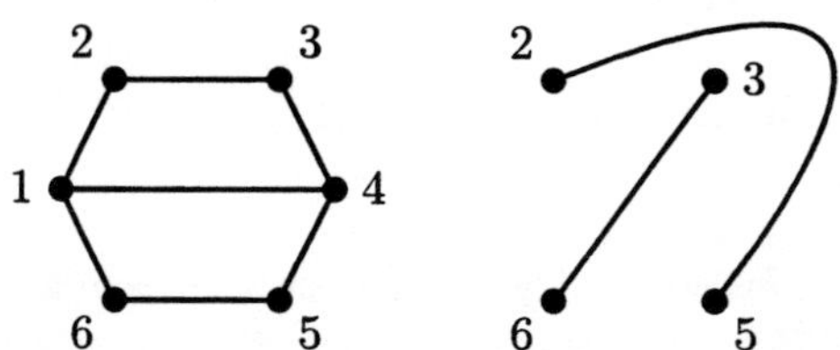

Figure 10: An optimal planar splitting of K_{10} (top) and its representation as a map (bottom). $\sigma(K_{10}) = 7$.

Figure 11: Two planar subgraphs of $K_{3,3}$ whose union is $K_{3,3}$.

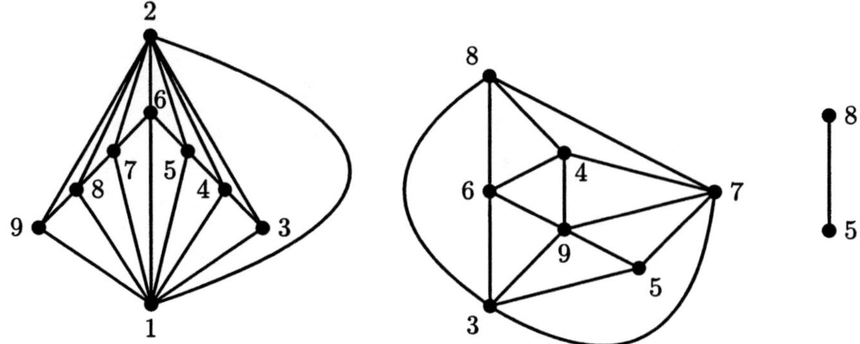

Figure 12: Three planar subgraphs G_1, G_2, and G_3 of K_9 whose union is K_9.

labeled i in the second subgraph. So we do not only have two *subgraphs* whose union is $K_{3,3}$, but we have two *embeddings* of two planar graphs so that the union of the embeddings yields a drawing of $K_{3,3}$. Kainen [Kai73] showed that this observation can be generalized:

Theorem 36 [Kai73] *Given a graph G with thickness $\theta(G)$, there exists a drawing of G, and there exist subgraphs $G_1, \ldots G_{\theta(G)}$ whose union is G, such that the drawing of G restricted to G_i is a planar embedding of G_i, for $1 \leq i \leq \theta(G)$.*

Note that the three subgraphs of K_9 in Figure 12 are drawn in a way so that the union of their embeddings does not yield a drawing of K_9.

Knowing the thickness of a given graph can be helpful in some application problems. [AKS91] proposes two new multilayer grid models for VLSI layout and shows for one of them that a graph with n vertices and thickness 2 can be embedded in two layers in an area of size $O(n^2)$. Furthermore, another algorithm embeds a graph with n vertices and thickness t in t layers in $O(n^3)$ area, respecting some additional constraints. [RL92, RL93] give approximate algorithms for scheduling multihop radio networks. They find a schedule whose length is a function of the thickness of the network.

The thickness of graphs has been widely studied as part of topological graph theory, but few algorithmic results for finding the thickness of a graph seem to be available. Early work about thickness and the introduction of the study of thickness into graph theory is described in detail by Hobbs [Hob69]. In particular, Tutte [Tut63b] establishes many results about the thickness of graphs in one of the earliest papers about this topic. Surveys about thickness are [WB78], [Bei88], and [MOS98].

The following sections give a brief summary of the known results about thickness: Section 5.1 describes the result of Mansfield [Man83] that says that determining the thickness of a graph is NP-hard, and mentions heuristic approaches for finding the thickness. Thickness-minimal graphs are discussed in Section 5.2, and Section 5.3 lists results about the thickness of graphs belonging to particular classes of graphs. Finally, Section 5.4 mentions two variations of the thickness.

5.1   Finding the Thickness of a Graph

Mansfield [Man83] defines the following problem (that was already mentioned in [GJ79, Problem OPEN3]):

Problem 37 (Thickness [Man83]) *Given a graph G and a positive integer K, does the thickness of G satisfy $\theta(G) \leq K$?*

Mansfield shows that this problem is NP-complete for the fixed value $K = 2$, thus establishing the NP-completeness of Thickness. The proof uses a reduction from Planar 3-SAT [GJ79, Problem LO1]. Before we state this problem, recall that given a set $U = \{u_1, \ldots, u_m\}$ of Boolean variables, the set $L = \{u_1, \overline{u_1}, \ldots, u_m, \overline{u_m}\}$ is the set of *literals* over U. A subset of literals $c \subseteq L$ is called a *clause* over U. A clause c is said to be *satisfied* if the disjunction of the literals in c has the Boolean value "true" (for some truth assignment for U). Given a set U of Boolean variables and a collection C of clauses over U, consider the bipartite graph $G_{U,C} = (U \cup C, E)$ with $E = \{uc \mid (u \in c$ or $\overline{u} \in c)$ and $u \in U$ and $c \in C\}$.

Problem 38 (Planar 3-SAT [GJ79, Problem LO1]) *Given a set U of Boolean variables and a collection C of clauses over U with $|c| \leq 3$ for all $c \in C$, and given that the graph $G_{U,C}$ is planar, is there a truth assignment for U that satisfies all clauses in C simultaneously?*

Lichtenstein [Lic82] showed that Planar 3-SAT is NP-complete. Mansfield first shows that Planar 3-SAT remains NP-complete if each clause contains *exactly* three literals, and then reduces this restricted version of Planar 3-SAT to Thickness with $K = 2$.

So it is unlikely that a polynomial time algorithm that determines the thickness of a given graph will be found. A heuristic approach for finding an upper bound on the thickness of a graph $G = (V, E)$ is to find a planar subgraph $G' = (V, E')$ of G, to form the difference graph $H = (V, E \setminus E')$, to then find a planar subgraph of H and so on until the difference graph itself is planar. This approach is studied in [OS94, MOS98] and, independently, in [Cim95b]. [MOS98] reports on using three different algorithms to find a planar subgraph: The maximal planar subgraph algorithm [CHT93], the maximal planar subgraph algorithm [JST89, JTS89], and the branch and cut heuristic [Mut94, JM96]. Computational studies are carried out for 19 complete graphs with 10 to 100 vertices, for 9 complete bipartite graphs with 20 to 100 vertices, and for 14 further graphs with 28 to 680 vertices originating in VLSI design. The two thickness heuristics using the maximal planar subgraph algorithms perform very similarly throughout. For the complete and complete bipartite graphs, their results are reported to be on average 38 and 24 percent, respectively, off the optimal solution, while the heuristic using the branch and cut approach is only off by 20 and

A. Liebers, *Planarizing Graphs*, JGAA, 5(1) 1–74 (2001)   36

12 percent, respectively. For the other 14 graphs, the thickness is not known. The results of all three heuristics are very similar for these graphs, with a small advantage for the branch and cut heuristic. But as with the maximum planar subgraph heuristics discussed in Section 3.3, the branch and cut heuristic often needs more than 100 times the run time of the heuristics based on maximal planar subgraphs.

[Cim95b] also reports on a thickness heuristic based on extracting maximal planar subgraphs. Heuristic improvements are made to increase the size of the planar subgraphs obtained, and computational studies on complete, complete bipartite, and random graphs with 10 to 115 vertices are carried out. For the complete and complete bipartite graphs that were also used in [MOS98], the performance of the heuristics in [Cim95b] is similar to the performance of the heuristics using maximal planar subgraphs reported in [MOS98].

5.2   Thickness-Minimal Graphs

The following facts about thickness and the concept of thickness-minimal graphs (also called θ-minimal graphs) are due to Tutte [Tut63b]: If a graph G has thickness $\theta(G) = t$, then every subgraph of G has thickness at most t. Furthermore, if a subgraph G' of G has exactly one edge less than G or exactly one vertex (and all its incident edges) less than G, then either $\theta(G') = t$ or $\theta(G') = t - 1$. In other words, deleting one edge or deleting one vertex decreases the thickness of a graph by at most one. These facts motivate the following definition:

Definition 39 (thickness-minimal graphs)[11] *If a graph G has thickness t and if every proper subgraph of G has thickness less than t, then G is called a* thickness-minimal *(or θ-minimal) graph. If G is thickness-minimal with $\theta(G) = t$, we also call G t-minimal.*

The 2-minimal graphs are exactly the subdivisions of K_5 and $K_{3,3}$. Note that if a graph G has thickness $t \geq 2$, then there exists a t-minimal subgraph of G. For $t \geq 2$, every t-minimal graph is 2-connected and has minimum vertex degree at least t and maximum vertex degree at least $2t - 1$. Tutte then establishes the following important theorem:

Theorem 40 [Tut63b] *For each integer $t \geq 2$ there exist infinitely many pairwise nonisomorphic t-minimal graphs with maximum vertex degree $2t - 1$, and of girth greater than any specified integer N.*

This theorem establishes the *existence* of infinitely many t-minimal graphs. But given an integer $t \geq 2$, it does not provide an explicit construction of t-minimal graphs. Beineke [Bei67] showed that for any integer $t \geq 2$, the complete bipartite graph $K_{2t-1,4t^2-10t+7}$ is t-minimal. Hobbs and Grossman [HG68a], and, independently, Bouwer and Broere [BB68] showed that $K_{4t-5,4t-5}$ is t-minimal for any integer $t \geq 2$. Hobbs and Grossman [HG68b] also showed that any t-minimal graph is t-edge-connected.

---

[11][Bei67, Wes83b, Wes89]use the term *critical* instead of *minimal.*

A. Liebers, *Planarizing Graphs*, JGAA, 5(1) 1–74 (2001)   37

Since $\theta(K_9) = 3$ [BHK62, Tut63a, Wes86], K_9 is a candidate for being 3-minimal. Figure 12 displays three subgraphs G_1, G_2, and G_3 of K_9 whose union is K_9, where G_3 consists of a single edge. Thus any proper subgraph of K_9 is the union of a subgraph of G_1 and a subgraph of G_2, and has therefore thickness at most 2. So K_9 is 3-minimal. K_9 appears to be the only θ-minimal complete graph.

Wessel [Wes83b, Wes89], and, independently, Širáň and Horák [HŠ87] finally give, for each integer $t \geq 2$, an explicit construction of an infinite number of t-minimal graphs. Širáň and Horák show that the bounds established by [Tut63b] and [HG68b] on connectedness and minimum vertex degree are actually tight: Their graphs are 2-connected, but not 3-connected, they are t-edge-connected, but not $(t+1)$-edge-connected, and they have minimum vertex degree t.

5.3 Results for Particular Classes of Graphs

There are few classes of graphs for which the thickness is known. For the complete graphs, the thickness was settled in a long process described in detail by White and Beineke [WB78, Section 9]. It is clear that $\theta(K_1) = \theta(K_2) = \theta(K_3) = \theta(K_4) = 1$, and it is easily seen that $\theta(K_5) = \theta(K_6) = \theta(K_7) = \theta(K_8) = 2$. Figure 12 shows that $\theta(K_9) \leq 3$. Battle, Harary, and Kodama [BHK62] were the first to show that indeed $\theta(K_9) = 3$. Alternative proofs were given by Tutte [Tut63a] and Wessel [Wes86]. Beineke and Harary [BH65] showed the formula for $\theta(K_n)$ for most cases, and Alekseev and Goncakov [AG76], and, independently, Vasak [Vas76], completed the result:

$$\theta(K_n) = \left\lfloor \frac{n+7}{6} \right\rfloor \qquad \text{for } n \geq 1,\, n \neq 9,\, n \neq 10$$

$$\theta(K_9) = \theta(K_{10}) = 3$$

For the complete bipartite graph, the thickness is still not settled for all cases. Beineke, Harary, and Moon [BHM64] found the following result:

$$\theta(K_{n_1,n_2}) = \left\lceil \frac{n_1 \cdot n_2}{2(n_1 + n_2 - 2)} \right\rceil$$

except possibly when n_1 and n_2 are both odd, and, assuming $n_1 \leq n_2$, there is an integer k such that $n_2 = \left\lfloor \frac{2k(n_1-2)}{n_1-2k} \right\rfloor$. In [Bei67], Beineke gives a more detailed description of the proof than in [BHM64].

The thickness of the hypercube of dimension n is $\theta(Q_n) = \lceil \frac{n+1}{4} \rceil$ [Kle67].

For a graph G and a general surface S, let the *thickness of G on S*, denoted $\theta(G, S)$, be the smallest number of subgraphs of G so that the subgraphs are all embeddable in S and so that their union is G. When S is the torus (also denoted S_1), $\theta(G, S_1)$ is also called the *toroidal thickness* of G. To avoid confusion, the thickness of a graph is sometimes called the *planar thickness*. [WB78] reviews known results about the thickness on other surfaces.

A. Liebers, *Planarizing Graphs*, JGAA, 5(1) 1–74 (2001)

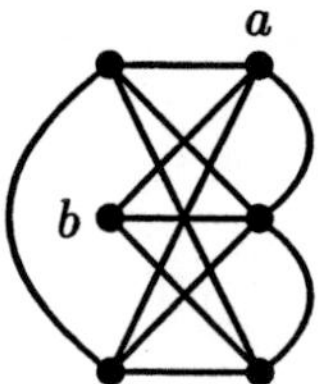

Figure 13: The graph G_{12}. G_{12} clearly has $K_{3,3}$ as a subgraph, so it has $K_{3,3}$ as a minor. To see that it also has K_5 as a minor, contract the edge ab.

[Rin65] and, independently, [Bei69] give the toroidal thickness of the complete graphs. [Bei69] also discusses the complete bipartite graphs, as well as some other surfaces. Further results about the toroidal thickness of graphs are given in [And82b, And82a].

Jünger et al. showed that the thickness of graphs that do not contain K_5 as a minor is at most 2 [JMOS94, Ode94], using a decomposition theorem of Truemper [Tru92, Theorem 10.5.24]. They were able to extend their result to graphs that do not contain the graph G_{12} as a minor [JMOS98] (G_{12} is depicted in Figure 13). Since G_{12} contains K_5 as well as $K_{3,3}$ as a minor, the result for G_{12} implies that the thickness of graphs that do not contain K_5 as a minor and the thickness of graphs that do not contain $K_{3,3}$ as a minor is at most two. For such a graph the thickness can be determined in linear time then, since it can be tested in linear time whether the graph is planar. If it is, its thickness is 1, otherwise it is 2.

For some graph classes, bounds on the thickness are known: A graph of orientable genus 1 (i.e. a nonplanar graph embeddable in the torus) has thickness 2 [Asa87], and a graph of orientable genus 2 has thickness either 2 or 3 [Asa94]. Every graph $G = (V, E)$ has thickness at most $\left\lfloor \sqrt{\frac{|E|}{3}} + \frac{3}{2} \right\rfloor$ [DHS91]. If δ and Δ are the minimum and maximum vertex degree of G, respectively, then $\left\lceil \frac{\delta+1}{6} \right\rceil \leq \theta(G) \leq \left\lceil \frac{\Delta}{2} \right\rceil$ [Wes84]. Independently, [Hal91] presents similar results about the relation between the minimum and maximum vertex degrees of a graph and its thickness.

5.4 Variations of Thickness

Bernhart and Kainen [BK79, Kai90] introduced the *book thickness* of a graph. A *book* B with $n \geq 0$ pages consists of a line L in 3-dimensional space, called the *spine*, together with n distinct half-planes (called the *pages*) with L as their common boundary. A graph G is embeddable in B if the vertices of G can be placed on L and if each edge can be embedded in at most one page of B. The *book thickness* (also called *pagenumber*) of a graph G is the smallest number n so that G can be embedded in a book with n pages. The book thickness

A. Liebers, *Planarizing Graphs*, JGAA, 5(1) 1–74 (2001)   39

has been studied for several classes of graphs, see for example [CLR87, Hea87, MLW88, HI92, Obr93, Mal94, SGB95, SS96].

[BS84] showed that any planar graph can be embedded in a 9-page book. [Hea84] lowered this bound to 7 pages and also gave an $O(n^2)$ algorithm to actually find an embedding. Yannakakis [Yan86, Yan89] showed that any planar graph can be embedded in a book with 4 pages. Yannakakis also gives a linear time algorithm to find such an embedding.

If a graph G has a straight line drawing and two subgraphs G_1 and G_2 whose union is G, and if the straight line drawing of G restricted to G_i is a planar embedding of G_i, for $1 \leq i \leq 2$, then G is called *doubly linear*. Clearly any doubly linear graph has thickness at most two. Hutchinson et al. [HSV96, HSV99] study doubly linear graphs. They show that a doubly linear graph with n vertices has at most $6n - 18$ edges.

Other variations of thickness are discussed in [Hob69, HŠ82, Hor83, Wes83a, PCK89, DEH00], for example.

6   Crossing Number

In graph drawing, but also in other application areas such as VLSI layout, we are interested in a drawing of a given graph with as few edge crossings as possible. Here, we do not allow drawings of graphs where a point in the plane belongs to more than two curves representing edges (unless the point represents a common end vertex of those edges). The crossing number problem goes back to Turán, who describes how he had first dealt with it as a brick factory problem, and how Zarankiewicz conjectured a solution that was later disproved [Tur77, Guy69].

Definition 41 (crossing number) *The crossing number of a graph G, denoted $\nu(G)$, is the smallest number k so that G can be drawn in the plane with at most k edge crossings.*

Clearly the crossing number of a graph is 0 if and only if the graph is planar, and the crossing number of a graph is bounded from below by the skewness of the graph. Surveys on the crossing number can be found in [SSV95, Sch95], and Vrt'o maintains a chronological bibliography [Vrt].

A graph G with n vertices, m edges, and with crossing number $\nu(G) > 0$ (and a given drawing with $\nu(G)$ edge crossings) can be transformed into a planar graph by introducing $\nu(G)$ new vertices and placing them at the edge crossings of the drawing. The new graph G' has $n + \nu(G)$ vertices and $m + 2 \cdot \nu(G)$ edges. Figure 14 illustrates this process.

This planarization technique is used in graph drawing: Introduce dummy vertices to planarize a graph, then apply a graph drawing algorithm to the planar graph, and afterwards eliminate the dummy vertices and re-introduce edge crossings into the drawing [DETT99, Sections 2.3, 2.5, and 2.6].

[SSV95] contains a survey on lower bounds for the crossing number, starting with the following observation: Equation 2 for the planar graph G' says that

A. Liebers, *Planarizing Graphs*, JGAA, 5(1) 1–74 (2001)   40

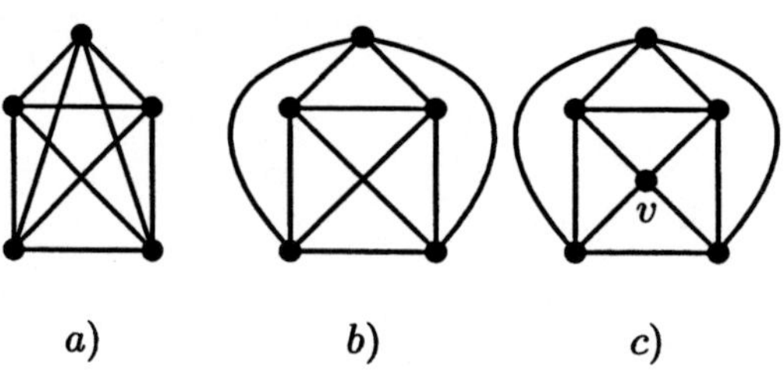

Figure 14: *a)* K_5 in a drawing with 5 edge crossings. *b)* K_5 with only one edge crossing. *c)* Vertex v is introduced instead of the edge crossing to planarize K_5.

$m + 2 \cdot \nu(G) \leq 3 \cdot (n + \nu(G)) - 6$. This immediately yields the following lower bound for the crossing number of G:

$$\nu(G) \geq m - 3 \cdot n + 6 \tag{42}$$

If the girth g of G is known, we have a more precise lower bound [Kai72a]:

$$\nu(G) \geq m - \frac{g}{g-2}(n-2) \qquad \text{for} \quad g \geq 3 \tag{43}$$

Note that Equation 42 follows from Equation 43 with $g = 3$. We may assume that $g \geq 3$ since otherwise G has no cycles and is planar. Kainen [Kai72a] actually generalizes Equation 43 to the crossing number of a graph on an orientable surface with given genus, and Kainen and White [KW78] provide the corresponding result for nonorientable surfaces.

Another general lower bound for the crossing number was found by Ajtai et al. [ACNS82] and, independently, by Leighton [Lei83, p. 108]. It was improved by Pach and Toth [PT97]:[12]

Theorem 44 [ACNS82], [Lei83, p. 108], [PT97] *If G is a graph with n vertices and m edges, and if $m \geq 7.5 \cdot n$, then*

$$\nu(G) \geq \frac{m^3}{33.75 \cdot n^2} \ .$$

[PT97] shows that this lower bound is asymptotically tight, and that if there is no restriction on m, then we still have $\nu(G) \geq \frac{m^3}{33.75 n^2} - 0.9n$. [SSSV94, SSSV96b]% give corresponding lower bounds for the crossing number of graphs on orientable surfaces with given genus. [SSV95] reports further results on lower bounds for the crossing number, involving the bisection width of the graph ([Lei83][Lei81, Lei84][SV93a, SV94][PSS94, PSS96]) as well as embedding a graph into another one ([Lei83][SS92, SS94][SSSV94, SSSV96b]). In particular, we have:

---

[12][ACNS82] actually shows that if $m \geq 4n$ then $\nu(G) \geq \frac{m^3}{100n^2}$. [Lei83, p. 108] shows that if $m \geq 4n$ then $\nu(G) \geq \frac{m^3}{375n^2}$.

Theorem 45 [SV93a] *If G is a graph with n vertices, maximum degree Δ, and with bisection width b, then*

$$\nu(G) \geq \frac{b^2}{75\Delta} - n \ .$$

The bisection width of a graph $G = (V, E)$ with n vertices is defined to be $b = \min\{|E(V_1, V_2)|\}$ where the minimum is taken over all partitions of V into two sets V_1 and V_2 with $|V_1|, |V_2| \geq n/3$, and where $E(V_1, V_2)$ denotes the set of edges with one endpoint in V_1 and the other in V_2.

See [Vrt] for other recent results on lower bounds.

Section 6.1 deals with the problem of finding the crossing number for a general graph. Section 6.2 introduces the notion of crossing-critical graphs. Section 6.3 lists papers that examine the crossing number for particular graph classes. The partial results that are known for complete and complete bipartite graphs are also listed. Finally, Section 6.4 mentions some variations of the crossing number.

6.1 Finding the Crossing Number of a Graph

We are interested in the following problem:

Problem 46 (Crossing Number) *Given a graph G and a positive integer K, is there a drawing of G with K or less edge crossings?*

The complexity status of this problem was mentioned as being open in [GJ79, Problem OPEN3]. Then Garey and Johnson [GJ83] showed that Crossing Number is NP-complete. They use a two-step reduction, starting with the following NP-complete problem [GJS76]:

Problem 47 (Optimal Linear Arrangement [GJ79, Problem GT42]) *Given a graph $G = (V, E)$ and a positive integer K, is there a bijection $f : V \to \{1, 2, \ldots, |V|\}$ such that $\sum_{uv \in E} |f(u) - f(v)| \leq K$?*

First, Optimal Linear Arrangement is reduced to a problem introduced as Bipartite Crossing Number:

Problem 48 (Bipartite Crossing Number [GJ83]) *Given a connected bipartite graph $G = (V_1, V_2, E)$ with multiple edges allowed, and given a positive integer K, can G be drawn in the unit square so that all vertices in V_1 are on the northern boundary, all vertices in V_2 are on the southern boundary, all edges are within the square, and there are at most K edge crossings?*

Then, Bipartite Crossing Number (on graphs with multiple edges allowed) is reduced to Crossing Number (on simple graphs).

So it is unlikely that a polynomial time algorithm that determines the crossing number of a given graph will be found. Note that any algorithm for drawing a graph in the plane can trivially be seen as a heuristic for the crossing number: We simply count the edge crossings of the resulting drawing. [OS94] presents a study on a concrete heuristic designed specifically for drawing a graph with few crossings. It is based on finding a maximal planar subgraph. Two variations are tested on complete graphs with 5 to 14 vertices, on complete bipartite graphs $K_{n,n}$ with 6 to 16 vertices, and on 11 other graphs for which the crossing number is known or conjectured. One of the two variants performs rather well on the graphs tested, but it incurs high running times.

6.2   Crossing-Critical Graphs

If a graph G has crossing number k, then clearly any subgraph of G has crossing number at most k. We are interested in the "smallest" graph with crossing number k:

Definition 49 (crossing-critical graphs) *If a graph G has crossing number k and if every proper subgraph of G has crossing number less than k, then G is said to be.*

[Koc87] gives, for any $k \geq 2$, a construction of an infinite family of 3-connected crossing-critical graphs with crossing number k. This improves the result in [Šir84].

Note the analogy of the above definition to thickness-minimal graphs discussed in Section 5.2. But this analogy does not carry over to structural results about the crossing number and crossing-critical graphs. While deleting an edge from a graph G can decrease the thickness of G by at most one, the same is not true for the crossing number:

Theorem 50 [Koc91] *For any positive integer k there is a 3-connected graph $G = (V, E)$ with crossing number $\nu(G) = 4k$ and with an edge $e \in E$ so that $\nu(G' = (V, E \setminus \{e\})) = 3k$, and so that e belongs to no Kuratowski subgraph of G.*

Further results about crossing-critical graphs can be found in [Šir83], [Ric88, MR94, RT93].

6.3   Results for Particular Classes of Graphs

There seem to be few results about the exact crossing number of particular graph classes. Some crossing numbers are known exactly, but often, only lower or upper bounds are known.

The crossing number of complete graphs is not known exactly:

Theorem 51

$$\nu(K_n) \leq \frac{1}{4} \left\lfloor \frac{n}{2} \right\rfloor \left\lfloor \frac{n-1}{2} \right\rfloor \left\lfloor \frac{n-2}{2} \right\rfloor \left\lfloor \frac{n-3}{2} \right\rfloor$$

and equality holds for $n \leq 10$.

A. Liebers, *Planarizing Graphs*, JGAA, 5(1) 1–74 (2001)   43

See for example [Guy71, Guy72] or [WB78] for a description of the history of this result. [Guy71, Guy72] also contains the proof of equality for $n \leq 10$. Leighton [Lei84] gives the lower bound $\nu(K_n) \geq \frac{1}{120}n(n-1)(n-2)(n-3)$ for $n \geq 5$, which is better than the general lower bounds discussed above.

The crossing number of complete bipartite graphs is not known exactly either:

Theorem 52

$$\nu(K_{n_1,n_2}) \leq \left\lfloor \frac{n_1}{2} \right\rfloor \left\lfloor \frac{n_1-1}{2} \right\rfloor \left\lfloor \frac{n_2}{2} \right\rfloor \left\lfloor \frac{n_2-1}{2} \right\rfloor$$

and equality holds for $\min(n_1, n_2) \leq 6$.

The upper bound was pointed out by Zarankiewicz [Zar54], and the most recent result that contributed to the equality for $\min(n_1, n_2) \leq 6$ is [Kle70]. [Woo93] was able to extend the proof of equality to K_{7,n_2} with $n_2 \leq 10$. See also [WB78] or [Wes96, Section 7.3] for the history of this result, and for further results towards the crossing numbers of complete and complete bipartite graphs, see also [RT97].

The crossing number of the hypercube of dimension n, Q_n, is studied in [Kai72a, Kai72b][Mad91][SV92, SV93b][DR95][FF00].

The crossing numbers of many Cartesian product graphs have been studied, see [KW78, RB78, JŠ82, PPV86, Kle91, Kle94, Kle95, Kle96, DR95, RT95, KRS96, SSSV98], for example. Further classes of graphs were investigated in [GH73, Asa86, Fio86, MR92][SV92, SV93b][RŠ96].

6.4   Variations of Crossing Number

The following restricted version of the crossing number is studied intensely within the field of graph drawing:

Definition 53 (rectilinear crossing number) *The rectilinear crossing number of a graph G, denoted $\overline{\nu}(G)$, is the smallest number k so that G can be drawn in the plane with at most k edge crossings, and so that each edge of G is drawn as a straight-line segment.*

Clearly $\overline{\nu}(G) \geq \nu(G)$. For graphs with bounded degree, the crossing number and the rectilinear crossing number are bounded functions of one another [BD92][SSSV95, SSSV96a]. But for every $m > k \geq 4$, there exists a graph G with $\nu(G) = k$ and $\overline{\nu}(G) \geq m$ [BD93]. So the rectilinear crossing number can be arbitrarily large in comparison to the crossing number.

A further restriction of the rectilinear crossing number yields the following problem: The vertices of the graph under consideration are partitioned into $k \geq 2$ classes (usually called *layers* in this context) numbered from 1 to k such that edges only exist between vertices of consecutive layers. All vertices of a particular layer have to be drawn on a horizontal line, while edges still have to be drawn as straight-line segments, with as few edge crossings as possible. This

A. Liebers, *Planarizing Graphs*, JGAA, 5(1) 1–74 (2001)   44

is a much studied problem, see for example [JM97, JLMO97][YS99][SSSV97, SSSV00].

Still further variations of the crossing number are studied in [MKNF87, MNKF90][Bie91][SSSV95, SSSV96a][PT98], for example.

7 Coarseness

Apparently by accident, Paul Erdös introduced the notion of coarseness of a graph [Har69, p. 121]:

Definition 54 (coarseness) *The* coarseness *of a graph G, denoted $\xi(G)$, is the largest number of pairwise edge disjoint nonplanar subgraphs contained in G.*

We only mention the coarseness for the sake of completeness, and refer the interested reader to the literature: The coarseness (and variations thereof) of some graph classes have been studied in [GB68, BG69, Kai73, Har79, Mic83, AP93], for example.

Acknowledgment

The author would like to thank Dorothea Wagner for her generous support and encouragement, as well as for her advice in preparing this manuscript. The author would also like to thank the anonymous referee for a detailed report and valuable comments.

List of Figures

A. Liebers, *Planarizing Graphs*, JGAA, 5(1) 1–74 (2001)   45

Author Index

A. Liebers, *Planarizing Graphs*, JGAA, 5(1) 1–74 (2001)   46

A. Liebers, *Planarizing Graphs*, JGAA, 5(1) 1–74 (2001)   47

A. Liebers, *Planarizing Graphs*, JGAA, 5(1) 1–74 (2001)   48

Subject Index

A. Liebers, *Planarizing Graphs*, JGAA, 5(1) 1–74 (2001)   53

References

[AAP+96] Pankaj K. Agarwal, Boris Aronov, János Pach, Richard Pollack, and Micha Sharir. Quasi-Planar Graphs Have a Linear Number of Edges. In Franz J. Brandenburg, editor, *Graph Drawing. Proceedings of the DIMACS International Workshop, GD'95*, pages 1–7. Springer-Verlag, Lecture Notes in Computer Science, vol. 1027, 1996.

[ABL95] Dan Archdeacon, C. Paul Bonnington, and Charles H.C. Little. An Algebraic Characterization of Planar Graphs. *J. of Graph Theory*, 19:237–250, 1995.

[ACNS82] M. Ajtai, V. Chvátal, M.N. Newborn, and E. Szemerédi. Crossing-free subgraphs. In Alexander Rosa and Anton Kotzig, editors, *Theory and practice of combinatorics: a collection of articles honoring Anton Kotzig on the occasion of his 60th birthday*, pages 9–12. North-Holland, Annals of Discrete Mathematics, vol. 12, 1982.

[AG76] V.B. Alekseev and V.S. Goncakov. The thickness of an arbitrary complete graph. *Mat. Sbornik*, 101(143):212–230, 1976. In Russian.

[AKS91] Alok Aggarwal, Maria Klawe, and Peter Shor. Multilayer Grid Embeddings for VLSI. *Algorithmica*, 6:129–151, 1991.

[AMO93] Ravindra K. Ahuja, Thomas L. Magnanti, and James B. Orlin. *Network Flows*. Prentice Hall, 1993.

[And82a] Ian Anderson. On the toroidal thickness of graphs. *J. of Graph Theory*, 6:177–184, 1982.

[And82b] Ian Anderson. The Toroidal Thickness of the Symmetric Quadripartite Graph. *J. of Combinatorial Theory, Series B*, 33:57–59, 1982.

[AP61] L. Auslander and S.V. Parter. On Imbedding Graphs in the Sphere. *J. of Mathematics and Mechanics*, 10(3):517–523, 1961.

[AP93] D.G. Akka and S.V. Panshetty. The edge-outercoarseness of complete graphs. *Bull. Cal. Math. Soc.*, 85:505–512, 1993.

[Arc84] Dan Archdeacon. Face Colorings of Embedded Graphs. *J. of Graph Theory*, 8:387–398, 1984.

[Asa86] Kouhei Asano. The Crossing Number of $K_{1,3,n}$ and $K_{2,3,n}$. *J. of Graph Theory*, 10:1–8, 1986.

[Asa87] Kouhei Asano. On the genus and thickness of graphs. *J. of Combinatorial Theory, Series B*, 43:287–292, 1987.

[Asa94] Kouhei Asano. On the thickness of graphs with genus 2. *Ars Combinatoria*, 38:87–95, 1994.

[Bar82] D.W. Barnette. Generating the triangulations of the projective plane. *J. of Combinatorial Theory, Series B*, 33:222–230, 1982.

[Bar83] Francisco Barahona. The max-cut problem on graphs not contractible to K_5. *Operations Research Letters*, 2:107–111, 1983.

[Bar87] D.W. Barnette. Generating closed 2-cell embeddings in the torus and the projective plane. *Discrete Comput. Geom.*, 2:233–247, 1987.

A. Liebers, *Planarizing Graphs*, JGAA, 5(1) 1–74 (2001)   54

[Bar90]   D.W. Barnette. Generating the 4-connected and strongly 4-connected triangulations on the torus and projective plane. *Discrete Mathematics*, 85(1):1–16, 1990.

[Bar91]   D.W. Barnette. The minimal projective plane polyhedral maps. In Peter Gritzmann and Victor Klee, editors, *Applied geometry and discrete mathematics: The Victor Klee Festschrift*, volume 4 of *DIMACS Series in Discrete Mathematics and Computer Science*, pages 63–70. American Mathematical Society, 1991.

[BB68]   I.Z. Bouwer and I. Broere. Note on t-minimal complete bipartite graphs. *Canad. Math. Bull.*, 11:729–732, 1968.

[BD92]   Daniel Bienstock and Nathaniel Dean. New Results on Rectilinear Crossing Numbers and Plane Embeddings. *J. of Graph Theory*, 16:389–398, 1992.

[BD93]   Daniel Bienstock and Nathaniel Dean. Bounds for rectilinear crossing numbers. *J. of Graph Theory*, 17:333–348, 1993.

[BE88]   D.W. Barnette and A.L. Edelson. All Orientable 2-Manifolds Have Finitely Many Minimal Triangulations. *Israel J. of Mathematics*, 62:90–98, 1988.

[BE89]   D.W. Barnette and A.L. Edelson. All 2-Manifolds Have Finitely Many Minimal Triangulations. *Israel J. of Mathematics*, 67:123–128, 1989.

[Bei67]   Lowell W. Beineke. Complete Bipartite Graphs: Decomposition into Planar Subgraphs. In Frank Harary, editor, *A Seminar on Graph Theory*. Holt, Rinehart and Winston, 1967.

[Bei69]   Lowell W. Beineke. Minimal decompositions of complete graphs. *Can. J. Math.*, 21:992–1000, 1969.

[Bei88]   Lowell W. Beineke. Fruited planes. In *Graph theory, 250th Anniv. Conf., Lafayette/Indiana 1986*, volume 63 of *Congressus Numerantium*, pages 127–138, 1988.

[Ber73]   Claude Berge. *Graphs and Hypergraphs*. North–Holland, 1973.

[Ber89]   Claude Berge. *Hypergraphs*. North–Holland, 1989.

[BG69]   Lowell W. Beineke and Richard K. Guy. The coarseness of the complete bipartite graph. *Can. J. Math.*, 21:1086–1096, 1969.

[BH65]   Lowell W. Beineke and Frank Harary. The thickness of the complete graph. *Can. J. Math.*, 17:850–859, 1965.

[BHK62]   Joseph Battle, Frank Harary, and Yukihiro Kodama. Every planar graph with nine points has a nonplanar complement. *Bulletin of the American Mathematical Society*, 68:569–571, 1962.

[BHM64]   Lowell W. Beineke, Frank Harary, and John W. Moon. On the thickness of the complete bipartite graph. *Proc. Camb. Phil. Soc.*, 60:1–5, 1964.

[Bie91]   Daniel Bienstock. Some Provably Hard Crossing Number Problems. *Discrete Comput. Geom.*, 6:443–459, 1991.

[BK79]   Frank Bernhart and Paul C. Kainen. The Book Thickness of a Graph. *J. of Combinatorial Theory, Series B*, 27:320–331, 1979.

A. Liebers, *Planarizing Graphs*, JGAA, 5(1) 1–74 (2001)   55

[BL76]   Kellogg S. Booth and George S. Lueker. Testing for the consecutive ones property, interval graphs, and planarity using PQ-tree algorithms. *J. Computer and System Sciences*, 13:335–379, 1976.

[BL95]   C. Paul Bonnington and Charles H.C. Little. *The Foundations of Topological Graph Theory*. Springer, 1995.

[BLW76]   Norman L. Biggs, E. Keith Lloyd, and Robin J. Wilson. *Graph Theory: 1736–1936*. Clarendon Press, Oxford, 1976.

[BM76]   John A. Bondy and Uppaluri R.S. Murty. *Graph theory with applications*. North–Holland, 1976.

[BM99]   John Boyer and Wendy Myrvold. Stop Minding Your P's and Q's: A Simplified $O(n)$ Planar Embedding Algorithm. In *Proceedings of the 10th Annual ACM-SIAM Symposium on Discrete Algorithms, SODA'99*, pages 140–146, 1999.

[BS84]   Jonathan F. Buss and Peter Shor. On the pagenumber of planar graphs. In *Proceedings of the 16th Annual ACM Symposium on Theory of Computing, STOC'84*, pages 98–100, 1984.

[BS93]   Graham R. Brightwell and Edward R. Scheinerman. Representations of planar graphs. *SIAM J. Discrete Math.*, 6:214–229, 1993.

[CF96]   Gruia Călinescu and Cristina G. Fernandes. Finding Large Planar Subgraphs and Large Subgraphs of a Given Genus. In Jin-Yi Cai and Chak Kuen Wong, editors, *Proceedings 2nd Annual International Conference on Computing and Combinatorics, COCOON'96*, pages 152–161. Springer-Verlag, Lecture Notes in Computer Science, vol. 1090, 1996.

[CFFK96]   Gruia Călinescu, Cristina G. Fernandes, Ulrich Finkler, and Howard Karloff. A Better Approximation Algorithm for Finding Planar Subgraphs. In *Proceedings of the 7th Annual ACM-SIAM Symposium on Discrete Algorithms, SODA'96*, pages 16–25, 1996.

[CFFK98]   Gruia Călinescu, Cristina G. Fernandes, Ulrich Finkler, and Howard Karloff. A Better Approximation Algorithm for Finding Planar Subgraphs. *J. Algorithms*, 27:269–302, 1998.

[CGP98]   Zhi-Zhong Chen, Michelangelo Grigni, and Christos H. Papadimitriou. Planar Map Graphs. In *Proceedings of the 30th Annual ACM Symposium on Theory of Computing, STOC'98*, pages 514–523, 1998.

[Che96]   Zhi-Zhong Chen. Practical Approximation Schemes for Maximum Induced-Subgraph Problems on $K_{3,3}$-free or K_5-free Graphs. In Friedhelm Meyer auf der Heide and Burkhard Monien, editors, *Proceedings of the 23rd International Colloquium on Automata, Languages and Programming ICALP'96*, pages 269–279. Springer-Verlag, Lecture Notes in Computer Science, vol. 1099, 1996.

[Che98]   Zhi-Zhong Chen. Efficient Approximation Schemes for Maximization Problems on $K_{3,3}$-Free Graphs. *J. Algorithms*, 26:166–187, 1998.

[CHT93]   Leizhen Cai, Xiaofeng Han, and Robert E. Tarjan. An $O(m \log n)$-time algorithm for the maximal planar subgraph problem. *SIAM J. Comput.*, 22:1142–1162, 1993.

A. Liebers, *Planarizing Graphs*, JGAA, 5(1) 1–74 (2001)   56

[Cim92]   Robert J. Cimikowski. Graph Planarization and Skewness. In *Proc. 23rd Southeastern Conference on Combinatorics, Graph Theory and Computing, Boca Raton, Florida, USA, February 3–7, 1992, part 1*, volume 88 of *Congressus Numerantium*, pages 21–32, 1992.

[Cim94]   Robert J. Cimikowski. Branch-and-bound techniques for the maximum planar subgraph problem. *Intern. J. Computer Math.*, 53:135–147, 1994.

[Cim95a]   Robert J. Cimikowski. An Analysis of Some Heuristics for the Maximum Planar Subgraph Problem. In *Proceedings of the 6th Annual ACM-SIAM Symposium on Discrete Algorithms, SODA'95*, pages 322–331, 1995.

[Cim95b]   Robert J. Cimikowski. On Heuristics for Determining the Thickness of a Graph. *Information Sciences*, 85:87–98, 1995.

[Cim97]   Robert J. Cimikowski. An analysis of heuristics for graph planarization. *J. Inf. Optimization Sci.*, 18:49–73, 1997.

[CK93]   Jianer Chen and Arkady Kanevsky. On Assembly of Four-Connected Graphs. In Ernst W. Mayr, editor, *Proceedings 18th International Workshop on Graph-Theoretic Concepts in Computer Science, WG'92*, pages 158–169. Springer-Verlag, Lecture Notes in Computer Science, vol. 657, 1993. Extended Abstract.

[CL96]   Gary Chartrand and Linda Lesniak. *Graphs & Digraphs*. Chapman & Hall, 3rd edition, 1996.

[CLR87]   Fan R.K. Chung, Frank Thomson Leighton, and Arnold L. Rosenberg. Embedding graphs in books: A layout problem with applications to VLSI design. *SIAM J. Alg. Disc. Meth.*, 8:33–58, 1987.

[CLR94]   Thomas H. Cormen, Charles E. Leiserson, and Ronald L. Rivest. *Introduction to Algorithms*. MIT Press and McGraw–Hill, 1994.

[CNAO85]   Norishige Chiba, Takao Nishizeki, Shigenobu Abe, and Takao Ozawa. A Linear Algorithm for Embedding Planar Graphs Using PQ-Trees. *J. Computer and System Sciences*, 30:54–76, 1985.

[CNS79]   Toru Chiba, Ikuo Nishioka, and Isao Shirakawa. An algorithm of maximal planarization of graphs. In *Proceedings IEEE Symposium on Circuits & Systems, ISCAS, 1979*, pages 649–652, 1979.

[Com92]   F. Comellas. Using genetic algorithms for planarization problems. In *Computational and applied mathematics, I. Algorithms and theory, Sel. Rev. Pap. IMACS 13th World Congr., Dublin/Irel. 1991*, pages 93–100, 1992.

[DEH00]   Michael B. Dillencourt, David Eppstein, and Daniel S. Hirschberg. Geometric Thickness of Complete Graphs. *Journal of Graph Algorithms and Applications*, 4:5–17, 2000.

[DETT94]   Giuseppe Di Battista, Peter Eades, Roberto Tamassia, and Ioannis G. Tollis. Algorithms for Drawing Graphs: an Annotated Bibliography. *Computational Geometry*, 4:235–282, 1994.

[DETT99]   Giuseppe Di Battista, Peter Eades, Roberto Tamassia, and Ioannis G. Tollis. *Graph Drawing. Algorithms for the Visualization of Graphs*. Prentice Hall, 1999.

A. Liebers, *Planarizing Graphs*, JGAA, 5(1) 1–74 (2001)   57

[dFdM96] Hubert de Fraysseix and Patrice Ossona de Mendez. Planarity and edge poset dimension. *Europ. J. Combinatorics*, 17:731–740, 1996.

[DFF85] M.E. Dyer, L.R. Foulds, and A.M. Frieze. Analysis of heuristics for finding a maximum weight planar subgraph. *European J. of Operational Research*, 20:102–114, 1985.

[dFPP90] Hubert de Fraysseix, János Pach, and Richard Pollack. How to draw a planar graph on a grid. *Combinatorica*, 10:41–51, 1990.

[DHS91] Alice M. Dean, Joan P. Hutchinson, and Edward R. Scheinerman. On the Thickness and Arboricity of a Graph. *J. of Combinatorial Theory, Series B*, 52:147–151, 1991.

[Dji84] Hristo N. Djidjev. On some properties of nonplanar graphs. *Comptes rendus de l'Académie Bulgare des Sciences: Sciences mathématiques et naturelles*, 37:1183–1184, 1984.

[Dji85] Hristo N. Djidjev. A linear algorithm for partitioning graphs of fixed genus. *SERDICA Bulgaricae mathematicae publications*, 11:369–387, 1985.

[Dji95] Hristo N. Djidjev. A Linear Algorithm for the Maximal Planar Subgraph Problem. In Selim G. Akl, Frank Dehne, Jörg-Rüdiger Sack, and Nicola Santoro, editors, *Proceedings 4th International Workshop on Algorithms and Datastructures, WADS'95*, pages 369–380. Springer-Verlag, Lecture Notes in Computer Science, vol. 955, 1995.

[DR91] Hristo N. Djidjev and John H. Reif. An Efficient Algorithm for the Genus Problem with Explicit Construction of Forbidden Subgraphs. In *Proceedings of the 23th Annual ACM Symposium on Theory of Computing, STOC'91*, pages 337–347, 1991.

[DR95] Alice M. Dean and R. Bruce Richter. The Crossing Number of $C_4 \times C_4$. *J. of Graph Theory*, 19:125–129, 1995.

[DT89] Giuseppe Di Battista and Roberto Tamassia. Incremental Planarity Testing. In *Proceedings of the 30th Annual Symposium on Foundations of Computer Science, FOCS'89*, pages 436–441, 1989.

[DT90] Giuseppe Di Battista and Roberto Tamassia. On-Line Graph Algorithms with SPQR-Trees. In Michael S. Paterson, editor, *Proceedings of the 17th International Colloquium on Automata, Languages and Programming ICALP'90*, pages 598–611. Springer-Verlag, Lecture Notes in Computer Science, vol. 443, 1990.

[DT96a] Giuseppe Di Battista and Roberto Tamassia. On-Line Maintenance of Triconnected Components with SPQR-Trees. *Algorithmica*, 15:302–318, 1996.

[DT96b] Giuseppe Di Battista and Roberto Tamassia. On-Line Planarity Testing. *SIAM J. Comput.*, 25(5):956–997, 1996.

[dV90] Yves Colin de Verdière. Sur un Nouvel Invariant des Graphes et un Critère de Planarité. *J. of Combinatorial Theory, Series B*, 50:11–21, 1990.

[dV93] Yves Colin de Verdière. On a New Graph Invariant and a Criterion for Planarity. In Neil Robertson and Paul Seymour, editors, *Graph Structure Theory, Proceedings of the AMS-IMS-SIAM Joint Summer Research Conference on Graph Minors, Seattle, Washington, USA, June 22 – July 5,*

A. Liebers, *Planarizing Graphs*, JGAA, 5(1) 1–74 (2001)   58

1991, volume 147 of *Contemporary Mathematics*, pages 137–147. American Mathematical Society, 1993.

[DV95]   Hristo N. Djidjev and Shankar M. Venkatesan. Planarization of graphs embedded on surfaces. In Manfred Nagl, editor, *Proceedings 21st International Workshop on Graph-Theoretic Concepts in Computer Science, WG'95*, pages 62–72. Springer-Verlag, Lecture Notes in Computer Science, vol. 1017, 1995.

[EFG82]   Peter Eades, L.R. Foulds, and J.W. Giffin. An efficient heuristic for identifying a maximum weight planar subgraph. In *Combinatorial Mathematics IX, Proceedings 9th Australian Conference, Brisbane, Australia, 1981*, pages 239–251. Springer-Verlag, Lecture Notes in Mathematics, vol. 952, 1982.

[ELMM95]   Peter Eades, Cláudio L. Lucchesi, João Meidanis, and Candido Ferreira Xavier de Mendonça Neto. NP-hardness Results for Tension-Free Layout. Technical Report DCC-95-15, Departamento de Ciência da Computação, Instituto de Matemaática, Estatística e Ciência da Computação, Universidade Estadual de Campinas, Brazil, 1995.

[EM93]   Peter Eades and Candido Ferreira Xavier de Mendonça Neto. Heuristics for Planarization by Vertex Splitting. In Giuseppe Di Battista, Peter Eades, Hubert de Fraysseix, Pierre Rosenstiehl, and Roberto Tamassia, editors, *ALCOM International Workshop on Graph Drawing, GD'93, Paris, France, September 25–29, 1993*, pages 83–85. Available electronically at ftp://ftp.cs.brown.edu/pub/gd94/gd-92-93/gd93-v2.ps.Z, 1993.

[EM96]   Peter Eades and Candido Ferreira Xavier de Mendonça Neto. Vertex Splitting and Tension-Free Layout. In Franz J. Brandenburg, editor, *Graph Drawing. Proceedings of the DIMACS International Workshop, GD'95*, pages 202–211. Springer-Verlag, Lecture Notes in Computer Science, vol. 1027, 1996.

[ET76]   Shimon Even and Robert E. Tarjan. Computing an *st*-numbering. *Theoretical Computer Science*, 2:339–344, 1976.

[Eve79]   Shimon Even. *Graph algorithms*. Computer Software Engineering Series. Computer Science Press, 1979.

[Fár48]   István Fáry. On straight line representation of planar graphs. *Acta Scientiarum Mathematicarum (Institutum Bolyainum, Universitatis Szegediensis)*, 11:229–233, 1948.

[FCE95]   Qing-Wen Feng, Robert F. Cohen, and Peter Eades. Planarity for Clustered Graphs. In Paul Spirakis, editor, *Third European Symposium on Algorithms, ESA'95*, pages 213–226. Springer-Verlag, Lecture Notes in Computer Science, vol. 979, 1995.

[FF00]   Luérbio Faria and Celina Miraglia Herrera de Figueiredo. On Eggleton and Guy conjectured upper bound for the crossing number of the n-cube. *Math. Slovaca*, 50:271–287, 2000.

[FFM98a]   Luérbio Faria, Celina Miraglia Herrera de Figueiredo, and Candido Ferreira Xavier de Mendonça Neto. Splitting Number is NP-Complete. In Juraj Hromkovič and Ondrej Sýkora, editors, *Proceedings 24th International Workshop on Graph-Theoretic Concepts in Computer Science,*

A. Liebers, *Planarizing Graphs*, JGAA, 5(1) 1–74 (2001)   59

WG'98, pages 285–297. Springer-Verlag, Lecture Notes in Computer Science, vol. 1517, 1998.

[FFM98b] Luérbio Faria, Celina Miraglia Herrera de Figueiredo, and Candido Ferreira Xavier de Mendonça Neto. The Splitting Number of the 4-Cube. In Cláudio L. Lucchesi and Arnaldo V. Moura, editors, *Theoretical Informatics. Proceedings of the 3rd Latin American Symposium, LATIN'98*, pages 141–150. Springer-Verlag, Lecture Notes in Computer Science, vol. 1380, 1998.

[FFM01] Luérbio Faria, Celina Miraglia Herrera de Figueiredo, and Candido Ferreira Xavier de Mendonça Neto. Splitting number is NP-complete. *Discrete Applied Math.*, 108:65–83, 2001.

[FGG85] L.R. Foulds, P.B. Gibbons, and J.W. Giffin. Facilities Layout Adjacency Determination: An Experimental Comparison of Three Graph Theoretic Heuristics. *Operations Research*, 33:1091–1106, 1985.

[Fio86] S. Fiorini. On the crossing number of generalized Petersen graphs. In *Proceedings Int. Conf. Finite Geom. Comb. Struct., Bari, Italy, 1984, Combinatorics '84*, pages 225–241. North-Holland, Annals of Discrete Mathematics, vol. 30, 1986.

[FMN94] Steve Fisk, Bojan Mohar, and Roman Nedela. Minimal Locally Cyclic Triangulations of the Projective Plane. *J. of Graph Theory*, 18:25–35, 1994.

[Fou92] L.R. Foulds. *Graph Theory Applications*. Springer, 1992.

[FR78] L.R. Foulds and David F. Robinson. Graph Theoretic Heuristics for the Plant Layout Problem. *International J. of Production Research*, 16:27–37, 1978.

[FR95] T. Feo and Mauricio G.C. Resende. Greedy randomized adaptive search procedures. *J. of Global Optimization*, 6:109–133, 1995.

[Gav73] Fanika Gavril. Algorithms for a Maximum Clique and a Maximum Independent Set of a Circle Graph. *Networks*, 3:261–273, 1973.

[GB68] Richard K. Guy and Lowell W. Beineke. The coarseness of the complete graph. *Can. J. Math.*, 20:888–894, 1968.

[GH73] Richard K. Guy and Anthony Hill. The crossing number of the complement of a circuit. *Discrete Mathematics*, 5:335–344, 1973.

[Gib85] Alan Gibbons. *Algorithmic Graph Theory*. Cambridge University Press, 1985.

[GJ79] Michael R. Garey and David S. Johnson. *Computers and Intractability: A Guide to the Theory of NP-Completeness*. W H Freeman & Co Ltd, 1979.

[GJ83] Michael R. Garey and David S. Johnson. Crossing Number is NP-Complete. *SIAM J. Alg. Disc. Meth.*, 4:312–316, 1983.

[GJS76] Michael R. Garey, David S. Johnson, and Larry Stockmeyer. Some simplified NP-complete graph problems. *Theoretical Computer Science*, 1:237–267, 1976.

[GM84] Michel Gondran and Michel Minoux. *Graphs and algorithms*. Wiley, 1984.

A. Liebers, *Planarizing Graphs*, JGAA, 5(1) 1–74 (2001)   60

[Gol63]   A.J. Goldstein. An Efficient and Constructive Algorithm for Testing Wether a Graph Can Be Embedded in a Plane. In *Graph and Combinatorics Conference, Contract No. NONR 1858-(21), Office of Naval Research Logistics Proj., Dept. of Mathematics, Princeton University, May 16-18, 1963*, 1963.

[Gol80]   Martin C. Golumbic. *Algorithmic graph theory and perfect graphs.* Academic Press, 1980.

[Grü67]   Banko Grünbaum. *Convex Polytopes.* Interscience Publishers, a divsion of John Wiley & Sons, 1967.

[GT87]   Jonathan L. Gross and Thomas W. Tucker. *Topological graph theory.* Wiley, 1987.

[GT94]   Oliver Goldschmidt and Alexan Takvorian. An efficient graph planarization two-phase heuristic. *Networks*, 24:69–73, 1994.

[Gub93]   Bradley S. Gubser. Planar graphs with no 6-wheel minor. *Discrete Mathematics*, 120:59–73, 1993.

[Guy69]   Richard K. Guy. The decline and fall of Zarankiewicz's theorem. In Frank Harary, editor, *Proof Techniques in Graph Theory. Proceedings of the Second Ann Arbor Graph Theory Conference, February 1968*, pages 63–69. Academic Press, New York, 1969.

[Guy71]   Richard K. Guy. Latest results on crossing numbers. In Michael Capobianco, Joseph B. Frechen, and M. Krolik, editors, *Recent Trends in Graph Theory, Proceedings of the First New York City Graph Theory Conference, June 11–13, 1970*, pages 143–156. Springer-Verlag, Lecture Notes in Mathematics, vol. 186, 1971.

[Guy72]   Richard K. Guy. Crossing Numbers of Graphs. In Y. Alavi, D.R. Lick, and Arthur T. White, editors, *Graph Theory and Applications*, pages 111–124. Proceedings of the Conference at Western Michigan University, May 10–13, 1972, Springer-Verlag, Lecture Notes in Mathematics, vol. 303, 1972.

[Hal43]   Dick Wick Hall. A Note on Primitive Skew Curves. *Bulletin of the American Mathematical Society*, 49:935–936, 1943.

[Hal91]   John H. Halton. On the Thickness of Graphs of Given Degree. *Information Sciences*, 54:219–238, 1991.

[Har69]   Frank Harary. *Graph theory.* Addison–Wesley Publishing Company, 1969.

[Har79]   Jehuda Hartman. Bounds on the Coarseness of the n-Cube. *Canad. Math. Bull.*, 22:171–175, 1979.

[Har86]   Nora Hartsfield. The toroidal splitting number of the complete graph K_n. *Discrete Mathematics*, 62:35–47, 1986.

[Har87]   Nora Hartsfield. The splitting number of the complete graph in the projective plane. *Graphs and Combinatorics*, 3:349–356, 1987.

[Hea90]   P.J. Heawood. Map Colour Theorem. *Quart. J. Math.*, 24:332–338, 1890.

[Hea84]   Lenny Heath. Embedding Planar Graphs in Seven Pages. In *Proceedings of the 25th Annual Symposium on Foundations of Computer Science, FOCS'84*, pages 74–83, 1984.

A. Liebers, *Planarizing Graphs*, JGAA, 5(1) 1–74 (2001)   61

[Hea87]   Lenwood S. Heath. Embedding outerplanar graphs in small books. *SIAM J. Alg. Disc. Meth.*, 8:198–218, 1987.

[HG68a]   Arthur M. Hobbs and J.W. Grossman. A Class of Thickness-Minimal Graphs. *J. of Research of the National Bureau of Standards – B. Mathematical Sciences*, 72B:145–153, 1968.

[HG68b]   Arthur M. Hobbs and J.W. Grossman. Thickness and connectivity in graphs. *J. of Research of the National Bureau of Standards – B. Mathematical Sciences*, 72B:239–244, 1968.

[HHW88]   Frank Harary, John P. Hayes, and Horng-Jyh Wu. A survey of the theory of hypercube graphs. *Comput. Math. Applic.*, 15:277–289, 1988.

[HI92]   Lenwood S. Heath and Sorin Istrail. The pagenumber of genus g graphs is $O(g)$. *J. of the Association for Computing Machinery*, 39:479–501, 1992.

[HJR85]   Nora Hartsfield, Brad Jackson, and Gerhard Ringel. The splitting number of the complete graph. *Graphs and Combinatorics*, 1:311–329, 1985.

[HM87]   Joan P. Hutchinson and Gary L. Miller. On deleting vertices to make a graph of positive genus planar. In David S. Johnson, editor, *Discrete algorithms and complexity. Proc. of the Japan-US joint seminar, Kyoto, Japan, June 4–6, 1986*, pages 81–98, 1987.

[Hob69]   Arthur M. Hobbs. A survey of thickness. In William T. Tutte, editor, *Recent progress in combinatorics. Proceedings of the 3rd Waterloo Conference on Combinatorics, May 20–31, 1968*, pages 255–264. Academic Press, New York, 1969.

[Hor83]   Peter Horák. Solution of four problems from Eger, 1981. I. In Miroslav Fiedler, editor, *Graphs and Other Combinatorial Topics. Proceedings of the Third Czechoslovak Symposium on Graph Theory, Prague, August 24–27, 1982*, pages 115–117. Teubner, Texte zur Mathematik, vol. 59, 1983.

[HŠ82]   Peter Horák and Jozef Širáň. On a Modified Concept of Thickness of a Graph. *Math. Nachr.*, 108:305–306, 1982.

[HŠ87]   Peter Horák and Jozef Širáň. A construction of thickness-minimal graphs. *Discrete Mathematics*, 64:263–268, 1987.

[Hsu95]   Wen-Lian Hsu. A linear time algorithm for finding maximal planar subgraphs. In John Staples, Peter Eades, Naoki Katoh, and Alistair Moffat, editors, *Algorithms and Computation, 6th International Symposium, ISAAC'95*, pages 352–361. Springer-Verlag, Lecture Notes in Computer Science, vol. 1004, 1995.

[HSV96]   Joan P. Hutchinson, Thomas Shermer, and Andrew Vince. On Representations of Some Thickness-Two Graphs. In Franz J. Brandenburg, editor, *Graph Drawing. Proceedings of the DIMACS International Workshop, GD'95*, pages 324–332. Springer-Verlag, Lecture Notes in Computer Science, vol. 1027, 1996.

[HSV99]   Joan P. Hutchinson, Thomas Shermer, and Andrew Vince. On representations of some thickness-two graphs. *Computational Geometry*, 13:161–171, 1999.

[HT65]   Frank Harary and William T. Tutte. A dual form of Kuratowski's theorem. *Canad. Math. Bull.*, 8:17–20, 1965.

[HT74]   John E. Hopcroft and Robert E. Tarjan. Efficient planarity testing. *J. of the Association for Computing Machinery*, 21:549–568, 1974.

[Hut89]   Joan P. Hutchinson. On genus-reducing and planarizing algorithms for embedded graphs. In R. Bruce Richter, editor, *Graphs and algorithms, Proceedings of the AMS-IMS-SIAM Joint Summer Research Conference, Boulder, Colorado, USA, June 28 – July 4, 1987*, volume 89 of *Contemporary Mathematics*, pages 19–26. American Mathematical Society, 1989.

[JLM97]   Michael Jünger, Sebastian Leipert, and Petra Mutzel. Pitfalls of Using PQ-Trees in Automatic Graph Drawing. In Giuseppe Di Battista, editor, *Proceedings of the Symposium on Graph Drawing, GD'97*, pages 193–204. Springer-Verlag, Lecture Notes in Computer Science, vol. 1353, 1997.

[JLMO97]   Michael Jünger, Eva K. Lee, Petra Mutzel, and Thomas Odenthal. A Polyhedral Approach to the Multi-Layer Crossing Minimization Problem. In Giuseppe Di Battista, editor, *Proceedings of the Symposium on Graph Drawing, GD'97*, pages 13–24. Springer-Verlag, Lecture Notes in Computer Science, vol. 1353, 1997.

[JM96]   Michael Jünger and Petra Mutzel. Maximum Planar Subgraphs and Nice Embeddings: Practical Layout Tools. *Algorithmica*, 16:33–59, 1996.

[JM97]   Michael Jünger and Petra Mutzel. 2-Layer Straightline Crossing Minimization: Performance of Exact and Heuristic Algorithms. *Journal of Graph Algorithms and Applications*, 1:1–25, 1997.

[JMOS94]   Michael Jünger, Petra Mutzel, Thomas Odenthal, and Mark Scharbrodt. The Thickness of Graphs without K_5-Minors. Technical Report 94.168, Universität zu Köln, 1994. Available electronically at ftp://ftp.ZPR.uni-koeln.de/pub/paper/zpr94-168.ps.gz.

[JMOS98]   Michael Jünger, Petra Mutzel, Thomas Odenthal, and Mark Scharbrodt. The Thickness of a Minor-Excluded Class of Graphs. *Discrete Mathematics*, 182:169–176, 1998.

[Joh85]   David S. Johnson. The NP-Completeness Column: An Ongoing Guide. *J. Algorithms*, 6:434–451, 1985.

[JR84a]   Brad Jackson and Gerhard Ringel. Plane constructions for graphs, networks, and maps measurements of planarity. In *Selected topics in operations research and mathematical economics, Proc. 8th Symp., Karlsruhe 1983*, pages 315–324. Springer-Verlag, Lecture Notes in Economics and Mathematical Systems, vol. 226, 1984.

[JR84b]   Brad Jackson and Gerhard Ringel. The splitting number of complete bipartite graphs. *Arch. Math.*, 42:178–184, 1984.

[JR85]   Brad Jackson and Gerhard Ringel. Splittings of graphs on surfaces. In Frank Harary, editor, *Graphs and applications, Proceedings of the 1st Colorado Symposium on Graph Theory, Boulder, Colorado, 1982*, pages 203–219, 1985.

[JŠ82]   Stanislav Jendrol and Mária Ščerbová. On the crossing numbers of $S_m \times P_n$ and $S_m \times C_n$. *Casopis pro pestování matematiky*, 107:225–230, 1982.

[JST89]   R. Jayakumar, Madisetti N.S. Swamy, and K. Thulasiraman. $O(n^2)$ Algorithms for Graph Planarization. In Jan v. Leeuwen, editor, *Proceedings*

A. Liebers, *Planarizing Graphs*, JGAA, 5(1) 1–74 (2001) 63

14th International Workshop on Graph-Theoretic Concepts in Computer Science, WG'88, pages 352–377. Springer-Verlag, Lecture Notes in Computer Science, vol. 344, 1989.

[JT95] Tommy R. Jensen and Bjarne Toft. *Graph Coloring Problems*. Wiley, 1995.

[JTS89] R. Jayakumar, K. Thulasiraman, and Madisetti N.S. Swamy. $O(n^2)$ Algorithms for Graph Planarization. *IEEE Transactions on Computer-Aided Design of Integrated Circuits and Systems*, 8:257–267, 1989.

[Kai72a] Paul C. Kainen. A Lower Bound for Crossing Numbers of Graphs with Application to K_n, $K_{p,q}$, and $Q(d)$. *J. of Combinatorial Theory, Series B*, 12:287–298, 1972.

[Kai72b] Paul C. Kainen. On the stable crossing number of cubes. *Proceedings of the American Mathematical Society*, 36:55–62, 1972.

[Kai73] Paul C. Kainen. Thickness and Coarseness of Graphs. *Abhandlungen aus dem Mathematischen Seminar der Universität Hamburg*, 39:88–95, 1973.

[Kai90] Paul C. Kainen. The book thickness of a graph, II. In *Proc. 20th Southeastern Conference on Combinatorics, Graph Theory, and Computing, Boca Raton, Florida, USA, 1989*, volume 71 of *Congressus Numerantium*, pages 127–132, 1990.

[Kan92] Goos Kant. An $O(n^2)$ Maximal Planarization Algorithm based on PQ-trees. Technical Report RUU-CS-92-03, Dept. of Comp. Science, Utrecht University, 1992.

[Kan93] Goos Kant. *Algorithms for Drawing Planar Graphs*. University of Utrecht, 1993. PhD Thesis.

[Kar72] Richard M. Karp. Reducibility Among Combinatorial Problems. In Raymond E. Miller and J.W. Thatcher, editors, *Complexity of computer computations. Proceedings of a Symposium on the Complexity of Computer Computations, IBM Thomas J.Watson Research Center, Yorktown Heights, New York, March 20–22, 1972*, pages 85–103. Plenum Press, New York, 1972.

[KD79] M.S. Krishnamoorthy and Narsingh Deo. Node-deletion NP-complete problems. *SIAM J. Comput.*, 8:619–625, 1979.

[Kel93] Alexander K. Kelmans. Graph planarity and related topics. In Neil Robertson and Paul Seymour, editors, *Graph Structure Theory, Proceedings of the AMS-IMS-SIAM Joint Summer Research Conference on Graph Minors, Seattle, Washington, USA, June 22 – July 5, 1991*, volume 147 of *Contemporary Mathematics*, pages 635–667. American Mathematical Society, 1993.

[KH78] Bernhard Korte and Dirk Hausmann. An Analysis of the Greedy Heuristic for Independence Systems. In Brian R. Alspach, editor, *Algorithmic Aspects of Combinatorics*, pages 65–74. North-Holland, Annals of Discrete Mathematics, vol. 2, 1978.

[Khu90] Samir Khuller. Extended Planar Graph Algorithms to $K_{3,3}$-Free Graphs. *Information and Computation*, 84:13–25, 1990.

A. Liebers, *Planarizing Graphs*, JGAA, 5(1) 1–74 (2001)   64

[KJ83]   Kazimierz Kuratowski and Jan Jaworowski. On the problem of skew curves in topology. In M. Borowiecki, J.W. Kennedy, and Maciej M. Sysło, editors, *Graph Theory, Proc. of a Conference held in Lagów, Poland, February 10–13, 1981*, pages 1–13. Springer-Verlag, Lecture Notes in Mathematics, vol. 1018, 1983. Translation of [Kur30] by Jan Jaworowski.

[Kle67]   Michael Kleinert. Die Dicke des n-dimensionalen Würfel-Graphen. *J. of Combinatorial Theory*, 3:10–15, 1967.

[Kle70]   Daniel J. Kleitman. The Crossing Number of $K_{5,n}$. *J. of Combinatorial Theory*, 9:315–323, 1970.

[Kle91]   Marián Klešč. On the crossing numbers of Cartesian products of stars and paths or cycles. *Math. Slovaca*, 41(2):113–120, 1991.

[Kle94]   Marián Klešč. The Crossing Numbers of Products of Paths and Stars with 4-Vertex Graphs. *J. of Graph Theory*, 18:605–614, 1994.

[Kle95]   Marián Klešč. The crossing numbers of certain Cartesian products. *Discussiones Mathematicae – Graph Theory*, 15:5–10, 1995.

[Kle96]   Marián Klešč. The crossing number of $K_{2,3} \times P_n$ and $K_{2,3} \times S_n$. *Tatra Mountains Math. Publ.*, 9:51–56, 1996.

[KM92]   André Kézdy and Patrick McGuinness. Sequential and Parallel Algorithms to Find a K_5 minor. In *Proceedings of the 3rd Annual ACM-SIAM Symposium on Discrete Algorithms, SODA'92*, pages 345–356, 1992.

[Koc87]   Martin Kochol. Construction of crossing-critical graphs. *Discrete Mathematics*, 66:311–313, 1987.

[Koc91]   Martin Kochol. Linear Jump of Crossing Number for Non Kuratowski Edge of a Graph. *Radovi Matematički*, 7:177–184, 1991.

[Kön36]   Dénes König. *Theorie der endlichen und unendlichen Graphen. Kombinatorische Topologie der Streckenkomplexe.* Akademische Verlagsgesellschaft, Leipzig, 1936.

[Kön90]   Dénes König. *Theory of Finite and Infinite Graphs.* Birkhäuser, 1990. Originally published in 1936 as *Theorie der endlichen und unendlichen Graphen.* With a commentary by William T. Tutte.

[KRS96]   Marián Klešč, R. Bruce Richter, and Ian Stobert. The crossing number of $C_5 \times C_n$. *J. of Graph Theory*, 22:239–243, 1996.

[Kur30]   Kazimierz Kuratowski. Sur le problème des courbes gauches en topologie. *Fundamenta Mathematicae*, 15:271–283, 1930.

[KW78]   Paul C. Kainen and Arthur T. White. On Stable Crossing Numbers. *J. of Graph Theory*, 2:181–187, 1978.

[La 94]   J.A. La Poutré. Alpha-Algorithms for Incremental Planarity Testing. In *Proceedings of the 26th Annual ACM Symposium on Theory of Computing, STOC'94*, pages 706–715, 1994.

[LEC67]   A. Lempel, Shimon Even, and I. Cederbaum. An Algorithm for Planarity Testing of Graphs. In Pierre Rosenstiehl, editor, *Theory of Graphs. International Symposium, Rome, Italy, July 1966*, pages 215–232, New York, 1967. Gordon and Breach.

A. Liebers, *Planarizing Graphs*, JGAA, 5(1) 1–74 (2001)   65

[Lee90]   Jan v. Leeuwen. Graph Algorithms. In Jan v. Leeuwen, editor, *Algorithms and Complexity*, volume A of *Handbook of Theoretical Computer Science*, pages 525–631. Elsevier Science Pub., Amsterdam, 1990.

[Lei81]   Frank Thomson Leighton. New Lower Bound Techniques for VLSI. In *Proceedings of the 22th Annual Symposium on Foundations of Computer Science, FOCS'81*, pages 1–12, 1981.

[Lei83]   Frank Thomson Leighton. *Complexity Issues in VLSI*. MIT Press, 1983.

[Lei84]   Frank Thomson Leighton. New Lower Bound Techniques for VLSI. *Math. Systems Theory*, 17:47–70, 1984.

[Lei94]   Sebastian Leipert. The Problem of Computing a Maximal Planar Subgraph Using *PQ*-Trees is Still Not Solved. In *Student Proceedings 8th Conference of the European Consortium for Mathematics in Industry, ECMI'94, Kaiserslautern, Germany, 1994*, pages 95–109, 1994.

[Len89]   Thomas Lengauer. Hierarchical Planarity Testing Algorithms. *J. of the Association for Computing Machinery*, 36:474–509, 1989.

[Leu92]   Janny Leung. A new graph-theoretic heuristic for facility layout. *Management Science*, 38:594–605, 1992.

[LG79]   P.C. Liu and R.C. Geldmacher. On the Deletion of Nonplanar Edges of a Graph. In *Proc. of the 10th Southeastern Conference on Combinatorics, Graph Theory, and Computing, Boca Raton, Florida, USA, 1979, part 2*, volume 24 of *Congressus Numerantium*, pages 727–738, 1979.

[Lic82]   D. Lichtenstein. Planar Formulae and their Uses. *SIAM J. Comput.*, 11:329–343, 1982.

[LY80]   John M. Lewis and Mihalis Yannakakis. The Node-Deletion Problem for Hereditary Properties is NP-Complete. *J. Computer and System Sciences*, 20:219–230, 1980.

[Mac37]   S. MacLane. A combinatorial condition for planar graphs. *Fundamenta Mathematicae*, 28:22–32, 1937.

[Mad91]   Tom Madej. Bounds for the Crossing Number of the N-Cube. *J. of Graph Theory*, 15:81–97, 1991.

[Mal94]   Seth M. Malitz. Graphs with E Edges Have Pagenumber $O(\sqrt{E})$. *J. Algorithms*, 17:71–84, 1994.

[Man83]   Anthony Mansfield. Determining the thickness of graphs is NP-hard. *Math. Proc. Camb. Phil. Soc.*, 93:9–23, 1983.

[ME93a]   Matthias Mayer and Fikret Ercal. Genetic Algorithms For Vertex Splitting in DAGs. Technical Report CSC-93-02, Computer Science Department, University of Missouri-Rolla, MO 65401, USA, January 1993. Presented as a poster at *The Fifth Int'l Conf. on Genetic Algorithms (ICGA-93)*. Available electronically at http://www.cs.umr.edu/techrpt/postscript/93-02.ps.

[ME93b]   Matthias Mayer and Fikret Ercal. Parallel Genetic Algorithms For the DAG Vertex Splitting Problem. Technical Report CSC-93-10, Computer Science Department, University of Missouri-Rolla, MO 65401, USA, May 1993.

[Meh84]   Kurt Mehlhorn. *Data Structures and Algorithms 2: Graph Algorithms and NP-Completeness.* Springer, 1984.

[Men94]   Candido Ferreira Xavier de Mendonça Neto. *A Layout System for Information System Diagrams.* PhD thesis, University of Queensland, Australia, March 1994. Technical Report No. 94-01, Department of Computer Science, The University of Newcastle, Australia.

[Mic83]   Danuta Michalak. On middle and total graphs with coarseness number equal 1. In M. Borowiecki, J.W. Kennedy, and Maciej M. Sysło, editors, *Graph Theory, Proc. of a Conference held in Lagów, Poland, February 10–13, 1981*, pages 139–150. Springer-Verlag, Lecture Notes in Mathematics, vol. 1018, 1983.

[MKNF87]   Sumio Masuda, Toshinobu Kashiwabara, Kazuo Nakajima, and Toshio Fujisawa. On the NP-completeness of a computer network layout problem. In *Proceedings of the 1987 IEEE International Symp. on Circuits and Systems*, pages 292–295, 1987.

[MLW88]   Douglas Muder, Margaret Lefevre Weaver, and Douglas B. West. Pagenumber of Complete Bipartite Graphs. *J. of Graph Theory*, 12:469–489, 1988.

[MM92]   Aleksander Malnič and Bojan Mohar. Generating locally-cyclic triangulations of surfaces. *J. of Combinatorial Theory, Series B*, 56:147–164, 1992.

[MM96]   Kurt Mehlhorn and Petra Mutzel. On the Embedding Phase of the Hopcroft and Tarjan Planarity Testing Algorithm. *Algorithmica*, 16:233–242, 1996.

[MMN93]   Kurt Mehlhorn, Petra Mutzel, and Stefan Näher. An Implementation of the Hopcroft and Tarjan Planarity Test and Embedding Algorithm. Technical Report MPI-I-93-151, Max-Planck-Institut für Informatik, Im Stadtwald, 66123 Saarbrücken, Germany, 1993. Available electronically at http://www.mpi-sb.mpg.de/~mutzel/mpireports/MPI-I-93-151.ps.gz.

[MNKF90]   Sumio Masuda, Kazuo Nakajima, Toshinobu Kashiwabara, and Toshio Fujisawa. Crossing Minimization in Linear Embeddings of Graphs. *IEEE Transactions on Computers*, 39(1):124–127, 1990.

[Moh96]   Bojan Mohar. Embedding Graphs in an Arbitrary Surface in Linear Time. In *Proceedings of the 28th Annual ACM Symposium on Theory of Computing, STOC'96*, 1996.

[MOS98]   Petra Mutzel, Thomas Odenthal, and Mark Scharbrodt. The Thickness of Graphs: A Survey. *Graphs and Combinatorics*, 14:59–73, 1998.

[MP95]   Raanon Manor and Michal Penn. A Fast Algorithm for Integral Two-Flow in $K_{3,3}$-free Graphs. Technical report, Technion – Israel Institute of Technology, September 1995.

[MR92]   Dan McQuillan and R. Bruce Richter. On the crossing numbers of certain generalized Petersen graphs. *Discrete Mathematics*, 104:311–320, 1992.

[MR94]   Dan McQuillan and R. Bruce Richter. On 3-Regular Graphs Having Crossing Number at least 2. *J. of Graph Theory*, 18:831–839, 1994.

A. Liebers, *Planarizing Graphs*, JGAA, 5(1) 1–74 (2001)   67

[Mut94]   Petra Mutzel. *The Maximum Planar Subgraph Problem.* PhD thesis, Universität zu Köln, 1994. The Introduction and the Table of Contents are available electronically at http://www.mpi-sb.mpg.de/˜mutzel/koelnreports/.

[NC88]   Takao Nishizeki and Norishige Chiba. *Planar Graphs: Theory and Algorithms*, volume 32 of *Annals of Discrete Mathematics*. North–Holland, 1988.

[Nis90]   Takao Nishizeki. Planar graph problems. In Gottfried Tinhofer, Ernst W. Mayr, Hartmut Noltemeier, and Maciej M. Sysło, editors, *Computational Graph Theory*, volume 7 of *Computing Supplementum*, pages 53–68. Springer-Verlag Wien New York, 1990.

[NP94]   Zeev Nutov and Michal Penn. Minimum Feedback Arc Set and Maximum Integral Dicycle Packing Algorithms in $K_{3,3}$-free Digraphs. Technical report, Technion – Israel Institute of Technology, June 1994.

[NW79]   Crispin St.J.A. Nash-Williams. Acyclic detachments of graphs. In Robin J. Wilson, editor, *Proceedings of the Conference in Combinatorics and Graph Theory, The Open University, England, May 12, 1978*, Research Notes in Mathematics, pages 87–97. Pitman Publishing Limited, 1979.

[NW85a]   Crispin St.J.A. Nash-Williams. Connected detachments of graphs and generalized Euler trails. *J. London Math. Soc. (2)*, 31:17–29, 1985.

[NW85b]   Crispin St.J.A. Nash-Williams. Detachments of graphs and generalised Euler trails. In Ian Anderson, editor, *Surveys in Combinatorics 1985, Pap. 10th British Combinatorial Conference, Glasgow, Scotland, 1985*, volume 103 of *London Mathematical Society Lecture Note Series*, pages 137–151. Cambridge Univ. Press, Cambridge-New York, 1985.

[NW87]   Crispin St.J.A. Nash-Williams. Amalgamations of almost regular edge-colourings of simple graphs. *J. of Combinatorial Theory, Series B*, 43:322–342, 1987.

[Obr93]   Bojana Obrenić. Embedding de Bruijn and shuffle-exchange graphs in five pages. *SIAM J. Discrete Math.*, 6:642–654, 1993.

[Ode94]   Thomas Odenthal. Die Schichtzahl von Graphen. Master's thesis, Institut für Informatik, Universität zu Köln, 1994.

[OS94]   Thomas Odenthal and Mark Scharbrodt. Maximal Planarization as a Tool for Approximating Thickness and Crossing Number. In *Student Proceedings 8th Conference of the European Consortium for Mathematics in Industry, ECMI'94, Kaiserslautern, Germany, 1994*, pages 137–151, 1994.

[OT81]   Takao Ozawa and H. Takahashi. A graph-planarization algorithm and its application to random graphs. In N. Saito and Takao Nishizeki, editors, *Proceedings of the 17th Symposium, Research Institute of Electrical Communication, Tohoku University, Sendai, Japan, October 24–25, 1980*, pages 95–107. Springer-Verlag, Lecture Notes in Computer Science, vol. 108, 1981.

[Pap94]   Christos H. Papadimitriou. *Computational complexity.* Addison–Wesley Publishing Company, 1994.

[PCK89]   Y.H. Peng, C.C. Chen, and K.M. Koh. On the Factor-Thickness of Regular Graphs. *Graphs and Combinatorics*, 5:173–188, 1989.

[PPV86]   Giustina Pica, Tomaž Pisanski, and Aldo G.S. Ventre. Cartesian products of graphs and their crossing numbers. In *Proceedings Int. Conf. Finite Geom. Comb. Struct., Bari, Italy, 1984, Combinatorics '84*, pages 339–346. North-Holland, Annals of Discrete Mathematics, vol. 30, 1986.

[PSS94]   János Pach, Farhad Shahrokhi, and Mario Szegedy. Applications of the Crossing Number. In *Proceedings of the 10th ACM Symposium on Computational Geometry, SCG'94, Stony Brook, New York, USA, June 6–8, 1994*, pages 198–202. The Association for Computing Machinery, New York, 1994.

[PSS96]   János Pach, Farhad Shahrokhi, and Mario Szegedy. Applications of the Crossing Number. *Algorithmica*, 16:111–117, 1996.

[PT97]   János Pach and Géza Tóth. Graphs drawn with few crossings per edge. *Combinatorica*, 17:427–439, 1997.

[PT98]   János Pach and Géza Tóth. Which crossing number is it, anyway? In *Proceedings of the 39th Annual Symposium on Foundations of Computer Science, FOCS'98*, pages 617–626, 1998.

[RB78]   Richard D. Ringeisen and Lowell W. Beineke. The crossing number of $C_3 \times C_n$. *J. of Combinatorial Theory, Series B*, 24:134–136, 1978.

[Ric88]   R. Bruce Richter. Cubic Graphs with Crossing Number Two. *J. of Graph Theory*, 12:363–374, 1988.

[Rin65]   Gerhard Ringel. Die toroidale Dicke des vollständigen Graphen. *Math. Z.*, 87:19–26, 1965.

[RL92]   Subramanian Ramanathan and Errol L. Lloyd. Scheduling algorithms for multihop radio networks. *ACM SIGCOMM Computer Communication Review*, 22:211–222, 1992.

[RL93]   Subramanian Ramanathan and Errol L. Lloyd. Scheduling algorithms for multihop radio networks. *IEEE/ACM Transactions on Networking*, 1:166–177, 1993.

[RLWW97] Heike Ripphausen-Lipa, Dorothea Wagner, and Karsten Weihe. The Vertex-Disjoint Menger Problem in Planar Graphs. *SIAM J. Comput.*, 26:331–349, 1997.

[RR97]   Mauricio G.C. Resende and Celso C. Ribeiro. A GRASP for Graph Planarization. *Networks*, 29:173–189, 1997.

[RR98]   Mauricio G.C. Resende and Celso C. Ribeiro. Graph planarization, 1998. Available electronically at http://www.research.att.com/~mgcr/doc/splanar.ps.Z.

[RŠ96]   R. Bruce Richter and Jozef Širáň. The Crossing Number of $K_{3,n}$ in a Surface. *J. of Graph Theory*, 21:51–54, 1996.

[RSST96] Neil Robertson, Daniel P. Sanders, Paul Seymour, and Robin Thomas. A new proof of the four-colour theorem. *ERA Amer. Math. Soc.*, 2(1):17–25, 1996. Available electronically at http://www.ams.org/era/1996-02-01/.

[RT85]   J. Roskind and Robert E. Tarjan. A note on finding minimum–cost edge–disjoint spanning trees. *Mathematics of Operations Research*, 10:701–708, 1985.

A. Liebers, *Planarizing Graphs*, JGAA, 5(1) 1–74 (2001)   69

[RT93]   R. Bruce Richter and Carsten Thomassen. Minimal graphs with crossing number at least k. *J. of Combinatorial Theory, Series B*, 58:217–224, 1993.

[RT95]   R. Bruce Richter and Carsten Thomassen. Intersections of curve systems and the crossing number of $C_5 \times C_5$. *Discrete Comput. Geom.*, 13:149–159, 1995.

[RT97]   R. Bruce Richter and Carsten Thomassen. Relations Between Crossing Numbers of Complete and Complete Bipartite Graphs. *The American Mathematical Monthly*, 104:131–137, 1997.

[RZ95]   Jürgen Richter-Gebert and Günter M. Ziegler. Realization spaces of 4-polytopes are universal. *Bulletin (New Series) of the American Mathematical Society*, 32:403–412, 1995. Research report.

[SC89]   Ranjan Kumar Sen and Pranay Chaudhuri. A parallel approximate algorithm for minimum edge deletion bipartite subgraph problem. *J. of Combinatorics, Information & System Sciences*, 14:111–123, 1989.

[Sch86]   Karl Schaffer. *The Splitting Number and Other Topological Parameters of Graphs*. PhD thesis, University of California at Santa Cruz, January 1986.

[Sch89]   Walter Schnyder. Planar Graphs and Poset Dimension. *Order*, 5:323–343, 1989.

[Sch90a]   Walter Schnyder. Embedding Planar Graphs on the Grid. In *Proceedings of the 1st Annual ACM-SIAM Symposium on Discrete Algorithms, SODA'90*, pages 138–148, 1990.

[Sch90b]   H. Schumacher. On 2-Embeddable Graphs. In *Topics in Combinatorics and Graph Theory, Essys in Honour of Gerhard Ringel*, pages 651–661. Physica Verlag, Heidelberg, 1990.

[Sch91]   Haijo Schipper. Generating triangulations of 2-manifolds. In *Computational geometry – methods, algorithms and applications, Proceedings International Workshop, CG'91*, pages 237–248. Springer-Verlag, Lecture Notes in Computer Science, vol. 553, 1991.

[Sch95]   Mark Scharbrodt. Die Kreuzungszahl von Graphen. Master's thesis, Institut für Informatik, Universität zu Köln, 1995.

[Sch97]   Alexander Schrijver. Minor-monotone Graph Invariants. In Rosemary A. Bailey, editor, *Surveys in Combinatorics 1997, Papers presented by the invited lectures at the 16th British Combinatorial Conference, 1997*, volume 241 of *London Mathematical Society Lecture Note Series*. Cambridge Univ. Press, Cambridge-New York, 1997.

[Sel88]   S.M. Selim. Split vertices in vertex colouring and their application in developing a solution to the faculty timetable problem. *The Computer Journal*, 31(1):76–82, 1988.

[Sen90]   Ranjan Kumar Sen. An approximate algorithm for minimum edge deletion bipartite subgraph problem. In *Proc. 20th Southeastern Conference on Combinatorics, Graph Theory, and Computing, Boca Raton, Florida, USA, 1989*, volume 74 of *Congressus Numerantium*, pages 38–54, 1990.

[SGB95]   R. Swaminathan, D. Giriraj, and D.K. Bhatia. The pagenumber of the class of bandwidth-k graphs is $k - 1$. *Information Processing Letters*, 55:71–74, 1995.

[SH99]   Wei-Kuan Shih and Wen-Lian Hsu. A new planarity test. *Theoretical Computer Science*, 223:179–191, 1999.

[Šir83]   Jozef Širáň. Crossing-critical edges and Kuratowski subgraphs of a graph. *J. of Combinatorial Theory, Series B*, 35:83–92, 1983.

[Šir84]   Jozef Širáň. Infinite families of crossing-critical graphs with a given crossing number. *Discrete Mathematics*, 48:129–132, 1984.

[SL89]   Majid Sarrafzadeh and D.T. Lee. A new approach to topological via minimization. *IEEE Transactions on Computer-Aided Design of Integrated Circuits and Systems*, 8:890–900, 1989.

[Sla74]   Peter J. Slater. A Classification of 4-Connected Graphs. *J. of Combinatorial Theory, Series B*, 17:281–298, 1974.

[SR34]   Ernst Steinitz and Hans Rademacher. *Vorlesungen über die Theorie der Polyeder unter Einschluß der Elemente der Topologie*, volume 41 of *Die Grundlehren der Mathematischen Wissenschaften*. Springer, 1934. Aus dem Nachlaß von Ernst Steinitz herausgegeben und ergänzt von Hans Rademacher.

[SS92]   Farhad Shahrokhi and László A. Székely. Effective lower bounds for crossing number, bisection width and balanced vertex separators in terms of symmetry. In *Proceedings of the 2nd Conference on Integer Programming and Combinatorial Optimization, IPCO'92, Carnegie-Mellon University, USA, May 25–27, 1992*, pages 102–113, 1992.

[SS94]   Farhad Shahrokhi and László A. Székely. On Canonical Concurrent Flows, Crossing Number and Graph Expansion. *Combinatorics, Probability and Computing*, 3:523–543, 1994.

[SS96]   Farhad Shahrokhi and Weiping Shi. Efficient Deterministic Algorithms for Embedding Graphs on Books. In Jin-Yi Cai and Chak Kuen Wong, editors, *Proceedings 2nd Annual International Conference on Computing and Combinatorics, COCOON'96*, pages 162–168. Springer-Verlag, Lecture Notes in Computer Science, vol. 1090, 1996.

[SSSV94]   Farhad Shahrokhi, Ondrej Sýkora, László A. Székely, and Imrich Vrt'o. Improved Bounds for the Crossing Numbers on Surfaces of Genus g. In Jan v. Leeuwen, editor, *Proceedings 19th International Workshop on Graph-Theoretic Concepts in Computer Science, WG'93*, pages 388–395. Springer-Verlag, Lecture Notes in Computer Science, vol. 790, 1994.

[SSSV95]   Farhad Shahrokhi, Ondrej Sýkora, László A. Székely, and Imrich Vrt'o. Book Embeddings and Crossing Numbers. In Ernst W. Mayr, G. Schmidt, and Gottfried Tinhofer, editors, *Proceedings 20th International Workshop on Graph-Theoretic Concepts in Computer Science, WG'94*, pages 256–268. Springer-Verlag, Berlin, 1995.

[SSSV96a]   Farhad Shahrokhi, László A. Székely, Ondrej Sýkora, and Imrich Vrt'o. The Book Crossing Number of a Graph. *J. of Graph Theory*, 21:413–424, 1996.

A. Liebers, *Planarizing Graphs*, JGAA, 5(1) 1–74 (2001)   71

[SSSV96b] Farhad Shahrokhi, László A. Székely, Ondrej Sýkora, and Imrich Vrt'o. Drawings of Graphs on Surfaces with Few Crossings. *Algorithmica*, 16:118–131, 1996.

[SSSV97] Farhad Shahrokhi, Ondrej Sýkora, László A. Székely, and Imrich Vrt'o. On Bipartite Crossings, Largest Biplanar Subgraphs, and the Linear Arrangement Problem. In Frank Dehne, Andrew Rau-Chaplin, Jörg-Rüdiger Sack, and Roberto Tamassia, editors, *Proceedings 5th International Workshop on Algorithms and Data Structures, WADS'97, Halifax, Nova Scotia, Canada, August 6–8, 1997*, pages 55–68. Springer-Verlag, Lecture Notes in Computer Science, vol. 1272, 1997.

[SSSV98] Farhad Shahrokhi, Ondrej Sýkora, László A. Székely, and Imrich Vrt'o. Intersection of curves and crossing number of $C_m \times C_n$ on surfaces. *Discrete Comput. Geom.*, 19:237–247, 1998.

[SSSV00] Farhad Shahrokhi, Ondrej Sýkora, László A. Székely, and Imrich Vrt'o. A new lower bound for the bipartite crossing number with applications. *Theoretical Computer Science*, 245:281–294, 2000.

[SSV95] Farhad Shahrokhi, László A. Székely, and Imrich Vrt'o. Crossing Numbers of Graphs, Lower Bound Techniques and Algorithms: A Survey. In Roberto Tamassia and Ioannis G. Tollis, editors, *Graph Drawing. Proceedings of the DIMACS International Workshop, GD'94*, pages 131–142. Springer-Verlag, Lecture Notes in Computer Science, vol. 894, 1995.

[Ste51] S.K. Stein. Convex Maps. *Proceedings of the American Mathematical Society*, 2:464–466, 1951.

[SV92] Ondrej Sýkora and Imrich Vrt'o. On the crossing number of the hypercube and the cube connected cycles. In G. Schmidt and R. Berghammer, editors, *Proceedings 17th International Workshop on Graph-Theoretic Concepts in Computer Science, WG'91*, pages 214–218. Springer-Verlag, Lecture Notes in Computer Science, vol. 570, 1992.

[SV93a] Ondrej Sýkora and Imrich Vrt'o. Edge separators for graphs of bounded genus with applications. *Theoretical Computer Science*, 112:419–429, 1993.

[SV93b] Ondrej Sýkora and Imrich Vrt'o. On crossing numbers of hypercubes and cube connected cycles. *BIT*, 33:232–237, 1993.

[SV94] Ondrej Sýkora and Imrich Vrt'o. On VLSI layout of the star graph and related networks. *INTEGRATION The VLSI Journal*, 17:83–93, 1994.

[Tho89] Carsten Thomassen. The Graph Genus Problem Is NP-Complete. *J. Algorithms*, 10:568–576, 1989.

[TJS86] K. Thulasiraman, R. Jayakumar, and Madisetti N.S. Swamy. On Maximal Planarization of Nonplanar Graphs. *IEEE Transactions on Circuits and Systems*, 33:843–844, 1986.

[TL89] Yoshiyasu Takefuji and Kuo Chun Lee. A near-optimum parallel planarization algorithm. *Science*, 245:1221–1223, 1989.

[TLC91] Yoshiyasu Takefuji, Kuo Chun Lee, and Yong Beom Cho. Comments on "$O(n^2)$ Algorithms for Graph Planarization". *IEEE Transactions on Computer-Aided Design of Integrated Circuits and Systems*, 10:1582–1583, 1991.

A. Liebers, *Planarizing Graphs*, JGAA, 5(1) 1–74 (2001) 72

[Tru92]   Klaus Truemper. *Matroid decomposition*. Academic Press, 1992.

[TS92]   K. Thulasiraman and Madisetti N.S. Swamy. *Graphs: theory and algorithms*. Wiley, 1992.

[TT97]   Hisao Tamaki and Takeshi Tokuyama. A Characterization of Planar Graphs by Pseudo-Line Arrangements. In *Algorithms and Computation, 8th International Symposium, ISAAC'97*, pages 133–142. Springer-Verlag, Lecture Notes in Computer Science, vol. 1350, 1997.

[Tur77]   Paul Turán. A Note of Welcome. *J. of Graph Theory*, 1:7–9, 1977.

[Tut61]   William T. Tutte. A Theory of 3-Connected Graphs. *Indagationes Mathematicae: ex actis quibus titulus "Proceedings of the Section of Science" Koninklijke Nederlandse Akademie van Wetenschappen*, 23:441–455, 1961.

[Tut63a]   William T. Tutte. The non-biplanar character of the complete 9-graph. *Canad. Math. Bull.*, 6:319–330, 1963.

[Tut63b]   William T. Tutte. The thickness of a graph. *Indagationes Mathematicae: ex actis quibus titulus "Proceedings of the Section of Science" Koninklijke Nederlandse Akademie van Wetenschappen*, 25:567–577, 1963.

[Tut66]   William T. Tutte. *Connectivity in Graphs*, volume 15 of *Mathematical Expositions*. University of Toronto Press, 1966.

[Tut84]   William T. Tutte. *Graph theory*, volume 21 of *Encyclopedia of Mathematics and its Applications*. Addison–Wesley Publishing Company, 1984.

[Val97]   Pavel Valtr. Graph Drawing with no k Pairwise Crossing Edges. In Giuseppe Di Battista, editor, *Proceedings of the Symposium on Graph Drawing, GD'97*, pages 205–218. Springer-Verlag, Lecture Notes in Computer Science, vol. 1353, 1997.

[Val98]   Pavel Valtr. On Geometric Graphs with No k Pairwise Parallel Edges. *Discrete Comput. Geom.*, 19:461–469, 1998.

[Vas76]   J.M. Vasak. The thickness of the complete graph. *Notices of the American Mathematical Society*, 23:A–479, 1976. Abstract.

[Vrt]   Imrich Vrt'o. Crossing Numbers of Graphs: A Bibliography. Available electronically at ftp://ifi.savba.sk/pub/imrich/crobib.ps.gz.

[Wag36]   K. Wagner. Bemerkungen zum Vierfarbenproblem. *Jber. d. Dt. Math.-Verein.*, 46:26–32, 1936.

[Wag37a]   K. Wagner. Über eine Eigenschaft der ebenen Komplexe. *Math. Ann.*, 114:570–590, 1937.

[Wag37b]   K. Wagner. Über eine Erweiterung eines Satzes von Kuratowski. *Deutsche Mathematik*, 2:280–285, 1937.

[WAN83]   Toshimasa Watanabe, Tadashi Ae, and Akira Nakamura. On the NP-hardness of edge-deletion and -contraction problems. *Discrete Applied Math.*, 6:63–78, 1983.

[WB78]   Arthur T. White and Lowell W. Beineke. Topological Graph Theory. In Lowell W. Beineke and Robin J. Wilson, editors, *Selected Topics in Graph Theory*, pages 15–49. Academic Press, New York, 1978.

[Wei97]   Karsten Weihe. Edge–disjoint (s, t)-paths in undirected planar graphs in linear time. *J. Algorithms*, 23:121–138, 1997.

A. Liebers, *Planarizing Graphs*, JGAA, 5(1) 1–74 (2001) 73

[Wes83a] Walter Wessel. On some variations of the thickness of a graph connected with colouring. In Miroslav Fiedler, editor, *Graphs and Other Combinatorial Topics. Proceedings of the Third Czechoslovak Symposium on Graph Theory, Prague, August 24–27, 1982*, pages 344–348. Teubner, Texte zur Mathematik, vol. 59, 1983.

[Wes83b] Walter Wessel. Thickness-critical graphs — a generalization of Kuratowski's topic. In M. Borowiecki, J.W. Kennedy, and Maciej M. Sysło, editors, *Graph Theory, Proc. of a Conference held in Lagów, Poland, February 10–13, 1981*, pages 266–277. Springer-Verlag, Lecture Notes in Mathematics, vol. 1018, 1983.

[Wes84] Walter Wessel. Über die Abhängigkeit der Dicke eines Graphen von seinen Knotenpunktvalenzen. In *Geometrie und Kombinatorik, 2. Kolloq., Karl-Marx-Stadt, GDR, 1983, part 2*, pages 235–238, 1984.

[Wes86] Walter Wessel. The non-biplanar character of the graph K_9. In *Algebra und Graphentheorie, Beitr. Jahrestag. Algebra Grenzgeb., Siebenlehn, GDR, 1985*, pages 123–126, 1986.

[Wes89] Walter Wessel. Construction of Critical Graphs by Replacing Edges. In Lars Døvling Andersen and Gabriel A. Dirac, editors, *Graph theory in memory of G. A. Dirac*, pages 473–486. North-Holland, Annals of Discrete Mathematics, vol. 41, 1989.

[Wes96] Douglas B. West. *Introduction to Graph Theory*. Prentice Hall, 1996.

[Wes01] Douglas B. West. *Introduction to Graph Theory*. Prentice Hall, 2nd edition, 2001.

[Whi33] Hassler Whitney. Planar graphs. *Fundamenta Mathematicae*, 21:245–254, 1933.

[Whi84] Arthur T. White. *Graphs, Groups and Surfaces*, volume 8 of *North-Holland Mathematics Studies*. North-Holland, 1984.

[Wil80] S.G. Williamson. Embedding graphs in the plane — algorithmic aspects. In J. Srivastava, editor, *Combinatorial Mathematics, Optimal Designs and their Applications*, pages 349–384. North-Holland, Annals of Discrete Mathematics, vol. 6, 1980.

[Wil86] Robin J. Wilson. An Eulerian Trail through Königsberg. *J. of Graph Theory*, 10:265–275, 1986.

[Woo93] D.R. Woodall. Cyclic-Order Graphs and Zarankiewicz's Crossing-Number Conjecture. *J. of Graph Theory*, 17:657–671, 1993.

[Yan78] Mihalis Yannakakis. Node- and Edge-Deletion NP-complete problems. In *Proceedings of the 10th Annual ACM Symposium on Theory of Computing, STOC'78*, pages 253–264, 1978.

[Yan81] Mihalis Yannakakis. Edge-deletion problems. *SIAM J. Comput.*, 10:297–309, 1981.

[Yan86] Mihalis Yannakakis. Four pages are necessary and sufficient for planar graphs. In *Proceedings of the 18th Annual ACM Symposium on Theory of Computing, STOC'86*, pages 104–108, 1986.

[Yan89] Mihalis Yannakakis. Embedding Planar Graphs in Four Pages. *J. Computer and System Sciences*, 38:36–67, 1989.

A. Liebers, *Planarizing Graphs*, JGAA, 5(1) 1–74 (2001)   74

[Yap81]   Hian Poh Yap. A construction of chromatic index critical graphs. *J. of Graph Theory*, 5:159–163, 1981.

[Yap83]   Hian Poh Yap. A class of chromatic index critical graphs. *Bull. Malays. Math. Soc.,II. Ser.*, 6:45–46, 1983.

[YS99]   Atsuko Yamaguchi and Akihiro Sugimoto. An Approximation Algorithms for the Two-Layered Graph Drawing Problem. In Takao Asano, Hiroshi Imai, D.T. Lee, Shin-ichi Nakano, and Takeshi Tokuyama, editors, *Proceedings 5th Annual International Conference on Computing and Combinatorics, COCOON'99*, pages 81–91. Springer-Verlag, Lecture Notes in Computer Science, vol. 1627, 1999.

[Zar54]   K. Zarankiewicz. On a problem of P. Turán concerning graphs. *Fundamenta Mathematicae*, 41:137–145, 1954.

This document was processed using LaTeX2e, TeX Version 3.14159, and Graph-TeX Version 1.0β. One of several ways to obtain LaTeX2e and TeX is by anonymous ftp from `ftp.dante.de` . Graph-TeX is currently available electronically at `http://www.ima.umn.edu/~pliam/gtht/gtht.html` .

For finding monographs and conference proceedings, heavy use was made of the database of the Bibliotheksservice-Zentrum Baden-Württemberg (through its current WWW interface at `http://www.bsz-bw.de/wwwroot/e.opac.html`). For finding individual papers, the bibliographic databases MATH and COMPUSCIENCE of the Fachinformationszentrum Karlsruhe in Germany (FIZ Karlsruhe) were searched.

Volume 5:2 (2001)

Journal of Graph Algorithms and Applications

http://www.cs.brown.edu/publications/jgaa/

vol. 5, no. 2, pp. 1–34 (2000)

Fully Dynamic 3-Dimensional Orthogonal Graph Drawing

M. Closson   S. Gartshore   J. Johansen   S. K. Wismath

Department of Mathematics and Computer Science,
University of Lethbridge,
Lethbridge, Alberta, T1K-3M4, Canada.
http://www.cs.uleth.ca/
wismath@cs.uleth.ca

Abstract

In a 3-dimensional orthogonal drawing of a graph, vertices are mapped to grid points on a 3-dimensional rectangular integer lattice and edges are routed along integer grid lines. In this paper, we present a technique that produces a 3D orthogonal drawing of any graph with n vertices of degree 6 or less, using at most 6 bends per edge route and in a volume bounded by $O(n^2)$. The advantage of our strategy over previous drawing methods is that our method is fully dynamic, allowing both insertion and deletion of vertices and edges, while maintaining the volume and bend bounds. The drawing can be obtained in $O(n)$ time and insertions/deletions are performed in $O(1)$ time. Multiple edges and self loops are permitted. Three related constructions are also presented: a more elaborate construction that uses only 5 bends per edge, a simpler, more balanced drawing that requires at most 7 bends per edge, and a technique for displaying directed graphs.

Communicated by D. Wagner; submitted June 2000; revised November 2000 and January 2001.

Financial support for this research was provided by N.S.E.R.C. and the University of Lethbridge via the ULRF and both are gratefully acknowledged. A preliminary version of this research was presented at Graph Drawing 99, Prague.

M. Closson et al., *Fully Dynamic 3-D Graph Drawing*, JGAA, 5(2) 1–34 (2000)2

1 Introduction and Previous Work

In this paper we describe a 3-dimensional orthogonal drawing strategy for graphs of maximum degree at most 6 (*i.e.* graphs containing no vertex of degree 7 or higher). The **3-dimensional orthogonal grid** consists of grid points whose coordinates are defined as (x', y', z'), where $x', y', z' \in \mathbf{Z}$, together with the axis-parallel grid lines determined by these grid points. A **3-dimensional orthogonal drawing** of a graph G is an embedding of G into the 3-dimensional orthogonal grid with the vertices located at grid points and each edge (u, v) represented as a sequence of contiguous orthogonal segments of grid lines and which join the 2 points corresponding to the vertices u and v. No pair of edge routes are permitted to intersect (except at common vertex endpoints). Thus as a consequence, each grid point corresponding to a vertex has 6 **ports** to which edge routes may be attached, namely along the 3 grid lines incident on it. It is common practice to use the same terminology for both the graph and its drawing.

The **bounding box** of a 3-dimensional orthogonal drawing is the minimum axis-parallel box which contains all vertices and edges of the drawing. The dimensions of such a bounding box are measured in terms of the number of grid points rather than the length. The **volume** of the bounding box is the product of the 3 dimensions which is then exactly equal to the number of grid points on or in the box.

For background information in graph drawing, the reader is referred to the text by Di Battista, Eades, Tamassia and Tollis [3].

There have been several 3-dimensional drawing strategies proposed in the graph drawing literature for graphs of maximum degree at most 6 — see for example [1, 4, 6, 8, 9]. Note that 3-dimensional drawings of graphs of higher degree have also been investigated in for example, [2, 8, 10]; however in this paper we consider only 3-dimensional orthogonal drawings of graphs of maximum degree 6.

For the most part, previous results have focussed on the trade-offs among: overall volume of the resulting drawing, the maximum number of bends on an edge, the maximum length of an edge, *etc.* Eades, Stirk and Whitesides [5], building on earlier work by Kolmogorov and Barzdin [7], presented an algorithm that produces a 3-d orthogonal drawing in $O(n\sqrt{n})$ time with at most 16 bends per edge, edge lengths bounded by $O(\sqrt{n})$ and in an asymptotically optimum volume of $O(n^{1.5})$. They also showed the following three problems are NP-Complete. Given a graph, determine a 3-d orthogonal drawing that:

- minimizes the total edge length;

- minimizes the area/volume;

- contains 0 bends.

These results follow from the related 2-dimensional cases (see also [3]). Experimental comparisons of several 3-d orthogonal drawing algorithms can be found in the paper by Di Battista, Patrignani, and Vargiu [4].

M. Closson et al., *Fully Dynamic 3-D Graph Drawing*, JGAA, 5(2) 1–34 (2000)3

Although there are some classes of graphs of maximum degree 6 that can be drawn with few bends (*e.g.* trees can be drawn with 0 bends), the introduction of multiple edges or self loops into the graph immediately implies that some edges will have at least three bends in the drawing. For a self loop, the necessity of three bends is evident. A simple but tedious case study can be used to demonstrate that a pair of vertices with 6 edges between them, cannot be drawn with at most two bends per edge. Whether all *simple* graphs can be drawn with 2 or fewer bends is still an important open problem; in [11] Wood established lower bounds on the total number of bends in any drawing of certain classes of graphs.

Our result is motivated primarily by a dynamic version of the drawing problem in which edges and vertices of the graph can be inserted or deleted over time. Papakostas and Tollis [8] introduced a semi-dynamic solution in which only vertices are permitted to be inserted, and at all stages the resulting drawing must correspond to a connected graph. No other algorithms in this area have discussed non-trivial dynamic graph drawing cases. For example, one of the classic drawing algorithms in this area is due to Eades, Symvonis and Whitesides [6]. Their algorithm achieves $O(n^{1.5})$ volume, uses 7 bends per edge, requires a matching and coloring pre-processing step, and is only suitable in a static context. The same general technique has recently been modified by the authors to examine volume/bend tradeoffs and in particular they achieve a drawing with at most 6 bends per edge and $O(n^2)$ volume.

Our main drawing strategy is *fully dynamic*, allowing insertion and deletion of vertices and edges. The volume of the drawing is bounded by $O(n^2)$, each edge has at most 6 bends, and insertions and deletions are accomplished in $O(1)$ time. The dynamic nature of our drawing algorithm relies on the fact that at any time, a free port on any vertex may safely be connected to a free port of any other vertex without disturbing the drawing, and this connection requires at most 6 bends. In one variant of our strategy, we show that the same volume bound can be achieved with at most 5 bends per edge.

Multiple edges and self-loops are also accommodated in our drawings. Multiple edges may prove to be an important consideration in applications that require some degree of "fault-tolerance". Some previous 3-dimensional orthogonal constructions such as [8] have been restricted to simple graphs.

One feature of our drawing technique that may be useful in some applications such as 3-dimensional VLSI design, is that the width of the drawing is constant; the entire drawing lies on 7 planes between the planes $Y = -3$ and $Y = +3$.

Finally, note that all four of our drawing techniques are practical and easily implemented. We supply a software package (OrthoPak) for constructing 3-dimensional orthogonal drawings. Detailed descriptions of the drawing techniques, two brief animations and several VRML worlds are also available via the web page associated with this paper.

M. Closson et al., *Fully Dynamic 3-D Graph Drawing*, JGAA, 5(2) 1–34 (2000)4

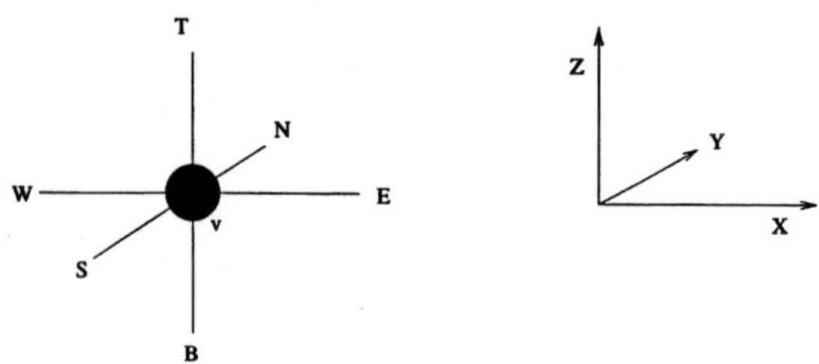

Figure 1: The 6 ports on a vertex

2 Drawing Graphs in $O(n^2)$ Volume with 6 Bends

2.1 Introduction and Definitions

Initially, we describe our drawing strategy in a static context, in which the entire graph (of maximum degree 6) is available. As will be seen, a dynamic version of the problem follows easily. We refer to this drawing strategy as the **staircase model**.

Let the vertices of the given graph be $v_1, v_2, v_3, ..., v_n$ (labeled in an arbitrary order). The vertices will be located at grid points in a staircase fashion, separated by an appropriate amount to permit the crossing-free embedding of the edges. More formally, vertex v_i is located at $(7i, 0, 5i)$. For convenience, we denote these coordinates as $(X(v_i), Y(v_i), Z(v_i))$. Each vertex may have at most 6 incident edges attached at **ports** which will be denoted as N, S, E, W, T, B — refer to Figure 1. When viewed from above (*i.e.* from $+\infty$ along the **Z** axis), the compass designations effectively describe the ports.

Containing each vertex v is a small box of dimensions $6 \times 7 \times 5$, called its **neighborhood** and denoted by $\mathbf{N(v)}$. The two diagonal corners of N(v) have coordinates: $(X(v) - 3, Y(v) + 3, Z(v) + 3)$ and $(X(v) + 2, Y(v) - 3, Z(v) - 1)$. Contained entirely inside N(v) are 6 **pedestals** — one for each port of v. Each pedestal consists of a line segment parallel to the Z axis. The **Bottom pedestal** is directly attached to v beneath it, and is of length 1. Similarly, the **Top pedestal** attaches to the top port of v and is of length 3. The **North pedestal** is of length 2 and extends from $(X(v), +1, Z(v))$ to $(X(v), +1, Z(v) + 2)$. The **South pedestal** is of length 1 and extends from $(X(v), -1, Z(v))$ to $(X(v), -1, Z(v) + 1)$. The **West** and **East pedestals** are degenerate points at $(X(v) - 1, 0, Z(v))$ and $(X(v) + 1, 0, Z(v))$ respectively; they are introduced for consistency in the description and will be modified in a subsequent section describing a related drawing strategy.

The Neighborhood N(v) consists of 5 layers in the **Z**-dimension, the:

- **Top layer** located on the plane $Z = Z(v) + 3$,

- **North layer** located on the plane $Z = Z(v) + 2$,

- **South layer** located on the plane $Z = Z(v) + 1$,

M. Closson et al., *Fully Dynamic 3-D Graph Drawing*, JGAA, 5(2) 1–34 (2000)5

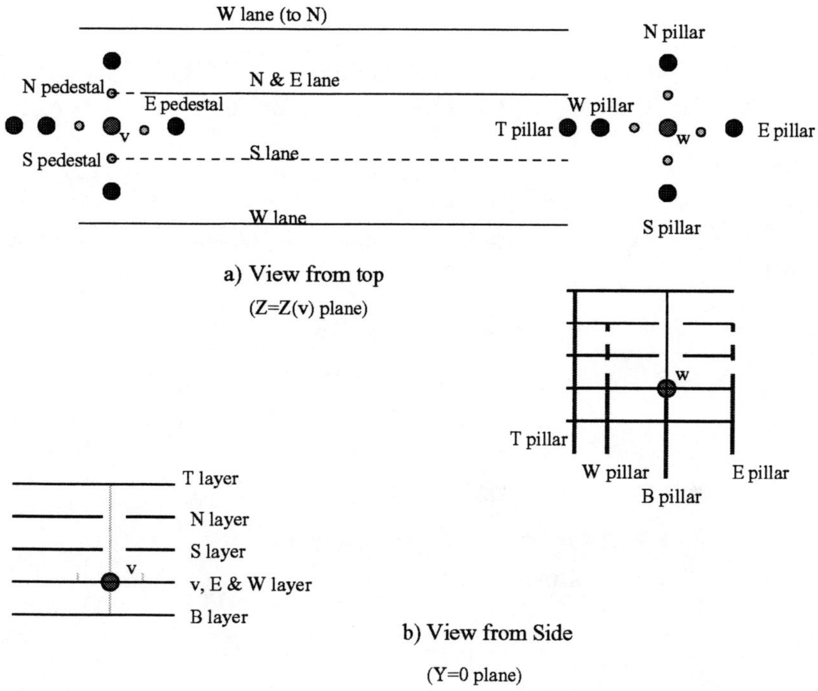

a) View from top

(Z=Z(v) plane)

b) View from Side

(Y=0 plane)

Figure 2: Two Neighborhoods

- **East, West, v layer** located on the plane $Z = Z(v)$,

- **Bottom layer** located on the plane $Z = Z(v) - 1$.

Edges are always considered as directed from the lower indexed vertex to the higher indexed vertex. Each pedestal is used only by *outgoing* edges from a vertex. *Incoming* edges are routed along **pillars**. The 6 pillars of a vertex v extend below the neighborhood of v to the plane $Z = -1$. More precisely, the

- **North pillar** extends from $(X(v), 2, -1)$ to $(X(v), 2, Z(v) + 2)$,

- **South pillar** extends from $(X(v), -2, -1)$ to $(X(v), -2, Z(v) + 2)$,

- **East pillar** extends from $(X(v) + 2, 0, -1)$ to $(X(v) + 2, 0, Z(v) + 2)$,

- **West pillar** extends from $(X(v) - 2, 0, -1)$ to $(X(v) - 2, 0, Z(v) + 2)$,

- **Bottom pillar** extends from $(X(v), 0, -1)$ to $(X(v), 0, Z(v))$,

- **Top pillar** extends from $(X(v) - 3, 0, -1)$ to $(X(v) - 3, 0, Z(v) + 3)$.

Note that the four side pillars (N, S, E, W) extend to $Z(v) + 2$ to accommodate self loops as will be described in a later section. Figure 2 displays two (perpendicular) views of the construction.

M. Closson et al., *Fully Dynamic 3-D Graph Drawing*, JGAA, 5(2) 1–34 (2000)6

We refer to the orthogonal box directly below a vertex v (and containing the pillars of v), as the **airspace of** v. An edge from v to w passes safely from the neighborhood of v to the airspace of w through intervening airspaces by traversing *lanes*, which are guaranteed to avoid pillars. In particular, note that any line segment lying in the planes $Y = 1, Y = 3, Y = -1$, or $Y = -3$ intersects no pillar of *any* vertex. Each layer contains one or two lanes on one of these four planes, extending from N(v), as far in the $+\mathbf{X}$ direction as necessary. Let M be $X(v_n) + 2$, where v_n is the last vertex in the staircase. The lanes are then defined as:

- **North lane** from $(X(v), +1, Z(v) + 2)$ to $(M, +1, Z(v) + 2)$,

- **South lane** from $(X(v), -1, Z(v) + 1)$ to $(M, -1, Z(v) + 1)$,

- **East lane** from $(X(v) + 1, +1, Z(v))$ to $(M, +1, Z(v))$,

- **West lanes**

    - to N: from $(X(v) - 1, 3, Z(v))$ to $(M, 3, Z(v))$,
    - to all other destinations: from $(X(v) - 1, -3, Z(v))$ to $(M, -3, Z(v))$,

- **Bottom lanes**

    - to N, E, T: from $(X(v), +1, Z(v) - 1)$ to $(M, +1, Z(v) - 1)$,
    - to S, W, B: from $(X(v), -1, Z(v) - 1)$ to $(M, -1, Z(v) - 1)$,

- **Top lanes**

    - to N, E, T: from $(X(v), +1, Z(v) + 3)$ to $(M, +1, Z(v) + 3)$,
    - to S, W, B: from $(X(v), -1, Z(v) + 3)$ to $(M, -1, Z(v) + 3)$.

Figure 3 displays the pedestals, pillars and layers for three vertices on the staircase and an edge routing from the north port of the first vertex to the east port of the third.

Description of edge-routes. Typically, an edge from a vertex v to a vertex w leaves v from an arbitrary free port, travels to the associated pedestal, climbs the pedestal to the appropriate layer, is routed *on that layer* from the pedestal to a lane, along the lane to the airspace under w, and (still on that layer) joins to the pillar associated with the destination port of w, climbs this pillar and finally enters w. The source and destination ports are chosen arbitrarily by the algorithm — any unused port can be specified. Although the routings differ slightly for each port to port pair, this general technique applies, the critical observation being that the routing from the neighborhood of v to the airspace of w is done at some *layer of v* (*i.e.* the lower vertex). By using one of the lanes, this section of the edge passes safely through the airspace of any intervening vertices without intersecting any of its pillars (which may have edges routed from below). Since the E and W ports of v share the same layer as v, special care is required to ensure routings from these ports do not intersect each other,

M. Closson et al., *Fully Dynamic 3-D Graph Drawing*, JGAA, 5(2) 1–34 (2000)7

Figure 3: Pedestals, pillars, and layers

$(X(v), 0, Z(v))$	- vertex v
$(X(v), +1, Z(v))$	
	N pedestal of v
$(X(v), +1, Z(v) + 2)$	
	N lane
$(X(w) + 2, +1, Z(v) + 2)$	
	connect to E pillar of w
$(X(w) + 2, 0, Z(v) + 2)$	
	climb E pillar
$(X(w) + 2, 0, Z(w))$	
$(X(w), 0, Z(w))$	- vertex w

Table 1: Route from North to East

M. Closson et al., *Fully Dynamic 3-D Graph Drawing*, JGAA, 5(2) 1–34 (2000)8

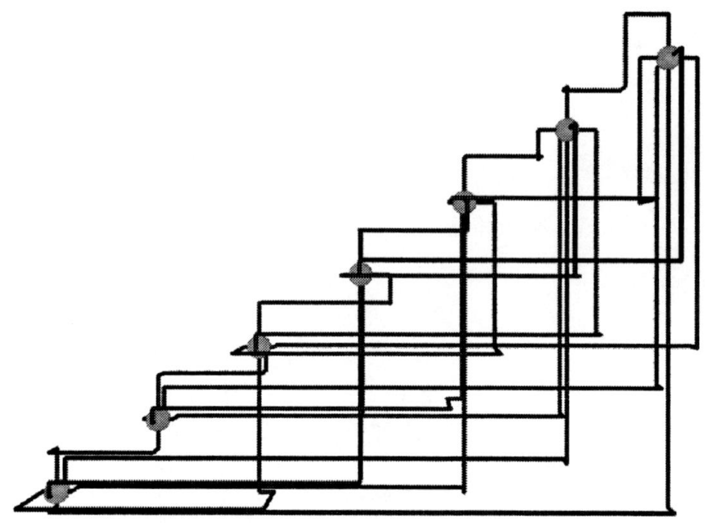

Figure 4: Drawing of K_7 using the Staircase method

or possible incoming edges to the N or S ports. A connection to the Top port of vertex w comes up the Top pillar, along w's Top layer, and then directly down to w, thus safely passing above w's West pillar.

For example, consider an edge routing from the N port of v to the E port of w. Table 1 shows the routing which consists of 6 segments and thus contains 5 bends.

Figure 4 displays a drawing of the graph K_7 using the staircase drawing method. For completeness, the 36 port-to-port routings are specified in Appendix A. Note that of these routings, 9 require six bends, 22 require five bends, and 5 require four bends.

Lemma 1 *No pair of edges in the construction intersect.*

Proof: Appendix A defines the paths for all port-to-port connections between a vertex v (at the lower level) to a vertex w ($\neq v$). To prove that edge (v, w) intersects no other edge in the drawing, note that the routing of each edge consists of three portions:

1. inside v's neighborhood

2. along a lane through intervening airspaces

3. in w's airspace.

Since the four lanes intersect no pillars, the second portion of edge (v, w) intersects no other edge.

Consider the portion of the edge (v, w) inside v's neighborhood. Each of the 6 possible routings out of v intersect no pillar of v and so cannot intersect any

M. Closson et al., *Fully Dynamic 3-D Graph Drawing*, JGAA, 5(2) 1–34 (2000)9

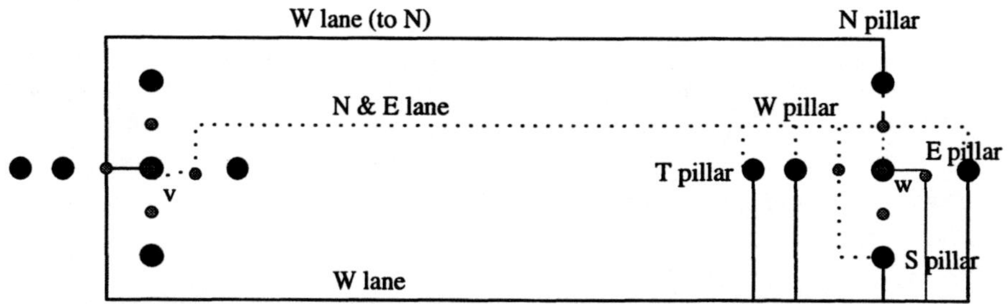

Figure 5: E and W routings do not intersect

incoming edge to v. And similarly, the portion in w's airspace can intersect no outgoing edge from w.

It remains to consider the intersection of (v, w) with some outgoing edge from v (inside v's neighborhood) or with some other incoming edge to w (inside w's airspace). Without loss of generality, we consider the case of (v, w) intersecting another edge between v and w but with different port assignments. Since the N,S,T, and B ports each have their own layers inside the neighborhood of v, it is simple to verify that no intersections are possible if either of the two edges involve these ports at v. However, the E and W ports share a plane with v itself, and it is therefore critical to check that none of the 30 possible pairs involving these two ports from v intersect. In Figure 5, the 6 routings from the West port are shown as solid lines and the 6 routings from the East port are displayed as dotted lines; note that only the routings to the pillars of w are shown. That no pair of solid and dotted paths intersect either in the neighborhood of v nor in the airspace of w is easily determined. □

2.2 Self-loops

The general drawing strategy for self loops is similar: from a port to a pedestal, up to the appropriate level, across to the specified pillar, down the pillar and into the vertex. A complete listing of the routes is presented in Appendix B, and on the web page.

2.3 Dynamic Case

The drawing strategy described in the previous section easily adapts to use in a dynamic context, in which both edges and vertices may be deleted or inserted. Insertion of a new vertex v_t is performed by placing it at $(7t, 0, 5t)$. It is clear that a new edge (u, v) can be added to the drawing in $O(1)$ time — any free port on u can be connected to a free port on v, and furthermore, the overall volume of the drawing remains the same. Deletion of a vertex is accomplished by replacing it by the top-most vertex and re-routing the (at most 6) incident edges. Let $v_1, v_2, ... v_t$ be the vertices in the drawing at time t. To delete vertex v_i, the

M. Closson et al., *Fully Dynamic 3-D Graph Drawing*, JGAA, 5(2) 1–34 (2000)10

edges on v_i are first deleted. Denote by $w_1, w_2, ...w_6$ the vertices adjacent to vertex v_t. These edges are all deleted in the drawing and vertex v_i is connected to $w_1, w_2, ...w_6$, using arbitrary ports on vertex v_i. Vertex v_t is then removed from the drawing. The resulting drawing has a bounding box of dimensions $6(t-1) \times 7 \times 5(t-1)$. Note that we measure the number of grid points rather than the length in each dimension.

Theorem 1 *For every graph G of maximum degree at most 6 the staircase algorithm produces a 3D orthogonal drawing contained in a bounding box with volume $210t^2$, where t is the number of vertices in the drawing. Insertion or deletion of vertices or edges can be accomplished in $O(1)$ update time.*

3   Variants

In this section, we consider three related constructions - a *spiral* drawing technique that produces a more balanced drawing, a drawing strategy for *directed* graphs, and a more elaborate version of the *staircase* model that requires at most 5 bends per edge.

3.1   Ortho-Spiral Layout

One possible criticism of the drawings produced by the staircase model is that the dimensions of the resulting drawing are *unbalanced* - the construction is very narrow, with a Y dimension consisting of exactly 7 planes. Although there may be practical applications of this feature of our strategy (for example in future VLSI design), from an aesthetic standpoint a more balanced drawing may be preferable. In this section, we outline a strategy that produces a drawing of dimensions $O(\sqrt{n}) \times O(\sqrt{n}) \times O(n)$, but with a slight penalty on the number of bends: 7 bends are required for some port-to-port connections. The dynamic nature of the previous strategy is preserved; an implementation and a short video describing the routing is available on the web page associated with the paper.

Instead of placing the vertices in a linear staircase fashion, the vertices are embedded in an orthogonal spiral manner. When viewed from above, the vertices appear in a $O(\sqrt{n}) \times O(\sqrt{n})$ grid as displayed in Figure 6.

As in the previous method, each vertex v, is assigned to a unique Z-plane, thus forming a spiral of linear height. Associated with each vertex are 7 planes (one for each port, and one for the vertex itself). The degenerate E and W pedestals described in Section 2.1 are extended from the v layer to the appropriate plane.

Figure 7 shows a cross section (from the Top) of the modified neighborhood through a vertex. The four side pedestals (for N, S, E, W) have their bases on this plane and extend to their respective layers. The pedestals for the Bottom and Top layers originate at the vertex itself.

M. Closson et al., *Fully Dynamic 3-D Graph Drawing*, JGAA, 5(2) 1–34 (2000)11

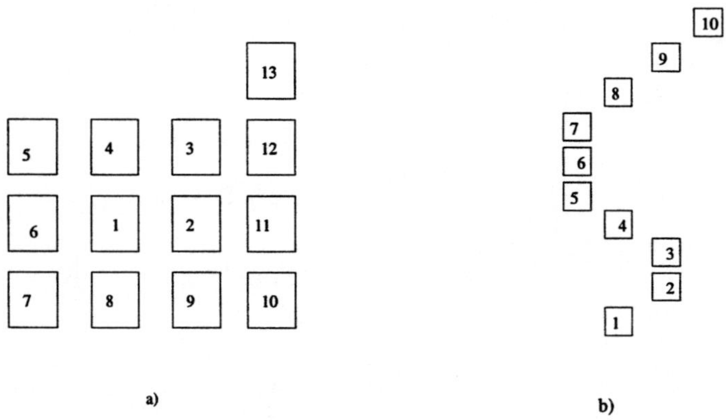

Figure 6: Spiral drawing: a) from above; b) from the side

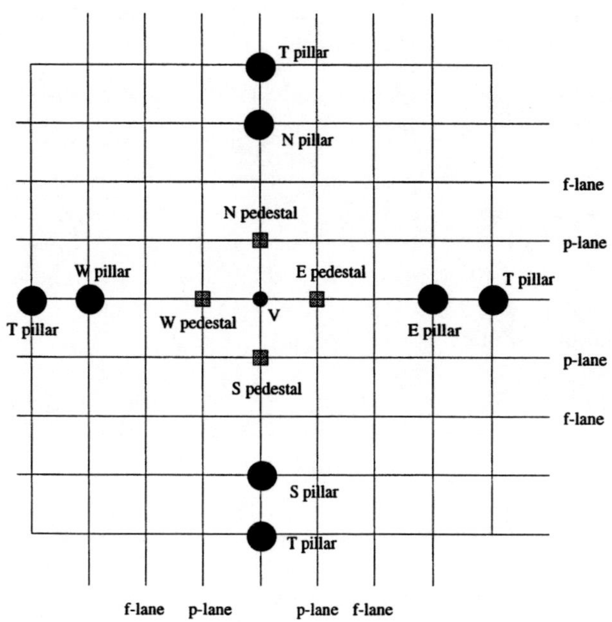

Figure 7: Spiral drawing: cross-section

M. Closson et al., *Fully Dynamic 3-D Graph Drawing*, JGAA, 5(2) 1–34 (2000)12

$(X(v), Y(v), Z(v))$	- vertex v
$(X(v), Y(v) + 1, Z(v))$	
	N pedestal of v
$(X(v), Y(v) + 1, Z(v) + 4)$	
	p-lane (correct in X)
$(X(w) + 2, Y(v) + 1, Z(v) + 4)$	
	f-lane (correct in Y)
$(X(w) + 2, Y(w), Z(v) + 4)$	
	connect to E pillar of w
$(X(w) + 3, Y(w), Z(v) + 4)$	
	climb E pillar
$(X(w) + 3, Y(w), Z(w))$	
$(X(w), Y(w), Z(w))$	- vertex w

Table 2: Route from North to East — Ortho Spiral

Side pillars and the Bottom pillar (for incoming routings) are spaced 1 unit further from v, but otherwise remain the same. A total of four pillars for connection to the Top port are provided.

On each particular layer, there are 4 pairs of lanes used for routing:

- **f-lanes** - two pairs (horizontal and vertical) that intersect no pillar or pedestal on any level (*i.e.* free of any intersections), and

- **p-lanes** - two pairs (horizontal and vertical) that intersect only a single pedestal on (or passing through) this level.

More precisely, each lane is defined as the intersection of 2 planes: one of the layer planes and one of the 8 lane planes. The 6 layers lie on the planes: $Z = Z(v) - 1$ (Bottom), $Z = Z(v) + 1$ (West), $Z = Z(v) + 2$ (East), $Z = Z(v) + 3$ (South), $Z = Z(v) + 4$ (North) and $Z = Z(v) + 5$ (Top). The 4 p-lanes lie in the planes $Y = Y(v) + 1$, $Y = Y(v) - 1$, $X = X(v) + 1$ and $X = X(v) - 1$, and the 4 f-lanes lie in the planes $Y = Y(v) + 2$, $Y = Y(v) - 2$, $X = X(v) + 2$ and $X = X(v) - 2$.

The general edge routing technique remains the same as in the staircase model, however note that unlike the staircase model, neighborhoods may differ in both their X and Y coordinates. An edge from a vertex v at a lower level, to a vertex w, exits a specific port on v, joins its pedestal, climbs the pedestal to the appropriate layer, traverses the p-lane to correct in the X-coordinate (N, S layers) or the Y-coordinate (E, W layers), then traverses a f-lane to correct in the Y-coordinate or X-coordinate (respectively) in order to enter w's airspace. Connection to the specified pillar of w then allows the edge to climb the appropriate pillar, and finally to enter w at the specified port. Edges from the Top or Bottom port climb/descend their pedestal and then safely join an f-lane, a second (perpendicular) f-lane and then connect to the specified pillar. As in the staircase model, connection into the Bottom port of w is directly from below

M. Closson et al., *Fully Dynamic 3-D Graph Drawing*, JGAA, 5(2) 1–34 (2000)13

		destination		
		side	Bottom	Top
	side	6	5	7
source	Bottom	6	5	7
	Top	6	5	7

Table 3: Bends per edge

and connection to the Top port of w requires two bends and is performed via w's Top layer. In each case, the connection from the pillar of w to a particular port of w itself intersects only the pedestal associated with that port which of course cannot be in use.

In Table 2 for example, the routing from the North port on vertex v to the East port of vertex w is described. Appendix C lists all 36 possible routings between an arbitrary pair of vertices. Table 3 summarizes the number of bends per edge for each routing in this spiral model.

The coordinate specification for the elements of this drawing is more elaborate than in the staircase model. For notational convenience assume the vertices are $v_0, v_1, ..., v_{n-1}$ and that vertex v_i is to be located at (x_i, y_i, z_i). Then $z_i := 7i$ since 7 layers are required for each vertex. Let $c = \lceil \sqrt{i} \rceil$ and $f = \lfloor \sqrt{i} \rfloor$. Then the X and Y coordinates for vertex v_i can be determined depending on whether f is even or odd:

- f is even:
  If $i \leq f^2 + f$ then $x_i := -\frac{9}{2}f$ and $y_i := 9(f^2 - i + \frac{f}{2})$.
  If $i \geq f^2 + f$ then $x_i := 9(i - c^2 + \frac{c+1}{2})$ and $y_i := -\frac{9}{2}f$.

- f is odd:
  If $i \leq f^2 + f$ then $x_i := 9\frac{f+1}{2}$ and $y_i := 9(i - f^2 - \frac{f-1}{2})$.
  If $i \geq f^2 + f$ then $x_i := 9(c^2 - i - \frac{c}{2})$ and $y_i := 9\frac{f+1}{2}$.

The proof that no pair of edges intersect in the drawing is simpler than that of the staircase model since associated with each port is a distinct layer. There are no intersections possible within v's neighborhood, within w's airspace, and p-lanes and f-lanes pass safely through any intervening pillars.

Note that for each vertex, 7 layers are required and that the X and Y extents of each neighborhood are 9×9 (see Figure 7).

Theorem 2 *Any graph of maximum degree 6 can be drawn orthogonally in 3-dimensions with at most 7 bends per edge in a bounding box with dimensions* $9\lceil \sqrt{n} \rceil \times 9\lceil \sqrt{n} \rceil \times 7n$.

Again insertions and deletions of vertices and edges are accomplished in $O(1)$ time, and self-loops and multiple edges are easily accommodated. A brief animation and the implementation details are provided on the web page.

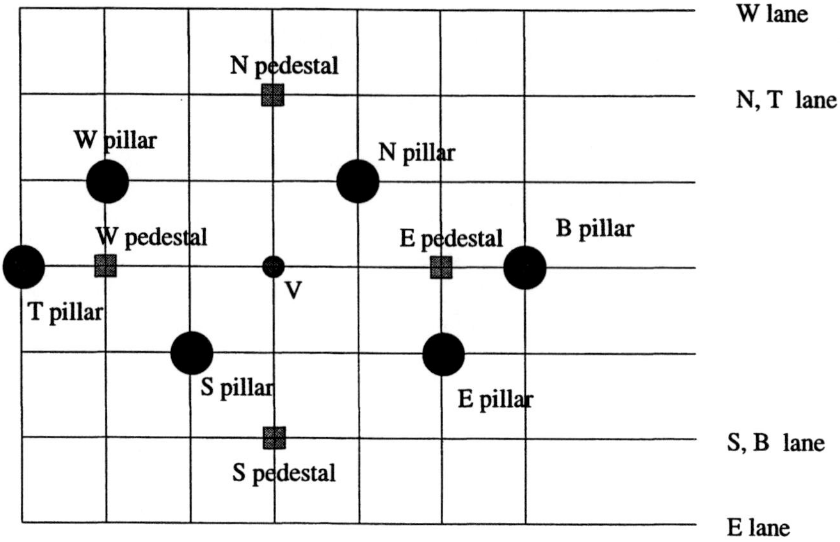

M. Closson et al., *Fully Dynamic 3-D Graph Drawing*, JGAA, 5(2) 1–34 (2000)14

Figure 8: Directed Case

3.2 Drawings of Directed Graphs

Typically, 3-dimensional orthogonal drawings of graphs are visually difficult for humans to work with, even with the addition of colored edges, and both the staircase and orthogonal spiral techniques suffer from this drawback. In this subsection, we outline a modified staircase drawing for *directed* graphs which allows immediate recognition of the direction of each edge.

We refer to the orthogonal box directly below the neighborhood of vertex v, as the **lower-airspace** of v. The orthogonal box directly above v's neighborhood is referred to as its **upper-airspace**.

In this variant, by repositioning the pillars and pedestals slightly (as illustrated in Figure 8), the number of layers required can be reduced to three and the overall volume is slightly smaller (70% of a staircase drawing), however *all* port to port routings require exactly six bends per edge. If v is lower than w on the staircase, then an edge directed from vertex v to vertex w is routed below the staircase, whereas an edge from vertex w to vertex v is routed above the staircase. Such a drawing allows for quick visual identification of the direction of an edge.

The staircase is formed with vertex v_i positioned at $(7i, 0, 3i)$. The neighborhood $N(v)$ has a size of $7 \times 7 \times 3$ and consists of three layers in the **Z**-dimension. The:

- Top, South layer located on the plane $Z = Z(v) + 1$,

- West, East layer located on the plane $Z = Z(v)$,

- North, Bottom layer located on the plane $Z = Z(v) - 1$,

M. Closson et al., *Fully Dynamic 3-D Graph Drawing*, JGAA, 5(2) 1–34 (2000)15

Outgoing edges are routed to the pedestal associated with the outgoing port, up or down the pedestal to the appropriate layer and then out to the lane. Let (v, w) be a directed edge to be drawn. If the vertex v is below the vertex w (on the staircase), the edge travels in the positive $\mathbf{X}$ direction along the lane. If the vertex v is above the vertex w, the edge is routed in the negative $\mathbf{X}$ direction. The North, South, Bottom and Top pedestals are each of length one; their extents are respectively: from $(X(v), +2, Z(v))$ to $(X(v), +2, Z(v) - 1)$, from $(X(v), -2, Z(v))$ to $(X(v), -2, Z(v)+1)$, from $(X(v), 0, Z(v))$ to $(X(v), 0, Z(v)-1)$, from $(X(v), 0, Z(v))$ to $(X(v), 0, Z(v)+1)$. The East and West pedestals are degenerate points at $(X(v) + 2, 0, Z(v))$ and $(X(v) - 2, 0, Z(v))$

An incoming edge ascends or descends a pillar. The six pillars of vertex v extend from the plane of $Z = -1$ through the neighborhood of v to the top of the drawing at plane $Z = Z(v_n) + 1$, where n is the index of the highest vertex in the drawing. The pillar extents are as follows:

- North: from $(X(v) + 1, 1, -1)$ to $(X(v) + 1, 1, Z(v_n) + 1)$
- South from $(X(v) - 1, -1, -1)$ to $(X(v) - 1, -1, Z(v_n) + 1)$
- East from $(X(v) + 2, -1, -1)$ to $(X(v) + 2, -1, Z(v_n) + 1)$
- West from $(X(v) - 2, 1, -1)$ to $(X(v) - 2, 1, Z(v_n) + 1)$
- Bottom from $(X(v) + 3, 0, -1)$ to $(X(v) + 3, 0, Z(v_n) + 1)$
- Top from $(X(v) - 3, 0, -1)$ to $(X(v) - 3, 0, Z(v_n) + 1)$

The routing of an edge from vertex v to w follows the standard pattern: from the specified port of v to the associated pedestal, up or down the pedestal to the associated layer, out to the assigned lane, along the lane to w's airspace, into the specified pillar, up (if v is before w) or down (if v is after w) that pillar a correction to align with the port, and finally into the specified port of w. A detailed listing of all port-to-port connections is contained in Appendix D.

Theorem 3 *The algorithm produces a 3D orthogonal drawing of any directed graph with n vertices of maximum degree at most 6, in a bounding box of $7n \times 7 \times 3n$ and with 6 bends per edge.*

Proof: Refer to vertex v in Figure 8 and imagine an arbitrary *incoming* edge to v. The edge runs along one of the four lanes (note that the North and Top lanes and the South and Bottom lanes are actually on different layers). The edge turns and joins the appropriate pillar, climbs the pillar and enters v. Note that a route using the Top pillar safely passes over the West pedestal and a route using the Bottom pillar safely passes under the East pedestal. No pair of incoming edges can intersect and no incoming edge intersects an outgoing edge.

Outgoing edges from v intersect no pillars (except West and East which safely intersect their own pillars) on their way to their assigned lanes. Note that three pairs share levels but each pair has separate lanes, and thus do not intersect. Since no lane intersects a pillar, there can be no intersections caused by intervening vertices.

$\square$

M. Closson et al., *Fully Dynamic 3-D Graph Drawing*, JGAA, 5(2) 1–34 (2000)16

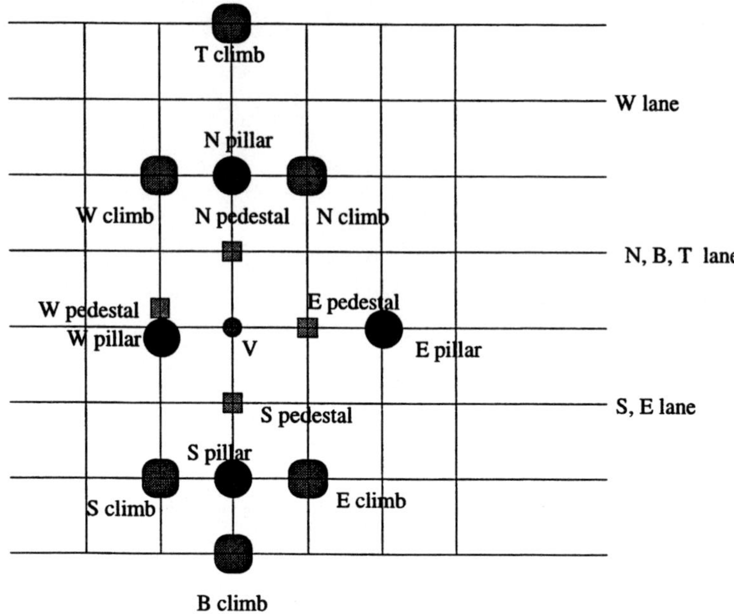

Figure 9: The 5 bend case

3.3 Achieving 5 Bends per Edge

The same basic format of the staircase models can be modified to produce a drawing with at most 5 bends per edge. In addition to using the region above the staircase, a new set of $2n$ planes called *under-planes* are introduced under the staircase. As a consequence of the under-plane construction, some of the dynamic nature of the drawing strategy is sacrificed. Although insertion and deletion of edges still takes $O(1)$ time, vertex insertions and deletions require $O(n)$ time. Note that in a *relative-change scenario* [8], such changes can be measured as $O(1)$; in this paper we assume the stronger *no-change* scenario.

In addition to the new construct of underplanes, we introduce **climbs** which extend only in the upper-airspace of v. Refer to Figure 9. These climbs are similar to the upper-airspace pillars of the digraph drawing; they are used in some routings instead of pedestals to climb to the level of w for routing at that level .

All routings are from a vertex v lower on the staircase to a vertex w that is higher. For the 36 possible cases there are three general types of routes in this model:

- using a pillar down to v's underplane, across to w's airspace and up its pillar (the 4 cases of: B to S, B to T, E to N, S to N),

- using a climb to a layer of w, and then across to w (the 10 cases of routes into T or S of w),

M. Closson et al., *Fully Dynamic 3-D Graph Drawing*, JGAA, 5(2) 1–34 (2000)17

- using a pedestal to a layer of v, across to w's airspace, up a pillar and into w (the remaining 22 cases).

The staircase is formed with vertex v_i placed at $(4i, 0, 4i)$. The size of each neighborhood $N(v)$ is $4 \times 8 \times 4$. As in the previous constructions, each vertex has a set of layers on which some local routing takes place. The four layers associated with a vertex v are: the **Top layer** located on the plane $Z = Z(v) + 2$, the **North, South layer** located on the plane $Z = Z(v) + 1$, the **East, West layer** located on the plane $Z = Z(v)$, and the **Bottom layer** located on the plane $Z = Z(v) - 1$.

The pedestals of vertex v are used by all routings into the W, E, B, and N ports of w except the 2 special cases of E to N and S to N. The **North pedestal** is of length 1 and extends from $(X(v), 1, Z(v))$ to $(X(v), 1, Z(v) + 1)$. The **South pedestal** is of length 1 and extends from $(X(v), -1, Z(v))$ to $(X(v), -1, Z(v) + 1)$. The **East and West pedestals** are degenerate points at $(X(v) + 1, 0, Z(v))$ and $(X(v) - 1, 0, Z(v))$ respectively. The **Bottom pedestal** is of length 1 and extends from $(X(v), 0, Z(v))$ to $(X(v), 0, Z(v) - 1)$. The **Top pedestal** is of length 2 and extends from $(X(v), 0, Z(v))$ to $(X(v), 0, Z(v) + 2)$.

The underplanes of the graph start at plane $Z = -2$ and extend downwards. There are two planes per vertex in the graph and are consecutive. The underplanes are arranged so that vertex v_n's (highest on the staircase) under-planes are at $Z = -2$ and $Z = -3$. The underplanes for vertex v_i lie at $Z = -2(n-i)-2$ and $Z = -2(n-i)-3$. Only a portion of each underplane is actually used and is termed the **under-layer**; it extends from $X(v_n)$ to $X(v)$. The under-layers also form a staircase shape with the longest under-layers being the bottom planes of the graph. The underplane (and thus under-layer) positions are dependent on the number of vertices in the graph and thus must change when a vertex is inserted or deleted. Since there are $2n$ underplanes, this vertical shifting of the planes and the resultant lengthening or shortening of the routings that use these underplanes results in the $O(n)$ time penalty for vertex insertion or deletion.

The **North climb** extends from $(X(v)+1, 2, Z(v))$ to $(X(v)+1, 2, Z(v_n)+2)$. The **South climb** extends from $(X(v)-1, -2, Z(v))$ to $(X(v)-1, -2, Z(v_n)+2)$. The **East climb** extends from $(X(v)+1, -2, Z(v))$ to $(X(v)+1, -2, Z(v_n)+2)$. The **West climb** extends from $(X(v)-1, 2, Z(v))$ to $(X(v)-1, 2, Z(v_n)+2)$. The **Bottom climb** extends from $(X(v), -3, -2n)$ to $(X(v), -3, Z(v_n) + 2)$; this is a special case that runs from the lowest underplane of vertex v to the top of the drawing. The **Top climb** extends from $(X(v), 3, Z(v) + 2)$ to $(X(v), 3, Z(v_n) + 2)$.

The climbs of vertex v are used for all routes into the T and S ports of w except the 2 special cases of B to T and B to S. For a route to the South port of w, the edge leaves v and with one bend joins to the climb associated with v's port, ascends the climb to the level of w, joins the S lane, traverses this lane to the base of w's pedestal and finally enters w at the South port. The routings from v to the Top port of w are similar except the lane used is on the plane $Y = 0$, which can be safely used since at w's level this lane intersects no intervening pillars.

M. Closson et al., *Fully Dynamic 3-D Graph Drawing*, JGAA, 5(2) 1–34 (2000)18

The pillars of v have two purposes, namely for some routes into v and also for the 4 special case routes that use the underplanes. This causes no conflict since a port can only be used in one direction at a time (either to, or from another port). Some pillars need to extend to the bottommost underplane (located at $Z = -2n - 1$), and others extend only to the lowest plane of the staircase $Z = -1$. The **North pillar** extends from $(X(v), 2, -1)$ to $(X(v), 2, Z(v))$ The **East pillar** extends from $(X(v) + 2, 0, -2n - 1)$ to $(X(v) + 2, 0, Z(v))$ and is used to reach the bottom underplane of v. The **West pillar** extends from $(X(v) - 1, 0, -1)$ to $(X(v) - 1, 0, Z(v))$. The **Bottom pillar** extends from $(X(v), 0, -2n-1)$ to $(X(v), 0, Z(v))$ and is used to reach the bottom underplane of v. The **South pillar** extends from $(X(v), -2, -2n)$ to $(X(v), -2, Z(v))$, and is only used to reach the topmost underplane associated with v. Note that the West pillar and West pedestal safely overlap. However, it is not possible to align the East pillar and pedestal in a similar fashion.

The edge routings are relatively elaborate and are listed in Appendix E. We also refer the interested reader to a series of web pages associated with this paper that describe the routings in more detail. The proof that no pair of edges intersects involves a lengthy case study since there are 3 general types of routes. Here we simply note that each edge has exactly 5 bends, but that self loops are not possible. A bounding box of dimensions $4n \times 8 \times 8n$ contains the drawing.

4 Implementation and Experimental Results

One important feature of the drawing strategy described in this paper is its simplicity. An implementation of the drawing is available in a package of 3-dimensional drawing tools, called OrthoPak, from:
`www.cs.uleth.ca/~wismath/packages`. This package was written in C++ and uses the LEDA library extensively. It runs under Solaris 2.6 or Linux, and produces either a VRML world that can be examined with an appropriate browser, or a PovRay file that can subsequently produce a static ray traced image of the drawing. OrthoPak is free for research and teaching purposes and is part of a larger suite of research tools developed at the University of Lethbridge.

A test suite of graphs for 3-D orthogonal drawing was described in [4]. Our time trials indicate that our staircase drawing strategy is very efficient: 7.013 seconds for the entire test suite of 1820 graphs containing from 5 to 95 vertices, running on a SPARC 5. On average, our staircase drawing had between 4.71 and 4.83 bends per edge over the entire distribution of graphs. Since our algorithm does not pre-preprocess the graph, most of the time is spent on simple I/O operations.

5 Conclusion and Open Problems

A 3-dimensional orthogonal drawing of graphs with n vertices of degree at most 6 was presented in which each edge is routed with at most 6 bends in a volume

M. Closson et al., *Fully Dynamic 3-D Graph Drawing*, JGAA, 5(2) 1–34 (2000)19

bounded by $O(n^2)$. The technique is fully dynamic, allowing insertion and deletion of edges and vertices in $O(1)$ time. Multiple edges and self-loops are permitted. Variants of the drawing strategy were also presented which may be more suitable in other contexts; for example the number of bends per edge can be reduced to 5 with the same volume bounds. It is an open problem to establish whether a volume of $O(n^2)$ is achievable with at most 4 bends per edge.

An implementation of the drawing strategy is available from the authors' web site. Demonstration VRML worlds, 2 short videos and several supporting web pages describing the techniques are supplied on the web page associated with this paper and maintained by the journal.

If the drawing strategy is not used in a dynamic setting then some pre-processing of the graph is possible and the number of bends can be reduced in the average case. Alternately, special properties of the graph may also be exploitable to reduce the number of bends. For example, trees of maximum degree 6 can be embedded with our staircase construction with no more than 5 bends. An interesting open problem is to determine if there are other classes of graphs that can be drawn using this strategy with a minimal amount of pre-processing.

Acknowledgment Detailed comments and suggestions by the anonymous referees substantially improved the presentation of the paper and are gratefully acknowledged.

References

[1] T. Biedl, Heuristics for 3D-Orthogonal Graph Drawings, *Proceedings of the 4th Twente Workshop on Graphs and Combinatorial Optimization*, Enshende, June 1995, pp. 41-44.

[2] T. Biedl, T. Shermer, S. Whitesides, S. Wismath, Bounds for Orthogonal 3-D Graph Drawing, *J. Graph Algorithms and Applications*, Vol. 3, No. 4, pp. 63-79, 1999.

[3] G. Di Battista, P. Eades, T. Tamassia, I. Tollis, Graph Drawing: Algorithms for the Visualization of Graphs, Prentice-Hall Inc., 1999.

[4] G. Di Battista, M. Patrignani, F. Vargiu, A Split & Push Approach to 3D Orthogonal Drawing, *Symposium on Graph Drawing 98*, Lecture Notes in Computer Science 1547, Springer Verlag, 1998. pp.87-101.

[5] P. Eades, C. Stirk, S. Whitesides, The Techniques of Komolgorov and Bardzin for Three Dimensional Orthogonal Graph Drawing, *Information Processing Letters*, Vol. 60, No 2, pp. 9-103, 1996.

[6] P. Eades, A. Symvonis, S. Whitesides, Three-Dimensional Orthogonal Graph Drawing Algorithms, *Discrete Applied Math.*, Vol. 103, pp. 55-87, 2000.

M. Closson et al., *Fully Dynamic 3-D Graph Drawing*, JGAA, 5(2) 1–34 (2000)20

[7] A. Kolmogorov, Y. Barzdin, About Realisation of Sets in 3-Dimensional Space, *Problems in Cybernetics*, Vol 8, pp. 261-268, March 1967.

[8] A. Papakostas, I. Tollis, Incremental Orthogonal Graph Drawing in Three Dimensions, *J. of Graph Algorithms and Applications*, Vol. 3, No. 4, pp. 81-115, 1999.

[9] D. Wood, An Algorithm for Three-Dimensional Orthogonal Graph Drawing, *Symposium on Graph Drawing 98*, Lecture Notes in Computer Science 1547, Springer Verlag, 1998. pp. 332-346.

[10] D. Wood, Multi-Dimensional Orthogonal Graph Drawing with Small Boxes, *Symposium on Graph Drawing 99*, Lecture Notes in Computer Science 1731, Springer Verlag, 1999, pp. 311-322.

[11] D. Wood, Lower Bounds for the Number of Bends in Three-Dimensional Orthogonal Graph Drawings, *Symposium on Graph Drawing 2000*, Lecture Notes in Computer Science 1984, Springer Verlag, Virginia, Sept. 2000, pp. 259-271.

A   Appendix: Routings - Staircase Model

All 36 port-to-port routings from a vertex v to vertex w will now be enumerated. In each case, the initial and terminating points of the route are $(X(v), Y(v) = 0, Z(v))$ and $(X(w), Y(w) = 0, Z(w))$ respectively.

- $N \to S$ (6 bends):
  $(X(v), 1, Z(v)) \to (X(v), 1, Z(v) + 2) \to (X(w) - 1, 1, Z(v) + 2) \to$
  $(X(w) - 1, -2, Z(v) + 2) \to (X(w), -2, Z(v) + 2) \to (X(w), -2, Z(w))$

- $N \to N$ (5 bends):
  $(X(v), 1, Z(v)) \to (X(v), 1, Z(v) + 2) \to (X(w), 1, Z(v) + 2) \to$
  $(X(w), +2, Z(v) + 2) \to (X(w), +2, Z(w))$

- $N \to E$ (5 bends):
  $(X(v), 1, Z(v)) \to (X(v), 1, Z(v) + 2) \to (X(w) + 2, 1, Z(v) + 2) \to$
  $(X(w) + 2, 0, Z(v) + 2) \to (X(w) + 2, 0, Z(w))$

- $N \to W$ (5 bends):
  $(X(v), 1, Z(v)) \to (X(v), 1, Z(v) + 2) \to (X(w) - 2, 1, Z(v) + 2) \to$
  $(X(w) - 2, 0, Z(v) + 2) \to (X(w) - 2, 0, Z(w))$

- $N \to T$ (6 bends):
  $(X(v), 1, Z(v)) \to (X(v), 1, Z(v) + 2) \to (X(w) - 3, 1, Z(v) + 2) \to$
  $(X(w) - 3, 0, Z(v) + 2) \to (X(w) - 3, 0, Z(w) + 3) \to (X(w), 0, Z(w) + 3)$

- $N \to B$ (4 bends):
  $(X(v), 1, Z(v)) \to (X(v), 1, Z(v) + 2) \to (X(w), 1, Z(v) + 2) \to$
  $(X(w), 0, Z(v) + 2)$

M. Closson et al., *Fully Dynamic 3-D Graph Drawing*, JGAA, 5(2) 1–34 (2000)21

- $S \to N$ (6 bends):
  $(X(v), -1, Z(v)) \to (X(v), -1, Z(v) + 1) \to (X(w) - 1, -1, Z(v) + 1) \to$
  $(X(w) - 1, 2, Z(v) + 1) \to (X(w), 2, Z(v) + 1) \to (X(w), 2, Z(w))$

- $S \to S$ (5 bends):
  $(X(v), -1, Z(v)) \to (X(v), -1, Z(v) + 1) \to (X(w), -1, Z(v) + 1) \to$
  $(X(w), -2, Z(v) + 1) \to (X(w), -2, Z(w))$

- $S \to E$ (5 bends):
  $(X(v), -1, Z(v)) \to (X(v), -1, Z(v) + 1) \to (X(w) + 2, -1, Z(v) + 1) \to$
  $(X(w) + 2, 0, Z(v) + 1) \to (X(w) + 2, 0, Z(w))$

- $S \to W$ (5 bends):
  $(X(v), -1, Z(v)) \to (X(v), -1, Z(v) + 1) \to (X(w) - 2, -1, Z(v) + 1) \to$
  $(X(w) - 2, 0, Z(v) + 1) \to (X(w) - 2, 0, Z(w))$

- $S \to T$ (6 bends):
  $(X(v), -1, Z(v)) \to (X(v), -1, Z(v) + 1) \to (X(w) - 3, -1, Z(v) + 1) \to$
  $(X(w) - 3, 0, Z(v) + 1) \to (X(w) - 3, 0, Z(w) + 3) \to (X(w), 0, Z(w) + 3)$

- $S \to B$ (4 bends):
  $(X(v), -1, Z(v)) \to (X(v), -1, Z(v) + 1) \to (X(w), -1, Z(v) + 1) \to$
  $(X(w), 0, Z(v) + 1)$

- $E \to N$ (5 bends):
  $(X(v) + 1, 0, Z(v)) \to (X(v) + 1, 1, Z(v)) \to (X(w), 1, Z(v)) \to$
  $(X(w), 2, Z(v)) \to (X(w), 2, Z(w))$

- $E \to S$ (6 bends):
  $(X(v) + 1, 0, Z(v)) \to (X(v) + 1, 1, Z(v)) \to (X(w) - 1, 1, Z(v)) \to$
  $(X(w) - 1, -2, Z(v)) \to (X(w), -2, Z(v)) \to (X(w), -2, Z(w))$

- $E \to E$ (5 bends):
  $(X(v) + 1, 0, Z(v)) \to (X(v) + 1, 1, Z(v)) \to (X(w) + 2, 1, Z(v)) \to$
  $(X(w) + 2, 0, Z(v)) \to (X(w) + 2, 0, Z(w))$

- $E \to W$ (5 bends):
  $(X(v) + 1, 0, Z(v)) \to (X(v) + 1, 1, Z(v)) \to (X(w) - 2, 1, Z(v)) \to$
  $(X(w) - 2, 0, Z(v)) \to (X(w) - 2, 0, Z(w))$

- $E \to T$ (6 bends):
  $(X(v) + 1, 0, Z(v)) \to (X(v) + 1, 1, Z(v)) \to (X(w) - 3, 1, Z(v)) \to$
  $(X(w) - 3, 0, Z(v)) \to (X(w) - 3, 0, Z(w) + 3) \to (X(w), 0, Z(w) + 3)$

- $E \to B$ (4 bends):
  $(X(v) + 1, 0, Z(v)) \to (X(v) + 1, 1, Z(v)) \to (X(w), 1, Z(v)) \to$
  $(X(w), 0, Z(v))$

- $W \to N$ (5 bends):
  $(X(v) - 1, 0, Z(v)) \to (X(v) - 1, 3, Z(v)) \to (X(w), 3, Z(v)) \to$
  $(X(w), 2, Z(v)) \to (X(w), 2, Z(w))$

M. Closson et al., *Fully Dynamic 3-D Graph Drawing*, JGAA, 5(2) 1–34 (2000)22

- $W \to S$ (5 bends):
  $(X(v) - 1, 0, Z(v)) \to (X(v) - 1, -3, Z(v)) \to (X(w), -3, Z(v)) \to$
  $(X(w), -2, Z(v)) \to (X(w), -2, Z(w))$

- $W \to W$ (5 bends):
  $(X(v) - 1, 0, Z(v)) \to (X(v) - 1, -3, Z(v)) \to (X(w) - 2, -3, Z(v)) \to$
  $(X(w) - 2, 0, Z(v)) \to (X(w) - 2, 0, Z(w))$

- $W \to E$ (5 bends):
  $(X(v) - 1, 0, Z(v)) \to (X(v) - 1, -3, Z(v)) \to (X(w) + 2, -3, Z(v)) \to$
  $(X(w) + 2, 0, Z(v)) \to (X(w) + 2, 0, Z(w))$

- $W \to T$ (6 bends):
  $(X(v) - 1, 0, Z(v)) \to (X(v) - 1, -3, Z(v)) \to (X(w) - 3, -3, Z(v)) \to$
  $(X(w) - 3, 0, Z(v)) \to (X(w) - 3, 0, Z(w) + 3) \to (X(w), 0, Z(w) + 3)$

- $W \to B$ (5 bends):
  $(X(v) - 1, 0, Z(v)) \to (X(v) - 1, -3, Z(v)) \to (X(w) + 1, -3, Z(v)) \to$
  $(X(w) + 1, 0, Z(v)) \to (X(w), 0, Z(v))$

- $T \to N$ (5 bends):
  $(X(v), 0, Z(v) + 3) \to (X(v), +1, Z(v) + 3) \to (X(w), 1, Z(v) + 3) \to$
  $(X(w), 2, Z(v) + 3) \to (X(w), 2, Z(w))$

- $T \to S$ (5 bends):
  $(X(v), 0, Z(v) + 3) \to (X(v), -1, Z(v) + 3) \to (X(w), -1, Z(v) + 3) \to$
  $(X(w), -2, Z(v) + 3) \to (X(w), -2, Z(w))$

- $T \to E$ (5 bends):
  $(X(v), 0, Z(v) + 3) \to (X(v), +1, Z(v) + 3) \to (X(w) + 2, 1, Z(v) + 3) \to$
  $(X(w) + 2, 0, Z(v) + 3) \to (X(w) + 2, 0, Z(w))$

- $T \to W$ (5 bends):
  $(X(v), 0, Z(v) + 3) \to (X(v), -1, Z(v) + 3) \to (X(w) - 2, -1, Z(v) + 3) \to$
  $(X(w) - 2, 0, Z(v) + 3) \to (X(w) - 2, 0, Z(w))$

- $T \to B$ (4 bends):
  $(X(v), 0, Z(v) + 3) \to (X(v), -1, Z(v) + 3) \to (X(w), -1, Z(v) + 3) \to$
  $(X(w), 0, Z(v) + 3)$

- $T \to T$ (6 bends):
  $(X(v), 0, Z(v) + 3) \to (X(v), +1, Z(v) + 3) \to (X(w) - 3, 1, Z(v) + 3) \to$
  $(X(w) - 3, 0, Z(v) + 3) \to (X(w) - 3, 0, Z(w) + 3) \to (X(w), 0, Z(w) + 3)$

- $B \to N$ (5 bends):
  $(X(v), 0, Z(v) - 1) \to (X(v), +1, Z(v) - 1) \to (X(w), 1, Z(v) - 1) \to$
  $(X(w), 2, Z(v) - 1) \to (X(w), 2, Z(w))$

- $B \to S$ (5 bends):
  $(X(v), 0, Z(v) - 1) \to (X(v), -1, Z(v) - 1) \to (X(w), -1, Z(v) - 1) \to$
  $(X(w), -2, Z(v) - 1) \to (X(w), -2, Z(w))$

M. Closson et al., *Fully Dynamic 3-D Graph Drawing*, JGAA, 5(2) 1–34 (2000)23

- $B \to E$ (5 bends):
  $(X(v), 0, Z(v) - 1) \to (X(v), +1, Z(v) - 1) \to (X(w) + 2, 1, Z(v) - 1) \to$
  $(X(w) + 2, 0, Z(v) - 1) \to (X(w) + 2, 0, Z(w))$

- $B \to W$ (5 bends):
  $(X(v), 0, Z(v) - 1) \to (X(v), -1, Z(v) - 1) \to (X(w) - 2, -1, Z(v) - 1) \to$
  $(X(w) - 2, 0, Z(v) - 1) \to (X(w) - 2, 0, Z(w))$

- $B \to T$ (6 bends):
  $(X(v), 0, Z(v) - 1) \to (X(v), +1, Z(v) - 1) \to (X(w) - 3, 1, Z(v) - 1) \to$
  $(X(w) - 3, 0, Z(v) - 1) \to (X(w) - 3, 0, Z(w) + 3) \to (X(w), 0, Z(w) + 3)$

- $B \to B$ (4 bends):
  $(X(v), 0, Z(v) - 1) \to (X(v), -1, Z(v) - 1) \to (X(w), -1, Z(v) - 1) \to$
  $(X(w), 0, Z(v) - 1)$

B Appendix: Self-Loop Routings - Staircase Model

- $T \to W$ (3 bends):
  $(X(v), 0, Z(v) + 3) \to (X(v) - 2, 0, Z(v) + 3) \to (X(v) - 2, 0, Z(v))$

- Similarly, $T \to N, S, E$.

- $W \to E$ (4 bends):
  $(X(v) - 1, 0, Z(v)) \to (X(v) - 1, -3, Z(v)) \to (X(v) + 2, -3, Z(v)) \to$
  $(X(v) + 2, 0, Z(v))$

- $W \to S$ (3 bends):
  $(X(v) - 1, 0, Z(v)) \to (X(v) - 1, -3, Z(v)) \to (X(v), -3, Z(v))$

- $B \to W$ (3 bends):
  $(X(v), 0, Z(v) - 1) \to (X(v) - 2, 0, Z(v) - 1) \to (X(v) - 2, 0, Z(v))$

- Similarly, $B \to N, S, E$.

- $B \to T$ (6 bends):
  $(X(v), 0, Z(v) - 1) \to (X(v), -1, Z(v) - 1) \to (X(v) - 3, -1, Z(v) - 1) \to$
  $(X(v) - 3, 0, Z(v) - 1) \to (X(v) - 3, 0, Z(v) + 3) \to (X(v), 0, Z(v) + 3)$

- $N \to E$ (5 bends):
  $(X(v), +1, Z(v)) \to (X(v), +1, Z(v) + 2) \to (X(v) + 2, 1, Z(v) + 2) \to$
  $(X(v) + 2, 0, Z(v) + 2) \to (X(v) + 2, 0, Z(v))$

- Similarly, $N \to W$.

- $N \to S$ (6 bends):
  $(X(v), +1, Z(v)) \to (X(v), +1, Z(v) + 2) \to (X(v) + 1, 1, Z(v) + 2) \to$
  $(X(v) + 1, -2, Z(v) + 2) \to (X(v), -2, Z(v) + 2) \to (X(v), -2, Z(v))$

M. Closson et al., *Fully Dynamic 3-D Graph Drawing*, JGAA, 5(2) 1–34 (2000)24

- $S \to E$ (5 bends):
  $(X(v), -1, Z(v)) \to (X(v), -1, Z(v) + 1) \to (X(v) + 2, -1, Z(v) + 1) \to$
  $(X(v) + 2, 0, Z(v) + 1) \to (X(v) + 2, 0, Z(v))$

C Appendix: Routings - Orthogonal Spiral Model

The 36 port-to-port routings from a vertex v to vertex w will now be enumerated. The offsets for each of the 6 layers are denoted as: T', N', S', E', W', B' and have the values $5, 4, 3, 2, 1, -1$ respectively. In each case, the initial and terminating points of the route are $(X(v), Y(v), Z(v))$ and $(X(w), Y(w), Z(w))$; the computation of these coordinates is presented in Section 3.1.

- $N \to S$ (6 bends):
  $(X(v), Y(v) + 1, Z(v)) \to (X(v), Y(v) + 1, Z(v) + N') \to$
  $(X(w) + 2, Y(v) + 1, Z(v) + N') \to (X(w) + 2, Y(w) - 3, Z(v) + N') \to$
  $(X(w), Y(w) - 3, Z(v) + N') \to (X(w), Y(w) - 3, Z(w))$

- $N \to N$ (6 bends):
  $(X(v), Y(v) + 1, Z(v)) \to (X(v), Y(v) + 1, Z(v) + N') \to$
  $(X(w) + 2, Y(v) + 1, Z(v) + N') \to (X(w) + 2, Y(w) + 3, Z(v) + N') \to$
  $(X(w), Y(w) + 3, Z(v) + N') \to (X(w), Y(w) + 3, Z(w))$

- $N \to E$ (6 bends):
  $(X(v), Y(v) + 1, Z(v)) \to (X(v), Y(v) + 1, Z(v) + N') \to$
  $(X(w) + 2, Y(v) + 1, Z(v) + N') \to (X(w) + 2, Y(w), Z(v) + N') \to$
  $(X(w) + 3, Y(w), Z(v) + N') \to (X(w) + 3, Y(w), Z(w))$

- $N \to W$ (6 bends):
  $(X(v), Y(v) + 1, Z(v)) \to (X(v), Y(v) + 1, Z(v) + N') \to$
  $(X(w) - 2, Y(v) + 1, Z(v) + N') \to (X(w) - 2, Y(w), Z(v) + N') \to$
  $(X(w) - 3, Y(w), Z(v) + N') \to (X(w) - 3, Y(w), Z(w))$

- $N \to T$ (7 bends):
  $(X(v), Y(v) + 1, Z(v)) \to (X(v), Y(v) + 1, Z(v) + N') \to$
  $(X(w) + 2, Y(v) + 1, Z(v) + N') \to (X(w) + 2, Y(w) + 4, Z(v) + N') \to$
  $(X(w), Y(w) + 4, Z(v) + N') \to (X(w), Y(w) + 4, Z(w) + T') \to$
  $(X(w), Y(w), Z(w) + T')$

- $N \to B$ (5 bends):
  $(X(v), Y(v) + 1, Z(v)) \to (X(v), Y(v) + 1, Z(v) + N') \to$
  $(X(w) + 2, Y(v) + 1, Z(v) + N') \to (X(w) + 2, Y(w), Z(v) + N') \to$
  $(X(w), Y(w), Z(v) + N')$

- $S \to N$ (6 bends):
  $(X(v), Y(v) - 1, Z(v)) \to (X(v), Y(v) - 1, Z(v) + S') \to$
  $(X(w) - 2, Y(v) - 1, Z(v) + S') \to (X(w) - 2, Y(w) + 3, Z(v) + S') \to$
  $(X(w), Y(w) + 3, Z(v) + S') \to (X(w), Y(w) + 3, Z(w))$

M. Closson et al., *Fully Dynamic 3-D Graph Drawing*, JGAA, 5(2) 1–34 (2000)25

- $S \to S$ (6 bends):
  $(X(v), Y(v) - 1, Z(v)) \to (X(v), Y(v) - 1, Z(v) + S') \to$
  $(X(w) - 2, Y(v) - 1, Z(v) + S') \to (X(w) - 2, Y(w) - 3, Z(v) + S') \to$
  $(X(w), Y(w) - 3, Z(v) + S') \to (X(w), Y(w) - 3, Z(w))$

- $S \to E$ (6 bends):
  $(X(v), Y(v) - 1, Z(v)) \to (X(v), Y(v) - 1, Z(v) + S') \to$
  $(X(w) + 2, Y(v) - 1, Z(v) + S') \to (X(w) + 2, Y(w), Z(v) + S') \to$
  $(X(w) + 3, Y(w), Z(v) + S') \to (X(w) + 3, Y(w), Z(w))$

- $S \to W$ (6 bends):
  $(X(v), Y(v) - 1, Z(v)) \to (X(v), Y(v) - 1, Z(v) + S') \to$
  $(X(w) - 2, Y(v) - 1, Z(v) + S') \to (X(w) - 2, Y(w), Z(v) + S') \to$
  $(X(w) - 3, Y(w), Z(v) + S') \to (X(w) - 3, Y(w), Z(w))$

- $S \to T$ (7 bends):
  $(X(v), Y(v) - 1, Z(v)) \to (X(v), Y(v) - 1, Z(v) + S') \to$
  $(X(w) - 2, Y(v) - 1, Z(v) + S') \to (X(w) - 2, Y(w) + 4, Z(v) + S') \to$
  $(X(w), Y(w) + 4, Z(v) + S') \to (X(w), Y(w) + 4, Z(w) + T') \to$
  $(X(w), Y(w), Z(w) + T')$

- $S \to B$ (5 bends):
  $(X(v), Y(v) - 1, Z(v)) \to (X(v), Y(v) - 1, Z(v) + S') \to$
  $(X(w) - 2, Y(v) - 1, Z(v) + S') \to (X(w) - 2, Y(w), Z(v) + S') \to$
  $(X(w), Y(w), Z(v) + S')$

- $E \to N$ (6 bends):
  $(X(v) + 1, Y(v), Z(v)) \to (X(v) + 1, Y(v), Z(v) + E') \to$
  $(X(v) + 1, Y(w) + 2, Z(v) + E') \to (X(w), Y(w) + 2, Z(v) + E') \to$
  $(X(w), Y(w) + 3, Z(v) + E') \to (X(w), Y(w) + 3, Z(w))$

- $E \to S$ (6 bends):
  $(X(v) + 1, Y(v), Z(v)) \to (X(v) + 1, Y(v), Z(v) + E') \to$
  $(X(v) + 1, Y(w) - 2, Z(v) + E') \to (X(w), Y(w) - 2, Z(v) + E') \to$
  $(X(w), Y(w) - 3, Z(v) + E') \to (X(w), Y(w) - 3, Z(w))$

- $E \to E$ (6 bends):
  $(X(v) + 1, Y(v), Z(v)) \to (X(v) + 1, Y(v), Z(v) + E') \to$
  $(X(v) + 1, Y(w) + 2, Z(v) + E') \to (X(w) + 3, Y(w) + 2, Z(v) + E') \to$
  $(X(w) + 3, Y(w), Z(v) + E') \to (X(w) + 3, Y(w), Z(w))$

- $E \to W$ (6 bends):
  $(X(v) + 1, Y(v), Z(v)) \to (X(v) + 1, Y(v), Z(v) + E') \to$
  $(X(v) + 1, Y(w) + 2, Z(v) + E') \to (X(w) - 3, Y(w) + 2, Z(v) + E') \to$
  $(X(w) - 3, Y(w), Z(v) + E') \to (X(w) - 3, Y(w), Z(w))$

- $E \to T$ (7 bends):
  $(X(v) + 1, Y(v), Z(v)) \to (X(v) + 1, Y(v), Z(v) + E') \to$
  $(X(v) + 1, Y(w) + 2, Z(v) + E') \to (X(w) - 4, Y(w) + 2, Z(v) + E') \to$

M. Closson et al., *Fully Dynamic 3-D Graph Drawing*, JGAA, 5(2) 1–34 (2000)26

$$(X(w) - 4, Y(w), Z(v) + E') \to (X(w) - 4, Y(w), Z(w) + T') \to$$
$$(X(w), Y(w), Z(w) + T')$$

- $E \to B$ (5 bends):
  $$(X(v) + 1, Y(v), Z(v)) \to (X(v) + 1, Y(v), Z(v) + E') \to$$
  $$(X(v) + 1, Y(w) + 2, Z(v) + E') \to (X(w), Y(w) + 2, Z(v) + E') \to$$
  $$(X(w), Y(w), Z(v) + E')$$

- $W \to N$ (6 bends):
  $$(X(v) - 1, Y(v), Z(v)) \to (X(v) - 1, Y(v), Z(v) + W') \to$$
  $$(X(v) - 1, Y(w) + 2, Z(v) + W') \to (X(w), Y(w) + 2, Z(v) + W') \to$$
  $$(X(w), Y(w) + 3, Z(v) + W') \to (X(w), Y(w) + 3, Z(w))$$

- $W \to S$ (6 bends):
  $$(X(v) - 1, Y(v), Z(v)) \to (X(v) - 1, Y(v), Z(v) + W') \to$$
  $$(X(v) - 1, Y(w) - 2, Z(v) + W') \to (X(w), Y(w) - 2, Z(v) + W') \to$$
  $$(X(w), Y(w) - 3, Z(v) + W') \to (X(w), Y(w) - 3, Z(w))$$

- $W \to W$ (6 bends):
  $$(X(v) - 1, Y(v), Z(v)) \to (X(v) - 1, Y(v), Z(v) + W') \to$$
  $$(X(v) - 1, Y(w) - 2, Z(v) + W') \to (X(w) - 3, Y(w) - 2, Z(v) + W') \to$$
  $$(X(w) - 3, Y(w), Z(v) + W') \to (X(w) - 3, Y(w), Z(w))$$

- $W \to E$ (6 bends):
  $$(X(v) - 1, Y(v), Z(v)) \to (X(v) - 1, Y(v), Z(v) + W') \to$$
  $$(X(v) - 1, Y(w) - 2, Z(v) + W') \to (X(w) + 3, Y(w) - 2, Z(v) + W') \to$$
  $$(X(w) + 3, Y(w), Z(v) + W') \to (X(w) + 3, Y(w), Z(w))$$

- $W \to T$ (7 bends):
  $$(X(v) - 1, Y(v), Z(v)) \to (X(v) - 1, Y(v), Z(v) + W') \to$$
  $$(X(v) - 1, Y(w) - 2, Z(v) + W') \to (X(w) + 4, Y(w) - 2, Z(v) + W') \to$$
  $$(X(w) + 4, Y(w), Z(v) + W') \to (X(w) + 4, Y(w), Z(w) + T') \to$$
  $$(X(w), Y(w), Z(w) + T')$$

- $W \to B$ (5 bends):
  $$(X(v) - 1, Y(v), Z(v)) \to (X(v) - 1, Y(v), Z(v) + W') \to$$
  $$(X(v) - 1, Y(w) - 2, Z(v) + W') \to (X(w), Y(w) - 2, Z(v) + W') \to$$
  $$(X(w), Y(w), Z(v) + W')$$

- $T \to N$ (6 bends):
  $$(X(v), Y(v), Z(v) + T') \to (X(v) + 2, Y(v), Z(v) + T') \to$$
  $$(X(v) + 2, Y(w) + 2, Z(v) + T') \to (X(w), Y(w) + 2, Z(v) + T') \to$$
  $$(X(w), Y(w) + 3, Z(v) + T') \to (X(w), Y(w) + 3, Z(w))$$

- $T \to S$ (6 bends):
  $$(X(v), Y(v), Z(v) + T') \to (X(v) + 2, Y(v), Z(v) + T') \to$$
  $$(X(v) + 2, Y(w) - 2, Z(v) + T') \to (X(w), Y(w) - 2, Z(v) + T') \to$$
  $$(X(w), Y(w) - 3, Z(v) + T') \to (X(w), Y(w) - 3, Z(w))$$

M. Closson et al., *Fully Dynamic 3-D Graph Drawing*, JGAA, 5(2) 1–34 (2000)27

- $T \to E$ (6 bends):
  $(X(v), Y(v), Z(v) + T') \to (X(v), Y(v) + 2, Z(v) + T') \to$
  $(X(w) + 2, Y(v) + 2, Z(v) + T') \to (X(w) + 2, Y(w), Z(v) + T') \to$
  $(X(w) + 3, Y(w), Z(v) + T') \to (X(w) + 3, Y(w), Z(w))$

- $T \to W$ (6 bends):
  $(X(v), Y(v), Z(v) + T') \to (X(v), Y(v) + 2, Z(v) + T') \to$
  $(X(w) - 2, Y(v) + 2, Z(v) + T') \to (X(w) - 2, Y(w), Z(v) + T') \to$
  $(X(w) - 3, Y(w), Z(v) + T') \to (X(w) - 3, Y(w), Z(w))$

- $T \to B$ (5 bends):
  $(X(v), Y(v), Z(v) + T') \to (X(v), Y(v) + 2, Z(v) + T') \to$
  $(X(w) - 2, Y(v) + 2, Z(v) + T') \to (X(w) - 2, Y(w), Z(v) + T') \to$
  $(X(w), Y(w), Z(v) + T')$

- $T \to T$ (7 bends):
  $(X(v), Y(v), Z(v) + T') \to (X(v), Y(v) + 2, Z(v) + T') \to$
  $(X(w) - 2, Y(v) + 2, Z(v) + T') \to (X(w) - 2, Y(w) + 4, Z(v) + T') \to$
  $(X(w), Y(w) + 4, Z(v) + T') \to (X(w), Y(w) + 4, Z(w) + T') \to$
  $(X(w), Y(w), Z(w) + T')$

- $B \to N$ (6 bends):
  $(X(v), Y(v), Z(v) + B') \to (X(v) + 2, Y(v), Z(v) + B') \to$
  $(X(v) + 2, Y(w) + 2, Z(v) + B') \to (X(w), Y(w) + 2, Z(v) + B') \to$
  $(X(w), Y(w) + 3, Z(v) + B') \to (X(w), Y(w) + 3, Z(w))$

- $B \to S$ (6 bends):
  $(X(v), Y(v), Z(v) + B') \to (X(v) + 2, Y(v), Z(v) + B') \to$
  $(X(v) + 2, Y(w) - 2, Z(v) + B') \to (X(w), Y(w) - 2, Z(v) + B') \to$
  $(X(w), Y(w) - 3, Z(v) + B') \to (X(w), Y(w) - 3, Z(w))$

- $B \to E$ (6 bends):
  $(X(v), Y(v), Z(v) + B') \to (X(v), Y(v) - 2, Z(v) + B') \to$
  $(X(w) + 2, Y(v) - 2, Z(v) + B') \to (X(w) + 2, Y(w), Z(v) + B') \to$
  $(X(w) + 3, Y(w), Z(v) + B') \to (X(w) + 3, Y(w), Z(w))$

- $B \to W$ (6 bends):
  $(X(v), Y(v), Z(v) + B') \to (X(v), Y(v) - 2, Z(v) + B') \to$
  $(X(w) - 2, Y(v) - 2, Z(v) + B') \to (X(w) - 2, Y(w), Z(v) + B') \to$
  $(X(w) - 3, Y(w), Z(v) + B') \to (X(w) - 3, Y(w), Z(w))$

- $B \to T$ (7 bends):
  $(X(v), Y(v), Z(v) + B') \to (X(v), Y(v) - 2, Z(v) + B') \to$
  $(X(w) - 2, Y(v) - 2, Z(v) + B') \to (X(w) - 2, Y(w) - 4, Z(v) + B') \to$
  $(X(w), Y(w) - 4, Z(v) + B') \to (X(w), Y(w) - 4, Z(w) + T') \to$
  $(X(w), Y(w), Z(w) + T')$

- $B \to B$ (5 bends):
  $(X(v), Y(v), Z(v) + B') \to (X(v), Y(v) - 2, Z(v) + B') \to$

M. Closson et al., *Fully Dynamic 3-D Graph Drawing*, JGAA, 5(2) 1–34 (2000)28

$$(X(w) - 2, Y(v) - 2, Z(v) + B') \rightarrow (X(w) - 2, Y(w), Z(v) + B') \rightarrow$$
$$(X(w), Y(w), Z(v) + B')$$

D   Appendix: Routings - Directed Model

There are 72 port-to-port routings from a vertex v to vertex w in a directed drawing. Each of the following specifications represents 2 routings depending on whether v is above or below w on the staircase. In each case, the initial and terminating points of the route are $(X(v), Y(v) = 0, Z(v))$ and $(X(w), Y(w) = 0, Z(w))$ respectively and each edge contains exactly 6 bends.

- $N \rightarrow S$:
  $(X(v), 2, Z(v)) \rightarrow (X(v), 2, Z(v) - 1) \rightarrow (X(w) - 1, 2, Z(v) - 1) \rightarrow$
  $(X(w) - 1, -1, Z(v) - 1) \rightarrow (X(w) - 1, -1, Z(w)) \rightarrow (X(w), -1, Z(w))$

- $N \rightarrow N$:
  $(X(v), 2, Z(v)) \rightarrow (X(v), 2, Z(v) - 1) \rightarrow (X(w) + 1, 2, Z(v) - 1) \rightarrow$
  $(X(w) + 1, 1, Z(v) - 1) \rightarrow (X(w) + 1, 1, Z(w)) \rightarrow (X(w), 1, Z(w))$

- $N \rightarrow E$:
  $(X(v), 2, Z(v)) \rightarrow (X(v), 2, Z(v) - 1) \rightarrow (X(w) + 2, 2, Z(v) - 1) \rightarrow$
  $(X(w) + 2, -1, Z(v) - 1) \rightarrow (X(w) + 2, -1, Z(w)) \rightarrow (X(w) + 2, 0, Z(w))$

- $N \rightarrow W$:
  $(X(v), 2, Z(v)) \rightarrow (X(v), 2, Z(v) - 1) \rightarrow (X(w) - 2, 2, Z(v) - 1) \rightarrow$
  $(X(w) - 2, 1, Z(v) - 1) \rightarrow (X(w) - 2, 1, Z(w)) \rightarrow (X(w) - 2, 0, Z(w))$

- $N \rightarrow T$:
  $(X(v), 2, Z(v)) \rightarrow (X(v), 2, Z(v) - 1) \rightarrow (X(w) - 3, 2, Z(v) - 1) \rightarrow$
  $(X(w) - 3, 0, Z(v) - 1) \rightarrow (X(w) - 3, 0, Z(w) + 1) \rightarrow (X(w), 0, Z(w) + 1)$

- $N \rightarrow B$:
  $(X(v), 2, Z(v)) \rightarrow (X(v), 2, Z(v) - 1) \rightarrow (X(w) + 3, 2, Z(v) - 1) \rightarrow$
  $(X(w) + 3, 0, Z(v) - 1) \rightarrow (X(w) + 3, 0, Z(w) - 1) \rightarrow (X(w), 0, Z(w) - 1)$

- $S \rightarrow S$:
  $(X(v), -2, Z(v)) \rightarrow (X(v), -2, Z(v) + 1) \rightarrow (X(w) - 1, -2, Z(v) + 1) \rightarrow$
  $(X(w) - 1, -1, Z(v) + 1) \rightarrow (X(w) - 1, -1, Z(w)) \rightarrow (X(w), -1, Z(w))$

- $S \rightarrow N$:
  $(X(v), -2, Z(v)) \rightarrow (X(v), -2, Z(v) + 1) \rightarrow (X(w) + 1, -2, Z(v) + 1) \rightarrow$
  $(X(w) + 1, 1, Z(v) + 1) \rightarrow (X(w) + 1, 1, Z(w)) \rightarrow (X(w), 1, Z(w))$

- $S \rightarrow E$:
  $(X(v), -2, Z(v)) \rightarrow (X(v), -2, Z(v) + 1) \rightarrow (X(w) + 2, -2, Z(v) + 1) \rightarrow$
  $(X(w) + 2, -1, Z(v) + 1) \rightarrow (X(w) + 2, -1, Z(w)) \rightarrow (X(w) + 2, 0, Z(w))$

M. Closson et al., *Fully Dynamic 3-D Graph Drawing*, JGAA, 5(2) 1–34 (2000)29

- $S \to W$:
  $(X(v), -2, Z(v)) \to (X(v), -2, Z(v) + 1) \to (X(w) - 2, -2, Z(v) + 1) \to$
  $(X(w) - 2, 1, Z(v) + 1) \to (X(w) - 2, 1, Z(w)) \to (X(w) - 2, 0, Z(w))$

- $S \to T$:
  $(X(v), -2, Z(v)) \to (X(v), -2, Z(v) + 1) \to (X(w) - 3, -2, Z(v) + 1) \to$
  $(X(w) - 3, 0, Z(v) + 1) \to (X(w) - 3, 0, Z(w) + 1) \to (X(w), 0, Z(w) + 1)$

- $S \to B$:
  $(X(v), -2, Z(v)) \to (X(v), -2, Z(v) + 1) \to (X(w) + 3, -2, Z(v) + 1) \to$
  $(X(w) + 3, 0, Z(v) + 1) \to (X(w) + 3, 0, Z(w) - 1) \to (X(w), 0, Z(w) - 1)$

- $E \to N$:
  $(X(v) + 2, 0, Z(v)) \to (X(v) + 2, -3, Z(v)) \to (X(w) + 1, -3, Z(v)) \to$
  $(X(w) + 1, 1, Z(v)) \to (X(w) + 1, 1, Z(w)) \to (X(w), 1, Z(w))$

- $E \to S$:
  $(X(v) + 2, 0, Z(v)) \to (X(v) + 2, -3, Z(v)) \to (X(w) - 1, -3, Z(v)) \to$
  $(X(w) - 1, -1, Z(v)) \to (X(w) - 1, -1, Z(w)) \to (X(w), -1, Z(w))$

- $E \to E$:
  $(X(v) + 2, 0, Z(v)) \to (X(v) + 2, -3, Z(v)) \to (X(w) + 2, -3, Z(v)) \to$
  $(X(w) + 2, -1, Z(v)) \to (X(w) + 2, -1, Z(w)) \to (X(w) + 2, 0, Z(w))$

- $E \to W$:
  $(X(v) + 2, 0, Z(v)) \to (X(v) + 2, -3, Z(v)) \to (X(w) - 2, -3, Z(v)) \to$
  $(X(w) - 2, 1, Z(v)) \to (X(w) - 2, 1, Z(w)) \to (X(w) - 2, 0, Z(w))$

- $E \to T$:
  $(X(v) + 2, 0, Z(v)) \to (X(v) + 2, -3, Z(v)) \to (X(w) - 3, -3, Z(v)) \to$
  $(X(w) - 3, 0, Z(v)) \to (X(w) - 3, 0, Z(w) + 1) \to (X(w), 0, Z(w) + 1)$

- $E \to B$:
  $(X(v) + 2, 0, Z(v)) \to (X(v) + 2, -3, Z(v)) \to (X(w) + 3, -3, Z(v)) \to$
  $(X(w) + 3, 0, Z(v)) \to (X(w) + 3, 0, Z(w) - 1) \to (X(w), 0, Z(w) - 1)$

- $W \to N$:
  $(X(v) - 2, 0, Z(v)) \to (X(v) - 2, 3, Z(v)) \to (X(w) + 1, 3, Z(v)) \to$
  $(X(w) + 1, 1, Z(v)) \to (X(w) + 1, 1, Z(w)) \to (X(w), 1, Z(w))$

- $W \to S$:
  $(X(v) - 2, 0, Z(v)) \to (X(v) - 2, 3, Z(v)) \to (X(w) - 1, 3, Z(v)) \to$
  $(X(w) - 1, -1, Z(v)) \to (X(w) - 1, -1, Z(w)) \to (X(w), -1, Z(w))$

- $W \to E$:
  $(X(v) - 2, 0, Z(v)) \to (X(v) - 2, 3, Z(v)) \to (X(w) + 2, 3, Z(v)) \to$
  $(X(w) + 2, -1, Z(v)) \to (X(w) + 2, -1, Z(w)) \to (X(w) + 2, 0, Z(w))$

- $W \to W$:
  $(X(v) - 2, 0, Z(v)) \to (X(v) - 2, 3, Z(v)) \to (X(w) - 2, 3, Z(v)) \to$
  $(X(w) - 2, 1, Z(v)) \to (X(w) - 2, 1, Z(w)) \to (X(w) - 2, 0, Z(w))$

M. Closson et al., *Fully Dynamic 3-D Graph Drawing*, JGAA, 5(2) 1–34 (2000)30

- $W \to T$:
  $(X(v) - 2, 0, Z(v)) \to (X(v) - 2, 3, Z(v)) \to (X(w) - 3, 3, Z(v)) \to$
  $(X(w) - 3, 0, Z(v)) \to (X(w) - 3, 0, Z(w) + 1) \to (X(w), 0, Z(w) + 1)$

- $W \to B$:
  $(X(v) - 2, 0, Z(v)) \to (X(v) - 2, 3, Z(v)) \to (X(w) + 3, 3, Z(v)) \to$
  $(X(w) + 3, 0, Z(v)) \to (X(w) + 3, 0, Z(w) - 1) \to (X(w), 0, Z(w) - 1)$

- $T \to N$:
  $(X(v), 0, Z(v) + 1) \to (X(v), +2, Z(v) + 1) \to (X(w) + 1, 2, Z(v) + 1) \to$
  $(X(w) + 1, 1, Z(v) + 1) \to (X(w) + 1, 1, Z(w)) \to (X(w), 1, Z(w))$

- $T \to S$:
  $(X(v), 0, Z(v) + 1) \to (X(v), 2, Z(v) + 1) \to (X(w) - 1, 2, Z(v) + 1) \to$
  $(X(w) - 1, -1, Z(v) + 1) \to (X(w) - 1, -1, Z(w)) \to (X(w), -1, Z(w))$

- $T \to E$:
  $(X(v), 0, Z(v) + 1) \to (X(v), 2, Z(v) + 1) \to (X(w) + 2, 2, Z(v) + 1) \to$
  $(X(w) + 2, -1, Z(v) + 1) \to (X(w) + 2, -1, Z(w)) \to (X(w) + 2, 0, Z(w))$

- $T \to W$:
  $(X(v), 0, Z(v) + 1) \to (X(v), 2, Z(v) + 1) \to (X(w) - 2, 2, Z(v) + 1) \to$
  $(X(w) - 2, -1, Z(v) + 1) \to (X(w) - 2, -1, Z(w)) \to (X(w) - 2, 0, Z(w))$

- $T \to B$:
  $(X(v), 0, Z(v) + 1) \to (X(v), 2, Z(v) + 1) \to (X(w) + 3, 2, Z(v) + 1) \to$
  $(X(w) + 3, 0, Z(v) + 1) \to (X(w) + 3, 0, Z(w) - 1) \to (X(w), 0, Z(w) - 1)$

- $T \to T$:
  $(X(v), 0, Z(v) + 1) \to (X(v), 2, Z(v) + 1) \to (X(w) - 3, 2, Z(v) + 1) \to$
  $(X(w) - 3, 0, Z(v) + 1) \to (X(w) - 3, 0, Z(w) + 1) \to (X(w), 0, Z(w) + 1)$

- $B \to N$:
  $(X(v), 0, Z(v) - 1) \to (X(v), -2, Z(v) - 1) \to (X(w) + 1, -2, Z(v) - 1) \to$
  $(X(w) + 1, 1, Z(v) - 1) \to (X(w) + 1, 1, Z(w)) \to (X(w), 1, Z(w))$

- $B \to S$:
  $(X(v), 0, Z(v) - 1) \to (X(v), -2, Z(v) - 1) \to (X(w) - 1, -2, Z(v) - 1) \to$
  $(X(w) - 1, -1, Z(v) - 1) \to (X(w) - 1, -1, Z(w)) \to (X(w), -1, Z(w))$

- $B \to E$:
  $(X(v), 0, Z(v) - 1) \to (X(v), -2, Z(v) - 1) \to (X(w) + 2, -2, Z(v) - 1) \to$
  $(X(w) + 2, -1, Z(v) - 1) \to (X(w) + 2, -1, Z(w)) \to (X(w) + 2, 0, Z(w))$

- $B \to W$:
  $(X(v), 0, Z(v) - 1) \to (X(v), -2, Z(v) - 1) \to (X(w) - 2, -2, Z(v) - 1) \to$
  $(X(w) - 2, -1, Z(v) - 1) \to (X(w) - 2, -1, Z(w)) \to (X(w) - 2, 0, Z(w))$

- $B \to B$:
  $(X(v), 0, Z(v) - 1) \to (X(v), -2, Z(v) - 1) \to (X(w) + 3, -2, Z(v) - 1) \to$
  $(X(w) + 3, 0, Z(v) - 1) \to (X(w) + 3, 0, Z(w) - 1) \to (X(w), 0, Z(w) - 1)$

M. Closson et al., *Fully Dynamic 3-D Graph Drawing*, JGAA, 5(2) 1–34 (2000)31

- $B \to T$:
  $(X(v), 0, Z(v) - 1) \to (X(v), -2, Z(v) - 1) \to (X(w) - 3, -2, Z(v) - 1) \to$
  $(X(w) - 3, 0, Z(v) - 1) \to (X(w) - 3, 0, Z(w) + 1) \to (X(w), 0, Z(w) + 1)$

E   Appendix: Routings - 5 Bend Staircase Model

The 36 possible port-to-port routings from a vertex v to vertex w will now be enumerated. In each case, the initial and terminating points of the route are $(X(v), Y(v) = 0, Z(v))$ and $(X(w), Y(w) = 0, Z(w))$ respectively. We use $K(v)$ and $L(v)$ to denote the underlayers associated with vertex v: if v is the ith vertex, then $K(v) = -2(n - i) - 2$ and $L(v) = -2(n - i) - 3$, where n is the number of vertices in the staircase. All edge routes contain 5 bends except Top to Bottom, East to Bottom, North to Bottom, South to Bottom and Bottom to Bottom which have 4 bends.

- $N \to S$:
  $(X(v), 2, Z(v)) \to (X(v) + 1, 2, Z(v)) \to (X(v) + 1, 2, Z(w)) \to$
  $(X(v) + 1, -1, Z(w)) \to (X(w), -1, Z(w))$

- $N \to N$:
  $(X(v), 1, Z(v)) \to (X(v), 1, Z(v) + 1) \to (X(w), 1, Z(v) + 1) \to$
  $(X(w), +2, Z(v) + 1) \to (X(w), +2, Z(w))$

- $N \to E$:
  $(X(v), 1, Z(v)) \to (X(v), 1, Z(v) + 1) \to (X(w) + 2, 1, Z(v) + 1) \to$
  $(X(w) + 2, 0, Z(v) + 1) \to (X(w) + 2, 0, Z(w))$

- $N \to W$:
  $(X(v), 1, Z(v)) \to (X(v), 1, Z(v) + 1) \to (X(w) - 1, 1, Z(v) + 1) \to$
  $(X(w) - 1, 0, Z(v) + 1) \to (X(w) - 1, 0, Z(w))$

- $N \to T$:
  $(X(v), 2, Z(v)) \to (X(v) + 1, 2, Z(v)) \to (X(v) + 1, 2, Z(w) + 2) \to$
  $(X(v) + 1, 0, Z(w) + 2) \to (X(w), 0, Z(w) + 2)$

- $N \to B$ (4 bends):
  $(X(v), 1, Z(v)) \to (X(v), 1, Z(v) + 1) \to (X(w), 1, Z(v) + 1) \to$
  $(X(w), 0, Z(v) + 1)$

- $S \to N$:
  $(X(v), -2, Z(v)) \to (X(v), -2, K(v)) \to (X(w), -2, K(v)) \to$
  $(X(w), +2, K(v)) \to (X(w), 2, Z(w))$

- $S \to S$:
  $(X(v), -2, Z(v)) \to (X(v) - 1, -2, Z(v)) \to (X(v) - 1, -2, Z(w)) \to$
  $(X(v) - 1, -1, Z(w)) \to (X(w), -1, Z(w))$

M. Closson et al., *Fully Dynamic 3-D Graph Drawing*, JGAA, 5(2) 1–34 (2000)32

- $S \to E$:
  $(X(v), -1, Z(v)) \to (X(v), -1, Z(v) + 1) \to (X(w) + 2, -1, Z(v) + 1) \to$
  $(X(w) + 2, 0, Z(v) + 1) \to (X(w) + 2, 0, Z(w))$

- $S \to W$:
  $(X(v), -1, Z(v)) \to (X(v), -1, Z(v) + 1) \to (X(w) - 1, -1, Z(v) + 1) \to$
  $(X(w) - 1, 0, Z(v) + 1) \to (X(w) - 1, 0, Z(w))$

- $S \to T$:
  $(X(v), -2, Z(v)) \to (X(v) - 1, -2, Z(v)) \to (X(v) - 1, -2, Z(w) + 2) \to$
  $(X(v) - 1, 0, Z(w) + 2) \to (X(w), 0, Z(w) + 2)$

- $S \to B$ (4 bends):
  $(X(v), -1, Z(v)) \to (X(v), -1, Z(v) + 1) \to (X(w), -1, Z(v) + 1) \to$
  $(X(w), 0, Z(v) + 1)$

- $E \to N$:
  $(X(v) + 2, 0, Z(v)) \to (X(v) + 2, 0, L(v)) \to (X(w), 0, L(v)) \to$
  $(X(w), 2, L(v)) \to (X(w), 2, Z(w))$

- $E \to S$:
  $(X(v) + 1, 0, Z(v)) \to (X(v) + 1, -2, Z(v)) \to (X(v) + 1, -2, Z(w)) \to$
  $(X(v) + 1, -1, Z(w)) \to (X(w), -1, Z(w))$

- $E \to E$:
  $(X(v) + 1, 0, Z(v)) \to (X(v) + 1, -1, Z(v)) \to (X(w) + 2, -1, Z(v)) \to$
  $(X(w) + 2, 0, Z(v)) \to (X(w) + 2, 0, Z(w))$

- $E \to W$:
  $(X(v) + 1, 0, Z(v)) \to (X(v) + 1, -1, Z(v)) \to (X(w) - 1, -1, Z(v)) \to$
  $(X(w) - 1, 0, Z(v)) \to (X(w) - 1, 0, Z(w))$

- $E \to T$:
  $(X(v) + 1, 0, Z(v)) \to (X(v) + 1, -2, Z(v)) \to (X(v) + 1, -2, Z(w) + 2) \to$
  $(X(v) + 1, 0, Z(w) + 2) \to (X(w), 0, Z(w) + 2)$

- $E \to B$ (4 bends):
  $(X(v) + 1, 0, Z(v)) \to (X(v) + 1, -1, Z(v)) \to (X(w), -1, Z(v)) \to$
  $(X(w), 0, Z(v))$

- $W \to N$:
  $(X(v) - 1, 0, Z(v)) \to (X(v) - 1, 3, Z(v)) \to (X(w), 3, Z(v)) \to$
  $(X(w), 2, Z(v)) \to (X(w), 2, Z(w))$

- $W \to S$:
  $(X(v) - 1, 0, Z(v)) \to (X(v) - 1, 2, Z(v)) \to (X(v) - 1, 2, Z(w)) \to$
  $(X(v) - 1, -1, Z(w)) \to (X(w), -1, Z(w))$

- $W \to W$:
  $(X(v) - 1, 0, Z(v)) \to (X(v) - 1, 3, Z(v)) \to (X(w) - 1, 3, Z(v)) \to$
  $(X(w) - 1, 0, Z(v)) \to (X(w) - 1, 0, Z(w))$

M. Closson et al., *Fully Dynamic 3-D Graph Drawing*, JGAA, 5(2) 1–34 (2000)33

- $W \to E$:
  $(X(v) - 1, 0, Z(v)) \to (X(v) - 1, 3, Z(v)) \to (X(w) + 2, 3, Z(v)) \to$
  $(X(w) + 2, 0, Z(v)) \to (X(w) + 2, 0, Z(w))$

- $W \to T$:
  $(X(v) - 1, 0, Z(v)) \to (X(v) - 1, 2, Z(v)) \to (X(v) - 1, 2, Z(w) + 2) \to$
  $(X(v) - 1, 0, Z(w) + 2) \to (X(w), 0, Z(w) + 2)$

- $W \to B$:
  $(X(v) - 1, 0, Z(v)) \to (X(v) - 1, 3, Z(v)) \to (X(w) + 1, 3, Z(v)) \to$
  $(X(w) + 1, 0, Z(v)) \to (X(w), 0, Z(v))$

- $T \to N$:
  $(X(v), 0, Z(v) + 2) \to (X(v), +1, Z(v) + 2) \to (X(w), 1, Z(v) + 2) \to$
  $(X(w), 2, Z(v) + 2) \to (X(w), 2, Z(w))$

- $T \to S$:
  $(X(v), 0, Z(v) + 2) \to (X(v), 4, Z(v) + 2) \to (X(v), 4, Z(w)) \to$
  $(X(v), -1, Z(w)) \to (X(w), -1, Z(w))$

- $T \to E$:
  $(X(v), 0, Z(v) + 2) \to (X(v), +1, Z(v) + 2) \to (X(w) + 2, 1, Z(v) + 2) \to$
  $(X(w) + 2, 0, Z(v) + 2) \to (X(w) + 2, 0, Z(w))$

- $T \to W$:
  $(X(v), 0, Z(v) + 2) \to (X(v), 1, Z(v) + 2) \to (X(w) - 1, 1, Z(v) + 2) \to$
  $(X(w) - 1, 0, Z(v) + 2) \to (X(w) - 1, 0, Z(w))$

- $T \to B$ (4 bends):
  $(X(v), 0, Z(v) + 2) \to (X(v), 1, Z(v) + 2) \to (X(w), 1, Z(v) + 2) \to$
  $(X(w), 0, Z(v) + 3)$

- $T \to T$:
  $(X(v), 0, Z(v) + 2) \to (X(v), +4, Z(v) + 2) \to (X(v), 4, Z(w) + 2) \to$
  $(X(v), 0, Z(w) + 2) \to (X(w), 0, Z(w) + 2)$

- $B \to N$:
  $(X(v), 0, Z(v) - 1) \to (X(v), +1, Z(v) - 1) \to (X(w), 1, Z(v) - 1) \to$
  $(X(w), 2, Z(v) - 1) \to (X(w), 2, Z(w))$

- $B \to S$:
  $(X(v), 0, L(v)) \to (X(v), -3, L(v)) \to (X(v), -3, Z(w)) \to$
  $(X(v), -1, Z(w)) \to (X(w), -1, Z(w))$

- $B \to E$:
  $(X(v), 0, Z(v) - 1) \to (X(v), +1, Z(v) - 1) \to (X(w) + 2, 1, Z(v) - 1) \to$
  $(X(w) + 2, 0, Z(v) - 1) \to (X(w) + 2, 0, Z(w))$

- $B \to W$:
  $(X(v), 0, Z(v) - 1) \to (X(v), 1, Z(v) - 1) \to (X(w) - 1, 1, Z(v) - 1) \to$
  $(X(w) - 1, 0, Z(v) - 1) \to (X(w) - 1, 0, Z(w))$

M. Closson et al., *Fully Dynamic 3-D Graph Drawing*, JGAA, 5(2) 1–34 (2000)34

- $B \to T$:
  $$(X(v), 0, L(v)) \to (X(v), -3, L(v)) \to (X(v), -3, Z(w) + 2) \to$$
  $$(X(v), 0, Z(w) + 2) \to (X(w), 0, Z(w) + 2)$$

- $B \to B$ (4 bends):
  $$(X(v), 0, Z(v) - 1) \to (X(v), 1, Z(v) - 1) \to (X(w), 1, Z(v) - 1) \to$$
  $$(X(w), 0, Z(v) - 1)$$

Volume 5:3 (2001)

Journal of Graph Algorithms and Applications

http://www.cs.brown.edu/publications/jgaa/

vol. 5, no. 3, pp. 1–15 (2001)

1-Bend 3-D Orthogonal Box-Drawings: Two Open Problems Solved

Therese Biedl

Department of Computer Science
University of Waterloo
Waterloo, ON N2L 3G1, Canada
biedl@uwaterloo.ca

Abstract

This paper studies three-dimensional orthogonal box-drawings where edge-routes have at most one bend. Two open problems for such drawings are: (1) Does every drawing of K_n have volume $\Omega(n^3)$? (2) Is there a drawing of K_n for which additionally the vertices are represented by cubes with surface $O(n)$? This paper answers both questions in the negative, and provides related results concerning volume bounds as well.

Communicated by G. Liotta: submitted May 2000;
revised November 2000 and March 2001.

Research partially supported by NSERC. The results in this paper were presented at the 12th Canadian Conference on Computational Geometry, August 2000.

T. Biedl, *1-Bend 3-D Orthogonal Box-Drawings*, JGAA, 5(3) 1–15 (2001)   2

1   Background

A *3-D orthogonal box-drawing* of a graph is a drawing of the graph where vertices are represented by disjoint axis-parallel boxes and edges are represented by disjoint routes along an underlying three-dimensional rectangular grid. (Since no other type of drawings will be studied here, the term *drawing* is used to mean a 3-D orthogonal box-drawing from now on.)

The route of each edge thus consists of a sequence of contiguous *grid segments*, i.e., axis-parallel line segments for which the fixed coordinates are integers. The transition from one grid segment to another is called a *bend*. A drawing is called a *k-bend drawing* if all edge routes have at most k bends.

Every vertex is represented by an axis-parallel box with integral boundaries; such a box is called a *grid box*. An *X-plane* is a plane that is perpendicular to the X-axis. It is called an X-grid plane if its fixed coordinate is integral. *Y-planes* and *Z-planes* are defined similarly. For any vertex v, let $X(v)$ be the number of X-grid planes that intersect the box of v; $Y(v)$ and $Z(v)$ are defined similarly. The *surface of v* is $2(X(v)Y(v) + Y(v)Z(v) + Z(v)X(v))$. The *volume of v* is $X(v)Y(v)Z(v)$.

When no confusion arises, we will use graph-theoretic terms, such as "vertex" and "edge", to also mean the representation in a fixed drawing.

Given a drawing, denote by $X \times Y \times Z$ the size of the smallest enclosing rectangular box of the drawing. The *volume* of the drawing is $X \cdot Y \cdot Z$.

This paper studies bounds on the volume of drawings with very few bends per edge. Since not all graphs have a 0-bend drawing (also known as visibility representation) [BSWW99, FM99], the smallest applicable number of bends per edge is one.

1.1   Existing results for 1-bend drawings

In [BSWW99], it was shown that the complete graph K_n has a 1-bend drawing with $O(n^3)$ volume (more precisely, in an $n/2 \times n/2 \times n/2$-grid.) In the same paper, it was also shown that any drawing of K_n has volume $\Omega(n^{2.5})$. However, the lower bound does not take restrictions on the number of bends into account, and in particular, it was left as an open problem whether any 1-bend drawing of K_n needs $\Omega(n^3)$ volume.

One criticism of the drawings in [BSWW99] is that vertex boxes resemble "sticks", i.e., one dimension is very large while the other two dimensions are one unit each, hence there is no bound on the aspect ratio. A drawing is said to have *aspect-ratios at most r*, for some constants $r \geq 1$, if any vertex box has aspect ratio at most r. If $r = 1$, then the drawing is called a *cube-drawing*.

The construction in [BSWW99] can be modified to obtain a cube-drawing of K_n by "blowing up" every vertex (see also Figure 2). However, this increases the volume of the drawing to $O(n^4)$. Also, the surface of each vertex box then becomes $O(n^2)$, which seems excessive since every vertex has only $O(n)$ incident edges. A drawing is said to be *degree-restricted* if the surface of a vertex v is at most $\alpha \deg(v)$, for some constant $\alpha \geq 1$. The construction in [BSWW99] is

T. Biedl, *1-Bend 3-D Orthogonal Box-Drawings*, JGAA, 5(3) 1–15 (2001)   3

degree-restricted for K_n, but when converted to a cube-drawing, it is no longer degree-restricted. Hence, the question was posed whether K_n has a degree-restricted 1-bend cube-drawing.

In [BTW01], the lower bounds of [BSWW99] were extended to graphs other than the complete graph. More precisely, it was shown that there exist graphs with n vertices and m edges that have volume $\Omega(mn^{1/2})$ in any drawing. This lower bound also does not take restrictions on the number of bends into account.

Finally, in [Woo00], it was shown that every n-vertex m-edge graph with genus g has a 1-bend drawing of volume $O(nm\sqrt{g})$, which is $O(nm^{3/2})$ in the worst case.

1.2   Contributions of this paper

This paper settles the two open problems mentioned above, and provides other results for 1-bend drawings of *simple graphs*, i.e., graphs without loops and multiple edges. Specific results are as follows:

- Any 1-bend cube-drawing of a simple graph G with $\Omega(\Delta n)$ edges represents $\Omega(n)$ many vertices with an $\Omega(\Delta) \times \Omega(\Delta) \times \Omega(\Delta)$-box, where Δ is the maximum degree of G.[1]

  This has the following consequences:

  - Any such graph does not have a 1-bend degree-restricted cube-drawing. In particular, K_n does not have a 1-bend degree-restricted cube-drawing. (This settles the second open problem mentioned above.)

  - Any 1-bend cube-drawing of such a graph has volume $\Omega(\Delta^3 n)$. In particular, since K_n is $(n-1)$-regular, any 1-bend cube-drawing of K_n has volume $\Omega(n^4)$. (This bound is matched by a construction.)

- Other lower bounds are obtained using a so-called *Ramanujan-graph $G_{n,d}$*, which is a simple d-regular n-vertex graph with special cut-properties which will be reviewed in Section 3.1:

  - Any 1-bend drawing of $G_{n,d}$, for n and d sufficiently big, has a grid plane that intersects at least $\frac{1}{8}n$ vertices.

  - Any 1-bend drawing of $G_{n,d}$ has volume $\Omega(n^2 d)$.

    Since $K_n = G_{n,n-1}$, any 1-bend drawing of K_n has volume $\Omega(n^3)$, which answers the first open problem mentioned above.

2   Cube-Drawings

This section proves that K_n (or more generally, any graph with $\Omega(\Delta n)$ edges) does not have a degree-restricted 1-bend cube-drawing. As a preliminary result,

---

[1]Note that a graph with $\Omega(\Delta n)$ edges has asymptotically the maximum number of edges, since all graphs have at most $\frac{1}{2}\Delta n$ edges. However, there are graphs with $o(\Delta n)$ edges.

T. Biedl, *1-Bend 3-D Orthogonal Box-Drawings*, JGAA, 5(3) 1–15 (2001)   4

we first show that in any 1-bend drawing of such a graph many (i.e., $\Omega(n)$) vertices are intersected by many (i.e., $\Omega(\Delta)$) grid planes each.

Lemma 2.1 *If G is a simple graph with at least $\kappa\Delta n$ edges, for some $0 < \kappa \leq \frac{1}{2}$, then at least $\frac{1}{6}\kappa n$ vertices intersect at least $\frac{1}{6}\kappa\Delta$ grid planes each.*

Proof: Fix an arbitrary 1-bend drawing of G. For any edge e, the route of e has at most one bend, and hence is entirely contained within one grid plane P. We say that edge e *belongs* to P and P *owns* e. (If the route of e has no bend, then it is contained in two grid planes. Arbitrarily choose one of them to own e, so that each edge belongs to exactly one grid plane.)

Let $P_1, \ldots, P_l$ be the grid planes that own at least one edge. For $i = 1, \ldots, l$, let $n(P_i)$ be the number of vertices for which an incident edge belongs to P_i. See also Figure 1.

The crucial observation is that edges do not cross, hence the graph formed by the edges owned by P_i is planar. In particular, by simplicity of G at most $3n(P_i)$ edges can be owned by P_i. Since each of the m edges of G belongs to a grid plane,

$$\sum_{i=1}^{l} 3n(P_i) \geq m \geq \kappa\Delta n. \tag{1}$$

Now count $\sum_{i=1}^{l} n(P_i)$ in another way. For every vertex v, denote by $p(v)$ the number of grid planes that own an incident edge of v; see also Figure 1. Observe that $p(v) \leq X(v) + Y(v) + Z(v)$ because any grid plane that contributes to $p(v)$ must also intersect the box of v. Also, $\sum_{i=1}^{l} n(P_i) = \sum_{v \in V} p(v)$, because both sums count the incidences between a vertex v and a grid plane that owns an edge incident to v.

Let V_b be the set of vertices v with $p(v) \geq \frac{1}{6}\kappa\Delta$. The lemma holds if $|V_b| \geq \frac{1}{6}\kappa n$, because $X(v) + Y(v) + Z(v) \geq p(v) \geq \frac{1}{6}\kappa\Delta$ for every vertex $v \in V_b$. So assume for contradiction that fewer than $\frac{1}{6}\kappa n$ vertices belong to V_b. Observe that $p(v) \leq \Delta$ for all vertices (because for each grid plane there is at least one incident edge of v), and that at most n vertices could be in $V - V_b$. Therefore

$$\sum_{i=1}^{l} n(P_i) \;=\; \sum_{v \in V} p(v) = \sum_{v \notin V_b} p(v) + \sum_{v \in V_b} p(v)$$

$$<\; |V - V_b| \cdot \frac{1}{6}\kappa\Delta + |V_b| \cdot \Delta < n \cdot \frac{1}{6}\kappa\Delta + \frac{1}{6}\kappa n \cdot \Delta = \frac{1}{3}\kappa\Delta n.$$

This contradicts inequality (1), therefore $|V_b| \geq \frac{1}{6}\kappa\Delta$ and the lemma holds. $\square$

Note that the constants in the above lemma could be improved for bipartite graphs, because then at most $2n(P)$ edges could be owned by grid plane P. While this would improve some of the lower bounds to follow by a small fraction, we will not pursue this detail for simplicity's sake. Also note that the proof relies only on that any edge is routed entirely within one grid plane. While this

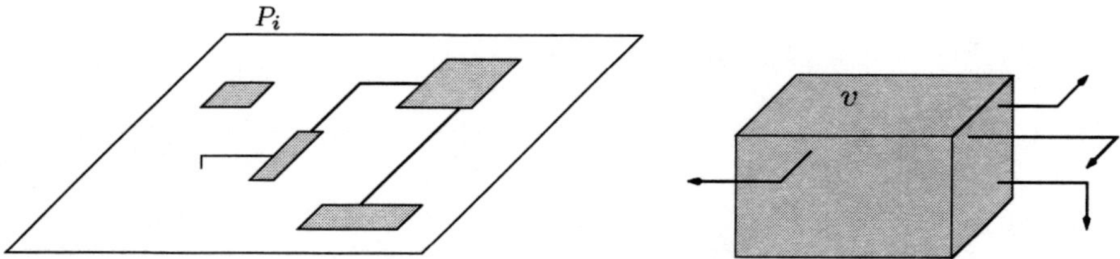

Figure 1: Illustration of $n(P_i)$ and $p(v)$. In the left picture $n(P_i) = 3$, because while four vertices intersect P_i, only three of them are incident to an edge that belongs to P_i. In the right picture, $p(v) = 2$.

certainly holds for any 1-bend drawing, it also holds for many other constructions (e.g., the ones of [BSWW99]). Hence, the lemma and its corollaries could be generalized to any so-called *co-planar* drawing in which each edge is routed within a grid plane.

Lemma 2.1 implies that any 1-bend cube-drawing contains many big vertex boxes. In fact, this result holds for any drawing with bounded aspect ratios.

Lemma 2.2 *Let G be a graph with $\Omega(\Delta n)$ edges. Then any 1-bend drawing of G with aspect ratios at most r contains $\Omega(n)$ vertices whose box has minimum dimension $\Omega(\Delta/r)$, surface $\Omega(\Delta^2/r)$ and volume $\Omega(\Delta^3/r^2)$.*

Proof: Assume that G has at least $\kappa\Delta n$ edges for some constant $0 < \kappa \leq \frac{1}{2}$. Fix an arbitrary drawing of G and let v be one of the at least $\frac{1}{6}\kappa n$ vertices whose box is intersected by at least $\frac{1}{6}\kappa\Delta$ grid planes; these exist by Lemma 2.1. Let the box representing v be an $X \times Y \times Z$-box; without loss of generality assume that $X \leq Y \leq Z$. The box of v intersects $X + Y + Z$ grid planes, so $X + Y + Z \geq \frac{1}{6}\kappa\Delta$ by assumption on v. Also, $Z \leq rX$ because the aspect ratio of v is at most r.

Minimizing the minimum dimension X of v under the constraints $X \leq Y \leq Z$, $X + Y + Z \leq \frac{1}{6}\kappa\Delta$ and $Z \leq rX$ yields $X \geq \frac{1}{6}\kappa\Delta/(1 + 2r) \in \Omega(\Delta/r)$. The surface of v is $2(XY + YZ + XZ)$ and the volume is XYZ. Minimizing each expression, subject to the above constraints, one obtains (for both of them) the solution $X = \frac{1}{6}\kappa\Delta/(r+2) = Y$ and $Z = rX = \frac{1}{6}\kappa r\Delta/(r+2)$. Hence the surface of v is

$$2(XY + YZ + XZ) \geq \frac{2}{36}\kappa^2\Delta^2(1 + 2r)/(r + 2)^2 \in \Omega(\Delta^2/r),$$

and the volume is

$$XYZ \geq \frac{1}{216}\kappa^3\Delta^3 r/(r + 2)^3 \in \Omega(\Delta^3/r^2).$$

$\square$

This lemma implies the answer for the open problem: K_n does not have a degree-restricted 1-bend cube-drawing.

T. Biedl, *1-Bend 3-D Orthogonal Box-Drawings*, JGAA, 5(3) 1–15 (2001)   6

Theorem 1 *Any simple graph G with $\Omega(\Delta n)$ edges does not have a degree-restricted 1-bend drawing with aspect ratios $o(\Delta)$.*

Proof: In any 1-bend drawing of G with aspect ratios at most r, there are $\Omega(n)$ vertices with surface $\Omega(\Delta^2/r)$ by Lemma 2.2. Unless $r \in \Omega(\Delta)$, the surface of these vertices is not proportional to their degrees, which is at most Δ. $\square$

This lemma can also be used for lower bounds on the volume of drawings with bounded aspect ratios.

Theorem 2 *If a simple graph G has $\Omega(\Delta n)$ edges, then any 1-bend drawing with aspect ratios at most r has volume $\Omega(n\Delta^3/r^2)$.*

Proof: In any 1-bend drawing of G with aspect ratios at most r, there are $\Omega(n)$ vertices with volume $\Omega(\Delta^3/r^2)$ by Lemma 2.2. Since vertex boxes are disjoint, these $\Omega(n)$ vertices together occupy an area of volume $\Omega(n\Delta^3/r^2)$. $\square$

Depending on the values of Δ and r, this theorem improves in some cases on the lower bound of $\Omega(m^{3/2}/\sqrt{r})$ for such drawings known from [BTW01].

The above lower bound is optimal for cube-drawings of K_n, because the lower bound states $\Omega(n^4)$ for K_n, and a construction with volume $O(n^4)$ can be obtained easily by "blowing up" the vertex boxes of the construction of [BSWW99]. See Figure 2.

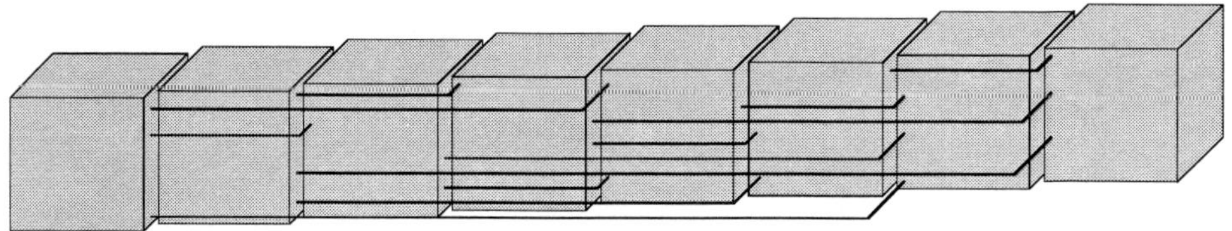

Figure 2: A cube-drawing of K_n with volume $O(n^4)$. Only half of the edges are shown; the other half is routed behind the cubes.

3   Lower Bounds

This section provides lower bounds on the volume of 1-bend drawings, and proves that the $O(n^3)$ volume drawing for K_n in [BSWW99] is asymptotically optimal.

The lower bound proof follows a scheme developed in [BSWW99] and also used in [BTW01]. For a given drawing there are three cases: (1) One grid line intersects "many" vertices; (2) one grid plane, but no grid line, intersects "many" vertices; (3) neither of the above is the case. In [BSWW99], it was shown that the volume of K_n is $\Omega(n^3)$ in the first and third case, but in the second case, only a bound of $\Omega(n^{2.5})$ was achieved. In [BTW01], it was shown

T. Biedl, *1-Bend 3-D Orthogonal Box-Drawings*, JGAA, 5(3) 1–15 (2001)   7

that the volume for so-called Ramanujan-graphs is $\Omega(\Delta n^2)$ in the first case, $\Omega(\Delta n^{1.5})$ in the second case and $\Omega((\Delta n)^{1.5})$ in the third case.

This paper shows a lower bound of $\Omega(\Delta n^2)$ for all 1-bend drawings of Ramanujan-graphs. If the drawing is in the first case, then this is proved exactly as in [BTW01] (the proof is repeated here for completeness). The proof in the second case uses the observation that every edge has at most one bend, and hence the two endpoints must "see" each other in some sense. Finally one can show that the third case cannot happen for sufficiently large n when edges have at most one bend.

This section is structured as follows: We first review the Ramanujan-graphs. Then we prove that the third case cannot happen. Finally, we proceed to prove lower bounds for all drawings.

3.1 Ramanujan-graphs

Ramanujan-graphs were introduced in [LPS88] and have already been used in [BTW01] for lower bounds for orthogonal graph drawing. They have the useful property that for any two disjoint subsets of size $\Omega(n)$, there are $\Omega(m)$ edges between the two subsets. This was first reported in [BTW01], we repeat and slightly modify their proof to obtain the statement for an arbitrary constant μ.

Lemma 3.1 *[BTW01] Let $0 < \mu < 1$ be a constant. If $p \neq q$ are primes, $p \equiv 1 \bmod 4$, $q \equiv 1 \bmod 4$, $p + 1 \geq 16/\mu^2$, then there exists a simple graph $G_{n,d}$ (called a* Ramanujan-graph*) with the following properties:*

- *$G_{n,d}$ is d-regular for $d = p + 1$,*

- *the number n of vertices of $G_{n,d}$ is at least $q(q-1)/2$ and at most $q(q-1)$.*

- *for any disjoint vertex sets S, T of $G_{n,d}$ with $|S| \geq \mu n$, $|T| \geq \mu n$, there are at least $\frac{1}{2}\mu^2 \cdot dn$ edges between S and T.*

Proof: Let $G_{n,d}$ be the graph $X^{p,q}$ defined in [LPS88]; the first two properties of the graph were shown in this paper. It was also shown that $\lambda \leq 2\sqrt{d-1}$, where λ denotes the second-largest eigenvalue of $G_{n,d}$. Assume S and T are as specified above. As shown in [AS92], the number of edges between S and T is at least $\frac{d|S||T|}{n} - \lambda\sqrt{|S||T|}$. Now,

$$\lambda\sqrt{|S||T|} \leq 2\sqrt{d-1}\sqrt{|S||T|} \cdot \underbrace{\sqrt{|S||T|/\mu^2 n^2}}_{\geq 1} \cdot \underbrace{\sqrt{d\mu^2/16}}_{\geq 1} \leq \frac{1}{2}d|S||T|/n.$$

Hence, the number of edges between S and T is at least $(1 - \frac{1}{2}) \cdot d|S||T|/n \geq \frac{1}{2}\mu^2 \cdot dn$. $\qquad\square$

It suffices to state lower bounds only for Ramanujan-graphs, because as was shown in [BTW01], graphs containing Ramanujan-graphs can be constructed for almost all values of m and n.

T. Biedl, *1-Bend 3-D Orthogonal Box-Drawings*, JGAA, 5(3) 1–15 (2001)   8

Lemma 3.2 *[BTW01] There exist constants n_0 and d_0 such that for any $n \geq n_0$ and any $m \geq d_0 n$ there exists a graph with n vertices and m edges that has a Ramanujan-graph with $\theta(n)$ vertices and $\theta(m)$ edges as a subgraph.*

In particular, using these graphs, the lower bounds can be transferred from Ramanujan-graphs to all values of n and m without affecting the order of magnitude, similarly as done in [BTW01].

3.2   Vertices in one grid plane

Now we prove that the "third case" mentioned above cannot happen for 1-bend drawings of Ramanujan-graphs, i.e., there always exists a grid plane intersecting $\Omega(n)$ vertices. For this and the lower bound proofs to come, we will often refer to positions of vertices relative to grid planes. A vertex is said to be *left (right)* of an $(X = X_0)$-plane if all the points in its box have X-coordinates less than X_0 (greater than X_0). A vertex is said to be *before (behind)* a $(Y = Y_0)$-plane if all the points in its box have Y-coordinates less than Y_0 (greater than Z_0). A vertex is said to be *below (above)* a $(Z = Z_0)$-plane if all the points in its box have Z-coordinates less than Z_0 (greater than Z_0).

Also, for the proofs to come, for ease of notation we neglect rounding issues, and assume that n is divisible as needed. This has no effect on the order of magnitude of the lower bounds, since for example in the next theorem, one could show a bound of $\frac{1}{8}n - o(n)$ vertices for all values of n.

Theorem 3 *Let $G_{n,d}$ be a Ramanujan-graph with $d \geq 2^{16}$ and n divisible by 8. Then any 1-bend drawing of $G_{n,d}$ has a grid plane that intersects at least $\frac{1}{8}n$ vertices.*

Proof: Assume to the contrary that no grid plane intersects as many as $\frac{1}{8}n$ vertices.

Informally, this leads to a contradiction because the drawing can be split into non-empty octants. Two of these octants have no grid-plane in common, and hence cannot have an edge with 0 or 1 bends between them. See Figure 3 for an illustration. The precise proof is as follows:

As an $(X = X_0)$-plane is swept from smaller to larger values, we encounter an integer X' where, for the last time, there are at most $\frac{7}{16}n$ vertices to the left of the $(X = X')$-plane. Thus, there are at least $\frac{7}{16}n$ vertices to the left of the $(X = X' + 1)$-plane. All these vertices, call them V_-, are also to the left of the $(X = X' + \frac{1}{2})$-plane. Also, since the $(X = X')$-plane intersects at most $\frac{1}{8}n$ vertices, and at most $\frac{7}{16}n$ vertices are to the left of it, there are at least $n - \frac{1}{8}n - \frac{7}{16}n = \frac{7}{16}n$ vertices to the right of the $(X = X')$-plane. All these vertices, call them V_+, are also to the right of the $(X = X' + \frac{1}{2})$-plane. Denote $X^* = X' + \frac{1}{2}$.

Note that no X-plane intersects both a vertex in V_+ and a vertex in V_-.

Apply the same argument to a sweep with a $(Y = Y_0)$-plane, considering only the vertices in V_-; recall that $|V_-| \geq \frac{7}{16}n$. Thus there is a value Y_-^* such that at least $\frac{5}{32}n$ vertices of V_- are before the $(Y = Y_-^*)$-plane, and at least

T. Biedl, *1-Bend 3-D Orthogonal Box-Drawings*, JGAA, 5(3) 1–15 (2001)   9

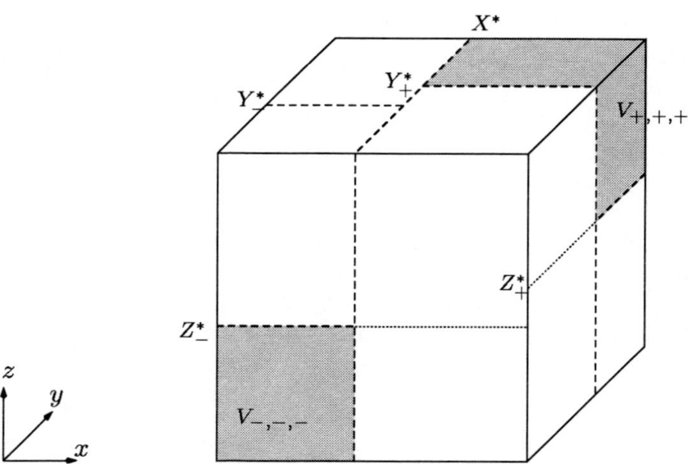

Figure 3: Two diagonally opposite octants yield two non-empty sets of vertices that cannot have an edge with 0 or 1 bends connecting them.

$\frac{7}{16}n - \frac{1}{8}n - \frac{5}{32}n = \frac{5}{32}n$ vertices of V_- are behind the $(Y = Y_-^*)$-plane. Denote these two sets of vertices as $V_{-,-}$ and $V_{-,+}$.

Apply the same argument to a sweep with a $(Y = Y_0)$-plane, considering only the vertices in V_+. Thus there is a value Y_+^* such that at least $\frac{5}{32}n$ vertices of V_+ are before the $(Y = Y_+^*)$-plane, and at least $\frac{5}{32}n$ vertices of V_+ are behind the $(Y = Y_+^*)$-plane. Denote these two sets of vertices as $V_{+,-}$ and $V_{+,+}$.

Without loss of generality, assume that $Y_-^* \leq Y_+^*$. In particular therefore, no Y-plane intersects both a vertex in $V_{-,-}$ and a vertex in $V_{+,+}$.

Apply the same argument to a sweep with a $(Z = Z_0)$-plane, considering only the vertices in $V_{-,-}$; recall that $|V_{-,-}| \geq \frac{5}{32}n$. Thus there is a value Z_-^* such that at least $\frac{1}{64}n$ vertices of $V_{-,-}$ are below the $(Z = Z_-^*)$-plane, and at least $\frac{5}{32}n - \frac{1}{8}n - \frac{1}{64}n = \frac{1}{64}n$ vertices of $V_{-,-}$ are above the $(Z = Z_-^*)$-plane. Denote these two sets of vertices as $V_{-,-,-}$ and $V_{-,-,+}$.

Apply the same argument to a sweep with a $(Z = Z_0)$-plane, considering only the vertices in $V_{+,+}$. Thus there is a value Z_+^* such that at least $\frac{1}{64}n$ vertices of $V_{+,+}$ are below the $(Z = Z_+^*)$-plane, and at least $\frac{1}{64}n$ vertices of $V_{+,+}$ are above the $(Z = Z_+^*)$-plane. Denote these two sets of vertices as $V_{+,+,-}$ and $V_{+,+,+}$.

Without loss of generality, assume that $Z_-^* \leq Z_+^*$. In particular therefore, no Z-plane intersects both a vertex in $V_{-,-,-}$ and a vertex in $V_{+,+,+}$.

Hence no grid plane intersects both a vertex in $V_{-,-,-}$ and $V_{+,+,+}$. These sets each contain at least $\frac{1}{64}n$ vertices. Since $G_{n,d}$ is a Ramanujan-graph with $d \geq 2^{16} = 16 \cdot 64^2$, there are edges between these two vertex sets. These edges cannot be drawn with at most one bend, a contradiction. $\square$

Remark: Any constant smaller than $\frac{1}{7}$ could take the role of $\frac{1}{8}$ in the theorem; the smaller the constant, the smaller also the lower bound on d. For example,

T. Biedl, *1-Bend 3-D Orthogonal Box-Drawings*, JGAA, 5(3) 1–15 (2001)   10

a bound $d \geq 82$ would suffice after replacing $\frac{1}{8}$ by $\frac{1}{407}$.

Also note that the above proof did not use that the drawing had no crossings, and hence would hold even if crossings were allowed.

3.3   1-bend drawings

Now we prove that any 1-bend drawing must have a large volume. The constants in the proof to follow are rather small and chosen for the convenience of a simple proof; they could be improved with a more detailed analysis.

Theorem 4 *Let $G_{n,d}$ be a Ramanujan-graph with $d \geq 2^{22}$ and n divisible by 512. Then any 1-bend drawing of $G_{n,d}$ has volume at least $2^{-27}dn^2$.*

Proof: There are two cases:

Case 1:   There exists a grid line that intersects at least $\frac{1}{256}n = 2^{-8}n$ many vertices. Assume that this grid line is an X-*line*, i.e., a line parallel to the X-axis; the other two directions are similar.

The argument in this case is exactly the same (except for a change of constants and directions) as in [BTW01]. Namely, let $v_1, \ldots, v_t$ be the vertices intersected by the X-line, listed in order of occurrence along the line. Let X_0 be a not necessarily integer X-coordinate such that the $(X = X_0)$-plane intersects none of these t vertices and separates the first $2^{-9}n$ of them from the remaining ones, of which there are at least $2^{-9}n$ many.

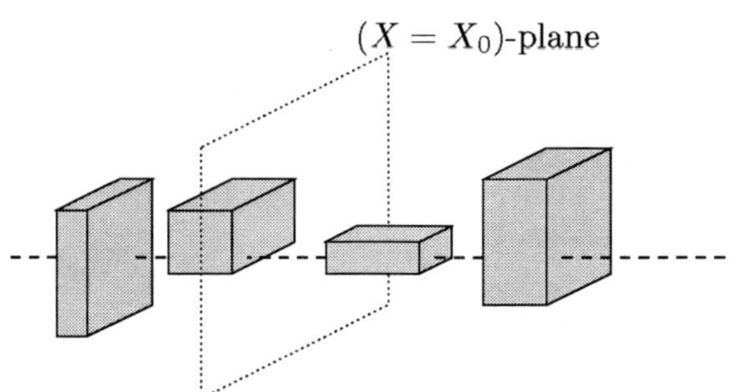

Figure 4: Illustration of case (1).

Because $G_{n,d}$ is a Ramanujan-graph and $d \geq 16 \cdot 2^{18}$, at least $2^{-19} \cdot dn$ edges connect these two vertex sets. Their edge routes cross the $(X = X_0)$-plane, which thus must contain at least $2^{-19} \cdot dn$ points having integer Y- and Z-coordinates. Hence $YZ \geq 2^{-19} \cdot dn$. Since the X-line intersects at least $2^{-8}n$ vertices, also $X \geq 2^{-8}n$, so $XYZ \geq 2^{-27} \cdot dn^2$.

Case 2: No grid line intersects many vertices.

By Theorem 3, there exists a grid plane, say the $(Z = Z')$-plane, that intersects at least $\frac{1}{8}n$ vertices; denote these vertices as V'. In all of the following argument, only vertices of V' are used.

T. Biedl, *1-Bend 3-D Orthogonal Box-Drawings*, JGAA, 5(3) 1–15 (2001)   11

As an $(X = X_0)$-plane is being swept from smaller to larger values, the intersection of the $(X = X_0)$-plane with the $(Z = Z')$-plane is a Y-line, which by assumption intersects at most $\frac{1}{256}n$ vertices at any one time. With an argument similar as in the proof of Theorem 3, we can thus obtain a value X^* such that at least $\frac{15}{256}n$ vertices of V' are to the left of the $(X = X^*)$-plane, and at least $\frac{1}{8}n - \frac{1}{256}n - \frac{15}{256}n \geq \frac{15}{256}n$ vertices of V' are to the right of the $(X = X^*)$-plane. Denote these two sets of vertices as V_- and V_+. Note that no X-plane intersects both a vertex in V_+ and a vertex in V_-.

Apply the same argument to a sweep with a $(Y = Y_0)$-plane, considering only the vertices in V_-. Thus there is a value Y_-^* such that at least $\frac{7}{256}n$ vertices of V_- are before the $(Y = Y_-^*)$-plane, and at least $\frac{15}{256}n - \frac{1}{256}n - \frac{7}{256}n = \frac{7}{256}n$ vertices of V_- are behind the $(Y = Y_-^*)$-plane. Denote these two sets of vertices as $V_{-,-}$ and $V_{-,+}$.

Apply the same argument to a sweep with a $(Y = Y_0)$-plane, considering only the vertices in V_+. Thus there is a value Y_+^* such that at least $\frac{7}{256}n$ vertices of V_+ are before the $(Y = Y_+^*)$-plane, and at least $\frac{7}{256}n$ vertices of V_+ are behind the $(Y = Y_+^*)$-plane. Denote these two sets of vertices as $V_{+,-}$ and $V_{+,+}$.

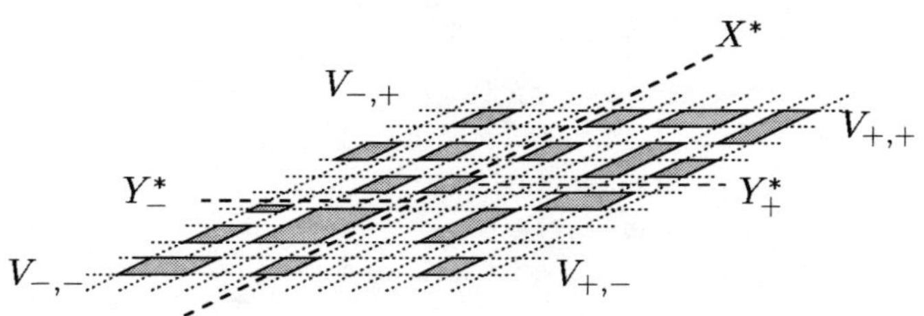

Figure 5: Illustration of case (2).

Without loss of generality, assume that $Y_-^* \leq Y_+^*$. In particular therefore, no Y-plane intersects both a vertex in $V_{-,-}$ and a vertex in $V_{+,+}$.

Since the graph is a Ramanujan-graph, there must be edges between $V_{-,-}$ and $V_{+,+}$. None of these edges can be routed within an X-plane or a Y-plane as observed above, hence they are all routed within a Z-plane.

Now we use the fact that every edge is drawn with one bend. Namely, let (v, w) be an edge with $v \in V_{-,-}$ and $w \in V_{+,+}$, and assume that it is routed in the $(Z = z)$-plane. The route of (v, w) consists of one X-segment and one Y-segment. If (say) its X-segment is incident to v, then no other vertex can be placed on the grid segment between v and the $(X = X^*)$-plane. This motivates the following definition illustrated in Figure 6.

Definition 1 *A vertex $v \in V_{-,-}$ ($v \in V_{+,+}$) is said to be* exposed at level z *if*

- *there exists an X-grid line in the $(Z = z)$-plane that intersects v and does not intersect any other vertex in $V_{-,-}$ ($V_{+,+}$) between v and the $(X = X^*)$-plane, or*

T. Biedl, *1-Bend 3-D Orthogonal Box-Drawings*, JGAA, 5(3) 1–15 (2001)   12

- *there exists a Y-grid line in the $(Z = z)$-plane that intersects v and does not intersect any other vertex in $V_{-,-}$ $(V_{+,+})$ between v and the $(Y = Y_-^*)$-plane $((Y = Y_+^*)$-plane).*

A vertex is called hidden *at level z if it is not exposed.*

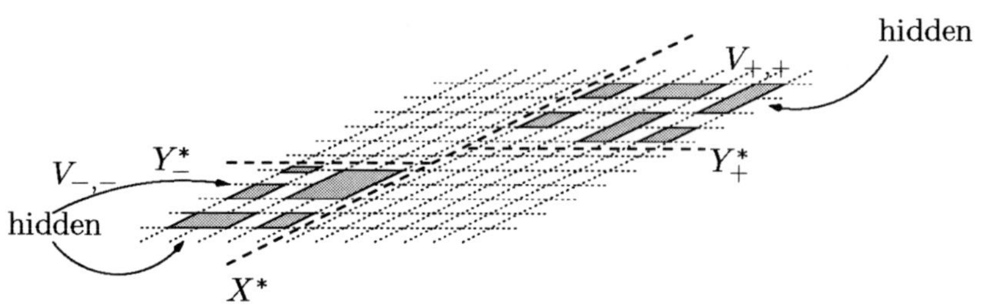

Figure 6: Examples of hidden vertices (we only show the cross-section with one Z-plane). All vertices not marked otherwise are exposed. Note that the top right vertex is hidden even though there is an X-line from it not intersecting other vertices, because this X-line is not a grid-line.

Hence, any edge (v, w) between $V_{-,-}$ and $V_{+,+}$ must be routed in a $(Z = z)$-plane such that both v and w are exposed at level z.

The crucial observation is now that if X and Y (the dimensions of the drawing) are small, then not very many vertices are exposed at any one level. This leads to a contradiction, because then not all edges can be routed. More precisely:

Claim: $X + Y > 2^{-8}n$.

To prove this claim, assume to the contrary that $X + Y \leq 2^{-8}n$. In particular therefore, at most $2^{-8}n$ vertices of $V_{-,-}$ can be exposed at any one given level, simply because there are at most $2^{-8}n$ possible grid lines, each of which can only intersect at most one vertex.

A vertex v is called *active* at level z if the $(Z = z)$-plane intersects the box of v, and *inactive* otherwise. Recall that all vertices in $V_{-,-}$ intersect the $(Z = Z')$-plane, so all vertex in $V_{-,-}$ are active on level Z'. If a vertex $v \in V_{-,-}$ is hidden on level $z - 1 \geq Z'$, but exposed on level z, then some other vertex $w \in V_{-,-}$ was "blocking" v at level $z - 1$, but not on level z, so w must have disappeared, i.e., w became inactive at level z. Hence, every time one vertex becomes exposed, another vertex must become inactive.

The precise argument is now as follows. Sweep a $(Z = Z_0)$-plane from smaller to larger values of Z, starting at $Z = Z'$. Initially, all vertices in $V_{-,-}$ are active (there are at least $\frac{7}{256}n$ many of them), and at most $2^{-8}n = \frac{1}{256}n$ of them are exposed.

During the sweep, more and more vertices become inactive, and hence more and more vertices become exposed. At some point, an integer $Z_{-,+}^*$ is encountered where for the first time at least $\frac{3}{256}n$ vertices of $V_{-,-}$ are inactive. Denote these vertices by $V_{-,-,-}$; hence none of them is exposed on any level $z \geq Z_{-,+}^*$.

T. Biedl, *1-Bend 3-D Orthogonal Box-Drawings*, JGAA, 5(3) 1–15 (2001)   13

At most $\frac{1}{256}n$ vertices were exposed on level Z', and at most $\frac{3}{256}n$ vertices became exposed on level $Z'+1,\ldots,Z^*_{-,+}-1$, because at most $\frac{3}{256}n$ vertices became inactive on these levels. Hence there are at least $|V_{-,-}|-\frac{1}{256}n-\frac{3}{256}n \geq \frac{3}{256}n$ vertices that are not exposed on any level between Z' and $Z^*_{-,+}-1$. Denote these vertices as $V_{-,-,+}$.

With a similar argument, find an integer $Z^*_{+,+}$ such that at least $\frac{3}{256}n$ vertices of $V_{+,+}$ are inactive on any level $z \geq Z^*_{+,+}$ (call them $V_{+,+,-}$), and at least $\frac{3}{256}n$ vertices of $V_{+,+}$ are not exposed on any level between Z' and $Z^*_{+,+}-1$ (call them $V_{+,+,+}$). See Figure 7.

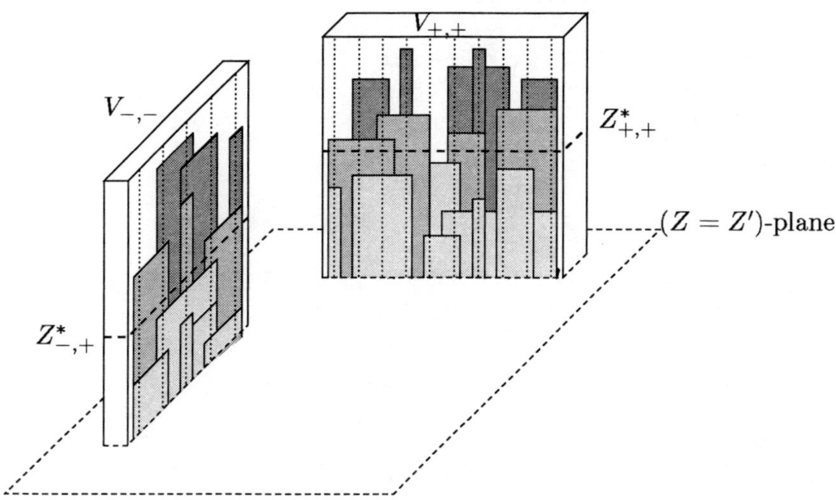

Figure 7: The darker vertices are hidden on all levels between Z' and $Z^*_{-,+}$ ($Z^*_{+,+}$), whereas the lighter vertices are inactive on all levels $z \geq Z^*_{-,+}$ ($z \geq Z^*_{+,+}$). (Only the projection of the vertices onto the bounding box is shown. Also, the picture is simplified in that only one of two possible directions of exposure is considered.)

Without loss of generality, assume that $Z_{-,+} \leq Z_{+,+}$. Therefore, there exists no level $z \geq Z'$ on which both a vertex in $V_{-,-,-}$ is active and a vertex in $V_{+,+,+}$ is exposed. Hence no edge between $V_{+,+,+}$ and $V_{-,-,-}$ can be routed on a $(Z = z)$-plane with $z \geq Z'$.

Now repeat the argument for the layers below Z', applied only to vertices in $V_{-,-,-}$ and $V_{+,+,+}$, respectively. Since $|V_{-,-,-}| \geq \frac{3}{256}n$, there is an integer $Z^*_{-,-} < Z'$ such that at least $\frac{1}{256}n$ vertices of $V_{-,-,-}$ are inactive on any level $z \leq Z^*_{-,-}$ (call them $V_{-,-,-,+}$), and at least $\frac{3}{256}n - \frac{1}{256}n - \frac{1}{256}n = \frac{1}{256}n$ vertices of $V_{-,-,-}$ (call them $V_{-,-,-,-}$) are hidden on any level between Z' and $Z^*_{-,-}+1$.

Also, there is a value $Z^*_{+,-} < Z'$ such that at least $\frac{1}{256}n$ vertices of $V_{+,+,+}$ are inactive on any level $z \leq Z^*_{+,-}$ (call them $V_{+,+,+,+}$), and at least $\frac{3}{256}n - \frac{1}{256}n - \frac{1}{256}n = \frac{1}{256}n$ vertices of $V_{+,+,+}$ (call them $V_{+,+,+,-}$) are hidden on any level between Z' and $Z^*_{+,-}+1$.

T. Biedl, *1-Bend 3-D Orthogonal Box-Drawings*, JGAA, 5(3) 1–15 (2001)   14

Without loss of generality, assume that $Z^*_{-,-} \leq Z^*_{+,-}$. Therefore, there exists no level $z \leq Z'$ on which both a vertex in $V_{-,-,-,-}$ is active and a vertex in $V_{+,+,+,+}$ is exposed. Hence none of the edges between $V_{+,+,+,+}$ and $V_{-,-,-,-}$ can be routed on a $(Z = z)$-plane with $z \leq Z'$. But as shown before, none of these edges can be routed in a $(Z = z)$-plane with $z \geq Z'$, and not in an X-plane or a Y-plane either. So if $X + Y \leq 2^{-8}n$, then the edges between $V_{+,+,+,+}$ and $V_{-,-,-,-}$ (which must exist because the graph is a Ramanujan-graph and $d \geq 16(256)^2 = 2^{20}$) cannot be routed with 0 or 1 bends, a contradiction.

Thus $X + Y > 2^{-8}n$, and if, say, $X = \max\{X, Y\}$, then $X > 2^{-9}n$. There are at least $\frac{1}{2}(\frac{7}{256})^2 \cdot dn$ edges between $V_{-,-}$ and $V_{+,+}$, since each of these sets contains at least $\frac{7}{256}n$ vertices. All their edges routes intersect the $(X = X^*)$-plane in a point with integer coordinates, therefore $YZ \geq \frac{1}{2}(\frac{7}{256})^2 dn = 49 \cdot 2^{-17}dn$. Combining this with $X > 2^{-9}n$, we obtain $XYZ > 49 \cdot 2^{-26}dn^2$, which gives the result. $\square$

4   Conclusion and Open Problems

This paper solved two open problems regarding three-dimensional orthogonal 1-bend drawings, namely, that any 1-bend drawing of K_n has volume $\Omega(n^3)$ and degree-restricted 1-bend cube-drawings are impossible for K_n, or more generally, for simple graphs with $\Omega(\Delta n)$ edges. Lower bounds for 1-bend cube-drawings were also established and hold for any graph for which any cut contains many edges, in particular for Ramanujan-graphs.

A number of open problems remain to be studied:

- Does every graph have a 1-bend drawing of volume $O(\Delta n^2)$? It is easy to construct such a drawing if crossings are allowed, by splitting the edges into $\Delta + 1$ matchings, and assigning a separate Z-plane to each matching, similarly as in [BSWW99]. Can a graph be split in $O(\Delta)$ matchings such that for a suitable vertex order all matchings are without crossing?

  If the answer is yes, does every graph have a 1-bend drawing of volume $O(mn)$?

- Does every graph have a 1-bend cube-drawing of volume $O(\Delta^3 n)$? If the answer is yes, does every graph have a 1-bend drawing of volume $O(\Delta^2 m)$?

- What is the correct lower bound for 2-bend drawings? There are drawings of size $O(n^3)$ for K_n [BSWW99] as well as $O(\Delta n^2)$ for all graphs [Bie98]. Is the lower bound $\Omega(\Delta n^2)$, as for the 1-bend case?

- What is the correct lower bound for 3-bend drawings? There are drawings of size $O(n^{2.5})$ for K_n [BSWW99], and this is optimal [BSWW99]. For Ramanujan-graphs, the lower bound is $\Omega(\Delta n^{1.5})$ [BTW01], but it is not known whether every graph has a 3-bend drawing of volume $O(\Delta n^{1.5})$. (Such drawings exist with 4 bends per edge [BTW01]; 3-bend drawings can be constructed with similar techniques if crossings are allowed.)

T. Biedl, *1-Bend 3-D Orthogonal Box-Drawings*, JGAA, 5(3) 1–15 (2001)   15

References

[AS92]   N. Alon and J. Spencer. *The Probabilistic Method.* John Wiley & Sons, 1992.

[Bie98]   T. Biedl. Three approaches to 3D-orthogonal box-drawings. In *Graph Drawing (GD'98)*, volume 1547 of *Lecture Notes in Computer Science*, pages 30–43. Springer-Verlag, 1998.

[BSWW99]   T. Biedl, T. Shermer, S. Whitesides, and S. Wismath. Bounds for orthogonal 3-D graph drawing. *J. Graph Alg. Appl,* 3(4):63–79, 1999.

[BTW01]   T. Biedl, T. Thiele, and D. Wood. Three-dimensional orthogonal graph drawing with optimal volume. In *Graph Drawing '00*, Lecture Notes in Computer Science, 2001. To appear.

[FM99]   S. Fekete and H. Meijer. Rectangle and box visibility graphs in 3D. *Internat. J. Comput. Geom. Appl.,* 9(1):1–27, 1999.

[LPS88]   A. Lubotzky, R. Phillips, and P. Sarnak. Ramanujan graphs. *Combinatorica,* 8:261–277, 1988.

[Woo00]   David R. Wood. *Three-Dimensional Orthogonal Graph Drawing.* PhD thesis, Monash University, March 2000.

Volume 5:4 (2001)

Journal of Graph Algorithms and Applications

http://www.cs.brown.edu/publications/jgaa/

vol. 5, no. 4, pp. 1–27 (2001)

Computing an Optimal Orientation of a Balanced Decomposition Tree for Linear Arrangement Problems

Reuven Bar-Yehuda

Computer Science Dept.
Technion
Haifa, Israel
http://www.cs.technion.ac.il/ reuven/
reuven@cs.technion.ac.il

Guy Even

Dept. of Electrical Engineering-Systems
Tel-Aviv University
Tel-Aviv, Israel
http://www.eng.tau.ac.il/ guy/
guy@eng.tau.ac.il

Jon Feldman

Laboratory for Computer Science
Massachusetts Institute of Technology
Cambridge, MA, USA
http://theory.lcs.mit.edu/ jonfeld/
jonfeld@theory.lcs.mit.edu

Joseph (Seffi) Naor

Computer Science Dept.
Technion
Haifa, Israel
http://www.cs.technion.ac.il/users/wwwb/cgi-bin/facultynew.cgi?Naor.Joseph
naor@cs.technion.ac.il

Abstract

Divide-and-conquer approximation algorithms for vertex ordering problems partition the vertex set of graphs, compute recursively an ordering of each part, and "glue" the orderings of the parts together. The computed ordering is specified by a decomposition tree that describes the recursive partitioning of the subproblems. At each internal node of the decomposition tree, there is a degree of freedom regarding the order in which the parts are glued together.

Approximation algorithms that use this technique ignore these degrees of freedom, and prove that the cost of every ordering that agrees with the computed decomposition tree is within the range specified by the approximation factor. We address the question of whether an optimal ordering can be efficiently computed among the exponentially many orderings induced by a binary decomposition tree.

We present a polynomial time algorithm for computing an optimal ordering induced by a binary balanced decomposition tree with respect to two problems: Minimum Linear Arrangement (MINLA) and Minimum Cutwidth (MINCW). For 1/3-balanced decomposition trees of bounded degree graphs, the time complexity of our algorithm is $O(n^{2.2})$, where n denotes the number of vertices.

Additionally, we present experimental evidence that computing an optimal orientation of a decomposition tree is useful in practice. It is shown, through an implementation for MINLA, that optimal orientations of decomposition trees can produce arrangements of roughly the same quality as those produced by the best known heuristic, at a fraction of the running time.

Communicated by T. Warnow; submitted July 1998;
revised September 2000 and June 2001.

Guy Even was supported in part by Intel Israel LTD and Intel Corp. under a grant awarded in 2000. Jon Feldman did part of this work while visiting Tel-Aviv University.

Bar-Yehuda et al., *Orientation of Decomposition Trees*, JGAA, 5(4) 1–27 (2001)3

1   Introduction

The typical setting in vertex ordering problems in graphs is to find a linear ordering of the vertices of a graph that minimizes a certain objective function [GJ79, A1.3,pp. 199-201]. These vertex ordering problems arise in diverse areas such as: VLSI design [HL99], computational biology [K93], scheduling and constraint satisfaction problems [FD], and linear algebra [R70]. Finding an optimal ordering is usually NP-hard and therefore one resorts to polynomial time approximation algorithms.

Divide-and-conquer is a common approach underlying many approximation algorithms for vertex ordering problems [BL84, LR99, Ha89, RAK91, ENRS00, RR98]. Such approximation algorithms partition the vertex set into two or more sets, compute recursively an ordering of each part, and "glue" the orderings of the parts together. The computed ordering is specified by a decomposition tree that describes the recursive partitioning of the subproblems. At each internal node of the decomposition tree, there is a degree of freedom regarding the order in which the parts are glued together. We refer to determining the order of the parts as assigning an *orientation* to an internal decomposition tree node since, in the binary case, this is equivalent to deciding which child is the left child and which child is the right child. We refer to orderings that can be obtained by assigning orientations to the internal nodes of the decomposition trees as orderings that *agree* with the decomposition tree.

Approximation algorithms that use this technique ignore these degrees of freedom, and prove that the cost of every ordering that agrees with the computed decomposition tree is within the range specified by the approximation factor. The questions that we address are whether an optimal ordering can be efficiently computed among the orderings that agree with the decomposition tree computed by the divide-and-conquer approximation algorithms, and whether computing this optimal ordering is a useful technique in practice.

Contribution.   We present a polynomial time algorithm for computing an optimal orientation of a balanced binary decomposition tree with respect to two problems: Minimum Linear Arrangement (MINLA) and Minimum Cutwidth (MINCW). Loosely speaking, in these problems, the vertex ordering determines the position of the graph vertices along a straight line with fixed distances between adjacent vertices. In MINLA, the objective is to minimize the sum of the edge lengths, and in MINCW, the objective is to minimize the maximum cut between a prefix and a suffix of the vertices.

Bar-Yehuda et al., *Orientation of Decomposition Trees*, JGAA, 5(4) 1–27 (2001)4

The corresponding decision problems for these optimization problems are NP-Complete [GJ79, problems GT42,GT44]. For binary 1/3-balanced decomposition trees of bounded degree graphs, the time complexity of our algorithm is $O(n^{2.2})$, and the space complexity is $O(n)$, where n denotes the number of vertices.

This algorithm also lends itself to a simple improvement heuristic for both MINLA and MINCW: Take some ordering π, build a balanced decomposition tree from scratch that agrees with π, and find its optimal orientation. In the context of heuristics, this search can be viewed as generalizing local searches in which only swapping of pairs of vertices is allowed [P97]. Our search space allows swapping of blocks defined by the global hierarchical decomposition of the vertices. Many local searches lack quality guarantees, whereas our algorithm finds the best ordering in an exponential search space.

The complexity of our algorithm is exponential in the depth of the decomposition tree, and therefore, we phrase our results in terms of balanced decomposition trees. The requirement that the binary decomposition tree be balanced does not incur a significant setback for the following reason. The analysis of divide-and-conquer algorithms, which construct a decomposition tree, attach a cost to the decomposition tree which serves as an upper bound on the cost of all orderings that agree with the decomposition tree. Even *et al.* [ENRS00] presented a technique for balancing binary decomposition trees. When this balancing technique is applied to MINLA the cost of the balanced decomposition tree is at most three times the cost of the unbalanced decomposition tree. In the case of the cutwidth problem, this balancing technique can be implemented so that there is an ordering that agrees with the unbalanced and the balanced decomposition trees.

Interestingly, our algorithm can be modified to find a *worst* solution that agrees with a decomposition tree. We were not able to prove a gap between the best and worst orientations of a decomposition tree, and therefore the approximation factors for these vertex ordering problems has not been improved. However, we were able to give experimental evidence that for a particular set of benchmark graphs the gap between the worst and best orientations is roughly a factor of two.

Techniques. Our algorithm can be interpreted as a dynamic programming algorithm. The "table" used by the algorithm has entries $\langle t, \alpha \rangle$, where t is a binary decomposition tree node, and α is a binary string of length $depth(t)$, representing the assignments of orientations to the ancestors of t in the decomposition tree. Note that the size of this table is exponential in the

Bar-Yehuda et al., *Orientation of Decomposition Trees*, JGAA, 5(4) 1–27 (2001)5

depth of the decomposition tree. If the decomposition tree has logarithmic depth (i.e. the tree is balanced), then the size of the table is polynomial.

The contents of a table entry $\langle t, \alpha \rangle$ is as follows. Let M denote the set of leaves of the subtree rooted at t. The vertices in M constitute a contiguous block in every ordering that agrees with the decomposition tree. Assigning orientations to the ancestors of t implies that we can determine the set L of vertices that are placed to the left of M and the set R of vertices that are placed to the right of M. The table entry $\langle t, \alpha \rangle$ holds the minimum *local cost* associated with the block M subject to the orientations α of the ancestors of t. This local cost deals only with edges incident to M and only with the cost that these edges incur within the block M. When our algorithm terminates, the local cost of the root will be the total cost of an optimal orientation, since M contains every leaf of the tree.

Our ability to apply dynamic programming relies on a locality property that enables us to compute a table entry $\langle t, \alpha \rangle$ based on four other table entries. Let t_1 and t_2 be the children of t in the decomposition tree, and let $\sigma \in \{0, 1\}$ be a possible orientation of t. We show that it is possible to easily compute $\langle t, \alpha \rangle$ from the four table entries $\langle t_i, \alpha \cdot \sigma \rangle$, where $i \in \{1, 2\}$, $\sigma \in \{0, 1\}$.

The table described above can viewed as a structure called an *orientations tree* of a decomposition tree. Each internal node $\widehat{t} = \langle t, \alpha \rangle$ of this orientations tree corresponds to a node t of the decomposition tree, and a string α of the orientations of the ancestors of t in the decomposition tree. The children of $\widehat{t}$ are the four children described above, so the value of each node of the orientations tree is locally computable from the value of its four children. Thus, we perform a depth-first search of this tree, and to reduce the space complexity, we do not store the entire tree in memory at once.

Relation to previous work. For MINLA, Hansen [Ha89] proved that decomposition trees obtained by recursive α-approximate separators yields an $O(\alpha \cdot log n)$ approximation algorithm. Since Leighton and Rao presented an $O(\log n)$-approximate separator algorithm, an $O(\log^2 n)$ approximation follows. Even *et al.* [ENRS00] gave an approximation algorithm that achieves an approximation factor of $O(\log n \log \log n)$. Rao and Richa [RR98] improved the approximation factor to $O(\log n)$. Both algorithms rely on computing a spreading metric by solving a linear program with an exponential number of constraints. For the cutwidth problem, Leighton and Rao [LR99] achieved an approximation factor of $O(\log^2 n)$ by recursive separation. The approximation algorithms of [LR99, Ha89, ENRS00] compute binary 1/3-

Bar-Yehuda et al., *Orientation of Decomposition Trees*, JGAA, 5(4) 1–27 (2001)6

balanced decomposition trees so as to achieve the approximation factors.

The algorithm of Rao and Richa [RR98] computes a non-binary non-balanced decomposition tree. Siblings in the decomposition tree computed by the algorithm of Rao and Richa are given a linear order which may be reversed (i.e. only two permutations are allowed for siblings) . This means that the set of permutations that agree with such decomposition trees are obtained by determining which "sibling orderings" are reversed and which are not. When the depth of the decomposition tree computed by the algorithm of Rao and Richa is super-logarithmic and the tree is non-binary, we cannot apply the orientation algorithm since the balancing technique of [ENRS00] will create dependencies between the orientations that are assigned to roots of disjoint subtree.

Empirical Results.   Our empirical results build on the work of Petit [P97]. Petit collected a set of benchmark graphs and experimented with several heuristics. The heuristic that achieved the best results was Simulated Annealing. We conducted four experiments as follows. First, we computed decomposition trees for the benchmark graphs using the HMETIS graph partitioning package [GK98]. Aside from the random graphs in this benchmark set, we showed a gap of roughly a factor of two between worst and best orientations of the computed decomposition trees. This suggests that finding optimal orientations is practically useful. Second, we computed several decomposition trees for each graph by applying HMETIS several times. Since HMETIS is a random algorithm, it computes a different decomposition each time it is invoked. Optimal orientations were computed for each decomposition tree. This experiment showed that our algorithm could be used in conjunction with a partitioning algorithm to compute somewhat costlier solutions than Simulated Annealing at a fraction of the running time. Third, we experimented with the heuristic improvement algorithm suggested by us. We generated a random decomposition tree based on the ordering computed in the second experiment, then found the optimal orientation of this tree. Repeating this process yielded improved the results that were comparable with the results of Petit. The running time was still less that Simulated Annealing. Finally, we used the best ordering computed as an initial solution for Petit's Simulated Annealing algorithm. As expected, this produced slightly better results than Petit's results but required more time due to the platform we used.

Bar-Yehuda et al., *Orientation of Decomposition Trees*, JGAA, 5(4) 1–27 (2001)7

Organization. In Section 2, we define the problems of MINLA and MINCW as well as decomposition trees and orientations of decomposition trees. In Section 3, we present the orientation algorithm for MINLA and MINCW. In Section 4 we propose a design of the algorithm with linear space complexity and analyze the time complexity of the algorithm. In Section 5 we describe our experimental work.

2 Preliminaries

2.1 The Problems

Consider a graph $G(V, E)$ with non-negative edge capacities $c(e)$. Let $n = |V|$ and $m = |E|$. Let $[i..j]$ denote the set $\{i, i + 1, \ldots, j\}$. A one-to-one function $\pi : V \to [1..n]$ is called an *ordering* of the vertex set V. We denote the cut between the first i nodes and the rest of the nodes by $\mathsf{cut}_\pi(i)$, formally,

$$\mathsf{cut}_\pi(i) = \{(u, v) \in E \; : \; \pi(u) \le i \text{ and } \pi(v) > i\}.$$

The capacity of $\mathsf{cut}_\pi(i)$ is denoted by $c(\mathsf{cut}_\pi(i))$. The *cutwidth* of an ordering π is defined by

$$cw(G, \pi) = \max_{i \in [1..n-1]} c(\mathsf{cut}_\pi(i)).$$

The goal in the *Minimum Cutwidth Problem* (MINCW) is to find an ordering π with minimum cutwidth.

The goal in the *Minimum Linear Arrangement Problem* (MINLA) is to find an ordering that minimizes the weighted sum of the edge lengths. Formally, the the weighted sum of the edge lengths with respect an ordering π is defined by:

$$la(G, \pi) = \sum_{(u,v) \in E} c(u, v) \cdot |\pi(u) - \pi(v)|.$$

The weighted sum of edge lengths can be equivalently defined as:

$$la(G, \pi) = \sum_{1 \le i < n} c(\mathsf{cut}_\pi(i))$$

2.2 Decomposition Trees

A *decomposition tree* of a graph $G(V, E)$ is a rooted binary tree with a mapping of the tree nodes to subsets of vertices as follows. The root is mapped to V, the subsets mapped to every two siblings constitute a partitioning of

Bar-Yehuda et al., *Orientation of Decomposition Trees*, JGAA, 5(4) 1–27 (2001)8

the subset mapped to their parent, and leaves are mapped to subsets containing a single vertex. We denote the subset of vertices mapped to a tree node t by $V(t)$. For every tree node t, let $T(t)$ denote the subtree of T, the root of which is t. The *inner cut* of an internal tree node t is the set of edges in the cut $(V(t_1), V(t_2))$, where t_1 and t_2 denote the children of t. We denote the inner cut of t by $in\_cut(t)$.

Every DFS traversal of a decomposition tree induces an ordering of V according to the order in which the leaves are visited. Since in each internal node there are two possible orders in which the children can be visited, it follows that 2^{n-1} orderings are induced by DFS traversals. Each such ordering is specified by determining for every internal node which child is visited first. In "graphic" terms, if the first child is always drawn as the left child, then the induced ordering is the order of the leaves from left to right.

2.3 Orientations and Optimal Orientations

Consider an internal tree node t of a decomposition tree. An *orientation* of t is a bit that determines which of the two children of t is considered as its left child. Our convention is that when a DFS is performed, the left child is always visited first. Therefore, a decomposition tree, all the internal nodes of which are assigned orientations, induces a unique ordering. We refer to an assignment of orientations to all the internal nodes as an *orientation* of the decomposition tree.

Consider a decomposition tree T of a graph $G(V, E)$. An orientation of T is optimal with respect to an ordering problem if the cost associated with the ordering induced by the orientation is minimum among all the orderings induced by T.

3 Computing An Optimal Orientation

In this section we present a dynamic programming algorithm for computing an optimal orientation of a decomposition tree with respect to MINLA and MINCW. We first present an algorithm for MINLA, and then describe the modifications needed for MINCW.

Let T denote a decomposition of $G(V, E)$. We describe a recursive algorithm $orient(t, \alpha)$ for computing an optimal orientation of T for MINLA. A say that a decomposition tree is *oriented* if all its internal nodes are assigned orientations. The algorithm returns an oriented decomposition tree isomorphic to T. The parameters of the algorithm are a tree node t and an assignment of orientations to the ancestors of t. The orientations of the

Bar-Yehuda et al., *Orientation of Decomposition Trees*, JGAA, 5(4) 1–27 (2001)9

ancestors of t are specified by a binary string α whose length equals $depth(t)$. The ith bit in α signifies the orientation of the ith node along the path from the root of T to t.

The vertices of $V(t)$ constitute a contiguous block in every ordering that is induced by the decomposition tree T. The sets of vertices that appear to the left and right of $V(t)$ are determined by the orientations of the ancestors of t (which are specified by α). Given the orientations of the ancestors of t, let L and R denote the set of vertices that appear to the left and right of $V(t)$, respectively. We call the partition $(L, V(t), R)$ an *ordered partition* of V.

Consider an ordered partition $(L, V(t), R)$ of the vertex set V and an ordering π of the vertices of $V(t)$. Algorithm $orient(t, \alpha)$ is based on the *local cost* of edge (u, v) with respect to $(L, V(t), R)$ and π. The local cost applies only to edges incident to $V(t)$ and it measures the length of the projection of the edge on $V(t)$. Formally, the local cost is defined by

$$
local\_cost_{L,V(t),R,\pi}(u, v) = \begin{cases} c(u, v) \cdot |\pi(u) - \pi(v)| & \text{if } u, v \in V(t) \\ c(u, v) \cdot \pi(u) & \text{if } u \in V(t) \text{ and } v \in L \\ c(u, v) \cdot (|V(t)| - \pi(u)) & \text{if } u \in V(t) \text{ and } v \in R \\ 0 & \text{otherwise.} \end{cases}
$$

Note that the contribution to the cut corresponding to $L \cup V(t)$ is not included.

Algorithm $orient(t, \alpha)$ proceeds as follows:

1. If t is a leaf, then return $T(t)$ (a leaf is not assigned an orientation).

2. Otherwise (t is not a leaf), let t_1 and t_2 denote children of t. Compute optimal oriented trees for $T(t_1)$ and $T(t_2)$ for both orientations of t. Specifically,

   (a) $T^0(t_1) = orient(t_1, \alpha \cdot 0)$ and $T^0(t_2) = orient(t_2, \alpha \cdot 0)$.

   (b) $T^1(t_1) = orient(t_1, \alpha \cdot 1)$ and $T^1(t_2) = orient(t_2, \alpha \cdot 1)$.

3. Let π_0 denote the ordering of $V(t)$ obtained by concatenating the ordering induced by $T^0(t_1)$ and the ordering induced by $T^0(t_2)$. Let $cost_0 = \sum_{e \in E} local\_cost_{L,V(t),R,\pi_0}(e)$.

4. Let π_1 denote the ordering of $V(t)$ obtained by concatenating the ordering induced by $T^1(t_2)$ and the ordering induced by $T^1(t_1)$. Let $cost_1 = \sum_{e \in E} local\_cost_{L,V(t),R,\pi_1}(e)$. (Note that here the vertices of $V(t_2)$ are placed first.)

Bar-Yehuda et al., *Orientation of Decomposition Trees*, JGAA, 5(4) 1–27 (2001)10

5. If $cost_0 < cost_1$, the orientation of t is 0. Return the oriented tree T' the left child of which is the root of $T^0(t_1)$ and the right child of which is the root of $T^0(t_2)$.

6. Otherwise ($cost_0 \geq cost_1$), the orientation of t is 1. Return the oriented tree T' the left child of which is the root of $T^1(t_2)$ and the right child of which is the root of $T^1(t_1)$.

The correctness of the *orient* algorithm is summarized in the following claim which can be proved by induction.

Claim 1: Suppose that the orientations of the ancestors of t are fixed as specified by the string α. Then, Algorithm *orient*(t, α) computes an optimal orientation of $T(t)$. When t is the root of the decomposition tree and α is an empty string, Algorithm *orient*(t, α) computes an optimal orientation of T.

An optimal orientation for MINCW can be computed by modifying the local cost function. Consider an ordered partition $(L, V(t), R)$ and an ordering π of $V(t)$. Let $V_i(t)$ denote that set $\{v \in V(t) : \pi(v) \leq i\}$. Let $E(t)$ denote the set of edges with at least one endpoint in $V(t)$. Let $i \in [0..|V(t)|]$. The ith local cut with respect to $(L, V(t), R)$ and an ordering π of $V(t)$ is the set of edges defined by,

$$local\_cut_{(L,V(t),R),\pi}(i) = \{(u, v) \in E(t) : u \in L \cup V_i(t) \text{ and } v \in (V(t) - V_i(t)) \cup R\}.$$

The local cutwidth is defined by,

$$local\_cw((L, V(t), R), \pi) = \max_{i \in [0..|V(t)|]} \sum_{e \in local\_cut_{(L,V(t),R),\pi}(i)} c(u, v).$$

The algorithm for computing an optimal orientation of a given decomposition tree with respect to MINCW is obtained by computing $cost_\sigma = local\_cw((L, V(t), R), \pi_\sigma)$ in steps 3 and 4 for $\sigma = 0, 1$.

4 Designing The Algorithm

In this section we propose a design of the algorithm that has linear space complexity. The time complexity is $O(n^{2.2})$ if the graph has bounded degree and the decomposition tree is 1/3-balanced.

Bar-Yehuda et al., *Orientation of Decomposition Trees*, JGAA, 5(4) 1–27 (2001)11

4.1 Space Complexity: The Orientation Tree

We define a tree, called an *orientation tree*, that corresponds to the recursion tree of Algorithm $orient(t, \alpha)$. Under the interpretation of this algorithm as a dynamic program, this orientation tree represents the table. The orientation tree $\widehat{T}$ corresponding to a decomposition tree T is a "quadary" tree. Every orientation tree node $\widehat{t} = \langle t, \alpha \rangle$ corresponds to a decomposition tree node t and an assignment α of orientations to the ancestors of t. Therefore, every decomposition tree node t has $2^{depth(t)}$ "images" in the orientations tree. Let t_1 and t_2 be the children of t in the decomposition tree, and let $\sigma \in \{0, 1\}$ be a possible orientation of t. The four children of $\langle t, \alpha \rangle$ are $\langle t_i, \alpha \cdot \sigma \rangle$, where $i \in \{1, 2\}$, $\sigma \in \{0, 1\}$. Figure 1 depicts a decomposition tree and the corresponding orientation tree.

The time complexity of the algorithm presented in Section 3 is clearly at least proportional to the size of the orientations tree. The number of nodes in the orientations tree is proportional to $\sum_{v \in V} 2^{depth(v)}$, where the $depth(v)$ is the depth of v in the decomposition tree. This implies the running time is at least quadratic if the tree is perfectly balanced. If we store the entire orientations tree in memory at once, our space requirement is also at least quadratic. However, if we are a bit smarter about how we use space, and never store the entire orientations tree in memory at once, we can reduce the space requirement to linear.

The recursion tree of Algorithm $orient(root(T), \phi)$ is isomorphic to the orientation tree $\widehat{T}$. In fact, Algorithm $orient(root(T), \alpha)$ assigns local costs to orientation tree nodes in DFS order. We suggest the following linear space implementation. Consider an orientation tree node $\widehat{t} = \langle t, \alpha \rangle$. Assume that when $orient(t, \alpha)$ is called, it is handed a "workspace" tree isomorphic to $T(t)$ which it uses for workspace as well as for returning the optimal orientation. Algorithm $orient(t, \alpha)$ allocates an "extra" copy of $T(t)$, and is called recursively for each of it four children. Each call for a child is given a separate "workspace" subtree within the two isomorphic copies of $T(t)$ (i.e. within the workspace and extra trees). Upon completion of these 4 calls, the two copies of $T(t)$ are oriented; one corresponding to a zero orientation of t and one corresponding to an orientation value of 1 for t. The best of these trees is copied into the "workspace" tree, if needed, and the "extra" tree is freed. Assuming that one copy of the decomposition tree is used throughout the algorithm, the additional space complexity of this implementation satisfies the following recurrence:

$$space(\widehat{t}) = sizeof(T(t)) + \max_{t' \text{ child of } \widehat{t}} space(t').$$

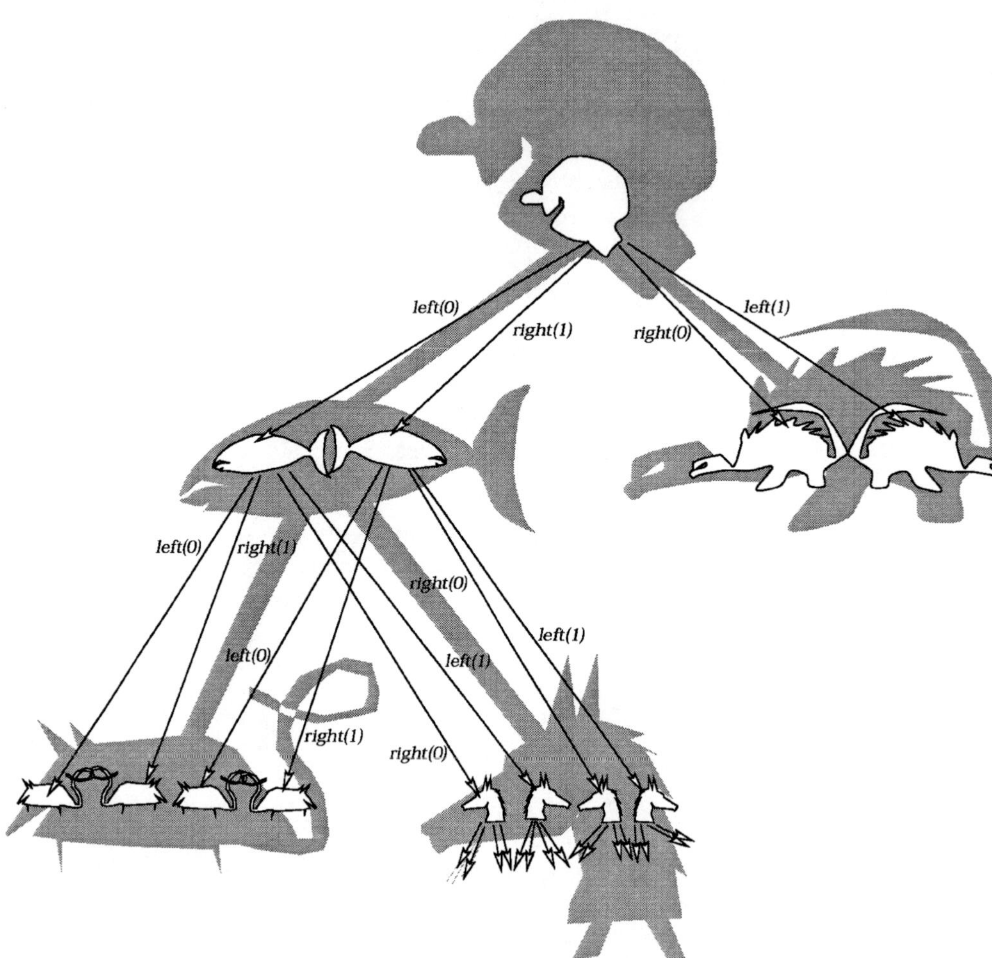

Figure 1: A decomposition tree and the corresponding orientation tree. The four node decomposition tree is depicted as a gray shadow. The corresponding orientation tree is depicted in the foreground. The direction of each non-root node in the orientation tree corresponds to the orientation of its parent (i.e. a node depicted as a fish swimming to the left signifies that its parent's orientation is left). The label of an edge entering a node in the orientation tree signifies whether the node is a left child or a right child and the orientation of its parent.

Bar-Yehuda et al., *Orientation of Decomposition Trees*, JGAA, 5(4) 1–27 (2001)13

Since $sizeof(T(t)) \leq 2^{depth(T(t))}$, it follows that

$$space(\widehat{t}) \leq 2 \cdot 2^{depth(T(t))}.$$

The space complexity of the proposed implementation of $orient(T, \alpha)$ is summarized in the following claim.

Claim 2: The space complexity of the proposed implementation is $O(2^{depth(T)})$. If the decomposition tree is balanced, then $space(root(T)) = O(n)$.

4.2   Time Complexity

Viewing Algorithm $orient(t, \alpha)$ as a DFS traversal of the orientation tree $\widehat{T}(t)$ corresponding to $T(t)$ also helps in designing the algorithm so that each visit of an internal orientation tree node requires only constant time. Suppose that each child of t is assigned a local cost (i.e. $cost(left(\sigma))$, $cost(right(\sigma))$, for $\sigma = 0, 1$). Let t_1 and t_2 denote the children of t. Let $(L, V(t), R)$ denote the ordered partition of V induced by the orientations of the ancestors of t specified by α. The following equations define $cost_0$ and $cost_1$:

$$
\begin{aligned}
cost_0 \;=\; & cost(left(0)) + cost(right(0)) \\
& + |V(t_2)| \cdot c(V(t_1), R) + |V(t_1)| \cdot c(L, V(t_2)) \\
cost_1 \;=\; & cost(left(1)) + cost(right(1)) \\
& + |V(t_1)| \cdot c(V(t_2), R) + |V(t_2)| \cdot c(L, V(t_1))
\end{aligned}
\tag{1}
$$

We now describe how Equation 1 can be computed in constant time. Let $in\_cut(t)$ denote the cut $(V(t_1), V(t_2))$. The capacity of $in\_cut(t)$ can be pre-computed by scanning the list of edges. For every edge (u, v), update $in\_cut(lca(u, v))$ by adding $c(u, v)$ to it.

We now define *outer cuts*. Consider the ordered partition $(L, V(t), R)$ corresponding to an orientation tree node $\widehat{t}$. The left outer cut and the right outer cut of $\widehat{t}$ are the cuts $(L, V(t))$ and $(V(t), R)$, respectively. We denote the left outer cut by $left\_cut(\widehat{t})$ and the right out cut by $right\_cut(\widehat{t})$.

We describe how outer cuts are computed for leaves and for interior nodes. The capacity of the outer cuts of a leaf $\widehat{t}$ are computed by considering the edges incident to t (since t is a leaf we identify it with a vertex in V). For every edge (t, u), the orientation of the orientation tree node along the path from the root to $\widehat{t}$ that corresponds to $lca(t, u)$ determines whether the edge belongs to the left outer cut or to the right outer cut. Since the least common ancestors in the decomposition tree of the endpoints of every edge

Bar-Yehuda et al., *Orientation of Decomposition Trees*, JGAA, 5(4) 1–27 (2001)14

are precomputed when the inner cuts are computed, we can compute the outer cuts of a leaf $\hat{t}$ in $O(\deg(t))$ time.

The outer cuts of a non-leaf $\hat{t}$ can be computed from the outer cuts of its children as follows:

$$
\begin{aligned}
c(\textit{left\_cut}(\hat{t})) &= c(\textit{left\_cut}(\textit{left}(0))) + c(\textit{left\_cut}(\textit{right}(0))) - c(\textit{in\_cut}(t)) \\
c(\textit{right\_cut}(\hat{t})) &= c(\textit{right\_cut}(\textit{left}(0))) + c(\textit{right\_cut}(\textit{right}(0))) - c(\textit{in\_cut}(t))
\end{aligned}
$$

Hence, the capacities of the outer cuts can be computed while the orientation tree is traversed. The following reformulation of Equation 1 shows that $cost_0$ and $cost_1$ can be computed in constant time.

$$
\begin{aligned}
cost_0 &= cost(\textit{left}(0)) + cost(\textit{right}(0)) \qquad\qquad (2)\\
&\quad + |V(t_2)| \cdot c(\textit{right\_cut}(\hat{t_1}))) + |V(t_1)| \cdot c(\textit{left\_cut}(\hat{t_2}))) \\
cost_1 &= cost(\textit{left}(1)) + cost(\textit{right}(1)) \\
&\quad + |V(t_1)| \cdot c(\textit{right\_cut}(\hat{t_2})) + |V(t_2)| \cdot c(\textit{left\_cut}(\hat{t_1}))
\end{aligned}
$$

Minimum Cutwidth. An adaptation of Equation 1 for MINCW is given below:

$$
\begin{aligned}
cost_0 &= \max\ \{\textit{local\_cw}(L, V(t_1), V(T_2) \cup R) + c(L, V(t_2)), \qquad (3)\\
&\qquad\quad \textit{local\_cw}(L \cup V(t_1), V(T_2), R) + c(V(t_1), R)\} \\
cost_1 &= \max\ \{\textit{local\_cw}(L, V(t_2), V(T_1) \cup R) + c(L, V(t_1)), \\
&\qquad\quad \textit{local\_cw}(L \cup V(t_2), V(T_1), R) + c(V(t_2), R)\}.
\end{aligned}
$$

These costs can be computed in constant time using the same technique described for MINLA.

Now we analyze the time complexity of the proposed implementation of Algorithm $orient(t, \alpha)$. We split the time spent by the algorithm into two parts: **(1)** The pre-computation of the inner cuts and the least common ancestors of the edges, **(2)** the time spent traversing the orientation tree $\hat{T}$ (interior nodes as well as leaves).

Precomputing the inner cuts and least common ancestors of the edges requires $O(m \cdot depth(T))$ time, where $depth(T)$ is the maximum depth of the decomposition tree T. Assume that the least common ancestors are stored and need not be recomputed.

The time spent traversing the orientation tree is analyzed as follows. We consider two cases: Leaves - the amount of time spent in a leaf $\hat{t}$ is linear in

Bar-Yehuda et al., *Orientation of Decomposition Trees*, JGAA, 5(4) 1–27 (2001)15

the degree of t. Interior Nodes - the amount of time spent in an interior node $\widehat{t}$ is constant. This implies that the complexity of the algorithm is linear in

$$|\widehat{T} - leaves(\widehat{T})| + \sum_{\widehat{t} \in leaves(\widehat{T})} \deg(t).$$

Every node $t \in T$ has $2^{depth(t)}$ "images" in $\widehat{T}$. Therefore, the complexity in terms of the decomposition tree T equals

$$\sum_{t \in T - leaves(T)} 2^{depth(t)} + \sum_{t \in leaves(T)} 2^{depth(t)} \cdot \deg(t).$$

If the degrees of the vertices are bounded by a constant, then the complexity is linear in

$$\sum_{t \in T} 2^{depth(t)}.$$

This quantity is quadratic if T is perfectly balanced, i.e. the vertex subsets are bisected in every internal tree node. We quantify the balance of a decomposition tree as follows:

Definition 1: A binary tree T is ρ-*balanced* if for every internal node $t \in T$ and for every child t' of t

$$\rho \cdot |V(t)| \leq |V(t')| \leq (1 - \rho) \cdot |V(t)|$$

The following claim summarizes the time complexity of the algorithm.

Claim 3: If the decomposition tree T is ρ balanced, then

$$\sum_{t \in T} 2^{depth(t)} \leq n^{\beta},$$

where β is the solution to the equation

$$\frac{1}{2} = \rho^{\beta} + (1 - \rho)^{\beta} \tag{4}$$

Observe that if $\rho \in (0, 1/2]$, then $\beta \geq 2$. Moreover, β increases as ρ increases. **Proof:** The quantity which we want to bound is the size of the orientation tree. This quantity satisfies the following recurrence:

$$f(T(t)) \triangleq \begin{cases} 1 & \text{if } t \text{ is a leaf} \\ 2f(T(t_1)) + 2f(T(t_2)) & \text{otherwise.} \end{cases}$$

Bar-Yehuda et al., *Orientation of Decomposition Trees*, JGAA, 5(4) 1–27 (2001)16

Define the function $f_\rho^*(n)$ as follows

$$f_\rho^*(n) = \max\{f(T) : \text{T is } \alpha\text{-balanced and has } n \text{ leaves}\}.$$

The function $f_\rho^*(n)$ satisfies the following recurrence:

$$f_\rho^*(n) \triangleq \begin{cases} 1 & \text{if } n = 1 \\ \max\left\{2f_\rho^*(n') + 2f_\rho^*(n - n') : n' \in [\lceil\rho n\rceil..(n - \lceil\rho n\rceil)]\right\} & \text{otherwise} \end{cases}$$

We prove that $f_\rho^*(n) \le n^\beta$ by induction on n. The induction hypothesis, for $n = 1$, is trivial. The induction step is proven as follows:

$$\begin{aligned} f_\rho^*(n) &= \max\left\{2f_\rho^*(n') + 2f_\rho^*(n - n') : n' \in [\lceil\rho n\rceil..(n - \lceil\rho n\rceil)]\right\} \\ &\le \max\left\{2(n')^\beta + 2(n - n')^\beta : n' \in [\lceil\rho n\rceil..(n - \lceil\rho n\rceil)]\right\} \\ &\le \max\left\{2(x)^\beta + 2(n - x)^\beta : x \in [\rho n, n - \rho n]\right\} \\ &= 2(\rho n)^\beta + 2(n - \rho n)^\beta \\ &= 2n^\beta \cdot (\rho^\beta + (1 - \rho)^\beta) \\ &= n^\beta \end{aligned}$$

The first line is simply the recurrence that $f_\rho^*(n)$ satisfies; the second line follows from the induction hypothesis; in the third line we relax the range over which the maximum is taken; the fourth line is justified by the convexity of $x^\beta + (n - x)^\beta$ over the range $x \in [\rho n, n - \rho n]$ when $\beta \ge 2$; in the fifth line we rearrange the terms; and the last line follows from the definition of β. $\square$

We conclude by bounding the time complexity of the orientation algorithm for 1/3-balanced decomposition trees.

Corollary 4: If T is a 1/3-balanced decomposition tree of a bounded degree, then the orientation tree $\widehat{T}$ of T has at most $n^{2.2}$ leaves.

Proof: The solution of Equation (4) with $\rho = 1/3$ is $\beta < 2.2$. $\square$

For graphs with unbounded degree, this algorithm runs in time $O(m \cdot 2^{depth(T)})$ which is $O(m \cdot n^{1/\log_2(1/1-\rho)})$. Alternatively, a slightly different algorithm gives a running time of n^β in general (not just for constant degree), but the space requirement of this algorithm is $O(n \log n)$. It is based on computing the outer cuts of a leaf of the orientations tree on the way down the recursion. We omit the details.

Bar-Yehuda et al., *Orientation of Decomposition Trees*, JGAA, 5(4) 1–27 (2001)17

5   Experiments And Heuristics

Since we are not improving the theoretical approximation ratios for the MINLA or MINCW problems, it is natural to ask whether finding an optimal orientation of a decomposition tree is a useful thing to do in practice. The most comprehensive experimentation on either problem was performed for the MINLA problem by Petit [P97]. Petit collected a set of benchmark graphs and ran several different algorithms on them, comparing their quality. He found that Simulated Annealing yielded the best results. The benchmark consists of 5 random graphs, 3 "regular" graphs (a hypercube, a mesh, and a binary tree), 3 graphs from finite element discretizations, 5 graphs from VLSI designs, and 5 graphs from graph drawing competitions.

5.1   Gaps Between Orientations

The first experiment was performed to check if there is a significant gap between the costs of different orientations of decomposition trees. Conveniently, our algorithm can be easily modified to find the *worst* possible orientation; every time we compare two local costs to decide on an orientation, we simply take the orientation that achieves the *worst* local cost. In this experiment, we constructed decomposition trees for all of Petit's benchmark graphs using the balanced graph partitioning program HMETIS [GK98]. HMETIS is a fast heuristic that searches for balanced cuts that are as small as possible. For each decomposition tree, we computed four orientations:

1. Naive orientation - all the decomposition tree nodes are assigned a zero orientation. This is the ordering you would expect from a recursive bisection algorithm that ignored orientations.

2. Random orientation - the orientations of the decomposition tree nodes are chosen randomly to be either zero or one, with equal probability.

3. Best orientation - computed by our orientation algorithm.

4. Worst orientation - computed by a simple modification of our algorithm.

The results are summarized in Table 1. On all of the "real-life" graphs, the best orientation had a cost of about half of the worst orientation. Furthermore, the costs on all the graphs were almost as low as those obtained by Petit (Table 5), who performed very thorough and computationally-intensive experiments.

Bar-Yehuda et al., *Orientation of Decomposition Trees*, JGAA, 5(4) 1–27 (2001)18

Notice that the costs of the "naive" orientations do not compete well with those of Petit. This shows that in this case, using the extra degrees of freedom afforded by the decomposition tree was essential to achieving a good quality solution. These results also motivated further experimentation to see if we could achieve comparable or better results using more iterations, and the improvement heuristic alluded to earlier.

5.2   Experiment Design

We ran the following experiments on the benchmark graphs:

1. Decompose & Orient. Repeat the following two steps k times, and keep the best ordering found during these k iterations.

   (a) Compute a decomposition tree T of the graph. The decomposition is computed by calling the HMETIS graph partitioning program recursively [GK98]. HMETIS accepts a balance parameter ρ. HMETIS is a random algorithm, so different executions may output different decomposition trees.

   (b) Compute an optimal orientation of T. Let π denote the ordering induced by the orientation of T.

2. Improvement Heuristic. Repeat the following steps k' times (or until no improvement is found during 10 consecutive iterations):

   (a) Let π_0 denote the best ordering π found during the Decompose & Orient step.

   (b) Set $i = 0$.

   (c) Compute a random ρ-balanced decomposition tree T' based on π_i. The decomposition tree T' is obtained by recursively partitioning the blocks in the ordering π_i into two contiguous sub-blocks. The partitioning is a random partition that is ρ-balanced.

   (d) Compute an optimal orientation of T'. Let π_{i+1} denote the ordering induced by the orientation of T'. Note that the cost of π_{i+1} is not greater than the cost of the ordering π_i. Increment i and return to Step 2c, unless $i = k' - 1$ or no improvement in the cost has occurred for the last 10 iterations.

3. Simulated Annealing. We used the best ordering found by the Heuristic Improvement stage as an input to the Simulated Annealing program of Petit [P97].

Bar-Yehuda et al., *Orientation of Decomposition Trees*, JGAA, 5(4) 1–27 (2001)19

Preliminary experiments were run in order to choose the balance parameter ρ. In graphs of less than 1000 nodes, we simply chose the balance parameter that gave the best ordering. In bigger graphs, we also restricted the balance parameter so that the orientation tree would not be too big. The parameters used for the Simulated Annealing program were also chosen based on a few preliminary experiments.

5.3 Experimental Environment

The programs have been written in C and compiled with a gcc compiler. The programs were executed on a dual processor Intel P-III 600MHz computer running under Red Hat Linux. The programs were only compiled for one processor, and therefore only ran on one of the two processors.

5.4 Experimental Results

The following results were obtained:

1. Decompose & Orient. The results of the Decompose & Orient stage are summarized in Table 2. The columns of Table 2 have the following meaning: "UB Factor" - the balance parameter used when HMETIS was invoked. The relation between the balance parameter ρ and the UB Factor is $(1/2 - \rho) \cdot 100$. "Avg. OT Size" - the average size of the orientation tree corresponding to the decomposition tree computed by HMETIS. This quantity is a good measure of the running time without overheads such as reading and writing of data. "Avg. L.A. Cost" - the average cost of the ordering induced by an optimal orientation of the decomposition tree computed by HMETIS. "Avg. HMETIS Time (sec)" - the average time in seconds required for computing a decomposition tree, "Avg. Orienting Time (sec)" - the average time in seconds required to compute an optimal orientation, and "Min L.A. Cost" - the minimum cost of an ordering, among the computed orderings.

2. Heuristic Improvement. The results of the Heuristic Improvement stage are summarized in Table 3. The columns of Table 3 have the following meaning: "Initial solution" - the cost of the ordering computed by the Decompose & Orient stage, "10 Iterations" - the cost of the ordering obtained after 10 iterations of Heuristic Improvements, "Total Iterations" - The number iterations that were run until there was no improvement for 10 consecutive iterations, "Avg. Time per Iteration" - the average time for each iteration (random decomposition

Bar-Yehuda et al., *Orientation of Decomposition Trees*, JGAA, 5(4) 1–27 (2001)20

and orientation), and "Final Cost" - the cost of the ordering computed by the Heuristic Improvement stage.

The same balance parameter was used in the Decompose & Orient stage and in the Heuristic Improvement stage. Since the partition is chosen randomly in the Heuristic Improvement stage such that the balance is never worse than ρ, the decomposition trees obtained were shallower. This fact is reflected in the shorter running times required for computing an optimal orientation.

3. Simulated Annealing. The results of the Simulated Annealing program are summarized in Table 4.

5.5 Conclusions

Table 5 summarizes our results and compares them with the results of Petit [P97]. The running times in the table refer to a single iteration (the running time of the Decompose & Orient stage refers only to orientation not including decomposition using HMETIS). Petit ran 10-100 iterations on an SGI Origin 2000 computer with 32 MIPS R10000 processors.

Our main experimental conclusion is that running our Decompose & Orient algorithm usually yields results within 5-10% of Simulated Annealing, and is significantly faster. The results were improved to comparable with Simulated Annealing by using our Improvement Heuristic. Slightly better results were obtained by using our computed ordering as an initial solution for Simulated Annealing.

5.6 Availability of Programs and Results

Programs and output files can be downloaded from
http://www.eng.tau.ac.il/~guy/Projects/Minla/index.html.

Graph	Naive	Random	Worst	Best	Best/Worst
randomA1	1012523	1010880	1042034	980631	0.94
randomA2	6889176	6908714	6972740	6842173	0.98
randomA3	14780970	14767044	14826303	14709068	0.99
randomA4	1913623	1907549	1963952	1866510	0.95
randomG4	210817	220669	282280	157716	0.56
hc10	523776	523776	523776	523776	1.
mesh33x33	55727	55084	62720	35777	0.57
bintree10	4850	5037	9021	3742	0.41
3elt	720174	729782	887231	419128	0.47
airfoil1	521014	524201	677191	337237	0.5
whitaker3	2173667	2423313	3233350	1524974	0.47
c1y	91135	86362	141034	64365	0.46
c2y	122838	108936	172398	81952	0.48
c3y	152993	160114	262374	131612	0.5
c4y	146178	166764	246104	120799	0.49
c5y	121191	137291	203594	103888	0.51
gd95c	733	702	953	555	0.58
gd96a	132342	136973	177945	121140	0.68
gd96b	2538	2475	2899	1533	0.53
gd96c	695	719	868	543	0.63
gd96d	3477	3663	4682	2614	0.56

Table 1: A comparison of arrangement costs for different orientations of a single decomposition tree for each graph.

Graph Properties			Decompose & Orient Iterations					
Graph	Nodes	Edges	UB Factor	Avg. OT Size	Avg. L.A. Cost	Average HMETIS Time (sec)	Average Orienting Time (sec)	Min L.A Cost
randomA1	1000	4974	15	1689096	976788	15.08	5.32	956837
randomA2	1000	24738	10	1430837	6829113	31.46	17.89	6791813
randomA3	1000	49820	15	2682997	14677190	51.71	70.89	14621489
randomA4	1000	8177	15	1868623	1870265	17.41	8.22	1852192
randomG4	1000	8173	15	1952092	157157	18.07	8.28	156447
hc10	1024	5120	10	699051	523776	13.72	1.98	523776
mesh33x33	1089	2112	10	836267	35880	10.45	1.15	35728
bintree10	1023	1022	10	913599	3741	7.52	0.87	3740
3elt	4720	13722	15	34285107	423225	54.93	60.93	414051
airfoil1	4253	12289	16	30715196	328936	48.45	53.30	325635
whitaker3	9800	28989	10	110394027	1462858	114.50	192.50	1441887
c1y	828	1749	10	709463	68458	7.51	1.51	63803
c2y	980	2102	15	1456839	82615	9.40	5.63	81731
c3y	1327	2844	10	1902661	133027	12.46	4.24	130213
c4y	1366	2915	10	1954874	120899	13.24	3.97	119016
c5y	1202	2557	10	1528117	100990	12.39	3.59	99772
gd95c	62	144	20	7322	520	0.51	0	512
gd96a	1096	1676	10	1151024	117431	9.36	2.08	112551
gd96b	111	193	20	38860	1502	0.77	0.14	1457
gd96c	65	125	20	4755	544	0.49	0	533
gd96d	180	228	15	48344	2661	1.16	0.11	2521

Table 2: Results of the Decompose & Orient stage. 100 iterations were executed for all the graphs, except for 3elt and airfoil1 - 60 iterations, and whitaker3 - 25 iterations.

Graph Properties	Heuristic Improvement Iterations						
Graph	Initial solution	10 Iterations	20 Iterations	30 Iterations	Total Iterations	Avg. Time / Iteration (sec)	Final Cost
randomA1	956837	947483	941002	935014	1048	2.63	897910
randomA2	6791813	6759449	6736868	6717108	500	9.84	6604738
randomA3	14621489	14576923	14537348	14506843	200	23.73	14365385
randomA4	1852192	1833589	1821592	1812344	557	3.73	1761666
randomG4	156447	150477	149456	149243	76	3.53	149185
hc10	523776	523776	-	-	10	2.00	523776
mesh33x33	35728	35492	35325	35212	152	1.28	34845
bintree10	3740	3724	3720	3718	47	0.75	3714
3elt	414051	400469	395594	392474	724	46.49	372107
airfoil1	325635	313975	309998	307383	630	36.94	292761
whitaker3	1441887	1408337	1393045	1381345	35	149.43	1376511
c1y	63803	63283	63085	62973	80	0.73	62903
c2y	81731	81094	80877	80679	171	1.19	80109
c3y	130213	129255	128911	128650	267	1.91	127729
c4y	119016	118109	117690	117438	160	2.14	116621
c5y	99772	99051	98813	98661	182	1.72	98004
gd95c	512	506	-	-	14	0.00	506
gd96a	112551	110744	109605	108887	977	1.07	102294
gd96b	1457	1427	1424	1417	38	0.02	1417
gd96c	533	520	520	-	20	0.00	520
gd96d	2521	2463	2454	2449	62	0.03	2436

Table 3: Results of the Heuristic Improvement stage. Balance parameters equal the UB factors in Table 2.

Graph	SA results		
	Initial solution	Total Running Time (sec)	Final Cost
randomA1	897910	2328.14	884261
randomA2	6604738	191645	6576912
randomA3	14365385	58771.8	14289214
randomA4	1761666	10895.6	1747143
randomG4	149185	8375.88	146996
hc10	523776	2545.16	523776
mesh33x33	34845	278.037	33531
bintree10	3714	59.782	3762
3elt	372107	5756.52	363204
airfoil1	292761	4552.14	289217
whitaker3	1376511	29936.8	1200374
c1y	62903	204.849	62333
c2y	80109	298.844	79571
c3y	127729	553.064	127065
c4y	116621	579.751	115222
c5y	98004	445.604	96956
gd95c	506	1.909	506
gd96a	102294	268.454	99944
gd96b	1417	2.923	1422
gd96c	520	1.175	519
gd96d	2436	4.163	2409

Table 4: Results of the Simulated Annealing program. The parameters were t0=4, tl=0.1, alpha=0.99 except for whitacker3, airfoil1 and 3elt for which alpha=0.975 and randomA3 for which tl=1 and alpha=0.8

Graph	Petit's Results [P97]		Decompose & Orient stage			Heuristic Improvement stage			Simulated Annealing stage		
	cost	time (sec)	cost	% diff.	time (sec)	cost	% diff.	time (sec)	cost	% diff.	time (sec)
randomA1	900992	317	956837	6.20%	5	897910	-0.34%	3	884261	-1.86%	2328
randomA2	6584658	481	6791813	3.15%	18	6604738	0.30%	10	6576912	-0.12%	191645
randomA3	14310861	682	14621489	2.17%	71	14365385	0.38%	24	14289214	-1.51%	58771
randomA4	1753265	346	1852192	5.64%	8	1761666	0.48%	4	1747143	-0.35%	10896
randomG4	150490	346	156447	3.96%	8	149185	-0.87%	4	146996	-2.32%	8376
hc10	548352	335	523776	-4.48%	2	523776	-4.48%	2	523776	-4.48%	2545
mesh33x33	34515	336	35728	3.51%	1	34845	0.96%	1	33531	-2.85%	278
bintree10	4069	291	3740	-8.09%	1	3714	-8.72%	1	3762	-7.54%	60
3elt	375387	6660	414051	10.30%	61	372107	-0.87%	46	363204	-3.25%	5757
airfoil1	288977	5364	325635	12.69%	53	292761	1.31%	37	289217	0.08%	4552
whitaker3	1199777	3197	1441887	20.18%	193	1376511	14.73%	149	1200374	0.05%	29937
c1y	63854	196	63803	-0.08%	2	62903	-1.49%	1	62333	-2.38%	205
c2y	79500	277	81731	2.81%	6	80109	0.77%	1	79571	0.09%	299
c3y	124708	509	130213	4.41%	4	127729	2.42%	2	127065	1.89%	553
c4y	117254	535	119016	1.50%	4	116621	-0.54%	2	115222	-1.73%	580
c5y	102769	416	99772	-2.92%	4	98004	-4.64%	2	96956	-5.66%	446
gd95c	509	1	512	0.59%	0	506	-0.59%	0	506	-0.59%	2
gd96a	104698	341	112551	7.50%	2	102294	-2.30%	1	99944	-4.54%	268
gd96b	1416	3	1457	2.90%	0	1417	0.07%	0	1422	0.42%	3
gd96c	519	1	533	2.70%	0	520	0.19%	0	519	0.00%	1
gd96d	2393	8	2521	5.35%	0	2436	1.80%	0	2409	0.67%	4

Table 5: Comparison of results and running times . Note (a) Time of Decompose & Orient and Heuristic Improvement are given per iteration. (b) Decompose & Orient time does not include time for recursive decomposition.

Bar-Yehuda et al., *Orientation of Decomposition Trees*, JGAA, 5(4) 1–27 (2001)26

6   Acknowledgments

We would like to express our gratitude to Zvika Brakerski for his help with using HMETIS, running the experiments, generating the tables, and writing scripts.

References

[BL84]   S.N. Bhatt and F.T. Leighton, "A framework for solving VLSI graph layout problems", JCSS, Vol. 28, pp. 300-343, (1984).

[ENRS00] G. Even, J. Naor, S. Rao, and B. Schieber, "Divide-and-Conquer Approximation Algorithms via Spreading Metrics", Journal of the ACM, Volume 47, No. 4, pp. 585 - 616, Jul. 2000.

[ENRS99] G. Even, J. Naor, S. Rao, and B. Schieber, "Fast approximate graph partitioning algorithms", *SIAM Journal on Computing*. Volume 28, Number 6, pp. 2187-2214, 1999.

[FD]   Frost, D., and Dechter, R. "Maintenance scheduling problems as benchmarks for constraint algorithms" To appear in Annals of Math and AI.

[GJ79]   M.R. Garey and D.S. Johnson, "Computers and intractability: a guide to the theory of NP-completeness", W. H. Freeman, San Francisco, California, 1979.

[GK98]   G. Karypis and V. Kumar, "hMeTiS - a hypergraph partitioning package", http://www-users.cs.umn.edu/~karypis/metis/hmetis/files/manual.ps.

[Ha89]   M. Hansen, "Approximation algorithms for geometric embeddings in the plane with applications to parallel processing problems," *30th FOCS*, pp. 604-609, 1989.

[HL99]   Sung-Woo Hur and John Lillis, "Relaxation and clustering in a local search framework: application to linear placement", Proceedings of the 36th ACM/IEEE conference on Design Automation Conference, pp. 360 - 366, 1999.

[K93]   Richard M. Karp, "Mapping the genome: some combinatorial problems arising in molecular biology", Proceedings of the twenty-

Bar-Yehuda et al., *Orientation of Decomposition Trees*, JGAA, 5(4) 1–27 (2001)27

fifth annual ACM Symposium on Theory of Computing, pp. 278-285, 1993

[LR99] F.T. Leighton and S. Rao, "Multicommodity max-flow min-cut theorems and their use in designing approximation algorithms", Journal of the ACM, Volume 46, No. 6, pp. 787 - 832, Nov. 1999.

[P97]   J. Petit, "Approximation Heuristics and Benchmarkings for the MinLA Problem", Tech. Report LSI-97-41-R, Departament de Llenguatges i Sistemes Informàtics. Universitat Politècnica de Catalunya, http://www.lsi.upc.es/dept/techreps/html/R97-41.html, 1997.

[RR98] S. Rao and A. W. Richa, "New approximation techniques for some ordering problems", Proceedings of the 9th Annual ACM-SIAM Symposium on Discrete Algorithms, 1998, pp. 211-218.

[RAK91] R. Ravi, A. Agrawal and P. Klein, "Ordering problems approximated: single processor scheduling and interval graph completion", *18th ICALP*, pp. 751-762, 1991.

[R70]   D. J. Rose, "Triangulated graphs and the elimination process", J. Math. Appl., 32 pp. 597-609, 1970.

Volume 5:5 (2001)

Journal of Graph Algorithms and Applications

http://www.cs.brown.edu/publications/jgaa/

vol. 5, no. 5, pp. 1–1 (2001)

Special Issue on Selected Papers from the 1998 Dagstuhl Seminar on Graph Algorithms and Applications

Guest Editors' Foreword

Takao Nishizeki

Tohoku University
http://www.nishizeki.ecei.tohoku.ac.jp/nszk/nishi/nishi.html
nishi@ecei.tohoku.ac.jp

Roberto Tamassia

Brown University
http://www.cs.brown.edu/people/rt/
rt@cs.brown.edu

Dorothea Wagner

University of Konstanz
http://www.inf.uni-konstanz.de/wagner/
Dorothea.Wagner@uni-konstanz.de

This Special Issue is devoted to selected papers from Dagstuhl Seminar 98301 on "Graph Algorithms and Applications," which was held July 27–31, 1998 at Schloß Dagstuhl, Germany. This seminar promoted the collaboration between computer scientists, mathematicians, and applied researchers, both from academia and industry. It was a follow-up of a seminar on the same topic held in 1996.

The five papers published in this special issue went through a rigorous review process according to the standards of the *Journal of Graph Algorithms and Applications*. We would like to express our gratitude to the referees, for their invaluable help, and to the authors, whose contributions made this issue possible.

Journal of Graph Algorithms and Applications

http://www.cs.brown.edu/publications/jgaa/

vol. 5, no. 5, pp. 3–16 (2001)

Carrying Umbrellas:
An Online Relocation Game on a Graph

Jae-Ha Lee

Max-Planck-Institut für Informatik
Saarbrücken, Germany
http://www.mpi-sb.mpg.de
lee@mpi-sb.mpg.de

Chong-Dae Park   *Kyung-Yong Chwa*

Department of Computer Science
Korea Advanced Institute of Science and Technology
http://jupiter.kaist.ac.kr
{cdpark,kychwa}@jupiter.kaist.ac.kr

Abstract

We introduce an online relocation problem on a graph, in which a player that walks around the vertices makes decisions on whether to relocate mobile resources, while not knowing the future requests. We call it *Carrying Umbrellas*. This paper gives a necessary and sufficient condition under which a competitive algorithm exists. We also describe an online algorithm and analyze its competitive ratio.

Communicated by T. Nishizeki, R. Tamassia and D. Wagner:
submitted December 1998; revised March 2000 and July 2000.

Research supported in part by the KOSEF (Korea Science and Engineering Foundation) under grant 98-0102-07-01-3.

J.-H. Lee et al., *Carrying Umbrellas*, JGAA, 5(5) 3–16 (2001) 4

1 Introduction

"To carry an umbrella or not?" This is an everyday dilemma. This dilemma does not vanish even when correct short-term weather forecasts are available. For example, if it is sunny now but there is no umbrella at the current destination, one has to carry an umbrella so as not to get wet when leaving there later. To illustrate some of our concepts, we describe a detailed scenario.

Picture a person who walks around N places. At each place, he is told where to go next and then must go there. As a usual person, he dislikes being caught in the rain with no umbrella. Since today's weather forecast is correct, he carries an umbrella whenever it is rainy. However, he does not know the future destinations and weather, which might be controlled by the malicious adversary. We say a person is *safe* if he never gets wet. There exists a trivial safe strategy: 'Always carry an umbrella'. However, carrying an umbrella in sunny days is annoying. As an alternative, he has placed several umbrellas in advance and thinks about an *efficient* strategy; he hopes, through some cleverness, to minimize the number of sunny days on which he carries an umbrella. We say a person's strategy is *competitive* if he carries an umbrella in a small portion of sunny days (for details, see Section 1.1). The following questions are immediate: (1) What is the minimum number of umbrellas with which the person can be safe and competitive? (For example, is placing one umbrella per place sufficient or necessary?) (2) What is a safe and competitive strategy? We formally define the problem next.

1.1 A game

Let $G = (V, E)$ be a simple graph with N vertices and M edges. An integer $u(v)$ is associated with each vertex $v \in V$, indicating the number of umbrellas placed on the vertex v. We use $u(G)$ to represent the total number of umbrellas in G. Consider an on-line game between a player and the adversary, assuming that $u(G)$ is fixed in advance.

> **Game** Carrying-Umbrellas$(G, u(G))$
> 1. **Initialization**
>   Player determines the initial configuration of $u(G)$ umbrellas;
>   Adversary chooses the initial position $s \in V$ of Player;
> 2. **for** $i = 1$ **to** L **do**
>   (a) Adversary specifies (v, w), where w is a boolean value and
> $v \in V$ is adjacent to the current vertex of Player;
>   (b) Player goes to v; If $w = 1$, he must carry at least one umbrella;

As an initialization, we assume that the player first determines the initial configuration of $u(G)$ umbrellas and later the adversary chooses the initial position $s \in V$ of the player, although our results of the present paper also hold even if the initialization is done in reverse order. A *play* consists of L *phases*, where L is determined by the adversary and unknown to the player. In each phase, the adversary gives a request (v, w), where v is adjacent to the current vertex of

J.-H. Lee et al., *Carrying Umbrellas*, JGAA, 5(5) 3–16 (2001)   5

the player and w is a boolean value that represents the weather; $w=1$ indicates that it is *rainy*, and $w=0$ *sunny*. For this request (v, w), the player must go to the specified vertex v and decide whether to carry umbrellas or not. If $w=1$, he must carry at least one umbrella; only in sunny days, he may or may not carry umbrellas. Notice that the player can carry an arbitrary number of umbrellas, if available.

The player *loses* if he cannot find any umbrella at the current vertex in a rainy phase. A strategy of the player is said to be *safe* if it is guaranteed that whenever $w = 1$ the player carries an umbrella. Moreover, a strategy of the player should be *efficient*.

As a measure of efficiency, we adopt the *competitive ratio* [9], which has been widely used in analyzing the performance of online algorithms. Let $\sigma = \sigma_1 \cdots \sigma_L$ be a *request sequence* of the adversary, where $\sigma_i = (v_i, w_i)$ is the request in the i-th phase. The *cost* of a strategy $\mathcal{A}$ for σ, written $C_{\mathcal{A}}(\sigma)$, is defined to be the number of phases in which the player carries an umbrella by the strategy $\mathcal{A}$. Then, the *competitive ratio* of the strategy $\mathcal{A}$ is

$$\frac{C_{\mathcal{A}}(\sigma)}{C_{Opt}(\sigma)}$$

where *Opt* is the optimal off-line strategy of the player. (Since *Opt* knows the entire σ in advance, it pays the minimum cost. However, *Opt* cannot be implemented by any player and is used only for comparison.) A strategy whose competitive ratio is bounded by c is termed *c-competitive*; it is simply said to be *competitive*, if it is c-competitive for some bounded c irrespective of the length of σ. The player *wins*, if he has a safe and competitive strategy; he *loses*, otherwise.

Not surprisingly, whether the player has a winning strategy depends on the number of umbrellas. In this paper, we are interested in that number.

Definition 1 *For a graph G, we define $u^*(G)$ to be the minimum number of umbrellas with which the player has a winning strategy in G.*

1.2 Summary of our results

Let $G = (V, E)$ be a connected simple graph (having no parallel edges) with N vertices and M edges. We show that $u^*(G) = M+1$. That is, no strategy of the player is safe and competitive if $u(G) < M+1$ (Section 3) and there exists a competitive strategy of the player if $u(G) \geq M+1$ (Section 4). The competitive ratio of our strategy is $b(G)$, the number of vertices in the largest biconnected component in G. Moreover, the upper bound is attained by the *weak player* that carries at most one umbrella in a phase, and the competitive ratio of $b(G)$ is optimal for some graphs when $u(G) = M+1$. These results can be easily extended to the case that G is not connected. For a simple graph G with M edges and k connected components, $u^*(G) = M + k$. Throughout this paper, we assume G is connected and simple.

J.-H. Lee et al., *Carrying Umbrellas*, JGAA, 5(5) 3–16 (2001)

1.3 Related works

Every online problem can be viewed as a game between an online algorithm and the adversary [8, 3, 5]. In this prospect, Chrobak and Larmore [6] introduced online game as a general model of online problems. Clearly, the problem of the present paper is an example of the online game. Many researchers have studied online problems on a graph. Such examples include the k-server problem [1, 7] and graph coloring [2, 4].

2 Examples

In order to introduce the reader to the ideas, we explain two examples.

Example 1. Consider $K_2 = (\{v_1, v_2\}, \{(v_1, v_2)\})$ that consists of two vertices v_1 and v_2 and an edge (v_1, v_2). Let $u(K_2) = 1$. We assume without loss of generality that the unique umbrella is initially placed at v_1 (see Figure 1). We claim that no strategy of the player can be safe and competitive in K_2. Although the adversary can maliciously determine the initial position s of the player, we explain both cases.

Figure 1: K_2

Case A. If $s = v_2$, the adversary only has to choose $\sigma_1 = (v_1, rainy)$. The player at v_2 has no umbrella and so must get wet in the first phase. Thus it is not safe.

Case B. If $s = v_1$, the adversary first chooses $\sigma_1 = (v_2, sunny)$. For this request σ_1, the player *must* carry an umbrella because otherwise, the resulting configuration is isomorphic to that of Case A; afterwards, the request $\sigma_2 = (v_1, rainy)$ makes the player not safe. Therefore, at the end of the first phase, the player must be at v_2 and $u(v_1) = 0$ and $u(v_2) = 1$. Note that this is isomorphic to the initial configuration. Thus, in a similar fashion, the subsequent requests such as $\sigma_i = (v_1, sunny)$ if i is even and $\sigma_i = (v_2, sunny)$ if i is odd forces the player to always carry an umbrella. Therefore, $C_A(\sigma) = L$. However, since the weather is always *sunny*, *Opt* does not carry an umbrella and $C_{Opt}(\sigma) = 0$. Therefore, any safe strategy of the player is not competitive.

It is easily seen that if $u(K_2) = 2$, then the player has a simple 2-competitive strategy: when it is sunny, the player carries an umbrella if and only if he is going from a vertex with two umbrellas to one with zero.

J.-H. Lee et al., *Carrying Umbrellas*, JGAA, 5(5) 3–16 (2001)   7

Example 2. K_3 is a triangle, that is, $K_3 = (\{v_1, v_2, v_3\}, \{(v_1, v_2), (v_2, v_3), (v_3, v_1)\})$. Let $u(K_3) = 3$. A natural initial configuration of the umbrellas is that $u(v_i) = 1$ for $i = 1, 2, 3$ (Figure 2).

Though every vertex in K_3 initially has an umbrella, we claim that no strategy of the player is safe and competitive. Without loss of generality, let v_1 be the start vertex. The adversary first chooses $\sigma_1 = (v_2, rainy)$. For this input, the player must carry at least one umbrella, making $u(v_1) = 0$ and $u(v_2) = 2$. Next, let $\sigma_2 = (v_3, sunny)$. For this input, the player must carry an umbrella because otherwise, the resulting subgraph induced by $\{v_1, v_3\}$ contains only one umbrella and is isomorphic to K_2 in Example 1; afterwards, the adversary can make the player unsafe or not competitive in K_2. Hence, at the end of the phase 2, there are two cases depending on the number of umbrellas the player has carried: $u(v_2) = 1$ and $u(v_3) = 2$ or $u(v_2) = 0$ and $u(v_3) = 3$.

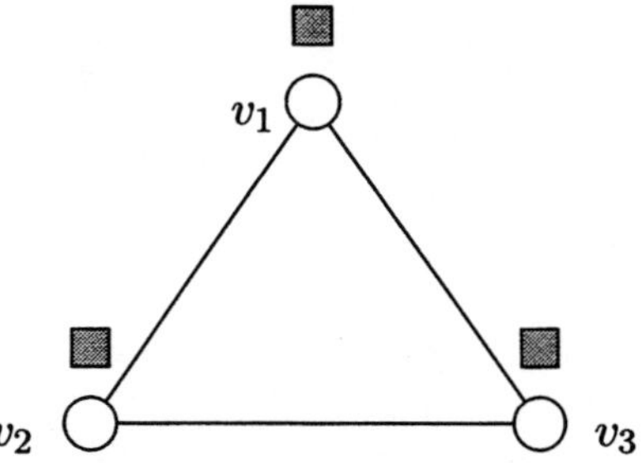

Figure 2: K_3

In either cases, the next request is $\sigma_3 = (v_2, sunny)$. Then the player must carry enough umbrellas to make $u(v_2) \geq 2$, because otherwise, K_2 consisting of v_1 and v_2 is deficient in umbrellas. Thus, the resulting configuration is isomorphic to that at the end of the phase 2. Similarly, the subsequent request sequence $\sigma_i = (v_3, sunny)$ if i is even and $\sigma_i = (v_2, sunny)$ if i is odd forces the player to always carry umbrellas. Therefore, $C_A(\sigma) = L$. However, since Opt only has to carry an umbrella in the first phase, the competitive ratio can be arbitrarily large as L becomes large.

As we shall see later, the player has a 3-competitive strategy, if $u(K_3) \geq 4$. It means $u^*(K_3) = 4$.

3   Lower Bound

In this section, we show the lower bound on $u^*(G)$. To simplify the explanation, we first consider the weak player that can carry at most one umbrella in a phase, and show that $u^*(G) \geq M{+}1$ for the weak player. Later the same bound is shown for the general player that can carry multiple umbrellas.

J.-H. Lee et al., *Carrying Umbrellas*, JGAA, 5(5) 3–16 (2001)

Theorem 1 *Let G be a simple connected graph with N vertices and M edges. Under the constraint that the player can carry at most one umbrella in a phase, $u^*(G) \geq M+1$.*

Proof: We show that if $u(G) \leq M$ then no strategy of the player is safe and competitive in G. Equivalently, it suffices to show that any safe strategy of the player is not c-competitive for any fixed c ($< \infty$). For convenience, imagine we are the adversary that would like to defeat the player.

Let $\sigma = \sigma_1 \sigma_2 \cdots \sigma_L$ denote the request sequence that we generate. Recall that the starting vertex s of the player is determined by the adversary. We fix s as a vertex such that $G - \{s\}$ is connected; we call such a vertex *non-cut vertex* of G. (A non-cut vertex of G is easily obtained by taking a leaf vertex in an arbitrary spanning tree of G.) We say that a graph G' is *deficient* in umbrellas, if $u(G')$ is no greater than the number of edges in it.

The basic idea is to force the player to go into a deficient subgraph of G and recursively defeat the player in it. Suppose we were somehow able to make the player be at s and at the same time, make $u(s) = d+1$, where d is the degree of s in G. This would tell us that $u(G-\{s\})$ is strictly less than the number of edges in $G-\{s\}$. In the next step, we make the player go into $G-\{s\}$. Even if the player has carried an umbrella, $G- \{s\}$ is deficient and the player is in it. Now we use recursion: after making the player go to a non-cut vertex in $G-\{s\}$, we can make the player not safe or not competitive in $G-\{s\}$ recursively. So our subgoal is to make $u(s)$ increase to $d+1$.

Let $v_1, \cdots, v_d$ be the adjacent vertices of s. We begin with explaining how to increase $u(s)$. Suppose that at the start of the $(2i-1)$-th phase, the player is located at s and $u(s) = j$. In the subsequent two phases, the adversary makes the player go to v_j and return to s. That is, $\sigma_{2i-1} = (v_j, sunny)$ and $\sigma_{2i} = (s, *)$. The $2i$-th weather w_{2i} depends on the player's decision in the $(2i-1)$-th phase; if the player carried an umbrella in the $(2i-1)$-th phase, then w_{2i} is set to *sunny*; otherwise *rainy*. Before the generation of σ_{2i}, the adversary tests if $G - \{s\}$ is deficient in umbrellas. If it is, the adversary makes the player move to a non-cut vertex of $G-\{s\}$ and calls the recursive procedure. The strategy of the adversary is summarized in Algorithm 1. Initially, the adversary calls $\mathbf{Adversary}(G, s, 1)$. The *lines 2–3* are the generation of σ_{2i-1} and the *lines 9–13* are that of σ_{2i}. The *lines 5–8* are the recursive procedure. In MOVE-TO-NON-CUT-VERTEX(G'), the adversary picks an arbitrary non-cut vertex s' in G', makes the player go to s' while setting the weather to *sunny*, and returns s'.

To see why this request sequence makes $u(s)$ increase, we define a weighted sum Φ of the umbrellas placed in vertex s and its neighbors, where j umbrellas in s weighs $0.5, 1.5, \cdots, j{-}0.5$, respectively and each umbrella in vertex v_i weighs i. Specifically,

$$\Phi = \sum_{i=1}^{d} i \cdot u(v_i) + \sum_{j=1}^{u(s)} (j - 0.5)$$

Let Φ_k denote Φ at the end of the k-th phase. We are interested in the change

J.-H. Lee et al., *Carrying Umbrellas*, JGAA, 5(5) 3–16 (2001) 9

Algorithm 1 Adversary(G, s, i)

1: **while** TRUE **do**
2:    $j = u(s)$;
3:    $\sigma_{2i-1} = (v_{\min(j,d)}, sunny)$;
4:    { Player's move for σ_{2i-1} }
5:    **if** $u(G-\{s\}) \leq |E(G-\{s\})|$ **then**
6:      $s' = \text{MOVE-TO-NON-CUT-VERTEX}(G-\{s\})$;
7:      Adversary$(G - \{s\}, s', i+1)$;
8:    **end if**
9:    **if** the player carried an umbrella in phase $(2i - 1)$ **then**
10:      $\sigma_{2i} = (s, sunny)$;
11:    **else**
12:      $\sigma_{2i} = (s, rainy)$;
13:    **end if**
14:    $i{+}{+}$;
15: **end while**

---

between Φ_{2k-2} and Φ_{2k}. Suppose that $u(s)$ is j at the end of the $(2k-2)$-th phase. Depending on the decision of the player, there are four cases to consider.

Case A. If the player carried an umbrella throughout the $(2k-1)$-th and the $2k$-th phase, we have $\Phi_{2k-2} = \Phi_{2k}$ because the number of umbrellas is unchanged. However, the weather must be *sunny* from Adversary(G, s, i). Therefore, the player did *useless carrying*.

Case B. If the player carried an umbrella from s to v_j only, $\Phi_{2k} = \Phi_{2k-2}+0.5$. This is because the umbrella moved had weight $j-0.5$ at s and has weight j at v_j.

Case C. If the player carried an umbrella from v_j to s only, $\Phi_{2k} = \Phi_{2k-2}+0.5$. This is because the weight of the umbrella moved changes from j to $j+0.5$.

Case D. The remaining case is that the player never carried an umbrella. However, if the player didn't carry an umbrella in the $(2k - 1)$-th phase, the adversary chooses the weather w_{2k} *rainy*. Since we are considering a safe strategy, the player must carry an umbrella. Hence this case is impossible.

In summary, over the two phases $2k-1$ and $2k$, Φ increases or is unchanged; if Φ is unchanged, the player did *useless carrying* in sunny days. Note that successive *useless carryings* make the player's strategy not competitive, because *Opt* would not carry an umbrella over two *sunny* days. Hence, in order to be competitive, the player must sometimes increase Φ.

How many times can Φ increase before $u(s)$ becomes $d+1$? Recall that once $u(s)$ becomes $d+1$, we can make $G-\{s\}$ deficient in umbrellas and start the recursive procedure. While $u(s) \leq d$, each umbrella can have $2d$ different weights. Moreover, since $u(G)$ is less than or equal to the number of edges in G, Φ can increase in at most $2d \times \frac{N^2}{2} < N^3$ phases before $u(s)$ becomes $d+1$. Thus, Φ can increase in at most N^3 phases, before the recursive procedure is called.

J.-H. Lee et al., *Carrying Umbrellas*, JGAA, 5(5) 3–16 (2001)   10

In $\mathtt{Adversary}(G-\{s\}, s', i+1)$, we recursively define the request sequence and weight sum Φ'. Thus, in $G-\{s\}$, Φ' can increase in at most $(N-1)^3$ phases, before the player moves into some deficient subgraph. In all recursive procedures, the weight sum can increase in at most $N^3+(N-1)^3+\cdots+2^3 < N^4$ phases. This means that the player carries an umbrella in all except at most N^4 phases, and that for the same input, *Opt* can carry an umbrella in at most N^4 phases. At the final recursive call, the graph becomes K_2 and the number of umbrellas in it is at most 1. If there is no umbrella, the player is not safe, which is a contradiction. Otherwise, the situation is the same as in Example 1 in the previous section; we can continue this game forever with no cost for *Opt* and the cost of 1 for the player in every phase.

One remaining step we have to consider is $\mathtt{MOVE\text{-}TO\text{-}NON\text{-}CUT\text{-}VERTEX}$, which is executed at the start of each recursive call. Since the i-th call of $\mathtt{MOVE\text{-}TO\text{-}NON\text{-}CUT\text{-}VERTEX}$ is done in G' with $(N-i)$ vertices, at most $(N-i)$ requests are given and thus the cost of *Opt* is at most $(N-i)$. Throughout all executions of $\mathtt{MOVE\text{-}TO\text{-}NON\text{-}CUT\text{-}VERTEX}$, the cost of *Opt* is at most N^2, because the total number of requests is at most N^2.

Since this game can continue forever as long as the player is safe, we can choose the length L of the input sequence σ to be arbitrarily large. Specifically, we let $L \geq (c+1) \cdot (N^4 + N^2)$. Then, the cost of the player's strategy $\mathcal{A}$ is

$$C_{\mathcal{A}}(\sigma) \geq L - N^4 - N^2 \geq c \cdot (N^4 + N^2)$$

And, the cost of *Opt* is

$$C_{Opt}(\sigma) < (N^4 + N^2)$$

Therefore, the competitive ratio is

$$\frac{C_{\mathcal{A}}(\sigma)}{C_{Opt}(\sigma)} > c$$

for an arbitrary $c(<\infty)$, completing the proof. $\qquad\qquad\square$

Now we present the main result of this section.

Theorem 2 *Let G be a simple connected graph with M edges. Then, $u^*(G) \geq M+1$.*

Proof: Here, the player can carry an arbitrary number of umbrellas in a phase. The algorithm of the adversary is exactly same as $\mathtt{Adversary}(G, s, i)$. However, the analysis is slightly more complex. The crucial fact in proving Theorem 1 was that if $u(s) \geq d+1$ then the adversary makes the player go to v_d, resulting that $G-\{s\}$ is deficient even if the player carries an umbrella, and calls the recursive procedure. In this theorem, however, the player can carry an arbitrary number of umbrellas and thus $u(s) \geq d+1$ does not imply that the recursive procedure is called in the next phase. In other words, $u(s)$ may fluctuate, even over $d+1$, without being trapped into the recursive procedure.

J.-H. Lee et al., *Carrying Umbrellas*, JGAA, 5(5) 3–16 (2001)   11

Φ is defined as in Theorem 1. We say Φ_{2k} is *stable* if $u(s) < d$ at the end of the $2k$-th phase; *unstable* otherwise. We divide the request sequence into a number of *stages*, each of which starts with stable Φ_{2k} and ends right before the next stable $\Phi_{2k'}$. We are interested in the change between Φ_{2k} and $\Phi_{2k'}$.

If $k' = k+1$ (i.e., no unstable Φ exists in this stage), the change of Φ is the same as in Theorem 1, except that Φ can increase by more than 0.5 when the player carries multiple umbrellas. Assume otherwise, that is, $\Phi_{2k''}$ ($k < k'' < k'$) is unstable. Unless the player carries enough umbrellas to make $G - \{s\}$ not deficient in phase $2k''+1$, the recursive procedure is called. Thus, the player must carry at least one umbrella in the $(2k''+1)$-th phase. Similarly, the player must carry at least one umbrella in phase $2k''$, to make Φ unstable. Therefore, the player always carries an umbrella right before and right after Φ is unstable and so the weather is always *sunny*. Comparing Φ_{2k} and $\Phi_{2k'}$ reveals that some umbrellas have moved from v_j to v_d (because the player went to some $v_j(j < d)$ in the $(2k+1)$-th phase, and returned to s with enough umbrellas to make Φ unstable and afterwards goes to and returns from v_d till Φ becomes stable). Therefore, $\Phi_{2k'} \geq \Phi_{2k} + 1$. Moreover, the player always carries at least one umbrella, except in the $(2k+1)$-th and the $2k'$-th phases.

Combining the cases of $k' = k + 1$ and $k' > k+1$, we can conclude that in a stage, either (1) Φ increases at least $0.5 \cdot x$ ($x = 1$ or 2) and the player does not carry umbrellas in at most x phases, or (2) Φ is unchanged and the player always does useless carrying. *Opt* still suffices to carry umbrellas only in x phases in case of (1), depending on the change of Φ. The remaining proof is same as that of Theorem 1. (Only one difference is that we concentrate on stable Φ. Before the recursive procedure is called, stable Φ can increase at most N^3 times, the player does not carry umbrellas in at most N^3 phases, and *Opt* carries umbrellas in at most N^3 phases.) By using the counting arguments in Theorem 1 (ignoring unstable Φ), this theorem is easily seen.   □

4   Upper Bound

In this section, we show the upper bound on $u^*(G)$.

Theorem 3 *Let G be a simple connected graph with N vertices and M edges. Then, $u^*(G) \leq M+1$.*

Proof: It suffices to show that if $u(G) = M+1$ then the player has a safe and competitive strategy. Throughout this section, we assume that $u(G) = M+1$. To show this theorem, we only consider the weak player that carries at most one umbrella in a phase. Imagine that we are the player that should be safe and competitive against the adversary.

The heart of our strategy is how to decide whether to carry an umbrella in sunny days. To help this decision, we maintain two things: labels of umbrellas and a subtree of G. First, let us explain the labels, recalling that $u(G) = M+1$. M umbrellas are labeled as $\gamma(e)$ for each edge $e \in E$, and one remaining umbrella

J.-H. Lee et al., *Carrying Umbrellas*, JGAA, 5(5) 3–16 (2001)   12

is labeled as *Current*. Under these constraints, labels are updated during the game. In addition to the labels, we also maintain a dynamic subtree S of G, called *skeleton*.

- S is a subtree of G whose root is the current position of the player.
- The root in S has the umbrella labeled as *Current*.
- For an edge $e \in S$, the child-vertex of e has the umbrella labeled as $\gamma(e)$.
- For an edge $e \notin S$, the umbrella labeled as $\gamma(e)$ lies in one endpoint of e.

Note that all umbrellas labeled as $\gamma(e)$ are placed on one endpoint of e. In an example of Figure 3, the current skeleton S is enclosed by a dotted line and every edge in S is directed towards the child, indicating the location of its umbrella.

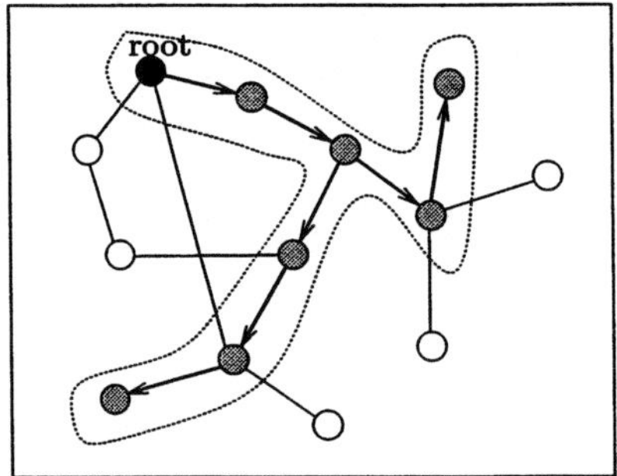

Figure 3: Example of a skeleton (enclosed in a dotted line).

S is initialized as follows: The player first chooses an arbitrary spanning tree T and places one umbrella in each vertex of T, while their labels undetermined. Then, the number of umbrellas used in T is N. For an edge e not in T, we place an umbrella on *any* endpoint of it; this umbrella is labeled as $\gamma(e)$. Then, the number of umbrellas labeled equals the number of edges not in T that is $M-N+1$ and so the total number of umbrellas used is $M+1$. After the adversary chooses the start vertex s, the umbrellas in T are labeled. The umbrella in s is labeled as *Current*, and the umbrella in v ($\neq s$) is labeled as $\gamma(e)$, where e connects v with its parent in T.

Now, we describe the strategy of the player. Suppose that currently, the player is at v_{i-1} and the current request is (v_i, w_i). If the weather is *rainy*, the player has no choice; it has to move to v_i with carrying the umbrella *Current*. In this case, the player resets S to a single vertex $\{v_i\}$. It is easily seen that the new skeleton satisfies the *invariants*, because only the umbrella *Current* is moved. The more difficult case is when the weather is *sunny*.

Suppose that $w_i = 0$, i.e., *sunny*. Depending on whether v_i belongs to S or not, there are two cases. If v_i belongs to S (Figure 4a), then the player moves to v_i without an umbrella. Though no umbrellas are moved, labels must be changed. Let $e_1, e_2, \cdots, e_r$ be the edges encountered when we traverse from v_{i-1} to v_i in S. The umbrella *Current* is relabeled as $\gamma(e_1)$ and the umbrella

J.-H. Lee et al., *Carrying Umbrellas*, JGAA, 5(5) 3–16 (2001)   13

$\gamma(e_k)$ is relabeled as $\gamma(e_{k+1})$ for $1 \le k \le r-1$ and finally, the umbrella $\gamma(e_r)$ is relabeled as the new *Current*. Observe that new skeleton $\mathcal{S}$ still satisfies the *invariants*.

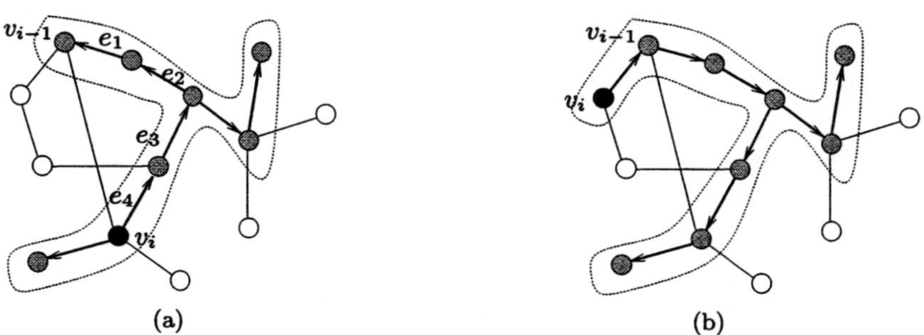

Figure 4: After Figure 3, $(v_i, sunny)$ is given. (a) v_i belongs to $\mathcal{S}$. (b) v_i does not belong to $\mathcal{S}$.

If v_i does not belong to $\mathcal{S}$ (Figure 4b), the edge $e = (v_{i-1}, v_i)$ does not lie in $\mathcal{S}$. Remember that the umbrella $\gamma(e)$ was at v_{i-1} or v_i from the *invariants*. If the umbrella $\gamma(e)$ was placed at v_{i-1}, the player moves to v_i carrying the umbrella *Current*, and adds the vertex v_i and the edge (v_{i-1}, v_i) to $\mathcal{S}$. If the umbrella $\gamma(e)$ was placed at v_i, the player moves to v_i without carrying an umbrella and adds the vertex v_i and the edge (v_{i-1}, v_i) to $\mathcal{S}$ and additionally, swaps the labels of *Current* and $\gamma(e)$. In both cases, it is easy to see that $\mathcal{S}$ still satisfies the *invariants* (see Figure 4b). The strategy of the player is summarized in Algorithm 2.

In order to complete the proof, it suffices to show that $\texttt{Player}(G)$ is a safe and competitive strategy. First, it is easily seen that the player is safe, because the player always has the umbrella *Current*. Next, we show that $\texttt{Player}(G)$ is an N-competitive strategy. Let us divide the request sequence into a number of *stages*, each of which contains exactly one 'rainy' and starts with 'rainy'. Remember that the player resets $\mathcal{S}$ to a single vertex at the start of every stage. The cost of $\texttt{Player}(G)$ in a stage is at most N, because the cost 1 is paid only when $\mathcal{S}$ is set to a single vertex or becomes larger. The cost of *Opt* in a stage is at least one, because each stage starts with 'rainy'. The cost of the initial part of the sequence until the first 'rainy' is zero for $\texttt{Player}(G)$ since $\mathcal{S}$ has size N. Therefore, the competitive ratio is at most N. □

Unfortunately, the competitive ratio of N we obtained above is best possible for some graph G when $u(G) = M+1$.

Theorem 4 *Suppose that G is the N-vertex ring and that $u(G) = N+1$. The competitive ratio of any strategy of the player is at least N.*

Proof: Assume N vertices in G are numbered from 0 to $N-1$. See Figure 5a. As an initial configuration, the player might evenly distribute $M + 1$ umbrellas

J.-H. Lee et al., *Carrying Umbrellas*, JGAA, 5(5) 3–16 (2001)   14

Algorithm 2   Player(G)

1: $i = 1$;
2: pick a spanning tree T of G; determine the initial configuration of $u(G)$ umbrellas;
3: receive s from the adversary;
4: transform T to the rooted tree $\mathcal{S}$ with root s; label the umbrellas;
5: **while** TRUE **do**
6:   $\{\sigma_i = (v_i, w_i)\}$
7:   **if** $w_i = rainy$ **then**
8:     carry an umbrella; $\mathcal{S} = (\{v_i\}, \emptyset)$;
9:   **else**
10:     **if** $v_i \in V(\mathcal{S})$ **then**
11:       do not carry an umbrella; new root is v_i; relabel;
12:     **else**
13:       **if** $\gamma(v_{i-1}, v_i)$ is placed at v_i **then**
14:         do not carry an umbrella; $V(\mathcal{S}) = V(\mathcal{S}) \cup \{v_i\}$; $E(\mathcal{S}) = V(\mathcal{S}) \cup \{(v_i, v_{i-1})\}$; relabel;
15:       **else**
16:         carry an umbrella; $V(\mathcal{S}) = V(\mathcal{S})\cup\{v_i\}$; $E(\mathcal{S}) = V(\mathcal{S})\cup\{(v_i, v_{i-1})\}$; relabel;
17:       **end if**
18:     **end if**
19:   **end if**
20:   $i{+}{+}$;
21: **end while**

(i.e., one umbrella per vertex and the remaining one to an arbitrary vertex). Otherwise, one vertex, say i, has no umbrella and the adversary forces the player to rotate the ring, setting the weather to *sunny*. When entering into i, the player must carry an umbrella, yet Opt never carries umbrellas, which makes the player not competitive.

Thus, we can assume that $u(i) = 2$ and $u(j) = 1$ for each j ($\neq i$). Then the adversary makes the player go to vertex $i-1$, setting the weather to *sunny*. The player must go to $i-1$ without carrying umbrellas. Next request is $(i, rainy)$, so the player must carry an umbrella from $i-1$ to i. As a result, $u(i) = 3$ and $u(i-1) = 0$ and $u(j) = 1$ for other j (Figure 5b).

In any subsequent k-th steps ($1 \leq k \leq N-2$), the adversary gives request $(i+k, sunny)$, where $+$ is taken modulo N. In spite of sunny days, the player must carry an umbrella because otherwise, the subgraph induced by the vertices $\{i+k, i+k+1, \cdots, i-1\}$ shall contain $N-k-1$ edges and $N-k-1$ umbrellas and the subgraph is deficient. The final request $(i-1, sunny)$ brings the player back to the starting point. Since v_{i-1} does not have any umbrella before this request, the player must bring one to be safe. For these requests, $C_A(\sigma)$ is at least N and C_{Opt} is 1 because only one day is rainy. Therefore, the competitive

J.-H. Lee et al., *Carrying Umbrellas*, JGAA, 5(5) 3–16 (2001)   15

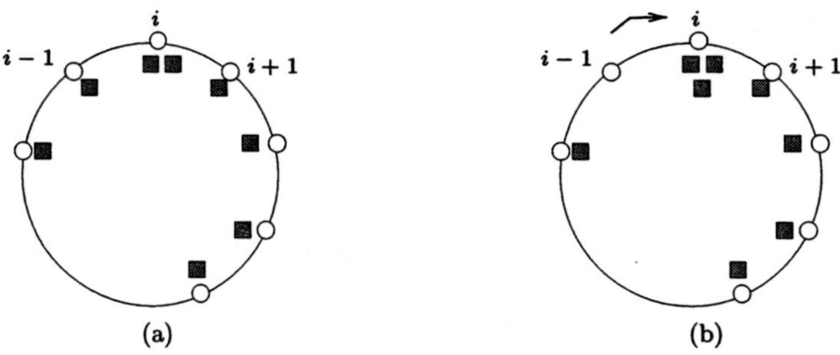

Figure 5: Theorem 4.

ratio is at least N. □

Though the competitive ratio of N is optimal for rings, we can easily design a 2-competitive algorithm for trees. Here we slightly modify the algorithm $\mathtt{Player}(G)$ and show that the new algorithm is $b(G)$-competitive, where $b(G)$ is the number of vertices in the largest biconnected component in G. Note that $b(G)$ is 2 for any tree G.

Theorem 5 *Let G be a simple connected graph with M edges and let $u(G) = M+1$. The player has a $b(G)$-competitive strategy, where $b(G)$ is the number of vertices in the largest biconnected component in G.*

Proof: It suffices to modify $\mathtt{Player}(G)$ for rainy phases. Suppose it is rainy when the player goes from u to v. In $\mathtt{Player}(G)$, the player resets S to a single vertex v (i.e. the current vertex) in every rainy phase.

In the new strategy, the player resets S in the biconnected component that contains u and v. More formally, suppose the edges in G are partitioned into l biconnected components $C_1, \cdots C_l$ and C_k contains the edge (u, v). Then the player resets S in C_k to a single vertex v and maintains other edges outside C_k. Note that the skeleton S may be disconnected during the game.

The analysis of the new strategy is nearly the same as that of $\mathtt{Player}(G)$. Consider the *sunny* phase i in which the player carries an umbrella from u to v. Let C_k be the biconnected component that contains the edge (u, v). This cost 1 is charged to the last rainy phase j $(< i)$ in which u and v are disconnected in S (i.e. the recent rainy phase in which the player was traversing an edge in C_k). The cost of the player in every rainy phase is charged to itself. In this way, all costs of the player are charged to rainy phases in the same biconnected component. It is easily seen that at most $b(G)$ is charged to each rainy phase, because the player resets S in the biconnected component in G containing the current edge, whose size is bounded by $b(G)$. □

J.-H. Lee et al., *Carrying Umbrellas*, JGAA, 5(5) 3–16 (2001)   16

5 Concluding Remarks

In this paper, we introduced an online game on a graph, *Carrying Umbrellas*. We showed that the player has a safe and competitive strategy if and only if $u(G)$ is at least $M+1$. We also presented an $b(G)$-competitive strategy of the player when $u(G) = M+1$. This paper is the first attempt on this problem, and many questions remain open. First, observe that $u^*(G)$ is independent of the *topology* of the graph, the starting position of the player, and the number of umbrellas that the player can carry in a phase. Finding this number $u^*(G)$ for digraphs seems to be interesting from a graph-theoretical viewpoint. Second, we believe that the problem *Carrying Umbrellas* can be extended to be of practical use. For now, however, it is mainly of theoretical interest. Finally, reducing the competitive ratio with more umbrellas is an open problem.

Acknowledgment We are grateful to Oh-Heum Kwon for motivating the problem of this paper, to Jung-Heum Park for helpful discussions and to anonymous referees for giving constructive comments.

References

[1] N. Alon, R. M. Karp, D. Peleg, and D. West. A graph-theoretic game and its application to the k-server problem. *SIAM Journal of Computing*, 24(1):78–100, 1995.

[2] R. Beigel and W. Gasarch. The mapmaker's dilemma. *Discrete Applied Mathematics*, 34(1):37–48, 1991.

[3] S. Ben-David, A. Borodin, R. Karp, G. Tardos, and A. Widgerson. On the power of randomization in online algorithms. In *Proc. of 22nd STOC*, pages 379–386, 1990.

[4] H. L. Bodlaender. On the complexity of some coloring games. *Int. J. of Foundation of Computer Science*, 2:133–147, 1991.

[5] A. Borodin and R. El-Yaniv. *Online computation and competitive analysis.* Cambridge University Press, 1998.

[6] M. Chrobak and L. Larmore. The server problem and online games. In *DIMACS Series in Discrete Mathematics and Theoretical Computer Science*, volume 7, pages 11–64. AMS-ACM, 1992.

[7] M. S. Manasse, L. A. McGeoch, and D. D. Sleator. Competitive algorithms for server problem. *Journal of Algorithms*, 24(1):78–100, 1990.

[8] P. Raghavan and M. Snir. Memory versus randomization in online algorithms. In *Proc. of 16th ICALP*, pages 687–703, 1987.

[9] D. Sleator and R. Tarjan. Amortized efficiency of list update and paging rules. *Communications of ACM*, 2(28):202–208, 1985.

Journal of Graph Algorithms and Applications

http://www.cs.brown.edu/publications/jgaa/

vol. 5, no. 5, pp. 17–38 (2001)

New Bounds for Oblivious Mesh Routing*

Kazuo Iwama [†]
School of Informatics
Kyoto University
Kyoto 606-8501, JAPAN
http://www.lab2.kuis.kyoto-u.ac.jp/~iwama/index.html
iwama@kuis.kyoto-u.ac.jp

Yahiko Kambayashi [‡]
School of Informatics
Kyoto University
Kyoto 606-8501, JAPAN
http://www.isse.kuis.kyoto-u.ac.jp/usr/yahiko/yahiko-e.html
yahiko@i.kyoto-u.ac.jp

Eiji Miyano [§]
Kyushu Institute of Design
Fukuoka 815-8540, JAPAN
http://www.kyushu-id.ac.jp/~miyano/index.html
miyano@kyushu-id.ac.jp

Abstract

We give two, new upper bounds for oblivious permutation routing on the mesh networks: Let N be the total number of processors in each mesh. One is an $O(N^{0.75})$ algorithm on the two-dimensional, $\sqrt{N} \times \sqrt{N}$ mesh with constant queue-size. This is the first algorithm which improves substantially the trivial $O(N)$ bound for oblivious routing in the mesh networks with constant queue-size. The other is a $1.16\sqrt{N} + o(\sqrt{N})$ algorithm on the three-dimensional, $N^{1/3} \times N^{1/3} \times N^{1/3}$ mesh with unlimited queue-size. This algorithm allows at most three bends in the path of each packet. If the number of bends is restricted to minimal, i.e., at most two, then the bound jumps to $\Omega(N^{2/3})$ as was shown in ESA'97.

Communicated by T. Nishizeki, R. Tamassia and D. Wagner:
submitted January 1999; revised April 2000 and October 2000.

---

*A preliminary version of this paper was presented at the 6th European Symposium on Algorithms (ESA'98).

[†]Supported in part by Scientific Research Grant, Ministry of Japan, 09480055 and 08244105, and Kayamori Foundation of Information Science Advancement, Japan

[‡]Supported in part by Scientific Research Grant, Ministry of Japan, 08244102

[§]Supported in part by Scientific Research Grant, Ministry of Japan, 10780198 and 12780234, and The Telecommunications Advancement Foundation, Japan.

K. Iwama et al., *Oblivious Mesh Routing*, JGAA, 5(5) 17–38 (2001)   18

1 Introduction

An algorithm for *packet routing* has to determine each packet's path through a network by using various information, such as source addresses, destinations, and the configuration of the network. So far a great deal of effort has been devoted to the design of efficient routing algorithms and there is a large amount of literature even if we focus our attention on deterministic, *permutation* routing, in which the destinations of packets are all different. The efficiency of a routing algorithm is generally measured by its *running time* and its *queue-size* of each processor, where the former is the total number of communication time-units the algorithm requires to route all packets to their destinations, and the latter is the maximum number of packets the processor temporally can hold at the same time during routing.

A very popular strategy for permutation routing is *oblivious* routing, in which the path of each packet is completely determined by its initial and final positions and is not affected by other packets it encounters. Hence, it is hard to avoid path-congestion in the worst case and it often takes much more time than it looks. Several lower-bounds which are rather surprising are known on the running time of oblivious permutation routing on standard *mesh* networks, each of which has N processors connected via point-to-point connections: For example: (i) An $\Omega(N)$ lower bound is known for oblivious permutation routing on any k-dimensional, constant queue-size mesh network including N processors, where k may be any constant [8]. (Note that an $O(N)$ upper bound can be achieved for permutation routing even on one-dimensional meshes (linear arrays) including N processors, i.e., increasing dimensions in meshes does not work in the worst case.) (ii) An $\Omega(N^{2/3})$ lower bound is known for oblivious permutation routing on *three-dimensional,* unbounded queue-size meshes including $N^{1/3} \times N^{1/3} \times N^{1/3}$ processors [5], which is much *worse* than the $O(\sqrt{N})$ bound for oblivious permutation routing on two-dimensional, unbounded queue-size meshes including $\sqrt{N} \times \sqrt{N}$ processors. (iii) An $\Omega(\sqrt{N})$ lower bound for oblivious permutation routing on any constant-degree, unbounded queue-size, N processor network [2, 3, 6]. It should be noted, however, that these lower bound proofs needed some supplementary conditions that might not seem so serious but are important for the proofs. In this paper, it is shown that the above lower bounds do not hold any more if those supplementary conditions are slightly relaxed.

More precisely, Krizanc needed the *pure condition* other than the oblivious condition to prove the $\Omega(N)$ lower bound in [8]. Roughly speaking, the pure condition requires that each packet must move if its next position is empty. Krizanc gave an open question, i.e., whether his linear bound can be improved by removing the pure condition. In this paper we give a positive answer to this question: It is shown that there is an $O(N^{0.75})$ oblivious algorithm for permutation on 2D, $\sqrt{N} \times \sqrt{N}$ meshes with constant queue-size, and there is an $O(N^{5/6})$ oblivious algorithm on 3D, $N^{1/3} \times N^{1/3} \times N^{1/3}$ meshes with constant queue-size. The oblivious condition used in [8] is a little more stronger than the normal one, called the *source-oblivious condition*. That is also satisfied by our new algorithm, i.e., we remove only the pure condition in this paper. Note that this $\Omega(N)$ lower bound is quite tough; it still holds even without the oblivious condition; the *destination-exchangeable* strategy also implies the same lower bound if the queue-size is bounded above by some constant [4]. Our new bound can be extended to the case of general queue-size k, namely, it is shown that there is an $O(N^{0.75}/\sqrt{k})$ algorithm for 2D, $\sqrt{N} \times \sqrt{N}$ meshes of queue-size k, and there is an $O(N^{5/6}/\sqrt{k})$ algorithm for 3D, $N^{1/3} \times N^{1/3} \times N^{1/3}$ meshes of queue-size k, while an $\Omega(N/k(8k)^{5k})$ lower bound was previously known for any constant degree, k-queue-size network including N processors under the pure condition [8]. For 2D meshes, if we set $k = \sqrt{N}$, then that is equivalent to unbounded queue-size. Our bound for this specific value of k is $O(N^{0.75}/N^{0.25}) = O(\sqrt{N})$, which matches the lower bound of [2, 3, 6].

Our second result concerns with 3D meshes: In [5] an important exception was proved against the well-known superiority of the 3D meshes over the 2D ones; oblivious permutation routing requires $\Omega(N^{2/3})$ steps over the 3D meshes including N processors under the following (not unusual, see the next paragraph) condition: The path must be shortest and be as straight as possible. In other words, each packet has to follow a path including at most two bends in the 3D case. [5] suggested that this lower bound may still hold even if the condition is removed; i.e., three-dimensional oblivious routing may be essentially inefficient. Fortunately this concern for 3D meshes was needless; we prove in this paper that any permutation can be routed over the 3D meshes including N processors in $1.16\sqrt{N} + o(\sqrt{N})$ steps by relaxing the condition a little bit: If we only allow the path of every packet to make one more bend, then the running time of the algorithm decreases from $\Omega(N^{2/3})$ to $\Theta(\sqrt{N})$. This upper bound is optimal within constant factor by [6] and does not change if we add the shortest-path condition.

For oblivious routing, there is a general lower bound, i.e., $\sqrt{N}/d$ for degree-d networks of any type [6]. This is tight for the hypercube including N processors, namely, $\Theta(\sqrt{N}/\log N)$ is both upper and lower bounds for oblivious routing over the hypercube [6]. This is also tight within constant factor for the 2D mesh including N processors, where $2\sqrt{N} - 2$ steps is an upper bound and also is a lower bound (without any supplementary condition) [1, 9, 10, 11]. Thus, tight bounds are known for two extremes, for the 2D mesh and for the $\log N$-dimensional mesh (= the hypercube). Furthermore, both upper bounds can be achieved rather easily, i.e., by using the most rigid, *dimension-order path strategy* [12, 2, 6]. However, for the 3D meshes, even the substantially weaker condition, i.e., the minimum-bending condition, provides the much worse bound as mentioned before [5]. Our second result now extends the family of meshes for which optimal oblivious routing is known. If randomization is allowed, then the bound decreases to $O(N^{1/3})$ [7, 13] for 3D meshes including N processors. A similar bound also holds for deterministic routing but for random permutations [9], including our new algorithm.

K. Iwama et al., *Oblivious Mesh Routing*, JGAA, 5(5) 17–38 (2001)   20

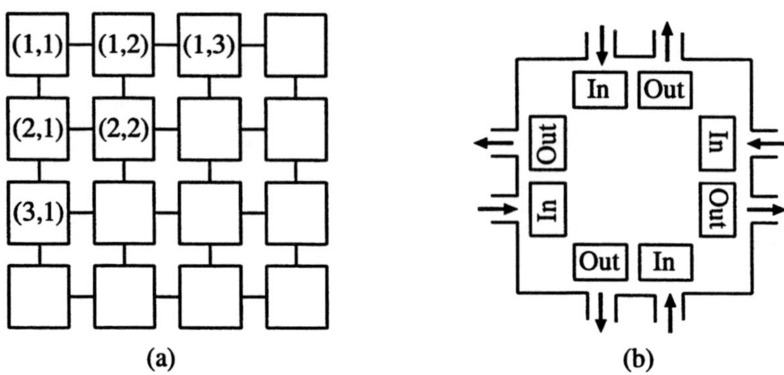

Figure 1: (a) 2D mesh   (b) Processor

2   Models and Problems

Two-dimensional meshes are illustrated in Figure 1-(a). The following definitions on the two-dimensional (2D for short) mesh can be naturally extended to
the three-dimensional (3D for short) mesh illustrated in Figure 2. A *position*
is denoted by (i,j), $1 \le i,j \le \sqrt{N}$ and a processor whose position is (i,j) is
denoted by $P_{i,j}$, i.e., the total number of processors is N. A connection between
the neighboring processors is called a (*communication*) *link*. A *packet* is denoted
by $[i,j]$, which shows that the destination of the packet is (i,j). (A real packet
includes more information besides its destination such as its original position
and body data, but they are not important within this paper and are omitted.)
So we have N different packets in total. An *instance* of *permutation routing*
consists of a sequence $\sigma_1\sigma_2\cdots\sigma_N$ of packets that is a permutation of the N
packets $[1,1],[1,2],\cdots,[\sqrt{N},\sqrt{N}]$, where σ_1 is originally placed in $P_{1,1}$, σ_2 in
$P_{1,2}$ and so on.

   Each processor has four input and four output queues (see Figure 1-(b)).
Each queue can hold up to k packets at the same time. The one-step computation consists of the following two steps: (i) Suppose that l (≥ 0) packets
remain, or there are $k - l$ spaces, in an output queue Q of processor P_i. Then
P_i selects at most $k - l$ packets from its input queues, and moves them to Q.
(ii) Let P_i and P_{i+1} be neighboring processors (i.e., P_i's right output queue Q_i
be connected to P_{i+1}'s left input queue Q_{i+1}). Then if the input queue Q_{i+1}
has space, then P_i selects at most one packet (at most one packet can flow on
each link in each time-step) from Q_i and send it to Q_{i+1}. Note that P_i makes
several decisions due to a specific algorithm in both steps (i) and (ii). When
making these decisions, P_i can use any information such as the information of
the packets now held in its queues. Other kind of information, such as how
many packets have moved horizontally in the recent t time-slots, can also be
used. Note that it may happen that a packet does not go out of its source until
a specific time in routing. In this paper, we assume that such an inactive packet

K. Iwama et al., *Oblivious Mesh Routing*, JGAA, 5(5) 17–38 (2001) 21

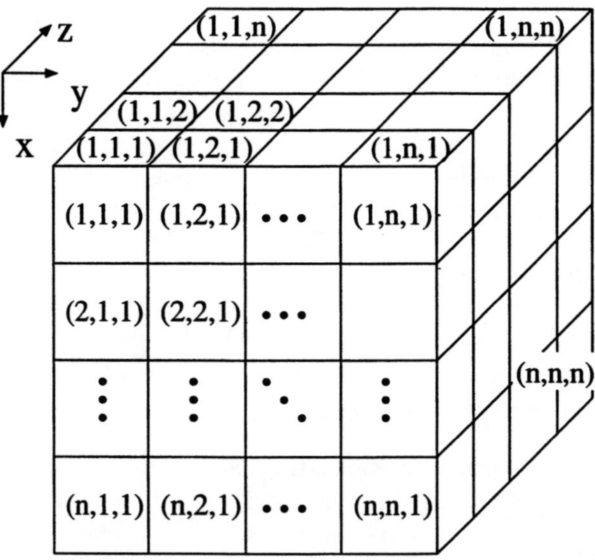

Figure 2: 3D mesh

is not included in the queue-size.

If we fix an algorithm and an instance, then the path R of each packet is determined, which is a sequence of processors, $P_1(=\text{source}), P_2, \cdots, P_j(=$ destination). R is said to be *b-bend* if R changes its direction at b positions. A routing algorithm, A , is said to be *b-bend* if the path of every packet is at most b-bend. A is said to be *oblivious* if the path of each packet is completely determined by its source and destination. Furthermore, A is said to be *source-oblivious* if the moving direction of each packet only depends on its current position and destination (regardless of its source position). A is said to be *minimal* if the path of every packet is the shortest one. A is said to be *pure* if a packet never stays at the current position when it is possible for the packet to advance.

The most rigid and typical oblivious scheme for routing on meshes (and hypercubes) is the so called *dimension-order* algorithm. In the two-dimensional case, a packet first moves horizontally to its destination column and then moves vertically to its destination row. If the queue-size is unbounded, then the dimension-order algorithm can route any permutation in $2\sqrt{N} - 2$ steps on 2D meshes including N processors [9, 12]. However, if the queue-size is bounded above by some constant, then we have to pay great attention to algorithm's behavior, especially to the *queuing discipline*. As an example, let us consider the following oblivious routing algorithm, say, A_0, for $k = 1$, which is based on the dimension-order strategy, i.e., all packets move horizontally first and then make turns at most once at the crossings of source rows and destination columns. (i) Suppose that the top output queue of processor P_i is empty. Then

K. Iwama et al., *Oblivious Mesh Routing*, JGAA, 5(5) 17–38 (2001)   22

P_i selects one packet whose destination is upward on this column. If there are more than one such packet, then the priority is given in the order of the left, right and bottom input queues. (Namely, if there is a packet that makes a turn in this processor, then it has a higher priority than a straight-moving packet.) Similarly for the bottom, left and right output queues, i.e., a turning packet has a priority if competition occurs. (ii) If an input queue is empty, then it is always filled by a packet from its neighboring output queue. Thus each queue of the processor never overflows under A_0. It is not hard to see that A_0 completes routing within roughly $c\sqrt{N}$ steps for many "usual" instances. Unfortunately, it is not always true.

Consider the following instance: Packets in the lower-left one-fourth plane are to move to the upper-right plane, and vice versa. The other packets in the lower-right and upper-left planes do not move at all. One can see that A_0 begins with moving (or shifting) packets in the lower-left plane to the right. Suppose that the flow of those packets looks like the following illustration: Here a shows a packet whose destination is on the rightmost column, b on the second rightmost column and so on. Note that the uppermost row includes a long sequence of a's. The second row includes five a's and a long b's, the third row includes five a's, five b 's and long c's and so on. We call such a sequence of packets which have the same destination column a *lump* of packets.

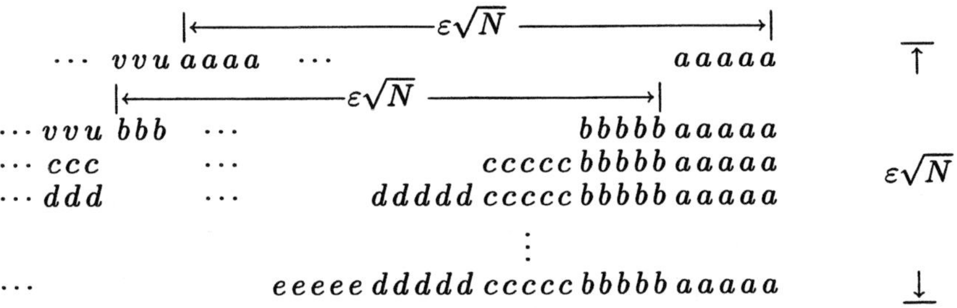

Now the lump of a's reaches the rightmost column. One can see that the a's in the uppermost row can move into the vertical line smoothly and the following packets can reach to their bending position smoothly also: Thus nothing happens against the uppermost row. However, the packet stream in the second row will encounter two different kinds of "blocking:" (1) The sequence of five a's in the second row is blocked at the upper-right corner and cannot move upward since the $\varepsilon\sqrt{N}$ a's in the uppermost row have privileges. (2) One can verify that the last (leftmost) a of these five a's stops at the left queue of the second rightmost processor, which blocks the next sequence of b's, namely, they cannot enter the second rightmost column even if it is empty (see Figure 3). Thus, we need $\varepsilon\sqrt{N}$ steps before the long b's start moving. After they start moving, those b's in turn block the five b's on the third row and below. This argument can continue until the $\varepsilon\sqrt{N}$th row, which means we need at least $(\varepsilon\sqrt{N})^2$ steps only to move those packets.

One might think that this congestion is due to the rule of giving a higher priority to the packets that turn to the top from the left. This is not necessarily

K. Iwama et al., *Oblivious Mesh Routing*, JGAA, 5(5) 17–38 (2001)   23

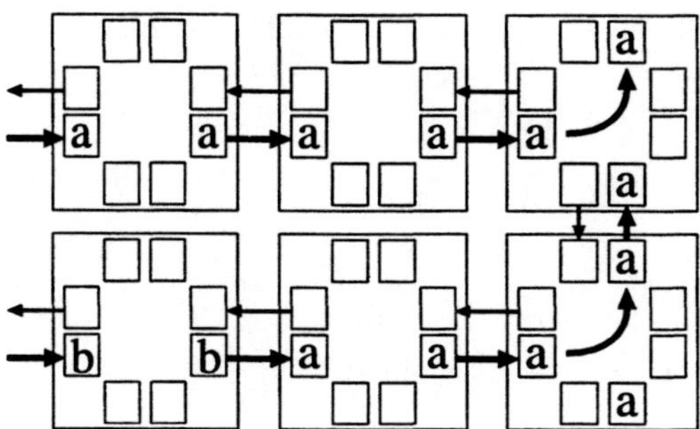

Figure 3: Blocking

true. Although details are omitted, we can create "adversaries" that imply a similar congestion against other resolution rules such as giving a priority to straight-moving packets. In the next section we will see how we can avoid this kind of bad blocking.

3   2D Oblivious Routing

3.1   Basic Ideas

Recall that the naive algorithm A_0 given in the previous section requires $\Omega(N)$ steps in the worst case. One can see that in the above example, serious delays do occur around at the crossings of source rows and destination columns, called the *critical zone*, and this inefficiency mainly comes from the co-existence of long and short lumps of packets; namely, short lumps of packets move ahead and prevent subsequent long lumps of packets from advancing. However, as shown in Lemma 3 later, the same algorithm runs quickly if there are only long lumps of packets in every row. On the other hand, if we have only short lumps of packets in every row, then we can also handle them efficiently (see Lemma 4). In the following, a lump of packets is said to be *long* if it includes at least d packets which have the same destination column for some positive integer d, which will be fixed later; otherwise, it is said to be *short*. So our basic strategy is quite simple: Before entering packets into their critical zone, (1) we first sort the sequence of packets in every row so that packets having nearer column destinations will go first (i.e., the packets move in nearest-first order), and then (2) count the size of each lump of packets which have the same destination column. As shown later, the operation (2) results in the sequence of the packets being changed into the reverse order. Note that it is only the purpose of these nearest-first and farthest-first orderings to count the number of packets going to the same column by grouping them. (3) We let only long lumps of packets go to the critical zone in farthest-first order. (4) The remaining short lumps of packets go afterwards but this time we adopt the so-called *shuffled order* as the sequence of packets. For example, consider the following sorted sequence of packets:

$$d_2 \quad d_1 \quad c_3 \quad c_2 \quad c_1 \quad b_2 \quad b_1 \quad a_3 \quad a_2 \quad a_1$$

Then, the shuffle order sequence looks like the following illustration:

$$c_3 \quad a_3 \quad d_2 \quad c_2 \quad b_2 \quad a_2 \quad d_1 \quad c_1 \quad b_1 \quad a_1$$

where a's have the farthest column destination, b's the second farthest, and so on as before. Namely, each of the rightmost four packets a_1, b_1, c_1 and d_1 comes from the right end of each lump, each of the next four packets a_2, b_2, c_2 and d_2 comes from the second right end of each lump and the remaining two packets are placed on the left end of the shuffle order sequence.

Recall that our main purpose of the sorting operation (1) is to gather packets heading for the same destination column, which makes it possible to execute the subsequent operations efficiently. Thus the key to implementing those ideas lies in designing an algorithm which can change any sequence of packets on each row to the sorted sequence according to the destination column without violating the oblivious condition:

Lemma 1 *Let $x = x_1 x_2 \cdots x_n$ be a sequence of n packets and $x_s = x_{s_1} x_{s_2} \cdots x_{s_n}$ be a sorted sequence of x such that x_{s_n} is the nearest packet among those n packets, and $x_{s_{n-1}}$ is the second nearest packet, and so on. Suppose that a linear array of $2n$ processors, P_1 through P_{2n}, is available and the sequence x of n packets is initially placed on the n processors of the left half of the linear array. Namely, P_1 through P_n hold x_1 through x_n in this order initially. Then there is an oblivious algorithm which runs in $2n - 1$ steps and needs queue-size $k = 2$ such that the sorted sequence x_s is finally placed on the right half of the linear array, i.e., P_{n+1} through P_{2n} finally hold x_{s_1} through x_{s_n} in this order.*

Proof: The basic idea of the following oblivious algorithm is very similar to the idea implemented to adaptive routing in [4]: (i) We first move those n packets to the right in nearest-first order. That means the leftmost processor P_{n+1} of the right-half linear array receives packets in nearest-first order, i.e., P_{n+1} first receives the nearest packet x_{s_n}, next the second nearest packet $x_{s_{n-1}}$ and so on. (ii) Then we keep moving each packet up to its correct position. Here is a more detailed description:

(i) For $1 \leq i \leq n$, each P_i selects a packet which should go to the nearer column out of the packets it currently holds in its queue (an arbitrary one if all packets P_i holds have the same destination column), and moves it to the right at each step. However, P_i starts this action at the ith step and does nothing until then. If P_i has no packet, then it does nothing again.

(ii) For $1 \leq i \leq n$, each P_{n+i} simply shifts its packet sent from P_{n+i-1} to the right at each step. Then P_{n+i} eventually receives its correct packet x_{s_i} exactly at the $(2n - 1)$th step.

Take a look at the following example:

	P_1	P_2	P_3	P_4	P_5	P_6	P_7	P_8	P_9	P_{10}
initial positions	a	c	b	a	c	c	a	a	b	a

which shows that the sequence of 10 packets are now included in the left half, P_1 through P_{10}, of a linear array of 20 processors. Note that a should go to the farthest column, b the second farthest and so on as before. At the first step, only the leftmost packet, a in this example, moves to the right and P_2 now holds a and c in its input queue (recall that $k = 2$). In the second step, c is selected among a and c since c should go out earlier than a, and it goes to P_3. In the third step, P_2 sends the remaining packet a to the right, and at the same time, P_3 sends c out since the destination column of c is nearer than b. For example, after the eighth step, the positions change as follows:

	P_1	P_2	P_3	P_4	P_5	P_6	P_7	P_8	P_9	P_{10}
after the 8th step						b	a	a	b	a
			a	a	c	c	c			

In the next step, P_{10}, P_9, P_8, P_7 and P_6 receive c, c, c, b and a from P_9, P_8, P_7, P_6 and P_5, respectively. In the next (10th) step, the packet c first goes out of the left half of the linear array and so on. The proof that the sorting operation works is based on the following claim, which can be shown by induction on m.

Claim 1 *For any $2 \leq m \leq n + 1$, the following statement is true: At the $(m-1)$th step P_m receives the nearest packet among the packets initially placed on P_1 through P_{m-1}, at the mth step P_m receives the second nearest among those packets, and so on. Finally, at the $(2m-3)$th step, P_m receives the $(m-1)$th nearest (i.e., farthest) packet.*

Proof: See [4]. □

By Claim 1, P_{n+1}, the leftmost processor of the right half array, can receive the jth nearest packet $x_{s_{n-j+1}}$ exactly at the $(n + j - 1)$th step for $1 \leq j \leq n$. In other words, P_{n+1} can receive the ith *farthest* packet x_{s_i} exactly at the $(2n - i)$th step for $1 \leq i \leq n$. Furthermore, the ith farthest packet x_{s_i} has to move rightward, and eventually, x_{s_i} arrives at its final position P_{n+i} exactly at the $(2n - 1)$th step since each P_{n+i} simply shifts each of the packets sent from P_{n+i-1} to the right and x_{s_i} requires $(i - 1)$ steps to travel from P_{n+1} to P_{n+i}. □

Note that the row routing algorithm in Lemma 1 works in linear time of n, but in order to obtain the sorted sequence of n packets we have to prepare a long path of (at least) $2n$ processors before entering the sequences into their critical zone.

3.2 Algorithms

Theorem 1 *There is an oblivious routing algorithm on the two-dimensional, $\sqrt{N} \times \sqrt{N}$ mesh of queue-size k ($2 \leq k \leq c\sqrt{N}$ for some constant c) which runs in $O(N^{0.75}/\sqrt{k})$ steps.*

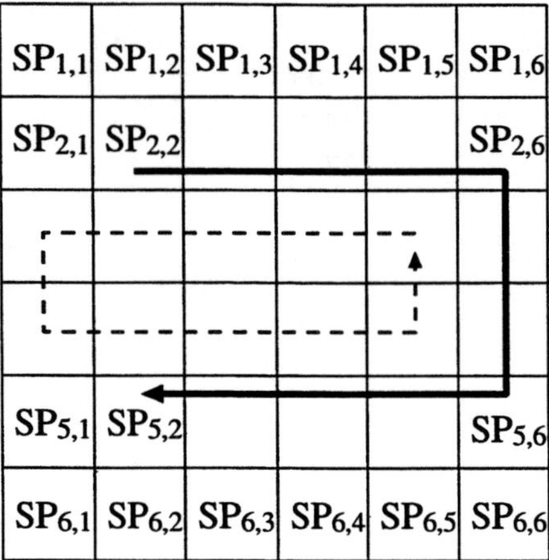

Figure 4: 36 subplanes

Proof: The whole plane is divided into 36 subplanes, $SP_{1,1}$ through $SP_{6,6}$ as shown in Figure 4. For simplicity, the total number of processors in 2D meshes is hereafter denoted by not N but $36n^2$, i.e., each subplane consists of $n \times n$ processors. (1) The entire algorithm is divided into 36×36 *phases*. In the first phase only packets whose sources and destinations are both in $SP_{1,1}$ move (i.e., they may be only a small portion of the n^2 packets in $SP_{1,1}$). In the second phase only packets from $SP_{1,1}$ to $SP_{1,2}$ move, and so on. (2) Suppose that it is now the phase where packets from $SP_{2,2}$ to $SP_{5,2}$ move. Then the paths of those packets are not shortest to make sorted sequences of the packets, but as shown by arrows in Figure 4: They first move rightward to $SP_{2,6}$, then downward to $SP_{5,6}$, and finally move leftward back to $SP_{5,2}$. Thus, the paths of the packets are denoted by the following sequence of subplanes, $SP_{2,2}$, $SP_{2,3}$, $\cdots$, $SP_{2,6}$, $SP_{3,6}$, $\cdots$, $SP_{5,6}$, $SP_{5,5}$, $\cdots$, $SP_{5,2}$. The general rule of path selection in each phase will be given soon. (3) These paths consist of three different zones, the *(parallel) shifting zone*, the *sorting zone* and the *critical zone*. In the above example, the sorting zone is composed of three consecutive subplanes $SP_{2,3}$, $SP_{2,4}$ and $SP_{2,5}$, the critical zone is $SP_{2,6}$ and the parallel shifting zone includes all the remaining portions.

We first describe the purpose of three different zones: First of all, before entering packets into the sorting zone, they move without changing their relative positions, i.e., they move in parallel just like a formation flight. In the above example again, packets initially placed on the top row in the source subplane $SP_{2,2}$ move through the top row without changing their order to the next subplane $SP_{2,3}$ of the sorting zone, packets on the second row in $SP_{2,2}$ move rightward through the second row without changing their order to $SP_{2,3}$, and so on. The sorting zone needs three consecutive subplanes where the flow of packets on each row is changed, i.e., from an arbitrary order to the farthest-first

K. Iwama et al., *Oblivious Mesh Routing*, JGAA, 5(5) 17–38 (2001)   27

order. More precisely, packets on each row in $SP_{2,3}$ of the sorting zone once move rightward to the next subplane $SP_{2,4}$ while sorting their order into the nearest-first order by using the row routing algorithm in Lemma 1. Furthermore, the packets move from $SP_{2,4}$ to the next subplane $SP_{2,5}$, while changing their order into the reverse order (i.e., farthest-first order) to calculate the size of each lump of packets, by using a new (but simple) idea described later. The critical zone is the most important zone where each packet enters its correct column position, but relatively within the subplane. Namely, if the destination of the packet is on the leftmost column in the final destination subplane $SP_{5,2}$, then the packet temporally enters the leftmost column in the critical zone $SP_{2,6}$, if the destination is on the second leftmost column in $SP_{5,2}$, then the packet temporally enters the second leftmost column in $SP_{2,6}$, and so on. Finally, all the packets move through the shifting zone towards their final positions without changing their relative positions.

Recall that our goal is to reduce path-congestion in the critical zone with a help of the sorting zone. To prepare the sorting zone before entering packets into the critical zone, the path of each packet from $SP_{2,2}$ to $SP_{5,2}$ takes such a long way. In general our oblivious routing algorithm has to determine the path independently for each of the 36^2 phases. Here is the rule: If the starting subplane is in the left side of the whole plane (as $SP_{2,2}$), then the packets are moved to the right end in the first stage of the algorithm, which allows us to prepare the sorting zone. If the starting subplane is on the right side, then the packets are moved to the left end in the first stage. If the starting and ending subplanes are on the same row, then the path goes like the one given by the dotted line in Figure 4. It should be noted that this path design is obviously source-oblivious in each time-step, but is not in the whole time-steps. To make it source-oblivious in the whole steps is not hard, see the remark at the end of this section.

Now we give more detailed description of a single phase of the algorithm Rout[2]. The size of each queue is two ($k = 2$) for a while, but it is extended to Rout[k] later. Suppose for exposition that it is now the phase where packets from a subplane in the left side of the upper half plane to another subplane in the lower half plane should move. We call those packets *active packets* on this phase. Also, suppose that the active packets move first through subplanes $PZ_1, \cdots, PZ_p$ as the parallel shifting zone, then move through subplanes SZ_1, SZ_2 and SZ_3 as the sorting zone, next move through subplane CZ as the critical zone, and finally move through $PZ'_1, \cdots, PZ'_q$ as the parallel shifting zone again for $p \geq 0$ and $q \geq 0$. (By $p = 0$ ($q = 0$) we mean that the algorithm needs no shifting subplanes before the sorting zone (after the critical zone).)

Algorithm: Rout[2]

A single phase of Rout[2] is divided into the following seven stages:

Stage 1: If $p = 0$, then go to Stage 2. Otherwise, the following is executed on every row in parallel: All the active packets in PZ_1 are routed horizontally into SZ_1 without changing their relative positions within every subplane. More precisely, at each step, each processor shifts rightward its packet it currently holds in its queue. Eventually, all the active packets simultaneously arrive at their temporal destinations in SZ_1 exactly at the pnth step since it is exactly pn positions long for every active packet to travel to its temporal destination in SZ_1.

Stage 2 (Sorting): The following is executed on every row in parallel: The same algorithm as the one described in the proof of Lemma 1 is executed by using $2n$ processors, P_1 through P_{2n}, on the row in SZ_1 and SZ_2.

However, the present situation is a little bit different. Now some processor may have no packet since only active packets on this phase have been routed here. So, the initial sequence of packets on P_1 through P_n may include spaces or "null packets," but it does not cause any problem by regarding the null packets as the farthest (going out latest) packets.

Stage 3 (Counting): The following is executed on every row in parallel: Suppose that P_{n+1} through P_{3n} are $2n$ processors in the row in SZ_2 and SZ_3. Then we change the current position of each packet in SZ_2 into the symmetrical position with respect to the boundary between SZ_2 and SZ_3 as follows: (i) Each of the processors P_{n+1} through P_{2n} in SZ_2 simply shifts every active packet one position to the right step by step. (ii) Also, at each step, each P_{2n+i} in SZ_3 shifts rightward its packet it currently holds in its queue except for the packet, say, y_{2n+i} which arrives there exactly at the $(2i-1)$th step (from the beginning of this stage) for $1 \leq i \leq n$. Simultaneously, each P_{2n+i} calculates how many packets whose destinations are the same column as its packet y_{2n+i} it has sent to the right. Then, at the $(2n-1)$th step of this stage, P_{2n+i} stores the number of the packets plus one (for y_{2n+i}) in N_{2n+i}.

Stage 4 (Moving Long Lumps in Farthest-First Order): The following is executed on every row in parallel: If $N_{2n+i} \geq d$, then every P_{2n+i} in SZ_3 $(1 \leq i \leq n)$ starts to move its active packet to the right at the first step. Then the packet keeps moving one position to the right at each step. Otherwise, the processor does not start and remains holding its packet in its queue. However, if each P_{2n+j} with $N_{2n+j} < d$ receives a packet whose destination is the same column as its packet y_{2n+j} during this stage, then it starts to move y_{2n+j} to the right by using the farthest-first contention resolution protocol.

Stage 5 (Moving Short Lumps in Shuffle Order): The following is executed on every row in parallel: Note that there remain only packets of short lumps in SZ_3. At the first step, all P_{2n+i}'s such that the values of N_{2n+i}'s are equal to one start to forward their active packets to the right (if they have the packets in their queues), and then those packets keep moving rightward. At the $(n+1)$th step, all P_{2n+i}'s with $N_{2n+i} = 2$ start to move their packets to the right, at the $(2n+1)$th step all P_{2n+i}'s with $N_{2n+i} = 3$ start to move their packets to the right, and so on. Namely, n steps are inserted between the first actions. Finally, at the $((d-2)n+1)$th step, every P_{2n+i} with $N_{2n+i} = d-1$ starts to forward its packet. All the packets keep being forwarded to the right.

K. Iwama et al., *Oblivious Mesh Routing*, JGAA, 5(5) 17–38 (2001)   29

Stage 6 (Critical Zone): Right after the sorting zone, the packets enter the critical zone. Recall that long lumps enter first in farthest-first order and then short lumps in shuffle order. Each packet changes the direction from row to column at the crossing of its correct destination column (relatively in the subplane). What Rout[2] does in this zone is exactly the same as A_0, i.e., it gives a higher priority to turning packets.

Stage 7 (Shifting Zone): All active packets move in the shifting zone towards their final goals without changing their relative positions within the subplane if $q \geq 1$. (Otherwise, the packets should have already been routed to their final positions).

Stage 1 and Stage 2 take pn steps and $(2n-1)$ steps, respectively, as shown before. Suppose for example that $n = 10$ and the sorted sequence of 10 packets be now placed on processors P_{11} through P_{20} in SZ_2 after Stage 2 (see the proof of Lemma 1 again):

$$P_{11}\, P_{12}\, P_{13}\, P_{14}\, P_{15}\, P_{16}\, P_{17}\, P_{18}\, P_{19}\, P_{20}\, P_{21}\, P_{22}\, P_{23}\, P_{24}\, P_{25}\, P_{26}\, P_{27}\, P_{28}\, P_{29}\, P_{30}$$
$$a_5\quad a_4\quad a_3\quad a_2\quad a_1\quad b_2\quad b_1\quad c_3\quad c_2\quad c_1$$

One can see that if every packet moves one position to the right step by step in Stage 3, then c_1 arrives at the temporal destination P_{21} in the first step, c_2 arrives at P_{22} in the third step, c_3 arrives at P_{23} in the fifth step, and so on. Finally, the leftmost packet a_5 arrives at the rightmost processor P_{30} in the 19th step of this stage (since $2 \times 10 - 1 = 19$):

$$P_{11}\, P_{12}\, P_{13}\, P_{14}\, P_{15}\, P_{16}\, P_{17}\, P_{18}\, P_{19}\, P_{20}\, P_{21}\, P_{22}\, P_{23}\, P_{24}\, P_{25}\, P_{26}\, P_{27}\, P_{28}\, P_{29}\, P_{30}$$
$$c_1\quad c_2\quad c_3\quad b_1\quad b_2\quad a_1\quad a_2\quad a_3\quad a_4\quad a_5$$

Thus, after Stage 3, the stream of packets is changed into the farthest-first order. More importantly, each of P_{21} through P_{30} now knows how many packets in the same lump exist on its right side by counting the number of the flowing packets. For example, P_{22} knows another c exists in P_{23}, i.e., P_{22} sets $N_{22} = 2$.

Suppose for a while that $d = 4$, i.e., a lump is defined to be long if it includes four or more packets and only the lump of a's is long in the above example. At the first step of Stage 4, P_{26} and P_{27} start to move a_1 and a_2 to the right, respectively, since $N_{26} = 5$ and $N_{27} = 4$. Since P_{28}, P_{29} and P_{30} eventually receive a_1 (or a_2) from the left, they move their packets, a_3, a_4 and a_5 to the right. Thus all a's are routed rightward. In similar ways, all the packets of long lumps move forward. Note that, for example, P_{21} who does not know whether the lump of c's is long (since $N_{21} = 3$) does not initiate the move of its packet, and thus this short lump remains in SZ_3.

In Stage 5, ROUT[2] works as follows: At the first step of Stage 5, P_{25} and P_{23} start to forward b_2 and c_3, respectively, to the right since $N_{25} = N_{23} = 1$. Then, at the $(n+1)$th step, b_1 and c_2 start to move since $N_{24} = N_{22} = 2$, and finally c_1 starts to move at the $(2n+1)$th step since $N_{21} = 3$. Namely, we change the farthest-first sequence to the following shuffle order sequence (which

K. Iwama et al., *Oblivious Mesh Routing*, JGAA, 5(5) 17–38 (2001)   30

may include spaces between some two packets):

$$c_1 \qquad c_2 \qquad b_1 \qquad c_3 \qquad b_2$$

Now we shall investigate the time complexity of Rout[2]. The following discussions are only about what happens in the critical zone since the time needed for the sorting and shifting zones is linear in n. Suppose that a packet σ is now in the left input queue of a processor P_i, where σ is to enter the destination column by making a turn. Then it is said that σ is *ready to turn*. We first show a simple but important lemma, which our following argument depends on.

Lemma 2 *Suppose that the rightmost packet of a lump L of packets is now ready to turn into its destination column. Then all the packets in L go out of the critical zone at most $3n$ steps regardless of the behavior of other packets.*

Proof: Suppose that the last (leftmost) packet α which is in the same lump L cannot move on. Then there must be a packet, β, which is ready to turn into the same column in some upper position of α and move on in the next time-step. Let us call such a packet β, a *blocking packet against* α. Note that β cannot be a blocking packet against α any longer in the next step. Thus, in each step from now on, the following (i) or (ii) must occur: (i) The leftmost packet α in L moves one position. (ii) One blocking packet against α disappears. Since there are at most n packets that are to turn into a single column, (ii) can occur at most n times. (i) can occur at most $2n$ times also since we have at most $2n$ processors on the path of each packet in the critical zone.   □

Recall that a lump is said to be long if it includes at least d packets for some positive constant d; otherwise, it is said to be short.

Lemma 3 *All the long lumps can go through the critical zone within $O(n^2/d)$ steps.*

Proof: The following argument holds for any row: Let L_i be a lump of packets in some row such that packets of L_i head for a farther column than packets of L_j for $i < j$. Since packets of long lumps move in the farthest-first order, they flow in the following form of lumps:

$$\cdots L_3 \cdots L_2 \cdots L_1$$

Since each long lump has at least d packets, there are at most n/d different lumps in each row. Now the rightmost packet of L_1 must be ready to turn within $2n$ steps. By Lemma 2, then, those packets in L_1 must go out of the critical zone within the next $2n$ steps, i.e., within $4n$ steps in total. After L_1 goes out of the critical zone, L_2 must go out within $4n$ steps and so on. Thus $4n \times n/d = 4n^2/d$ is the maximum amount of steps all the long lumps need to go out.   □

K. Iwama et al., *Oblivious Mesh Routing*, JGAA, 5(5) 17–38 (2001)   31

Lemma 4 *All the short lumps can go through the critical zone within $O(dn)$ steps.*

Proof: The flow of packets on each row looks as follows:

$$\cdots z_1 \cdots z_{l-1} \, z_l \, y_1 \, \cdots \, y_{j-1} \, y_j \, x_1 \, \cdots \, x_{i-1} \, x_i$$

Here $x_1 \cdots x_i$ is a sequence of different packets such that x_i is heading for a farther column than x_{i-1}. Let us call this sequence an *ordered string*. $y_1 \cdots y_j$ is the next ordered string (i.e., y_j is heading for a farther column than x_1) and so on. One can see that each packet in the first ordered string becomes ready to turn within $2n$ steps and must go out of the critical zone within the next $2n$ steps by Lemma 2 (regardless of possible spaces between some two packets). After that, the second ordered string must be ready to turn within $2n$ steps and then goes out as before. Note that we have only d ordered strings since each short lump has less than d packets. Hence, $4n \times d$ is the enough number of steps before all the short lumps go out.   □

Now we are almost done. When moving packets from the sorting zone to the critical zone, we first move only lumps whose size is at least $\sqrt{n}$. Then by Lemma 3, the routing for these "long" lumps will finish in $O(n\sqrt{n})$ steps. After that we move "short" lumps, which will be also finished, by Lemma 4, in $O(n^2/\sqrt{n}) = O(n\sqrt{n})$ steps.

Algorithm Rout[2] can be extended to Rout[k] for a general queue-size k: Suppose from now on that a lump is said to be long if it includes at least kd packets for some positive constant d; otherwise, it is said to be short. Then we can obtain similar lemmas, say Lemmas 2', 3' and 4' to Lemmas 2, 3 and 4, respectively for the general queue-size. Lemma 2' is exactly the same as Lemma 2. As for Lemma 3', we replace d by kd. Then the time complexity for Rout[k] is $O(n^2/kd)$. As for Lemma 4', we also replace d by kd and furthermore each ordered string $x_1 \cdots x_{i-1}x_i$ is replaced by $X_1 \cdots X_{i-1}X_i$. Here X_{i-m} consists of k packets of the same destination instead of only one packet for x_{i-m}. Since our queue-size is k this time, each X_{i-m} can fit into a single input queue, which means all the packets in X_{i-m} can get ready to turn within $2n$ steps. Also note that the number of the ordered strings is at most d. Hence the bound in Lemma 4, $O(nd)$, does not differ in Lemma 4', which is a benefit of the large queue-size.

If we set $d = \sqrt{n/k}$, then both $O(n^2/kd)$ and $O(nd)$ become $O(n^{1.5}/\sqrt{k})$.   □

Remark. Recall that the current movement of packets is not source-oblivious in the whole time-steps, which can be modified as follows: Suppose that the goal subplane is in the left-lower of the whole plane. Then all the packets but in the rightmost subplanes move upward first. Then they move to the right in the uppermost subplanes, and then go downward in the rightmost subplanes (here we place the sorting zone), and finally go to the left. (Exceptionally, packets in the rightmost-lower subplanes move to the left first and then follow the same path as above.) Thus there is only one direction set in each subplane, i.e., we can design a source-oblivious algorithm. However, here is one important point: Packets originally placed near the destination in this path cannot go through the sorting zone. Hence, we allow them once to go through the destinations and then join the standard path after that.

Remark. Extension to 3D case. Let the side-length of a 3D mesh be n (see Figure 2). Our algorithm consists of n sequential phases. In the first phase, only n^2 packets placed on the top horizontal plane are routed to their final positions. In the second phase, only n^2 packets placed on the second horizontal plane move, and so on. Here is an outline of each phase: (1) Each packet σ first moves up/down temporarily to the wth horizontal yz-plane if the destination of σ is on the wth vertical xy-plane. (2) The same algorithm as Rout$[k]$ is performed on every horizontal plane in parallel, however, packet $[u, v, w]$ is routed temporarily to position (w, v, u). (3) Every packet moves along z-dimension to its correct z-coordinate and finally moves along x-dimension to its destination. Hence each phase can be performed in $O(n^{1.5}/\sqrt{k})$ steps ($2 \leq k \leq cn$ for some constant c) as follows: We need at most $n-1$ steps to move each of the n^2 packets placed on some horizontal plane up to its correct horizontal one, and we need $O(n^{1.5}/\sqrt{k})$ steps to move the packets in one horizontal plane. It is not hard to move each packet from the horizontal to the vertical planes in $O(n)$ steps. Since there are n sequential phases, the total routing time is $O(n \cdot n^{1.5}/\sqrt{k}) = O(n^{2.5}/\sqrt{k})$ (or $O(N^{5/6}/\sqrt{k})$ for $2 \leq k \leq cN^{1/3}$). Although details are omitted, the similar ideas may provide us some $o(N/\sqrt{k})$ upper bounds for oblivious routing on higher dimensional, constant queue-size meshes including N processors, e.g., an $O(N^{7/8}/\sqrt{k})$ upper bound for 4D meshes, an $O(N^{9/10}/\sqrt{k})$ upper bound for 5D meshes, and so on.

4   3D Oblivious Routing

For oblivious routing on any d-degree network, Kaklamanis *et al.* [6] proved an $\Omega(\sqrt{N}/d)$ lower bound. It is known [6, 9, 12] that this bound is tight for the hypercube and the 2D meshes. However, in the case of 3D meshes, it was not known whether the lower bound is indeed tight. Here is our answer:

Theorem 2 *There exists a deterministic, oblivious routing algorithm on the three-dimensional, $N^{1/3} \times N^{1/3} \times N^{1/3}$ mesh that routes any permutation in $1.16\sqrt{N} + o(\sqrt{N})$ steps.*

Remark. Theorem 2 holds for minimal and 3-bend routing. However, if we impose the 2-bend condition, then any oblivious routing algorithm requires $\Omega(N^{2/3})$ steps [5]. Recall that in the case of 2D meshes, dimension-order routing, which is even more rigid than the 1-bend routing, can achieve the tight $O(\sqrt{N})$ bound. Thus there is a big gap between 2D and 3D meshes in the difficulty of oblivious routing.

K. Iwama et al., *Oblivious Mesh Routing*, JGAA, 5(5) 17–38 (2001)   33

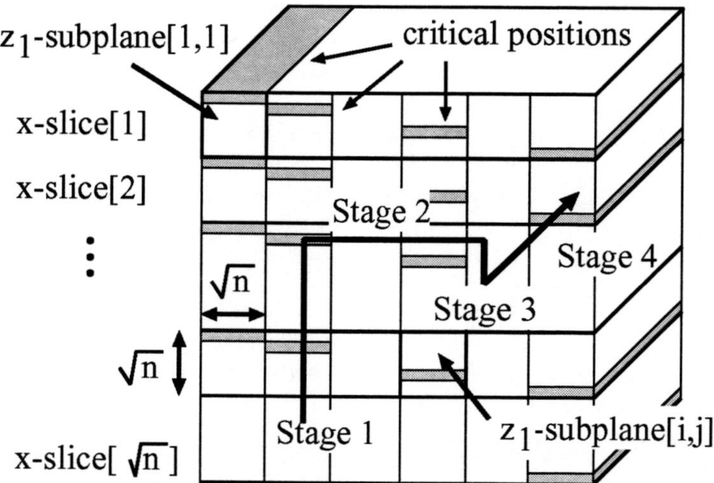

Figure 5: Slices, subplanes, critical positions

Proof: Let the total number of processors in 3D meshes be hereafter n^3, not N. We first present a deterministic, but non-minimal oblivious routing algorithm on the 3D mesh that routes any permutation in $O(n\sqrt{n})$ $(= O(\sqrt{N}))$ steps. At the end of this section, we will describe how to strengthen the algorithm to route every packet along its shortest path. See Figure 2 again. The three dimensions are denoted by *x-dimension*, *y-dimension* and *z-dimension*. x_i-*plane* $(1 \le i \le n)$ means the two-dimensional plane determined by $x = i$. Similarly for y_j-plane and z_k-plane. For example, the x_1-plane on the 3D mesh is the top, horizontal plane in Figure 2, which includes n^2 positions $(1,1,1), \cdots, (1,n,1)$, $(1,1,2), \cdots, (1,n,n)$. Let $x_i y_j$-*segment* $(1 \le i,j \le n)$ denote the linear array determined by $x = i$ and $y = j$, which consists of n processors $(i,j,1)$ through (i,j,n). Similarly for $y_j z_k$-segment and $z_k x_i$-segment.

The whole $n \times n \times n$ 3D mesh is partitioned into several submeshes. See Figure 5. *x-slice*[1] consists of the top $\sqrt{n}$ planes, the x_1-plane through the $x_{\sqrt{n}}$-plane. Similarly *x-slice*[2] consists of the next $\sqrt{n}$ planes. Each plane is divided into $\sqrt{n} \times \sqrt{n}$ groups, called *subplanes*. z_k-*subplane*$[i,j]$ on the z_k-plane $(1 \le k \le n)$ is the ith $\sqrt{n} \times \sqrt{n}$ 2D submesh from the top and jth from the left. Then $\sqrt{n}$ processors on the jth row from the top of each z_k-subplane$[i,j]$ are called *critical positions*, which play on an important role in the following algorithms. Note that the number of critical positions on a single $z_k x_i$-segment is exactly $\sqrt{n}$.

The following observation makes clear the reason why the elementary-path routing or dimension-order routing does not work efficiently: Suppose that n^2 packets placed on the z_1-plane should move to the x_1-plane. Also suppose that all of those n^2 packets first go up to the n processors on a single segment, $P_{1,1,1}$ through $P_{1,n,1}$, and then move along y-dimension. Then most of the n^2

K. Iwama et al., *Oblivious Mesh Routing*, JGAA, 5(5) 17–38 (2001)   34

paths probably go through one of the $n-1$ links between those n processors, which requires at least $\Omega(n^2)$ steps. Apparently we need to eliminate such heavy traffic on the single segment. The key strategy is as follows: The n^2 packets or paths (which go through a single segment in the worst case) are divided into $\sqrt{n}$ groups, $n\sqrt{n}$ paths each. Then the number of paths which a single segment has to handle can be reduced to those $n\sqrt{n}$ paths from the previous n^2 ones, where critical positions of each subplane play their roles. Here is our algorithm based on the dimension-order routing (see Figure 5 again):

Algorithm: DO-3-bend

Stage 1: Every packet moves along x-dimension to a critical position which is located in the same x-slice$[i]$ as its final destination. Namely, all of the packets go up or down into their correct x-slices.

Stage 2: Every packet placed now on the critical position moves along y-dimension to its correct y-coordinate, i.e., moves horizontally into its correct y_j-plane.

Stage 3: Every packet moves again along x-dimension to its correct x-coordinate, i.e., into its correct $x_i y_j$-segment.

Stage 4: Every packet moves along z-dimension to its final position.

One can see that the path of each packet is completely determined by its initial position and destination and that the algorithm can deliver all the packets for any permutation.

Lemma 5 DO-3-bend *can route any permutation in at most* $2n\sqrt{n} + o(n\sqrt{n})$ *steps.*

Proof: We shall analyze the running time of each stage: (1) Stage 1 requires at most n steps since every packet can move without delays. (2) Stage 2 requires at most $n\sqrt{n}+n$ steps. The reason is as follows: Recall that the number of critical positions on a single segment is exactly $\sqrt{n}$. Since each critical position holds at most n packets after Stage 1, the total number of packets that go through each $z_k x_i$-segment is at most $n\sqrt{n}$. Thus, it takes at most $n\sqrt{n}$ steps until those $n\sqrt{n}$ packets move out of the current subplane and it furthermore takes at most n steps until the packet which finally starts from the subplane arrives at its next intermediate position. (3) After Stage 2 every packet arrives at its correct sub-y_j-plane of the $\sqrt{n} \times n$ positions. Thus, the total number of packets that go through each link along x-dimension is at most $n\sqrt{n}$. Thus it takes at most $n\sqrt{n} + \sqrt{n}$ steps until all the packets arrive at their next positions. (4) Since each processor temporally holds at most n packets, $2n$ steps are sufficient for all the packets to arrive at their final positions. As a result, the whole algorithm requires at most $2n\sqrt{n} + o(n\sqrt{n})$ steps. $\square$

The above algorithm is based on the dimension-order routing, i.e., all the packets move in the same direction in each stage. However, by making the

K. Iwama et al., *Oblivious Mesh Routing*, JGAA, 5(5) 17–38 (2001)   35

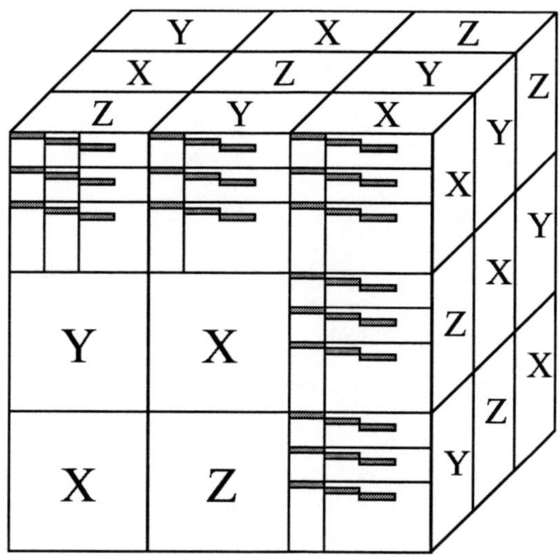

Figure 6: Three groups, X, Y, Z

packets move in different directions we can route more efficiently. The whole
3D mesh is partitioned into the three groups shown in Figure 6. Then, in the
first stage, the algorithm NDO-3-bend moves all the packets which are initially
placed on the groups marked "X", "Y" and "Z" along x-dimension, y-dimension
and z-dimension, respectively. In the remaining stages, dimensions are switched
adequately for each group. Note that each subplane can be viewed as a $\sqrt{n/3} \times
\sqrt{n/3}$ 2D mesh. Now the following lemma holds:

Lemma 6 NDO-3-bend *can route any permutation in at most* $1.16n\sqrt{n}+o(n\sqrt{n})$
steps.

Proof: It also takes at most n steps for each packet to arrive at its critical
position in the first stage. Note that there are $3 \times \sqrt{n/3}$ critical positions on a
single $z_k x_i$-segment and each critical position holds at most $n/3$ packets after
the first stage. Then it takes at most $n\sqrt{n/3}+n$ steps during the second stage.
In the third and fourth stages, it takes at most $n\sqrt{n/3} + \sqrt{n/3}$ and $2n$ steps,
respectively. Since $2/\sqrt{3} \leq 1.16$, the lemma holds.   □

    □

Finally we add the condition that all of the packets should follow their shortest routes: Suppose that a packet α which heads for the x_t-plane is originally
placed on the x_s-plane for $s > t$. Then, in the first stage, α should go up along
x-dimension to the closest but below critical position to the x_t-plane so as to
follow its shortest path. Similarly for $s < t$. However, as a special case, if
there is no critical position between α's original and destination planes, then α

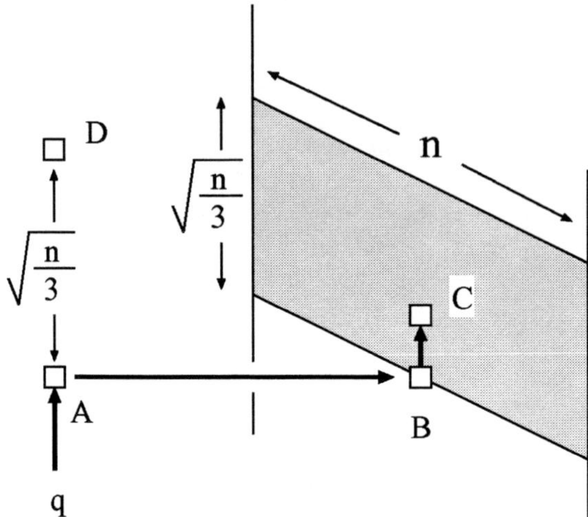

Figure 7: Minimal routing

does not move at all in the first stage. Let us count the total number of packets which temporally stay at each $z_k x_i$-segment (i.e., each horizontal segment) after the first stage. $\sqrt{3n}$ critical positions on a single segment hold at most $n\sqrt{n/3}$ packets, $n/3$ per critical position, and furthermore, at most n special packets may stay without moves. Thus it takes at most $n\sqrt{n/3} + 2n$ steps in the second stage. For the third stage, at most $n\sqrt{n/3} + \sqrt{n/3}$ steps suffice again. The reason is as follows: See Figure 7. Suppose that a packet q first moves up vertically to a critical position, A, and then it moves horizontally to a position, B. Then q's destination must belong to the upper $\sqrt{n/3} \times n$ positions of B, which is painted gray in the figure; otherwise, q must first arrive at the upper critical position, D. This implies that the number of packets which go through the link between B and its upper neighbor C is bounded by $n\sqrt{n/3}$. Implementation of this idea is not hard:

Theorem 3 *There exists a deterministic, oblivious, minimal, 3-bend routing algorithm on the three-dimensional, $N^{1/3} \times N^{1/3} \times N^{1/3}$ mesh that can route any permutation in at most $1.16n\sqrt{n} + o(n\sqrt{n})$ steps.*

Remark. Each algorithm can be modified so as to satisfy the source-oblivious condition by increasing the constant factor of its running time slightly.

Remark. It is not hard to show that our algorithm runs for random permutations in $3n + o(n)$ steps with high probability. Here is the reason: Since there are $\sqrt{n}$ critical positions on each vertical segment, each critical position receives $\sqrt{n}$ packets on average in the first stage. Then $\sqrt{n} \times \sqrt{n} = n$ packets temporally stay on each horizontal segment. In the second stage, those n packets move horizontally in n steps, and each processor receives one packet on average.

The third stage requires $\sqrt{n}$ steps. Finally, all the packets arrive at their final positions in n steps.

Acknowledgments

The authors wish to thank two anonymous referees for providing detailed suggestions for improving our original paper.

References

[1] A. Bar-Noy, P. Raghavan, B. Schieber and H. Tamaki, "Fast deflection routing for packets and worms," In *Proc. 1993 ACM Symposium on Principles of Distributed Computing* (1993) 75-86.

[2] A. Borodin and J.E. Hopcroft, "Routing, merging, and sorting on parallel models of computation," *J. Computer and System Sciences* 30 (1985) 130-145.

[3] A. Borodin, P. Raghavan, B. Schieber and E. Upfal, "How much can hardware help routing?," In *Proc. 1993 ACM Symposium on Theory of Computing* (1993) 573-582.

[4] D.D. Chinn, T. Leighton and M. Tompa, "Minimal adaptive routing on the mesh with bounded queue size," *J. Parallel and Distributed Computing* 34 (1996) 154-170.

[5] K. Iwama and E. Miyano, "Three-dimensional meshes are less powerful than two-dimensional ones in oblivious routing," In *Proc. 5th European Symposium on Algorithms* (1997) 284-295.

[6] C. Kaklamanis, D. Krizanc and A. Tsantilas, "Tight bounds for oblivious routing in the hypercube," *Mathematical Systems Theory 24* (1991) 223-232.

[7] C. Kaklamanis, D. Krizanc and S. Rao, "Simple path selection for optimal routing on processor arrays," In *Proc. 1992 ACM Symposium on Parallel Algorithms and Architectures* (1992) 23-30.

[8] D. Krizanc, "Oblivious routing with limited buffer capacity," *J. Computer and System Sciences* 43 (1991) 317-327.

[9] F.T. Leighton, *Introduction to Parallel Algorithms and Architectures: Arrays, Trees, Hypercubes*, Morgan Kaufmann (1992).

[10] F.T. Leighton, F. Makedon and I. Tollis, "A $2n - 2$ step algorithm for routing in an $n \times n$ array with constant queue sizes," *Algorithmica* 14 (1995) 291-304.

K. Iwama et al., *Oblivious Mesh Routing*, JGAA, 5(5) 17–38 (2001)   38

[11] J.F. Sibeyn, B.S. Chlebus and M. Kaufmann, "Deterministic permutation routing on meshes," *J. Algorithms* 22 (1997) 111-141.

[12] M. Tompa, *Lecture notes on message routing in parallel machines,* Technical Report # 94-06-05, Department of Computer Science and Engineering, University of Washington (1994).

[13] L.G. Valiant and G.J. Brebner, "Universal schemes for parallel communication," In *Proc. 1981 ACM Symposium on Theory of Computing* (1981) 263-277.

Journal of Graph Algorithms and Applications

http://www.cs.brown.edu/publications/jgaa/

vol. 5, no. 5, pp. 39–57 (2001)

Finding All the Best Swaps of a Minimum Diameter Spanning Tree Under Transient Edge Failures

Enrico Nardelli   *Guido Proietti*

Dipartimento di Matematica Pura ed Applicata
Università di L'Aquila, Italy, and
Istituto di Analisi dei Sistemi e Informatica, CNR, Roma, Italy
http://w3.dm.univaq.it/~{nardelli,proietti}
{nardelli,proietti}@univaq.it

Peter Widmayer

Institut für Theoretische Informatik
ETH Zentrum, Zürich, Switzerland
http://www.inf.ethz.ch/personal/widmayer
widmayer@inf.ethz.ch

Abstract

In network communication systems, frequently messages are routed along a minimum diameter spanning tree (MDST) of the network, to minimize the maximum delay in delivering a message. When a transient edge failure occurs, it is important to choose a temporary replacement edge which minimizes the diameter of a new spanning tree. Such an optimal replacement is called a *best swap*. As a natural extension, the *all-best-swaps (ABS) problem* is the problem of finding a best swap for every edge of the MDST. Given a weighted graph $G = (V, E)$, where $|V| = n$ and $|E| = m$, we solve the ABS problem in $O(n\sqrt{m})$ time and $O(m)$ space, thus improving previous bounds for $m = o(n^2)$. We also show that the diameter of the tree resulting from a best swap is at most 5/2 as long as the diameter of a MDST recomputed from scratch.

Communicated by Takao Nishizeki, Roberto Tamassia, and Dorothea Wagner:
submitted January 1999; revised February 2000 and December 2000.

Work supported by the EU TMR Grant CHOROCHRONOS and by the Swiss National Science Foundation. A preliminary version of this paper was presented to the 6th European Symposium on Algorithms (ESA'98), Venice, Italy, 1998.

E. Nardelli et al., *All Best Swaps of a MDST*, JGAA, 5(5) 39–57 (2001)   40

1   Introduction

For communication networks, it is important to remain operational even if individual network components fail. In the past few years, the ability of a network to survive a failure (its *survivability*) has been studied intensely (an excellent survey paper on survivable networks is [5]). From the practical side, this has largely been a consequence of the replacement of metal wire meshes by fiber optic networks: Their extremely high bandwidth makes it economically attractive to make networks as sparse as possible. A sparse network, however, is less likely to survive a link (or node) failure than a mesh with a multitude of connections that can be used as detours. Therefore, it is important for sparse networks to also take survivability into account from the very beginning.

In the extreme, a sparse network might coincide with a spanning tree of some underlying weighted graph G of n nodes and m edges. Such a spanning tree generally minimizes some objective function defined according to some network functionality that one wants to optimize. A typical example is that of the *minimum spanning tree* (MST), i.e., a spanning tree whose total weight is minimum among all the spanning trees of G. For tree-like communication networks, however, even a single link failure will affect the service. Notice that these failures should be considered *temporary* (or *transient*), since usually the disrupted link will be restored reasonably quickly. Hence, it makes sense to guarantee the network functionality during this short recovering time by simply reconnecting the two disconnected components through a *single* replacement link. The *emergency network* thus obtained will not be, in general, a best possible spanning tree (according to the chosen objective function) of the graph G now deprived of the failed link. However, it offers the considerable advantage of being readily operative at a low cost, since it will reuse almost all of both the physical network and the communication protocol previous to the failure. The alternative of rebuilding a new optimal spanning tree, which might possibly be totally different from the previous one, would indeed be excessive and disproportionate, considering its short life.

On the theoretical side, this scenario gave rise to an interesting family of problems on graphs. The fundamental question we want to answer to is the following: *Given a weighted, undirected and 2-edge connected graph* $G = (V, E)$, *and a spanning tree* $T = (V, E_T)$ *of* G *which minimizes some objective function* $\phi(G, T)$, *and given an edge* $e \in E_T$, *find an edge* $e' \in E$ *such that the* swap tree $T_{e/e'}$ *obtained by swapping* e *with* e' *is a spanning tree of* $G - e = (V, E \setminus \{e\})$, *and* $\phi(G - e, T_{e/e'})$ *is minimum over all the possible swap trees.* Such an edge e' will be referred to in the following as a *best swap edge* for e with respect to ϕ. To make things more concrete, let us go back to the sample case where T is a MST of G. Here, it is easy to see that a best swap edge is simply an edge of minimum weight connecting the two subtrees of T induced by the removal of an edge. Incidentally, in this particular case, the swap tree associated with a best swap edge, exhibits the property of being a MST of $G - e$. Another popular example arises when T is a *single source shortest paths tree* (SPT) rooted at a given node $r \in V$. In this case, the objective function minimized by T is the

E. Nardelli et al., *All Best Swaps of a MDST*, JGAA, 5(5) 39–57 (2001)   41

sum of the lengths of all the paths emanating from r, and it is not hard to see that a swap tree $T_{e/e'}$ *does not* coincide, in general, with a SPT rooted in r of $G - e$.

Concerning the algorithmic perspective, we observe that, from a network management point of view, it is desirable to know in advance, for each edge in the network, its respective best swap edge, since we can expect that sooner or later each edge will fail. This approach entails at least two advantages: First, as a failure happens, we are prepared to switch immediately to the emergency network; second, and perhaps most important, we can evaluate in advance the *vitality* of an edge, by measuring the degradation of the network functionality (as conveyed by the chosen optimization function) as soon as we switch to the emergency network. An *all-best-swaps* (ABS) problem is therefore the problem of finding a best swap edge–with respect to a given optimization function– *for every* edge in the network. In the past, the ABS problem has been solved both when the network is a MST and a SPT. In the first case, the fastest solution known to date is an $O(m)$ time algorithm [2], while in the second case, an $O(n^2)$ time algorithm has been presented in [9].

However, in several applications, the used spanning tree is neither an MST nor a SPT. Rather, many network architectures look for minimizing the *diameter* of the spanning tree, that is the length of the longest path in the tree between any two nodes, so that the maximum propagation delay in the network will be as low as possible. A tree exhibiting minimum diameter among all possible spanning trees is named a *minimum diameter spanning tree (MDST)*. A best swap edge for a failing edge of a MDST is a swap edge that keeps the diameter of the swap tree as low as possible among all possible swap trees, and the ABS problem for a MDST is defined accordingly.

In this paper we present an efficient solution precisely for this latter problem. Our approach makes use of some of the techniques presented in [8], in which the authors consider the related problem of computing a best swap edge in a fully dynamic context, where the original MDST evolves due to repeated insertions and deletions of edges in the graph. The solution presented in [8] computes a best swap in $O(n)$ time per update, using $O(m)$ space and preprocessing time. For the edge deletion case, that is of interest for this paper, the authors make use of a topology tree and a 2-dimensional topology tree [3], augmented with some extra information. This requires $O(m)$ time and space for initialization, and allows to compute the length of the diameter of any tree obtained as a consequence of a swap in $O(n)$ time. Hence, the approach in [8] is more general than what is needed for solving the ABS problem, and if we use it for solving the ABS problem, we spend $O(n^2)$ time and $O(m)$ space and preprocessing time. We get these bounds by computing a best swap in $O(n)$ time for each deleted edge. Recently, Alstrup et al. [1] improved the runtime for computing a best swap in an incremental context (i.e., when no deletions are allowed) to $O(\log^2 n)$. In the same paper, the authors claim that it is possible to maintain the diameters of trees in a dynamic forest under link and cut operations in $O(\log n)$ time and $O(n)$ space and preprocessing time. Using this and some of the results contained in [8], the ABS problem can be solved in the following way: we sequentially

E. Nardelli et al., *All Best Swaps of a MDST*, JGAA, 5(5) 39–57 (2001)   42

remove each edge in the MDST; when an edge e is removed, we initially compute in $O(\sqrt{m})$ time a set of $O(\sqrt{m})$ selected edges, among which a best swap is contained [8]; hence, to compute a best swap, we sequentially insert each of these selected edges, compute in $O(\log n)$ the new diameter and then remove it; finally, we re-insert e and move to the next edge in the MDST. Therefore, the obtained runtime is $O(n\sqrt{m}\log n)$, with $O(m)$ space and preprocessing time.

Our solution solves the ABS problem for a MDST in $O(n\sqrt{m})$ time and $O(m)$ space, thus strictly improving previous bounds for $m = o(n^2)$. This can be done by adapting the well-known *path halving* compression technique [10] for answering in $O(1)$ amortized time (over a total of $\Theta(n\sqrt{m})$ queries) the following query: *Given a rooted tree T and a pair of nodes y and v in T such that v belongs to the subtree of T rooted at y, what is the length of a longest simple undirected path in T, starting at v and staying within the subtree of T rooted at y?* Furthermore, we compare the diameter of $T_{e/e'}$ against that of a real MDST of $G-e$. Somewhat surprisingly, the ratio between them stays within a small constant factor in the worst case, that is $5/2$. Therefore, swapping can be viewed as both efficient and effective at the same time.

The paper is organized as follows: in Section 2 we define the problem more precisely and we propose the algorithm for solving it. In Section 3, we present the adaptation of the path halving compression technique which allows to efficiently solve the ABS problem. In Section 4, we compare the diameter of the tree obtained as a consequence of a best swap of a failing edge e as opposed to the diameter of a real new MDST. Finally, in Section 5, we present conclusions and list some open problems.

2 Solving the ABS problem

2.1 Problem statement

For basic definitions of graph terminologies, we refer to [6]. Let $T = (V, E_T)$ be a spanning tree of a 2-edge connected, undirected graph $G = (V, E)$, of n nodes and m edges, and with a nonnegative real length $|e|$ associated with each edge $e \in E$. Let $\mathcal{D}(T) = \langle d_1, d_2, \ldots, d_k \rangle$ denote a *diameter path* of T, that is a path whose length $|\mathcal{D}(T)|$ is maximum in T. We call $|\mathcal{D}(T)|$ the *diameter* of T. For simplicity, we will use the term diameter also for denoting the diameter path, whenever no confusion arises. T is said to be a *minimum diameter spanning tree* (MDST) of G if it has minimum diameter among all the spanning trees of G. If $e \in E_T$ is a tree edge, a *replacement edge* (or *swap edge*) for e is an edge $f \in E \setminus E_T$ such that $T_{e/f} = (V, (E_T \setminus \{e\}) \cup \{f\})$ is a spanning tree of $G - e = (V, E \setminus \{e\})$. Let $\mathcal{R}_e$ denote the set of replacement edges for e. A *best swap edge* for e with respect to the diameter is an edge $e' \in \mathcal{R}_e$ such that

$$|\mathcal{D}(T_{e/e'})| = \min_{f \in \mathcal{R}_e} \{|\mathcal{D}(T_{e/f})|\}.$$

The *all-best-swaps* (ABS) problem for a MDST is the problem of finding a best swap edge with respect to the diameter for every edge $e \in E_T$.

E. Nardelli et al., *All Best Swaps of a MDST*, JGAA, 5(5) 39–57 (2001)   43

2.2   The algorithm

To solve the ABS problem efficiently, we will exploit relationships among the original MDST T and the replacing ones. Let us denote the length of the path in T between any two nodes v, v' as $d(v, v')$. Let d_c, with $1 \leq c \leq k$, be the *center* of the diameter path, that is the node in $\mathcal{D}(T)$ for which $|d(d_1, d_c) - d(d_c, d_k)|$ is minimum. Let us denote by M this minimum. Notice that, in general, this node is not unique, since it might exist a sequence of edges of length 0 in $\mathcal{D}(T)$, say $(d_i, d_{i+1}), (d_{i+1}, d_{i+2}), \ldots, (d_{i+j-1}, d_{i+j})$, such that $M = |d(d_1, d_h) - d(d_h, d_k)|, h = i, \ldots, i+j$, or it might exist an edge (d_i, d_{i+1}) in $\mathcal{D}(T)$ of positive length and such that $d(d_1, d_i) = d(d_{i+1}, d_k)$. In the former case, to break the tie, we set $d_c = d_{i+j}$, while in the latter case we set $d_c = d_{i+1}$. Let $\widehat{T}$ denote a source directed tree obtained by rooting T in d_c and orienting the edges towards the leaves. Following [8], we maintain a topology tree and a 2-dimensional topology tree [3, 4], augmented with some extra-information, to efficiently retrieve only $O(\sqrt{m})$ *selected edges* among the $O(m)$ replacement edges, whenever an edge e in T is deleted. In fact, among the selected edges, a best swap is contained (for a proof of it, see [8]).

The general outline of our algorithm is the following:

Algorithm ABS(G, T);
Input: A weighted, 2-edge connected graph $G = (V, E)$ and a MDST $T = (V, E_T)$ of G.
Output: $\forall e = (x, y) \in E_T$, a swap edge $e' : |\mathcal{D}(T_{e/e'})| = \min\limits_{f \in \mathcal{R}_e} \{|\mathcal{D}(T_{e/f})|\}$.
Step 0: Perform preliminary computations.
Step 1: For each edge $e \in \widehat{T}$ as considered by any postorder visit **do** {
Step 2:   Delete e; update the topology and the 2-dim topology tree.
Step 3:   Compute the set of selected replacement edges $\mathcal{S}_e \subseteq \mathcal{R}_e$.
Step 4:   For each edge $f \in \mathcal{S}_e$, compute $|\mathcal{D}(T_{e/f})|$ and select a best swap.
Step 5:   Insert e and update the topology and the 2-dimensional topology tree.}

Step 0 requires $O(m)$ time and space, as we show next. Notice that in Step 1 we consider all the $O(n)$ edges of the tree, in the order they are generated by a postorder tree visit (this order is needed for a correct path halving compression, as is shown in Section 3). Steps 2, 3 and 5 can be accomplished in $O(\sqrt{m})$ time, and $|\mathcal{S}_e| = O(\sqrt{m})$ [8]. Concerning Step 4, we will show that $|\mathcal{D}(T_{e/f})|$ can be computed in $O(1)$ amortized time. This can be done using the *path halving compression* technique that will be presented and analyzed in detail in Section 3. This technique, given a rooted tree $\widehat{T}$ and a pair of nodes u and v in $\widehat{T}$ such that u belongs to the subtree of $\widehat{T}$ rooted at v, allows to find in $O(\log_{(1+Q/n)} n)$ amortized time per query (over a total of Q queries) the length $Findpath(u, v)$ of a longest simple undirected path starting from u and staying within the subtree rooted at v. Since for solving the ABS problem we will prove that $Q = \Theta(n\sqrt{m})$, $Findpath(u, v)$ can be computed in $O(1)$ amortized time. Thus, Step 4 costs $O(\sqrt{m})$ amortized time, and the global time for solving the ABS problem is $O(n\sqrt{m})$, using $O(m)$ space.

E. Nardelli et al., *All Best Swaps of a MDST*, JGAA, 5(5) 39–57 (2001)   44

2.3   Preliminary computations

Let $\mathcal{L}_{\mathcal{D}} = \langle d_1, \ldots, d_{c-1}, d_c \rangle$ and $\mathcal{R}_{\mathcal{D}} = \langle d_c, d_{c+1}, \ldots, , , d_k \rangle$. W.l.o.g., let us assume $|\mathcal{L}_{\mathcal{D}}| \geq |\mathcal{R}_{\mathcal{D}}|$. Let N_l (N_r) denote the subset of nodes of $\widehat{T}$ rooted in d_{c-1} (d_{c+1}), and let N_c denote the nodes of $\widehat{T}$ neither in N_l nor in N_r. Figure 1 depicts this notation.

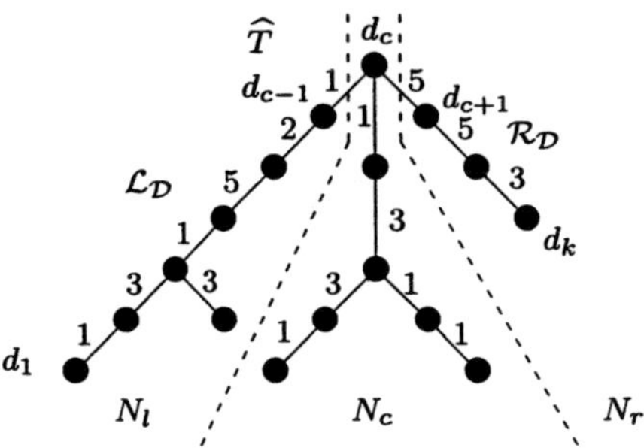

Figure 1: The oriented minimum diameter spanning tree.

We start by marking any node $v \in \widehat{T}$ with its *nearest ancestor* on $\mathcal{D}(T)$, that is the first node of $\mathcal{D}(T)$ we encounter on the path in $\widehat{T}$ from v to d_c. Notice that if v is on the diameter, then it is marked with itself. After, we associate with each node $v \in \widehat{T}$ its distance from d_c, and the lengths $h_1(v)$ and $h_2(v)$, with $h_1(v) \geq h_2(v)$, of the two longest directed paths in $\widehat{T}$ emanating from v and making use of two different subtrees of v (notice that if v is a leaf, then we set $h_1(v) = h_2(v) = 0$, while if there is only one subtree rooted at v, then we set $h_2(v) = 0$). Moreover, we also store with v the node $a(v)$ adjacent to v and belonging to the path of length $h_1(v)$ (notice that if $h_1(v) = 0$, then we set $a(v) = v$). With the root, we also associate a further value, say $h_3(d_c)$, corresponding to the length of a longest path in $\widehat{T}$ starting from d_c and not using d_{c-1} and d_{c+1} (notice that if such a path does not exist, then we set $h_3(d_c) = 0$). After, we associate with each node $d_j \in \mathcal{L}_{\mathcal{D}}$ ($d_j \in \mathcal{R}_{\mathcal{D}}$) a further value, say $\lambda(d_j)$, containing the length of a longest path in $\widehat{T}$ starting from d_c and containing neither d_{c+1} (d_{c-1}) nor the edge (d_j, d_{j-1}) ((d_j, d_{j+1})). Note that since d_c belongs to both $\mathcal{L}_{\mathcal{D}}$ and $\mathcal{R}_{\mathcal{D}}$, $\lambda(d_c)$ coincides with $h_3(d_c)$. We express $\lambda(d_j)$ recursively as follows:

$$\lambda(d_c) = h_3(d_c)$$

$$\lambda(d_j) = \max(\lambda(d_{j+1}), d(d_c, d_j) + h_2(d_j)) \qquad \text{for } j = c - 1, \ldots, 1$$

$$\lambda(d_j) = \max(\lambda(d_{j-1}), d(d_c, d_j) + h_2(d_j)) \qquad \text{for } j = c + 1, \ldots, k.$$

Next, we associate with each node $d_j \in \mathcal{L}_{\mathcal{D}}$ ($d_j \in \mathcal{R}_{\mathcal{D}}$) a further value, say $\mu(d_j)$, containing the length of a longest path in T starting from d_1 (d_k) and staying within the subtree of $\widehat{T}$ rooted at d_j. We express $\mu(d_j)$ recursively as

E. Nardelli et al., *All Best Swaps of a MDST*, JGAA, 5(5) 39–57 (2001)   45

follows:

$$\mu(d_1) = \mu(d_k) = 0$$
$$\mu(d_j) = \max(\mu(d_{j-1}), d(d_1, d_j) + h_2(d_j)) \qquad \text{for } j = 2, \ldots, c-1$$
$$\mu(d_j) = \max(\mu(d_{j+1}), d(d_k, d_j) + h_2(d_j)) \qquad \text{for } j = k-1, \ldots, c+1.$$

We also associate with each node d_j on the diameter a further value, say $\rho(d_j)$, containing the nearest node along the diameter of the path stored in $\mu(d_j)$ (this can be done during the computation of $\mu(d_j)$). It is easy to see that all the above computations cost $O(n)$ time.

Finally, we convert G into a graph $G' = (V', E')$ with maximum node degree 3, in the following way [6]: for each node $v \in V$ of degree $\delta(v) > 3$, having nodes $w_1, \ldots, w_{\delta(v)}$ adjacent to it, replace v with nodes $v_1, \ldots, v_{\delta(v)}$; then, add *dashed edges* $(v_i, v_{i+1}), i = 1, \ldots, \delta(v) - 1$, each of length 0, and replace edges $(v, w_i), i = 1, \ldots, \delta(v)$, with edges (v_i, w_i) of corresponding lengths. As a result of this transformation, the graph keeps its original edges, and has an additional $O(m)$ edges of length 0. It is not hard to see that $|V'| \leq 2m$ and $|E'| \leq 3m$, and therefore a MDST T' of G' can be derived from a MDST T of G in $O(m)$ time in the following way: we keep in T' all the edges in T, and for each node $v \in V$ we add the edges $(v_i, v_{i+1}), i = 1, \ldots, \delta(v) - 1$, of length 0. Then, we associate with T' a *topology tree* and a *2-dimensional topology tree* [3, 4], both augmented with some extra-information useful for solving our problem [8]. A topology tree Γ associated with T' is a hierarchical representation of G' based on T', while a 2-dimensional topology tree Γ' associated with Γ is a hierarchical representation of G' based on Γ. The structures of Γ and Γ' are quite involved, and we refer the reader to [3] for an exhaustive description of them. For our scopes, it suffices to mention that Γ and Γ', augmented with some extra-information [8], can be initialized in $O(m)$ time, and allow to find a best swap in a MDST undergoing to edge failures in $O(n)$ time. However, time needed to update Γ and Γ' is just $O(\sqrt{m})$, and since the approach in [8] is more general than what is needed for solving the ABS problem, we will show that this time is enough for computing a best swap in our case.

It is easy to see that to a swapping of an edge in T corresponds the same swap in T', and, vice versa, to a swapping of a non-dashed edge in T' corresponds the same swap in T. Therefore, in the following we will refer to the original spanning tree T, even though our algorithm makes use of the topology tree and the 2-dimensional topology tree associated with T'.

Summarizing, preliminary computations have an overall cost of $O(m)$ time and use $O(m)$ space.

2.4 Computing $|\mathcal{D}(T_{e/f})|$ in $O(1)$ amortized time

In the rest of the paper, two paths will be considered *adjacent* if they share the root d_c only. When the edge $e = (x, y)$ is removed, $\widehat{T}$ is split into two subtrees, say T_x and T_y, containing x and y, respectively, which will be later connected

E. Nardelli et al., *All Best Swaps of a MDST*, JGAA, 5(5) 39–57 (2001)   46

by means of a replacement edge $f = (u, v)$. As a consequence

$$|\mathcal{D}(T_{e/f})| = \max\{|\mathcal{D}(T_x)|, |\mathcal{D}(T_y)|, |\mathcal{P}_f|\} \tag{1}$$

where $|\mathcal{P}_f|$ is the length of a longest path in $T_{e/f}$ passing through f. Notice that (1) holds for any (possibly not MDST) spanning tree T of G, any edge e in T, and any swap edge f of e. Figure 2 shows the notations used.

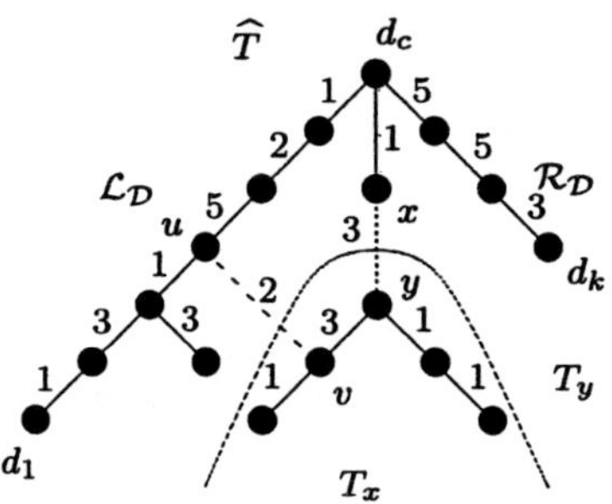

Figure 2: Edge $e = (x, y)$ is replaced by edge $f = (u, v)$.

We now analyze different cases that can arise in solving the ABS problem. For the sake of clarity, we perform a different analysis depending on whether the removed edge is located on the diameter or not.

Case 1: The removed edge is not on the diameter
Assume the edge $e = (x, y)$ is removed, where x is the parent of y in $\widehat{T}$, and $e \notin \mathcal{D}(T)$. In this case, neither $\mathcal{L_D}$ nor $\mathcal{R_D}$ are affected. Trivially, $\mathcal{D}(T_x) = \mathcal{D}(T)$. Moreover, $|\mathcal{D}(T_y)| \leq |\mathcal{D}(T)|$, since a path in T_y is a path in T as well. It then remains to compute $|\mathcal{P}_f|$, for any selected replacement edge $f \in \mathcal{S}_e$. Let $f = (u, v)$, where $u \in T_x$ and $v \in T_y$. It is clear that

$$|\mathcal{P}_f| = |\mathcal{L}_u| + |f| + |\mathcal{L}_v|$$

where $\mathcal{L}_u$ is a longest path in T_x starting from u and $\mathcal{L}_v$ is a longest path in T_y starting from v. Since v is a descendant of y in $\widehat{T}$, by using the path halving compression technique $|\mathcal{L}_v| = Findpath(v, y)$ can be computed in $O(1)$ amortized time, as we will see in Section 3, while $|f|$ is clearly available in $O(1)$ time. It remains to compute $|\mathcal{L}_u|$. The following claim is easy to prove:

Lemma 2.1 *At least one of the longest paths in T_x starting from u contains* d_c.

Proof: Suppose, for the sake of contradiction, that none of the longest paths in T_x starting from u contains d_c. Let us restrict our attention to any one of such longest paths, say $\mathcal{P}_u$. We will show that such a path can be modified into another path at least as long as $\mathcal{P}_u$ and passing through d_c, from which the claim will follow. Let w be the node in $\mathcal{P}_u$ nearest to d_c, and let z be the ending node of $\mathcal{P}_u$ other than u. Three cases are possible:

E. Nardelli et al., *All Best Swaps of a MDST*, JGAA, 5(5) 39–57 (2001)   47

1. $w \in N_l$: let $q \in \mathcal{L}_\mathcal{D}$ be the node on $\mathcal{D}(T)$ nearest to w (if w is on the diameter, then q coincides with w). It is trivial to see that in this case, being $q \neq d_c$ since $w \in N_l$, it must be

$$d(q, z) \leq d(q, d_k)$$

since otherwise $d(d_1, q) + d(q, z) > d(d_1, q) + d(q, d_k) = |\mathcal{D}(T)|$. Being $d(q, z) = d(q, w) + d(w, z)$, it then follows that $\mathcal{P}_u$ can be modified into the path $\mathcal{P}_u{}' = \langle u, \ldots, w, \ldots, q, \ldots, d_k \rangle$ containing d_c and such that

$$|\mathcal{P}_u{}'| = d(u, w) + d(w, q) + d(q, d_k) \geq d(u, w) + d(q, d_k) \geq$$

$$\geq d(u, w) + d(q, z) \geq d(u, w) + d(w, z) = |\mathcal{P}_u|$$

that is a contradiction (see Figure 3a).

2. $w \in N_r$: this case is symmetric to the first one.

3. $w \in N_c$: in this case, it must be clearly $d(w, z) \leq d(w, d_c) + d(d_c, d_1)$, since otherwise $d(d_c, d_1) + d(d_c, w) + d(w, z) > |\mathcal{D}(T)|$. It then follows that $\mathcal{P}_u$ can be modified into the path $\mathcal{P}_u{}' = \langle u, \ldots, w, \ldots, d_c, \ldots, d_1 \rangle$, containing d_c and such that

$$|\mathcal{P}_u{}'| = d(u, w) + d(w, d_c) + d(d_c, d_1) \geq d(u, w) + d(w, z) = |\mathcal{P}_u|$$

that is a contradiction (see Figure 3b).

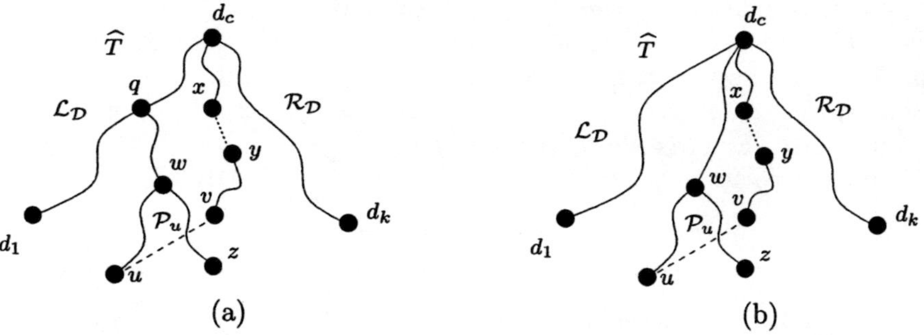

Figure 3: When edge $e = (x, y)$ is removed, there exists a longest path starting from u which passes through d_c (splines denote paths): (a) $w \in N_l$. (b) $w \in N_c$.

$\square$

From the above analysis and from the fact that $\mathcal{L}_\mathcal{D}$ is one of the longest paths emanating from d_c in $\widehat{T}$ and that $\mathcal{R}_\mathcal{D}$ is one of the longest paths emanating from d_c in $\widehat{T}$ which does not make use of d_{c-1} it follows that

E. Nardelli et al., *All Best Swaps of a MDST*, JGAA, 5(5) 39–57 (2001)   48

$$|\mathcal{L}_u| = \begin{cases} d(d_c, u) + |\mathcal{R}_\mathcal{D}| & \text{if } u \in N_l; \\ d(d_c, u) + |\mathcal{L}_\mathcal{D}| & \text{otherwise.} \end{cases} \qquad (2)$$

Therefore, $|\mathcal{L}_u|$ is available in $O(1)$ time. Summarizing, $|\mathcal{P}_f|$ can be computed in $O(1)$ amortized time, and

$$|\mathcal{D}(T_{e/f})| = \max(|\mathcal{D}(T)|, |\mathcal{P}_f|).$$

Once this value is computed for the $O(\sqrt{m})$ selected edges identified by the topology tree and the 2-dimensional topology tree, a best replacement edge is available. Therefore, the case $e \notin \mathcal{D}(T)$ can be managed in $O(\sqrt{m})$ amortized time.

Case 2: The removed edge is on the diameter
We will analyze the case in which $e = (d_i, d_{i-1}) \in \mathcal{L}_\mathcal{D}$ is removed, since the case $e \in \mathcal{R}_\mathcal{D}$ is symmetric. When e is removed, $\widehat{T}$ is split into two subtrees, say T_{d_i} and $T_{d_{i-1}}$, which will be later connected by means of a replacement edge $f = (u, v) \in \mathcal{S}_e$. Equation (1) becomes

$$|\mathcal{D}(T_{e/f})| = \max(|\mathcal{D}(T_{d_i})|, |\mathcal{D}(T_{d_{i-1}})|, |\mathcal{P}_f|).$$

Let us now analyze the value of these three terms.

- $|\mathcal{D}(T_{d_i})|$: We start by proving the following fact:

Lemma 2.2 *At least one of the diameters of T_{d_i} contains $d_k \in \mathcal{R}_\mathcal{D}$.*

Proof: For the sake of contradiction, suppose that none of the diameters of T_{d_i} contains d_k. Let us restrict our attention to any one of such diameters, say $\mathcal{P}$. We will show that in T_{d_i} there is a path containing d_k and at least as long as $\mathcal{P}$, from which the claim will follow. Let w be the node in $\mathcal{P}$ nearest to d_c. Let $\mathcal{P}_1$ and $\mathcal{P}_2$ be the two subpaths in which $\mathcal{P}$ splits with respect to w, with ending nodes z_1 and z_2, respectively, and suppose that z_1 precedes z_2 in a preorder traversal of $\widehat{T}$. Figure 4 shows the notations used.

Three cases are possible:

1. $w \in N_l$: this case is similar to the case 1 of the proof of Lemma 2.1. In fact, the modified path there built also contains d_k, apart from d_c.

2. $w \in N_r$: in this case, let $q \in \mathcal{R}_\mathcal{D}$ be the node on $\mathcal{D}(T)$ nearest to w (if w is on the diameter, then q coincides with w). Clearly, any path emanating from q in $\widehat{T}$ is no longer than $d(q, d_k)$. In particular

$$d(q, d_k) \geq d(q, z_2) \geq d(w, z_2).$$

It follows that $\mathcal{P}$ can be modified into the path $\mathcal{P}' = \langle z_1, \ldots, w, \ldots, q, \ldots, d_k \rangle$ containing d_k and such that

$$|\mathcal{P}'| = d(z_1, w) + d(w, q) + d(q, d_k) \geq d(z_1, w) + d(w, z_2) = |\mathcal{P}|$$

E. Nardelli et al., *All Best Swaps of a MDST*, JGAA, 5(5) 39–57 (2001)   49

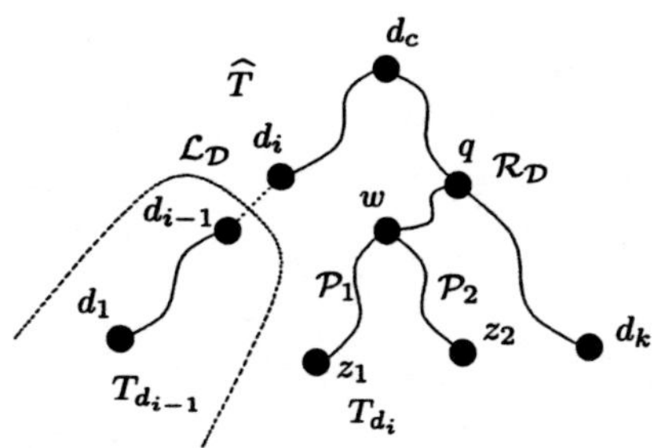

Figure 4: Edge $e = (d_i, d_{i-1})$ is removed: $\mathcal{P} = \mathcal{P}_1 \cup \mathcal{P}_2$ (splines denote paths).

3. $w \in N_c$: in this case, since z_2 descends from d_c in $\widehat{T}$, it must be clearly $d(w, z_2) \leq d(w, d_c) + d(d_c, d_k)$, since otherwise $d(d_c, d_1) + d(d_c, w) + d(w, z) > |\mathcal{D}(T)|$. It then follows that $\mathcal{P}$ can be modified into the path $\mathcal{P}' = \langle z_1, \ldots, d_c, \ldots, d_k \rangle$ containing d_k and such that

$$|\mathcal{P}'| = d(z_1, w) + d(w, d_c) + d(d_c, d_k) \geq d(z_1, w) + d(w, z_2) = |\mathcal{P}|$$

that is a contradiction.

$\square$

From the above result, it follows that a longest path starting from d_k and not using the edge e just removed can be computed in $O(1)$ time as

$$|\mathcal{D}(T_{d_i})| = \max(\mu(d_{c+1}), \lambda(d_i) + |\mathcal{R}_\mathcal{D}|).$$

• $|\mathcal{D}(T_{d_{i-1}})|$: analogously to the previous case, it can be proved that at least one of the diameters of $T_{d_{i-1}}$ must contain the node d_1. Therefore, it will be

$$|\mathcal{D}(T_{d_{i-1}})| = \mu(d_{i-1})$$

which can be computed in $O(1)$ time.

• $|\mathcal{P}_f|$: let be $f = (u, v)$, where $u \in T_{d_i}$ and $v \in T_{d_{i-1}}$, and $|\mathcal{P}_f| = |\mathcal{L}_u| + |f| + |\mathcal{L}_v|$. $|\mathcal{L}_v| = Findpath(v, d_{i-1})$ can be computed in $O(1)$ amortized time and $|f|$ is available in $O(1)$ time. It remains to analyze $|\mathcal{L}_u|$. Remember that with the node u is associated the nearest node q on the diameter from which it descends. The following three situations are possible for u:

1. $u \in N_l$: in this case, still using the same arguments as for the case 1 of the proof of Lemma 2.1, it can be easily proved that at least one of the longest paths in T_{d_i} starting from u must contain d_c, and then $|\mathcal{L}_u|$ can be obtained in $O(1)$ time as

$$|\mathcal{L}_u| = d(d_c, u) + |\mathcal{R}_\mathcal{D}|.$$

E. Nardelli et al., *All Best Swaps of a MDST*, JGAA, 5(5) 39–57 (2001)   50

2. $u \in N_r$: In this case, still using the same arguments as for the case 2 of the proof of Lemma 2.2, it can be proved that at least one of the longest paths in T_{d_i} starting from u must contain q. Therefore, it follows that $|\mathcal{L}_u|$ can be obtained in $O(1)$ time as (note that the following is equivalent to compute $Findpath(u, d_c)$)

$$|\mathcal{L}_u| = \max\left(d(u,q) + d\Big(q, \rho(d_{c+1})\Big) + \mu(d_{c+1}) - d\Big(d_k, \rho(d_{c+1})\Big),\right.$$
$$\left. d(u, d_k), d(u, d_c) + \lambda(d_i)\right).$$

3. $u \in N_c$: in this case, it is easy to see that at least one of the longest paths in T_{d_i} starting from u must contain d_c. In fact, for the sake of contradiction, suppose that none of the longest paths in T_{d_i} starting from u contains d_c. Let us restrict our attention to any one of such longest paths, say $\mathcal{P}_u$. Let w be the node in $\mathcal{P}_u$ nearest to d_c, and let z be the ending node of $\mathcal{P}_u$ other than u. Clearly, $d(w, z) \leq d(d_c, d_k)$, and therefore $\mathcal{P}_u$ can be modified into the path $\mathcal{P}_u' = \langle u, \ldots, w, \ldots, d_c, \ldots, d_k \rangle$, containing d_c and such that

$$|\mathcal{P}_u'| = d(u, w) + d(w, d_c) + d(d_c, d_k) \geq d(u, w) + d(w, z) = |\mathcal{P}_u|$$

that is a contradiction. Given that $\mathcal{L}_u$ contains d_c, it remains to compute the length of a longest path starting from d_c and not containing u, and this can be done by looking at $|\mathcal{R}_D|$ and at $\lambda(d_i)$ (note that if $\lambda(d_i)$ is exactly the length of a path passing through u, then it follows that $|\mathcal{R}_D| \geq \lambda(d_i)$). Thus, $|\mathcal{L}_u|$ can be computed in $O(1)$ time as

$$|\mathcal{L}_u| = \max(d(d_c, u) + |\mathcal{R}_D|, d(d_c, u) + \lambda(d_i)).$$

Summarizing, the case $e \in \mathcal{L}_D$ can be managed in $O(1)$ amortized time for any of the $O(\sqrt{m})$ selected edges. Since the case $e \in \mathcal{R}_D$ is symmetric to the previous one, it can be managed with the same amortized runtime. Repeating the above for all the $n - 1$ edges of T we therefore have the following:

Theorem 2.1 *The ABS problem for a minimum diameter spanning tree T of a graph G with n nodes and m edges can be solved in $O(n\sqrt{m})$ time, using $O(m)$ space.*

3   Constructing and using compressed paths

In this section we use an adaptation of the well-known *path halving* compression technique to prove that a $Findpath(v, y)$ operation can be satisfied in $O(1)$ amortized time, as required to solve efficiently the ABS problem.

E. Nardelli et al., *All Best Swaps of a MDST*, JGAA, 5(5) 39–57 (2001)   51

We start creating a *virtual forest* $\mathcal{F}$ of trees. Initially, $\mathcal{F}$ is composed of n singletons, one for each node $v \in V$. With each node v in $\mathcal{F}$ the following values are associated: $h_1(v), h_2(v), a(v)$, as defined in Section 2.3, and $up(v)$, which will contain an estimation of the length of a longest path emanating from v and ascending towards d_c in $\widehat{T}$, now considering the edges of the path from d_c to v as directed from v towards d_c. At the beginning, $up(v) = 0$, $\forall v \in \mathcal{F}$. The following instructions manipulate $\mathcal{F}$:

- *Link(u, v)*: combine the trees with roots u and v into a single tree rooted in u, adding the edge $e = (u, v)$ of length $|e|$;

- *Eval(v)*: Return the length of a longest path starting from v in the tree containing it and apply a suited path halving compression technique.

Note that *Eval(v)* assumes that a pointer to element v is obtained in constant time. The sequence of *Link()* operations in $\mathcal{F}$ is determined by the sequence of edge removals from $\widehat{T}$. Remember that we sequentially consider all the edges $e \in E_T$ in postorder fashion. When the edge $e = (x, y)$ is (temporarily) removed, we perform a sequence of *Link(y, z_i)* in $\mathcal{F}$, where $z_i, i = 1, \ldots, k$ are all the sons of y in $\widehat{T}$. Figure 5 illustrates the situation.

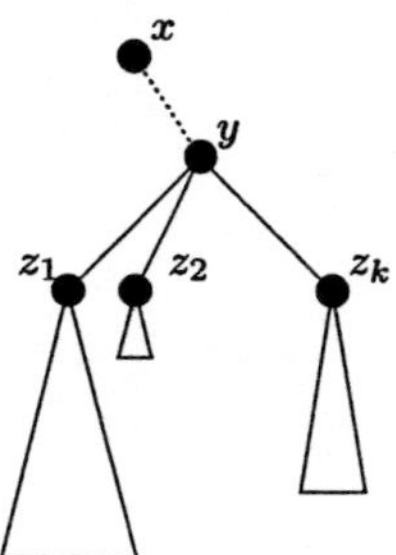

Figure 5: Edge $e = (x, y)$ is deleted and k *Link()* operations are applied.

This means that for any given node $v \in V$, whenever a *Findpath(v, y)* in $\widehat{T}$ occurs, the node y is exactly the root of v in $\mathcal{F}$. In fact, remember that we only ask *Findpath(v, y)* on nodes descending from the currently removed edge. This observation is crucial for the correctness of the method. We implement a *Findpath(v, y)* operation in $\widehat{T}$ by means of an *Eval(v)* operation in $\mathcal{F}$, that examines all the nodes along the path from v to the root y of the tree containing it and compresses such a path. The compression technique used is an adaptation of the *path halving* technique [10], which will guarantee the *associativity* of *Eval(v)*. Let us describe how the path halving works. Given a pair of nodes $v_1, v_2 \in V$, with v_1 ancestor of v_2 ($v_1 \prec v_2$) in $\widehat{T}$, we define the following function

$$h(v_1, v_2) = \begin{cases} h_2(v_1) & \text{if } a(v_1) \prec v_2 \text{ or } a(v_1) = v_2; \\ h_1(v_1) & \text{otherwise.} \end{cases}$$

E. Nardelli et al., *All Best Swaps of a MDST*, JGAA, 5(5) 39–57 (2001)   52

It is easy to see that $h(v_1, v_2)$ can be computed in $O(1)$ time, for any $v_1, v_2 \in V$, after $O(n)$ time of preprocessing of $\widehat{T}$. Assume an $Eval(v)$ operation occurs and this is the first time an $Eval()$ operation takes place. For the sake of simplicity, let us focus on a path of three nodes $\langle y, u, v \rangle$, with $y \prec u \prec v$ in $\widehat{T}$ and such that $|(u, v)| = \alpha$ and $|(y, u)| = \beta$. We have

$$Eval(v) = \max \Big(h_1(v), up(v), \alpha + h(u, v), \alpha + up(u), \alpha + \beta + h(y, v), \alpha + \beta + up(y) \Big).$$

The compression takes place as part of this operation, and replaces the edge (u, v) with an edge (y, v) of length $\alpha + \beta$ and the label $up(v)$ of v with

$$up(v) = \max \Big(up(v), \alpha + h(u, v), \alpha + up(u) \Big).$$

After the compression we hence have

$$Eval(v) = \max \Big(h_1(v), \max \big(up(v), \alpha + h(u, v), \alpha + up(u) \big),$$

$$\alpha + \beta + h(y, v), \alpha + \beta + up(y) \Big)$$

exactly as before the compression. Therefore, we can conclude that $Eval(v)$ compresses paths while correctly maintaining longest path information. Figure 6 illustrates the path halving procedure.

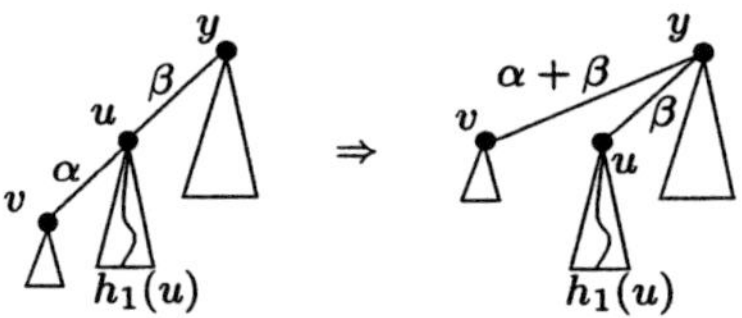

Figure 6: Path halving: Edge (y, v) of length $\alpha + \beta$ replaces edge (u, v) of length α. Note that v continues to take care of the path associated with $h_1(u)$, if needed, since this value is now stored in $up(v)$.

In the general case, let $\langle v_0 = v, v_1, \ldots, v_k = y \rangle$ be the path from v to the root y of the tree containing v in $\mathcal{F}$, having edges $e_i = (v_i, v_{i+1}), i = 0, \ldots, k-1$. We have

$$Eval(v) = \max_{i=1,\ldots,k} \left(h_1(v), up(v), h(v_i, v) + \sum_{j=0}^{i-1} |e_j|, up(v_i) + \sum_{j=0}^{i-1} |e_j| \right). \quad (3)$$

W.l.o.g., let k be even. The path halving technique makes every other node along the path (except the last and the next to last) point to its grandfather, i.e., replaces the edge $(v_i, v_{i+1}), i = 0, 2, \ldots, k - 2$, with the edge (v_i, v_{i+2}) of length $|e_i| + |e_{i+1}|$ and sets

E. Nardelli et al., *All Best Swaps of a MDST*, JGAA, 5(5) 39–57 (2001)   53

$$up(v_i) = \max\Big(up(v_i), |e_i| + h(v_{i+1}, v_i), |e_i| + up(v_{i+1})\Big), i = 0, 2, \ldots, k - 2. \quad (4)$$

Hence, after the halving, the path from v to y is $\langle v_0 = v, v_2, \ldots, v_{k-2}, v_k = y\rangle$, with edges $e'_i = (v_i, v_{i+2}), i = 0, 2, \ldots, k-2$, of length $|e_i| + |e_{i+1}|$. Therefore, after the halving, we have

$$Eval(v) = \max_{i=1,\ldots,k/2}\left(h_1(v), up(v), h(v_{2i}, v) + \sum_{j=0}^{2i-1} |e_j|, up(v_{2i}) + \sum_{j=0}^{2i-1} |e_j|\right)$$

which, from (4), is equivalent to (3). Therefore, it turns out that $Eval(v)$ before and after the halving is invariant. Remembering the order the edges are removed, it is clear that the compression of the paths works correctly (i.e., the compression incrementally proceeds towards the upper levels of $\widehat{T}$, according to the edge removals). Therefore, we conclude that $Findpath(v, y)$ is equivalent to $Eval(v)$.

Since naive linking in $\mathcal{F}$ must be applied to preserve paths of $\widehat{T}$, a sequence of $Q \geq n$ $Eval(v)$ queries in $\mathcal{F}$ can be satisfied in $O\left(Q \log_{(1+Q/n)} n\right)$ time [10]. Therefore, to establish that a single query can be satisfied in $O(1)$ amortized time, it remains to prove that $Q = \Omega(n^{1+\epsilon})$ for some constant $\epsilon > 0$. We now show that $Q = \Theta(n\sqrt{m})$, i.e., $\epsilon \geq 1/2$.

Let us distinguish the edges of T in *basic edges* (i.e., edges joining two nodes contained within the same basic cluster [8] and therefore associated with a node at level 0 in the corresponding topology tree) and *spanning edges* (i.e., edges joining two clusters at the same level $j \geq 0$ of the topology tree). If an edge removed from T is a basic edge, then we have $|\mathcal{S}_e| = \Theta(\sqrt{m})$. On the other hand, if an edge removed from T is a spanning edge and joins two clusters corresponding to nodes at level j in the topology tree, $0 \leq j \leq \lceil \log \sqrt{m} \rceil - 1$, then we have $|\mathcal{S}_e| = \Theta(\sqrt{m}/2^j)$. For each edge in $\mathcal{S}_e$ a $Findpath()$ query is issued. Hence the minimum overall number of queries is obtained when as many as possible of the $n-1$ failing edges of T are spanning edges at the highest possible level in the topology tree. Since there will be $\Theta(\sqrt{m}/2^{j+1})$ spanning edges for nodes at level j of the topology tree, i.e., a total of $\Theta(\sqrt{m})$ spanning edges, it follows that the minimum number of queries is posed when $\Theta(\sqrt{m})$ edges of T are spanning and the remaining $\Theta(n - \sqrt{m})$ are basic. Therefore, we have

$$Q = \Omega\left((n - \sqrt{m})\sqrt{m} + \sum_{j=0}^{\lceil \log \sqrt{m}\rceil - 1} \frac{\sqrt{m}}{2^{j+1}} \frac{\sqrt{m}}{2^j}\right) = \Omega(n\sqrt{m})$$

and since $Q = O(n\sqrt{m})$, it follows $Q = \Theta(n\sqrt{m})$, which implies $\epsilon \geq 1/2$.

E. Nardelli et al., *All Best Swaps of a MDST*, JGAA, 5(5) 39–57 (2001)   54

4 Swapping versus recomputing from scratch

In this section, we will show that even though swapping a failing edge in the best possible way does in general not yield a new MDST, it will not give a tree whose diameter differs all that much from a minimum one. Thus, swapping is quick and not too dirty. Let T_{G-e} be a MDST of $G - e$ and let $\sigma_e = |\mathcal{D}(T_{e/e'})|/|\mathcal{D}(T_{G-e})|$, where, as usual, e' is a best swap edge for e.

We start by observing that the MDST T must contain at least one path of length $\mathcal{D}(T)$ which is a shortest path between its two ending nodes [7]. W.l.o.g., let us assume that such a path is $\langle d_1, \ldots, d_k \rangle$, and therefore $\langle d_c, \ldots, d_1 \rangle$ and $\langle d_c, \ldots, d_k \rangle$ are shortest paths. The following holds:

Theorem 4.1 *For any edge $e = (x, y) \in E_T$, we have $\sigma_e \leq 5/2$. The bound is tight.*

Proof: Let $f = (u, v)$ be a replacement edge for e such that $f \in T_{G-e}$ and $|f|$ is minimum. As usual, let $\mathcal{L}_u$ be a longest path in T_x starting from u and $\mathcal{L}_v$ be a longest path in T_y starting from v. We have

$$|\mathcal{D}(T_{e/e'})| \leq |\mathcal{D}(T_{e/f})| \leq \max\left(|\mathcal{L}_u| + |f| + |\mathcal{L}_v|, |\mathcal{D}(T)|\right). \tag{5}$$

If there exists at least another edge $f' \in \mathcal{R}_e$ in T_{G-e}, then $\sigma_e \leq 5/2$ trivially follows from the fact that $|\mathcal{D}(T_{G-e})| \geq |f| + |f'| \geq 2|f|$, $|\mathcal{D}(T_{G-e})| \geq |\mathcal{L}_u|, |\mathcal{D}(T_{G-e})| \geq |\mathcal{L}_v|$ and $|\mathcal{D}(T_{G-e})| \geq |\mathcal{D}(T)|$.

Otherwise, let us assume that $f = (u, v)$ is the only edge in $\mathcal{R}_e$ belonging to T_{G-e}. We distinguish two cases: $e \in \mathcal{D}(T)$ and $e \notin \mathcal{D}(T)$.

1. $e = (x, y) \in \mathcal{D}(T)$: W.l.o.g., let $e \in \mathcal{L}_D$ and let x be the parent of y in $\widehat{T}$. Let $p(v, v')$ denote the length of a shortest path in $G - e$ between any two nodes v, v'. Trivially, $|\mathcal{L}_u| \leq |\mathcal{D}(T_x)|$ and $|\mathcal{L}_v| \leq |\mathcal{D}(T_y)|$. Moreover, in $G - e$, we have that $p(x, d_k) = d(x, d_k) \geq |\mathcal{R}_D| \geq |\mathcal{D}(T_x)|/2$. Therefore

$$\max\left(p(u, x), p(u, d_k)\right) \geq \frac{|\mathcal{D}(T_x)|}{4}.$$

    Similarly, we have that $p(d_1, y) = d(d_1, y) \geq |\mathcal{D}(T_y)|/2$, from which

$$\max\left(p(v, d_1), p(v, y)\right) \geq \frac{|\mathcal{D}(T_y)|}{4}.$$

    Hence, in $G - e$ there exists a shortest path passing through f at least as long as

$$|f| + \frac{|\mathcal{D}(T_x)|}{4} + \frac{|\mathcal{D}(T_y)|}{4}$$

    and whilst $|\mathcal{D}(T_x)| \leq |\mathcal{D}(T_{G-e})|$ and $|\mathcal{D}(T_y)| \leq |\mathcal{D}(T_{G-e})|$, (5) can be rewritten as

E. Nardelli et al., *All Best Swaps of a MDST*, JGAA, 5(5) 39–57 (2001)   55

$$|\mathcal{D}(T_{e/e'})| \leq \max\left(|f| + \frac{|\mathcal{D}(T_x)|}{4} + \frac{|\mathcal{D}(T_y)|}{4} + \frac{3}{4}\Big(|\mathcal{D}(T_x)| + |\mathcal{D}(T_y)|\Big),\right.$$

$$\left.|\mathcal{D}(T)|\right) \leq \max\left(|\mathcal{D}(T_{G-e})| + \frac{3}{2} \cdot |\mathcal{D}(T_{G-e})|, |\mathcal{D}(T_{G-e})|\right) \qquad (6)$$

from which $\sigma_e \leq 5/2$ follows. The bound is tight as shown in Figure 7.

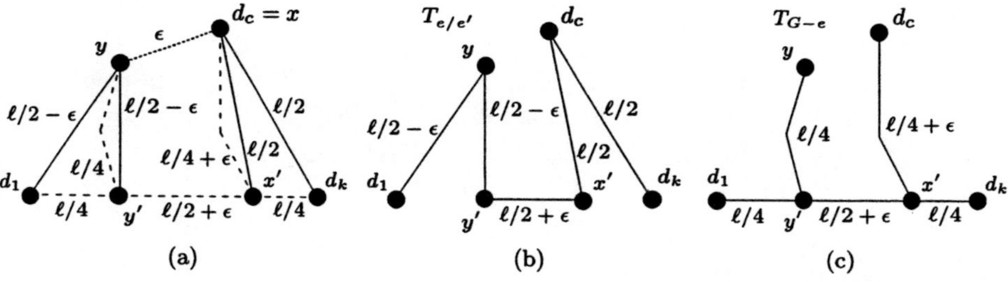

Figure 7: (a) The graph G with the MDST T, having $|\mathcal{D}(T)| = \ell$; non-tree edges are dashed, and edge $e = (x,y)$ is removed from T; (b) $T_{e/e'}$, with the swap edge $e' = (x',y')$, having $|\mathcal{D}(T_{e/e'})| = \frac{5}{2}\ell - \epsilon, \epsilon > 0$; (c) T_{G-e}, having $|\mathcal{D}(T_{G-e})| = \ell + 2\epsilon$.

2. $e = (x,y) \notin \mathcal{D}(T)$: W.l.o.g., let x be the parent of y in $\widehat{T}$. In this case, $\mathcal{D}(T) \in T_x$, and therefore in $G - e$ there exists a shortest path passing through f at least as long as

$$|f| + \frac{|\mathcal{D}(T_x)|}{2}$$

from which (5) can be rewritten as

$$|\mathcal{D}(T_{e/e'})| \leq \max\left(|f| + \frac{|\mathcal{D}(T_x)|}{2} + \frac{|\mathcal{D}(T_x)|}{2} + |\mathcal{D}(T_y)|, |\mathcal{D}(T)|\right) \leq$$

$$\leq \max\left(|\mathcal{D}(T_{G-e})| + \frac{1}{2} \cdot |\mathcal{D}(T_{G-e})| + |\mathcal{D}(T_{G-e})|, |\mathcal{D}(T_{G-e})|\right) \qquad (7)$$

from which $\sigma_e \leq 5/2$ follows. Also in this case, the bound is tight as shown in Figure 8.

$\square$

E. Nardelli et al., *All Best Swaps of a MDST*, JGAA, 5(5) 39–57 (2001)   56

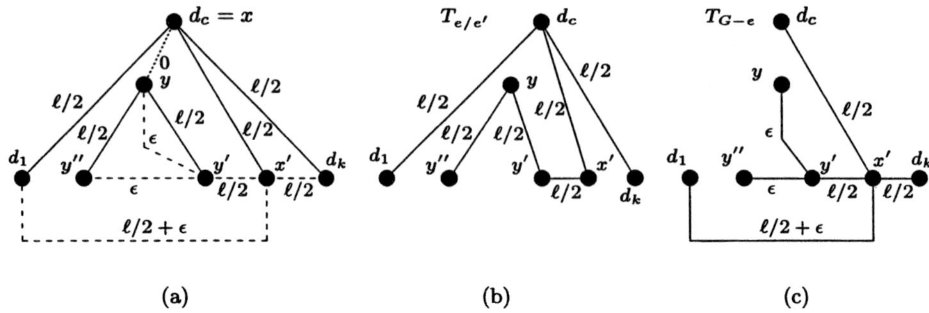

Figure 8: (a) The graph G with the MDST T, having $|\mathcal{D}(T)| = \ell$; non-tree edges are dashed, and edge $e = (d_c, y)$ is removed from T; (b) $T_{e/e'}$, with the swap edge $e' = (x', y')$, having $|\mathcal{D}(T_{e/e'})| = \frac{5}{2}\ell$; (c) T_{G-e}, having $|\mathcal{D}(T_{G-e})| = \ell + 2\epsilon, \epsilon > 0$.

5 Conclusions

In this paper we solved the ABS problem in $O(n\sqrt{m})$ time and $O(m)$ space, thus strictly improving previous bounds for $m = o(n^2)$. We have also shown that the diameter of the tree obtained as a consequence of a best swap is at most 5/2 times the diameter of the real new MDST. It might be interesting to study average values for the above ratio. Another open problem is the case of managing multiple simultaneous edge replacements. Clearly, it also remains to find out whether the algorithm we have proposed is optimal. Finally, the case of transient node failures deserves future study.

Acknowledgments

The authors would like to thank Dagmar Handke for asking the question treated in Section 4, and the referees for the suggestions that helped to improve the presentation of the paper.

E. Nardelli et al., *All Best Swaps of a MDST*, JGAA, 5(5) 39–57 (2001)   57

References

[1] S. Alstrup, J. Holm, K. de Lichtenberg, and M. Thorup. Minimizing diameters of dynamic trees. In P. Degano, R. Gorrieri, and A. Marchetti-Spaccamela, editors, *24th Int. Coll. on Automata, Languages and Programming (ICALP'97)*, pages 270–280. Lecture Notes in Computer Science Vol. 1256, Springer-Verlag, 1997.

[2] B. Dixon, M. Rauch, and R. Tarjan. Verification and sensitivity analysis of minimum spanning trees in linear time. *SIAM J. Computing*, 21(6):1184–1192, 1992.

[3] G. Frederickson. Data structures for on-line updating of minimum spanning trees. *SIAM J. Computing*, 14(4):781–798, 1985.

[4] G. Frederickson. Ambivalent data structures for dynamic 2-edge connectivity and k smallest spanning trees. In *Proc. 32nd IEEE Symp. on Foundations of Computer Science*, pages 632–641, 1991.

[5] M. Grötschel, C. Monma, and M. Stoer. Design of survivable networks. In *Handbooks in OR and MS*, chapter 7, pages 617–672. Elsevier, 1995.

[6] F. Harary. *Graph Theory*. Addison-Wesley, Reading, MA, 1969.

[7] R. Hassin and A. Tamir. On the minimum diameter spanning tree problem. *Information Processing Letters*, 53(2):109–111, 1995.

[8] G. Italiano and R. Ramaswami. Maintaining spanning trees of small diamete. *Algorithmica*, 22(3):275–304, 1998.

[9] E. Nardelli, G. Proietti, and P. Widmayer. How to swap a failing edge of a single source shortest paths tree. In T. Asano, H. Imai, D. Lee, S. Nakano, and T. Tokuyama, editors, *5th International Conference on Computing and Combinatorics (COCOON'99)*, pages 144–153. Lecture Notes in Computer Science Vol. 1627, Springer-Verlag, 1999.

[10] R. Tarjan and J. van Leeuwen. Worst-case analysis of set union algorithms. *J. of the ACM*, 31(2):245–281, 1984.

Journal of Graph Algorithms and Applications

http://www.cs.brown.edu/publications/jgaa/

vol. 5, no. 5, pp. 59–91 (2001)

Shelling Hexahedral Complexes for Mesh Generation

Matthias Müller–Hannemann

Technische Universität Berlin
Fachbereich Mathematik, Sekr. MA 6-1,
Straße des 17. Juni 136,
D 10623 Berlin, Germany
http://www.math.tu-berlin.de/~mhannema
mhannema@math.tu-berlin.de

Abstract

We present a new approach for the generation of hexahedral finite element meshes for solid bodies in computer-aided design. The key idea is to use a purely combinatorial method, namely a shelling process, to decompose a topological ball with a prescribed surface mesh into combinatorial cubes, so-called hexahedra. The shelling corresponds to a series of graph transformations on the surface mesh which is guided by the cycle structure of the combinatorial dual. Our method transforms the graph of the surface mesh iteratively by changing the dual cycle structure until we get the surface mesh of a single hexahedron. Starting with a single hexahedron and reversing the order of the graph transformations, each transformation step can be interpreted as adding one or more hexahedra to the so far created hex complex.

Given an arbitrary solid body, we first decompose it into simpler subdomains equivalent to topological balls by adding virtual 2-manifolds. Second, we determine a compatible quadrilateral surface mesh for all created subdomains. Then, in the main part we can use the shelling of topological balls to build up a hex complex for each subdomain independently. Finally, the combinatorial mesh(es) are embedded into the given solids and smoothed to improve quality.

Communicated by T. Nishizeki, R. Tamassia and D. Wagner:
submitted January 1999; revised August 2000 and July 2001.

The author was partially supported by the special program "Efficient Algorithms for Discrete Problems and Their Applications" of the Deutsche Forschungsgemeinschaft (DFG) under grant Mo 446/2-3.

M. Müller-Hannemann, *Shelling Hexahedral Complexes*, JGAA, 5(5) 59–91 (2001)60

1 Introduction

In recent years, global competition has led to an increasing demand to reduce the development time for new products. One step in this direction would be a more efficient computer simulation of the technical properties of prototypes for such products. For many years the finite element method has been successfully applied by engineers in simulations. As a prerequisite such a method needs a tool which converts a CAD model into a finite element mesh model suitable for a numerical analysis.

Therefore, various algorithms for the generation of meshes have been developed, mostly decomposing surfaces into triangles and solid bodies into tetrahedra, for surveys see [2, 3]. In many applications, however, quadrilateral and hexahedral meshes have numerical advantages. The potential savings gained from an all-hexahedral meshing tool compared to an analysis based on tetrahedral meshing may be enormous (with estimations in the range of 75% time and cost reductions [37]). On the other hand, the generation of hexahedral meshes turns out to be much more complex than for tetrahedral meshes. Recent years showed many research efforts and brought up several approaches, but up to now, hexahedral mesh generation for an arbitrary 3D solid is still a challenge.

In this paper, we propose a new method for all-hexahedral mesh generation which mainly exploits combinatorial properties of such a mesh. A preliminary description (without proofs) can be found in [21].

Geometric vs. combinatorial meshes. We distinguish between geometric and combinatorial meshes. A *geometric mesh* is a partition of some given domain into subdomains, in our context into hexahedra, i. e. regions combinatorially equivalent to cubes. In contrast, a *combinatorial hexahedral mesh*, is only a decomposition of the given domain into an abstract (cell) complex of combinatorial cubes but without an explicit embedding into space.

Fixed surface meshes. In many applications, in particular in structural mechanics simulations, high mesh quality is required near the boundary of the solid and is much more important than "deep inside the domain". Therefore, it is often preferred to start the volume meshing subject to a fixed quadrilateral surface mesh of an excellent quality. Moreover, the complete solid may consist of several components, for example, solid parts of different material or a solid and its complement within a larger box. Many subdomains may also arise for algorithmic reasons: we may want to divide the domain into smaller and simpler regions either to facilitate the meshing process for each part or to allow parallelism which can be crucial for some large-scale applications. In all such cases, it is usually essential to have a compatible mesh at the common boundary of adjacent components, that is the surface meshes must be compatible. The only solution for this problem we can envision is to prescribe the surface mesh for each component.

Thurston [36] and Mitchell [18] independently characterized the quadrilateral surface meshes which can be extended to hexahedral meshes. They showed

M. Müller-Hannemann, *Shelling Hexahedral Complexes*, JGAA, 5(5) 59–91 (2001)61

that for a volume which is topologically a ball and which is equipped with an all-quadrilateral surface mesh, there exists a combinatorial hexahedral mesh without further boundary subdivision if and only if the number of quadrilaterals in the surface mesh is even. Furthermore, Eppstein [10] used this existence result and proved that a linear number of hexahedra (in the number of quadrilaterals) are sufficient in such cases. Unfortunately, all these results are not completely constructive and it remains unclear whether they can be extended to constructive methods for geometric meshes. There are quite simple solids with natural looking quadrilateral surface meshes, for example the quadratic pyramid problem of Schneiders [30], where only rather complicated combinatorial meshes are known, but no geometric mesh with an acceptable quality is available.

Related work. We briefly review approaches to hexahedral mesh generation. For a more complete survey, online information and data bases on meshing literature see [29] and [26].

Most commercial systems rely on a mapping approach where the domain must be divided into simple shapes which are meshed separately. For example, *isoparametric mapping* [8] is a method for generating hexahedral meshes which is robust for block-type geometries, but does not work well for more complicated general volumes. Various techniques based on object feature recognition have been developed to automate the subdivision of an object into simpler parts. *Virtual decomposition* [40] separates the volume into mappable subvolumes by the creation of virtual surfaces (2-manifolds) inside the volume. Another method uses the *medial axis* of a surface and *midpoint subdivision* [28, 27]. Compatibility between adjacent subregions and mesh density is then modeled within an integer programming formulation [16]. Unfortunately, solving such integer programming problems is NP-hard, even for quadrilateral surface meshes [20], and such models often rely on a very restricted set of meshing primitives, so-called *templates.*

Grid-based [31] and *octree-based* [32] methods start with a perfect grid which is then adapted to the object's boundary by an isomorphism technique. The adaption step is difficult and often leads to badly shaped hexahedra near the boundary.

Plastering [7, 4] is an advancing front based method which starts from a quadrilateral surface mesh. It maintains throughout the algorithm the *meshing front*, that is a set of quadrilateral faces which represent the boundary of the region(s) yet to be meshed. The plasterer selects iteratively one or more quadrilaterals from the front, attaches a new hexahedron to them, and updates the front until the volume is completely meshed or the algorithm gives up and the remaining voids are filled with tetrahedra.

Whisker weaving [34, 35] also meshes from a quadrilateral surface mesh inward. But in contrast to plastering it first builds the combinatorial dual of a mesh and constructs the primal mesh and its embedding only afterwards. This method is based on the concept of the so-called *spatial twist continuum (STC)* [23, 24]. The STC is an interpretation of the geometric dual of a hexa-

M. Müller-Hannemann, *Shelling Hexahedral Complexes*, JGAA, 5(5) 59–91 (2001)62

hedral mesh as an arrangement of surfaces, the *sheets*. More precisely, the mesh dual is the cell complex induced by the intersection of the sheets. A fundamental data structure for an STC is a *sheet diagram* which represents the crossings of one sheet with other sheets. Whisker weaving starts with incomplete sheet diagrams based on the surface mesh. It seeks to complete the sheet diagrams by a set of rules which determine the local connectivity of the mesh. The creation of invalid mesh connectivity has been observed in the plastering and whisker weaving algorithms. Heuristic strategies have been developed to resolve such invalidities [35]. Calvo & Idelsohn [6] recently presented rough ideas of a recursive approach which inserts layers of hexahedra separating the whole domain into two subdomains.

We finally mention that some methods relax the meshing problem by allowing mixed elements, that is they try to mesh with mostly hexahedra, but include tetrahedra [37], or pyramids and wedges [17], as well.

A new approach. We suppose that a solid body is described by polygonal surface patches. For the reasons given above, the new approach presented in this paper uses an all-quadrilateral surface mesh as a starting point. But before doing the surface meshing, we first decompose a complex body into simpler subregions by adding internal 2-manifolds. Our method only requires that these subregions are topological balls (to avoid difficulties with holes and voids). It is not necessary, although desirable, that these regions are "almost convex." Roughly speaking, we mean by *almost convex* that a region should not deviate from a convex region by too much (in particular, the local dihedral angles are not too big).

The insertion of additional internal 2-manifolds creates *branchings*, i. e. edges which belong to more than two polygonal patches. All-quadrilateral surface meshing in the presence of branchings can be achieved by first solving a system of linear equations over $GF(2)$, and then using a network flows [22] or an advancing front based method like paving [5].

Because of the practical difficulties indicated by the pyramid example, our approach does not attempt to extend any quadrilateral surface mesh to a hexahedral volume mesh. It seems that self-intersecting cycles in the dual of the surface mesh are the main source for these difficulties. It is a disadvantage of advancing front based methods that they usually generate many such cycles. In contrast, our network flow based mesher [22] tends to produce very regular meshes with only few self-intersecting cycles. To get rid of the remaining self-intersecting cycles we use two strategies. First, we use additional heuristics to avoid them in the surface meshing. Second, we introduce a method which modifies the surface mesh such that all self-intersections disappear. This modification will be explained in Section 5. The important feature of this method is that one can still use any quadrilateral surface mesher (provided it can handle branchings consistently and mesh the virtual internal surfaces without an explicit geometric embedding).

After this preparation, the core of our method is to build up a compatible combinatorial cell complex of hexahedra for a solid body which is topologically

M. Müller-Hannemann, *Shelling Hexahedral Complexes*, JGAA, 5(5) 59–91 (2001)63

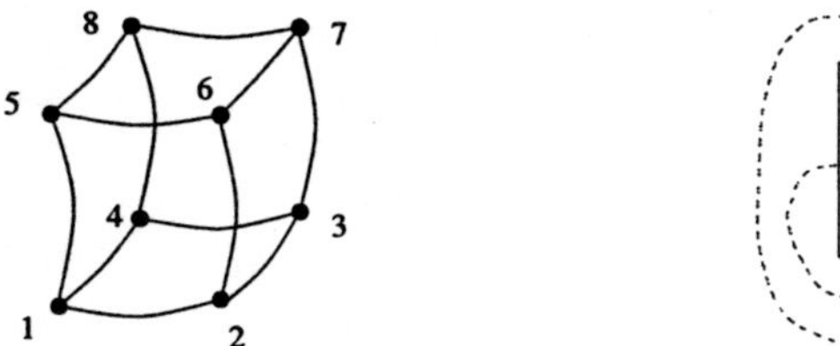

Figure 1: Hexahedron with curved quadrilateral facets (left), a planar surface graph embedding and its combinatorial dual (right).

a ball and for which a quadrilateral surface mesh is prescribed. Such a surface mesh, taken as a graph, is simple, planar, and 3-connected.

The step-wise creation of the hex complex is guided by the cycle structure of the combinatorial dual of the surface mesh. Our method transforms the graph of the surface mesh iteratively by changing the dual cycle structure until we get the surface mesh of a single hexahedron. During the transformation process we keep the invariant that the surface mesh remains simple, planar, and 3-connected. Our main strategy is a successive elimination of dual cycles. Hence, algorithmically, we try to determine a cycle elimination scheme transforming the given surface mesh to the mesh of a single hexahedron.

Starting with a single hexahedron and reversing the order of the graph transformations, each transformation step can be interpreted as adding one or more hexahedra to the so far created hex complex. The embedding and smoothing of the combinatorial mesh(es) finishes the mesh generation process.

Very recently, Folwell & Mitchell [11] changed the original whisker weaving algorithm and incorporated a strategy which is similar to ours. In contrast to us, they have no restrictions on the elimination of a cycle and thereby allow that the surface mesh (as well as the intermediate hex complex) becomes degenerated. Based on the constructions in [18], the new version of whisker weaving applies a number of heuristics to convert a degenerated hex complex into a well-defined one.

Organization. We first introduce some basic terminology in Section 2. Then, in Section 3, we present our new approach for the meshing of topological balls and show its relations to the shelling of cell complexes. In Section 4, we characterize a few classes of surface meshes which have a cycle elimination scheme corresponding to a shelling. A general meshing scheme for arbitrary domains will be given in Section 5. Finally, in Section 6 we summarize the main features of our approach and give directions for future work.

2 Basic Facts and Terminology

Planar graphs. We need some basic graph theory, see for example [25]. *Drawing* a graph in a given space means representing nodes as points and edges as curves. A graph can be *embedded* into some space if it can be drawn such

M. Müller-Hannemann, *Shelling Hexahedral Complexes*, JGAA, 5(5) 59–91 (2001)64

that no two edges intersect except at a common node. A graph is *planar* if it has an embedding in the plane, or equivalently, an embedding on a sphere in three-dimensional space. The curves representing the edges of a planar, embedded graph partition the plane or the sphere, respectively, into connected components, called *faces*. For a planar graph G, the *(geometric) dual* G^* is constructed as follows. A node v_i^* is placed in each face F_i of G; corresponding to each (primal) edge e of G we draw a dual edge e^* which crosses e but no other edge of G and joins the nodes v_i^* which lie in the faces F_i adjoining e.

A graph is *connected* if there is a path between any two distinct nodes of G, and the graph is *simple* if it has neither loops nor parallel edges. A graph G with at least 4 edges is 3-*connected* if it is simple and cannot be disconnected by removing 1 or 2 nodes from G. The reason for this version of the definitions (here, we follow Ziegler [41]) is that it is invariant under planar duality. That is, if we have a planar embedding of a graph G and construct its dual graph G^*, then G is 3-connected if and only if G^* is 3-connected.

A *(convex) polyhedron* is the intersection of finitely many halfspaces in some $\mathbb{R}^d$, and it is a *polytope* if it is bounded. The famous theorem of Steinitz relates planar graphs and polytopes.

Theorem 2.1 (Steinitz' theorem). *G is the graph of a 3-dimensional polytope if and only if it is simple, planar and 3-connected.*

Hex complex. The bodies we want to mesh have more general, curved surfaces (but, of course, orientable surfaces). Therefore we use the term *(geometric) cell* to mean a bounded region in 3-dimensional space, bounded by a finite number of orientable 2-manifolds. In the following, cells of different dimension will appear: 0-dimensional cells, i. e. single points, called *vertices*; 1-dimensional cells, i. e. segments of curves between two vertices, the *edges*, and 2-dimensional cells in the form of *quadrilateral facets*, i. e. smooth 2-manifolds bounded by a cycle of four distinct edges. A *hexahedron* is a 3-dimensional cell which is a combinatorial cube. It is bounded by 6 distinct quadrilateral facets, 12 distinct edges, and 8 distinct vertices. The quadrilaterals pairwise share edges as depicted in Fig. 1. Ideally, hexahedra are polyhedra, but in this abstract setting we do not require that the edges are straight line segments and that the quadrilateral facets are planar.

A geometric cell complex of hexahedra, called *hex complex* for short, is a finite, non-empty collection $\mathcal{C}$ of distinct (openly disjoint) hexahedra and all their lower dimensional cells such that the intersection of any two members of $\mathcal{C}$ is either empty or a cell of both of them. Two hexahedra are *neighbored* if they share a quadrilateral. By definition, a hexahedron of a geometric cell complex has at most one neighbored hexahedron for each quadrilateral face (*unique neighbor property*).

Combinatorial hex complex. If cells of a hex complex are not embedded but abstract, combinatorial entities, we regard a cell as composed by its lower-dimensional cells. For an abstract cell C, we define the *cell lattice* as the the set of all lower-dimensional cells including the cell C itself and the empty cell,

M. Müller-Hannemann, *Shelling Hexahedral Complexes*, JGAA, 5(5) 59–91 (2001)65

partially ordered by inclusion. Abstract cells are *combinatorially equivalent* if their cell lattices are isomorphic.

For a hex complex given by abstract cells, a *combinatorial hex complex*, we have to demand explicitly the unique neighbor property (which is no longer implied by definition) to avoid degeneracies. Hence, a combinatorial hex complex is *non-degenerated* if it has the unique neighbor property. A combinatorial hex complex yields a *surface compatible combinatorial mesh* if each quadrilateral is contained in exactly one hexahedron, if it belongs to the surface mesh, and in exactly two hexahedra, otherwise.

Valid geometric meshes. For a vertex of a hexahedron the Jacobian matrix is formed as follows. For that, let $x \in \mathbb{R}^3$ be the position of this vertex and $x_i \in \mathbb{R}^3$ for $i = 1, 2, 3$ be the position of its three neighbors in some fixed order. Using edge vectors $e_i = x_i - x$ with $i = 1, 2, 3$ the Jacobian matrix is then $A = [e_1, e_2, e_3]$. The determinant of the Jacobian matrix is usually called *Jacobian*. A hexahedron is said to be *inverted* if one of its Jacobians is less or equal to zero. As the sign of a determinant depends on the order of its column entries, the latter definition is only useful for checking the quality of a hexahedron if the order of its neighbors is carefully chosen for each vertex. However, a consistent and fixed ordering of the vertices can easily be derived from the combinatorial hex complex by a graph search from some hexahedron lying at the bounding surface.

A hex complex is *compatible* with the quadrilateral surface mesh of a body if this surface mesh is the union of all quadrilateral facets which belong to exactly one hexahedron of the complex. A surface compatible hex complex is a *non-degenerated geometric mesh* if all of its hexahedra are embedded inside the domain and are non-inverted. A quadrilateral surface mesh can also be seen as a cell complex of quadrilaterals. The *dimension $dim(\mathcal{C})$* of a cell complex is the largest dimension of a cell in $\mathcal{C}$. A cell complex is *pure* if all the inclusion-maximal cells have the same dimension. For example, hex complexes and quadrilateral surface meshes are pure. The *boundary complex $\partial\mathcal{C}$* of a cell complex is formed by the set of all cells of $\mathcal{C}$ with a dimension lower than $dim(\mathcal{C})$.

Shellings. Shellability of cell complexes has been widely studied in a very general context and is usually defined in a recursive way (on the dimension of the complex) [41]. A *shelling* of a pure d-dimensional cell complex $\mathcal{C}$ is a linear ordering $(F_1, F_2, \ldots, F_k)$ of the set of inclusion-maximal cells, which is arbitrary for $dim(\mathcal{C}) = 0$, but for $dim(\mathcal{C}) > 0$ has to satisfy the following two conditions:

(i) the boundary complex ∂F_1 has a shelling, and

(ii) for every $1 < j \leq k$, the boundary complex ∂F_j has a shelling $(G_1, G_2, \ldots, G_t)$ such that

$$F_j \cap \bigcup_{i=1}^{j-1} F_i = G_1 \cup G_2 \cup \cdots \cup G_r,$$

for some $1 \leq r \leq t$.

M. Müller-Hannemann, *Shelling Hexahedral Complexes*, JGAA, 5(5) 59–91 (2001)66

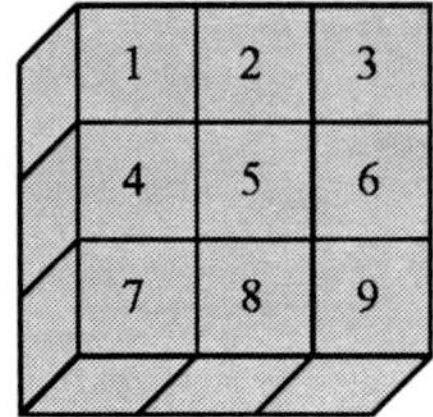 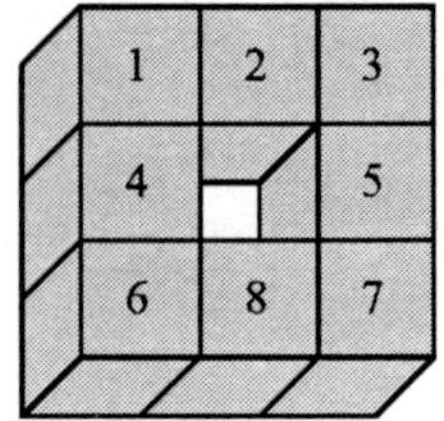

Figure 2: Example of a shellable(left) and a non-shellable (right) hex complex.

A complex is *shellable* if it has a shelling.

For the purpose of shelling hex complexes (or surface meshes) this can be simplified as the shellability of a single hexahedron and its lower-dimensional cells is obvious. Hence, for a hex complex, the concept of shellability can be defined as follows. A *shelling of a hex complex* C with k hexahedra is a linear ordering $H_1, H_2, \ldots, H_k$ of the hexahedra such that for $1 < j \leq k$ the intersection of the hexahedron H_j with the previous hexahedra is the non-empty union of quadrilateral faces of H_j and these quadrilaterals are connected with respect to the dual graph of H_j, that is

$$H_j \cap \bigcup_{i=1}^{j-1} H_i = Q_1 \cup Q_2 \cup \cdots \cup Q_r,$$

for some dually connected quadrilaterals $Q_1, \ldots Q_r$ of H_j, and $1 \leq r \leq 6$.

In Fig. 2, the left hex complex is shellable in the order of the numbered hexahedra, for example. However, the right hex complex is not shellable. To see this, just note that a beginning of a shelling sequence can never be completed. For example, a shelling sequence could start in the order $1, 2, \ldots, 7$, but the addition of hexahedron no. 8 would violate the condition that the quadrilaterals of the intersection of this hexahedron with the previous ones must be connected.

3   Shelling and Meshing Topological Balls

Our goal is to develop a purely combinatorial approach for hexahedral mesh generation starting from a fixed quadrilateral surface mesh. This implies that we can only use and therefore have to exploit the combinatorial structure of the surface mesh.

Throughout this section we restrict our discussion to the case that our input is a solid body which is topologically a ball. We further assume that a fixed all-quadrilateral surface mesh has been determined for this body. A non-degenerated surface mesh (which we assume) is combinatorially a simple, planar and 3-connected graph. Although the shape of the surfaces we consider is usually more general than that of a convex polytope, the combinatorial structure is not (by Steinitz' theorem).

M. Müller-Hannemann, *Shelling Hexahedral Complexes*, JGAA, 5(5) 59–91 (2001)67

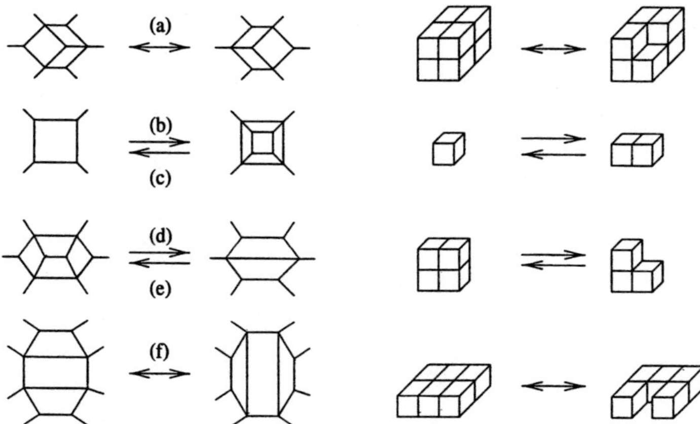

Figure 3: Local graph transformations (a) – (f) (left side) and examples of their application (right side).

Intuitive idea. Shelling a hex complex can be interpreted as building up the complex by successive additions of hexahedra. We would like to do just the opposite, that is to decompose a hex complex by taking away a hexahedron one after another. Of course, in the beginning no hex complex is available to us, all we have is the surface mesh. But if a compatible hex complex were available, then a reversed shelling order would give the desired decomposition. In general, for a given shelling order $H_1, H_2, \ldots, H_k$, the reversed order $H_k, H_{k-1}, \ldots, H_1$ is not again a shelling order (as, for example, in some step H_j may have empty intersection with $H_k, H_{k-1}, \ldots, H_{j+1}$), but it preserves that after j deletions the remaining hex complex is still topologically a ball.

The key observation is that each legal shelling step changes the planar surface graph of the so far created hex complex in a very restrictive way. Namely, such a shelling step corresponds to a local graph transformation which preserves that the surface graph is simple, planar, and 3-connected. The basic idea is to maintain this as an invariant also for the real problem where we do not know the hex complex.

What we are looking for is a *series of graph transformations* on the given surface graph which preserves that the surface mesh is a simple, planar and 3-connected graph. In other words, throughout the decomposition process the surface mesh corresponds to a topological ball. If this process ends up with the surface graph of a single hexahedron, the series of graph transformations yields a *"topology preserving reversed shelling"* of a hex complex. In the following paragraphs, we will make these general ideas more precise.

Shelling steps as graph transformations. Let us suppose for a while that a hex complex is already known for some body. Then we could try to shell this complex. The basic observation is that each shelling step, that is, the transition from a hex complex with k to $k + 1$ hexahedra, can be interpreted as a local

M. Müller-Hannemann, *Shelling Hexahedral Complexes*, JGAA, 5(5) 59–91 (2001)68

Figure 4: Quadrilaterals forming part of a self-intersecting dual cycle (left) and an example of a simple dual cycle (right) where the enclosed quadrilaterals a and b are disconnected with respect to the dual graph (although they share a primal vertex).

graph operation on the surface graph except for the very first shelling step with $k = 0$. Consider the six "local" transformations (a) - (f) shown in Fig. 3 on subgraphs of the surface graph. A transformation is applicable only if the subgraph has at least the edges shown in Fig. 3 (which ensures that each node has minimum degree 3 after the transformation). The operations represent the transformation of the current surface graph for all possible ways to add a single hexahedron to the shelling sequence in a legal way. The basic properties of these transformations are summarized in the next lemma.

Lemma 3.1. *Let G be a simple, planar and 3-connected graph whose faces are all quadrilaterals. Then any application of one of the six operations in Fig. 3 preserves the following invariants for the resulting graph G': G' remains simple, planar, 3-connected, all its faces are quadrilaterals, and the parity of the number of quadrilaterals remains unchanged.*

Observe that each graph transformation is reversible. It can either be interpreted as adding or as deleting of a hexahedron to a hex complex.

Reversed shelling orders as series of graph transformations. As mentioned above, the reversed order of a shelling is no shelling order in general. In contrast, not only all single graph transformations in Fig. 3 are reversible, a whole series is.

Lemma 3.2. *Suppose we are given a shellable hex complex with surface graph $G = G_0$. Then there is a series of graph transformations $g_1, g_2, \ldots, g_k$ from Fig. 3 which transform G successively to $G_1, G_2, \ldots, G_k$ such that G_k is isomorphic to the graph of a single hexahedron. The reversed order of these graph transformations corresponds to a shelling of the hex complex.*

In other words, topology preserving shelling of a hex complex can be seen as applying a series of graph operations on a planar graph. At first glance, this does not help too much as we usually do not know a hex complex compatible to the surface we want to mesh, and so cannot determine which operation we should apply and in what order. So we are faced with the problem of "shelling

M. Müller-Hannemann, *Shelling Hexahedral Complexes*, JGAA, 5(5) 59–91 (2001)69

an unknown complex," which seems to be intractable. However, we will combine single graph operations to larger units which will help to characterize classes of surface meshes where this concept is useful.

Dual cycle decomposition. Let G be the graph of a quadrilateral mesh and G^* its combinatorial dual. We say that two adjacent dual edges are *opposite to each other* if and only if they correspond to opposite sides of a quadrilateral, i. e. if they are not neighbored in the cyclic adjacency list of their common node. Hence, the four adjacent edges to each dual node can be partitioned into two pairs of opposite edges.

The dual graph $G^* = (V^*, E^*)$ can be decomposed in a canonical way into a collection of edge-disjoint cycles, say into $C_1, \ldots, C_k$, by putting a pair of adjacent edges e_1^*, e_2^* into the same cycle if they are opposite to each other. In other words, for each quadrilateral the edges which are dual to opposite sides are contained in the same cycle. With respect to the labeling in Fig. 1, the cycle decomposition for a single hexahedron yields the cycles $C_1 = \{a, b, c, d\}$, $C_2 = \{a, e, c, f\}$, and $C_3 = \{b, f, d, e\}$ (cycles taken as node sequences).

Observe that by transitivity two edges may belong to the same cycle even if they are neighbored in the cyclic adjacency list of some node. Hence, these dual cycles can be non-simple or *self-intersecting*, see Fig. 4.

Note that the set of dual cycles is well-defined and unique (in the sense that two cycles are equivalent if they have the same set of edges) and can be easily determined in linear time. We call this set the *canonical dual cycles* of G^*, and by a dual cycle we will henceforth always mean a canonical dual cycle.

Cycle elimination. Next we introduce the concept of a cycle elimination. To get an intuitive idea of the use of cycle eliminations for the meshing see Fig. 5. The elimination corresponds to the removal of a complete layer of hexahedra, that is in STC terminology to the removal of a complete sheet.

Let us suppose that the edges of a dual cycle are ordered within a cyclic list such that two edges are consecutive if and only if they are adjacent and opposite to each other with respect to their common node. The cyclic order of the edge list induces an orientation of the cycle. With respect to such an orientation, a simple dual cycle C separates the dual vertices $V^* \setminus V^*(C)$ into vertices $V^*_{C,\ell}$ on the "left hand side" of C and vertices $V^*_{C,r}$ on the "right hand side."

Given a planar graph G, its dual G^*, and a dual cycle C of G^*, the elimination of C transforms G to G' and G^* to G'^* in the following way. The new graph G' is obtained from G by contracting each primal edge corresponding to a dual edge contained in C, and removing parallel edges afterwards. The graph G'^* is then defined as the combinatorial dual of G'. An equivalent way to describe the elimination of a dual cycle is to remove all edges of C from G^*, and to replace for each node v^* of C the two remaining incident edges, say (u^*, v^*) and (w^*, v^*), by the new dual edge (u^*, w^*), and finally remove all vertices of C. In this second definition, dualization of G'^* gives the new primal graph G'.

In order to maintain the above mentioned invariants of the surface graph and to yield a connection to shellings (which will be explained below), cycle

M. Müller-Hannemann, *Shelling Hexahedral Complexes*, JGAA, 5(5) 59–91 (2001)70

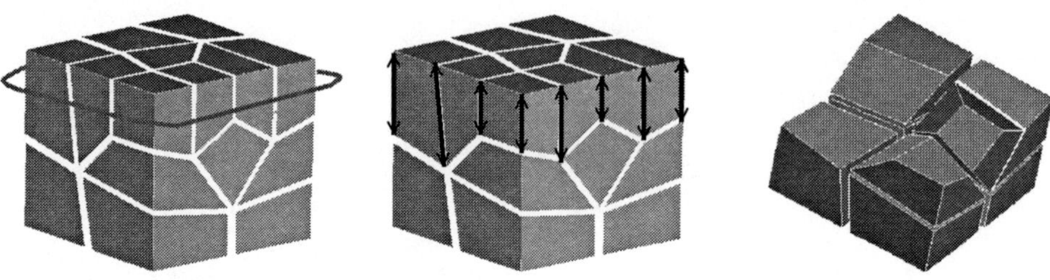

Figure 5: The elimination of a dual cycle (left) corresponds to the contraction of primal edges (middle) and leads to the removal of a sheet of hexahedra (right). In this example, we have taken the nine quadrilaterals on the top facet as the connected set of enclosed quadrilaterals.

eliminations have to be restricted. A *feasible elimination* of a dual cycle C from G^* requires that

1. C is a simple cycle,

2. at least one of the two subgraphs of G^* induced by $V^*_{C,\ell}$ and $V^*_{C,r}$, respectively, is connected, and

3. G'^* is 3-connected.

Note that 3-connectivity can be checked in linear time [13]. So we can test in linear time whether a cycle can be eliminated in a feasible way or not.

Cycle elimination corresponds to shelling operations. The concept of a feasible elimination is closely connected to shelling and graph transformations. It turns out that a feasible cycle elimination is equivalent to a series of graph operations and therefore also to shelling operations.

Suppose that G' is the surface mesh of some hex complex. If we (re)insert a dual cycle C which has been eliminated in a feasible way from graph G, this reinsertion can be interpreted as adding a complete layer (sheet) of hexahedra. Namely, if we assume that the subgraph G^* induced by the quadrilaterals $V^*_{C,\ell}$ on the left hand side of C is connected, then there is a one-to-one correspondence between the quadrilaterals belonging to the set $V^*_{C,\ell}$ and the hexahedra belonging to the added layer.

Lemma 3.3. *Let G be the graph of a quadrilateral mesh and G^* be its combinatorial dual. Let C be a dual cycle which can be eliminated in a feasible way from G^*. If $V^*_{C,\ell}$ ($V^*_{C,r}$) is connected then there is a sequence of exactly $|V^*_{C,\ell}|$ ($|V^*_{C,r}|$) graph operations from Fig. 3 which transforms G to G'.*

Proof: We first observe that the 2-dimensional cell complex formed by the union of the quadrilaterals contained in $V^*_{C,\ell}$ is shellable. This follows from a characterization of the shellability of 2-dimensional (pseudo)-manifolds [9]. In

M. Müller-Hannemann, *Shelling Hexahedral Complexes*, JGAA, 5(5) 59–91 (2001)71

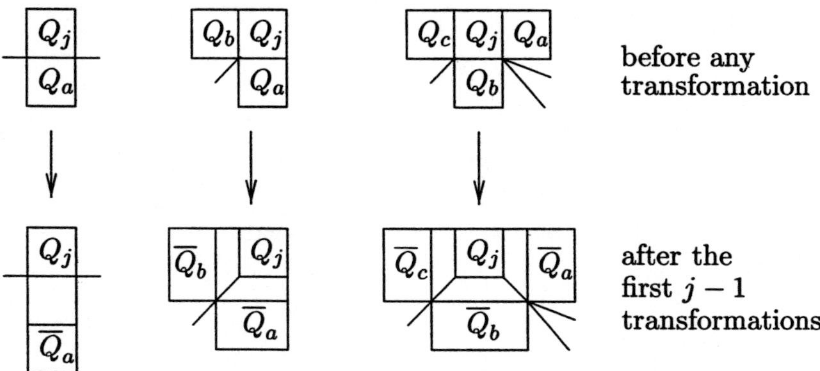

Figure 6: The different cases of local subgraphs before a hexahedron is placed on top of quadrilateral Q_j (Q_a, Q_b and Q_c come before Q_j in the shelling order). The number of quadrilaterals sharing a corner with Q_j and lying "between" Q_a and Q_b (or "between" Q_b and Q_c) can be an arbitrary non-negative number (including zero!).

our setting this characterization asserts that the cell complex is shellable if the planar primal graph G_ℓ induced by these quadrilaterals can be embedded in a 2-ball or 2-sphere M in such a way that $M \setminus G_\ell$ is the union of $|V^*_{C,\ell}|$ pairwise disjoint open 2-balls whose boundaries are the bounding cycles of the quadrilaterals. Clearly, these conditions are fulfilled as the complete quadrilateral mesh is embedded in a 2-manifold and so the quadrilaterals contained in $V^*_{C,\ell}$ are embedded in a 2-ball by our assumption that the dual subgraph induced by $V^*_{C,\ell}$ is connected.

We claim that any shelling order $(Q_1, Q_2, \ldots, Q_k)$ of quadrilaterals (with $k = |V^*_{C,\ell}|$) determines a possible series of graph operations to transform G' to G. The general idea is to place a hexahedron "on top" of each of quadrilaterals Q_j in the shelling order. At each step, we maintain the property that the newly created hexahedron contains the quadrilateral Q_j as the "bottom facet," and exactly one quadrilateral facet from each hexahedron which has been placed on a quadrilateral Q_i adjacent to Q_j with $i < j$. Thus, all these faces disappear from the surface graph. At the same time, the addition of a hexahedron creates new quadrilateral facets, namely the "top facet", denoted by $\overline{Q}_j$ and all the remaining facets of the hexahedron which are not shared with a previously added hexahedron.

Hence, the graph transformation starts with operation (b) of Fig. 3 applied to the first quadrilateral Q_1 in the shelling order. For the j-th element Q_j of the shelling order ($1 < j \leq k$), the shelling conditions guarantee that the intersection of Q_j with quadrilaterals which come earlier in the shelling order is a connected set of edges. Moreover, the following is a crucial observation about a correct shelling order. If Q_a and Q_b are quadrilaterals which come before Q_j in the shelling order and both share an edge with Q_j incident to a common

M. Müller-Hannemann, *Shelling Hexahedral Complexes*, JGAA, 5(5) 59–91 (2001)72

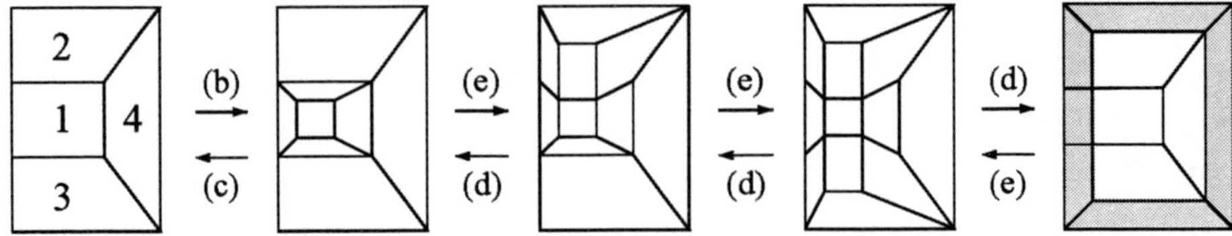

Figure 7: Graph transformations corresponding to a cycle elimination/reinsertion according to the shelling order (1,2,3,4).

vertex v, then all other quadrilaterals incident to this vertex v also precede Q_j in the given shelling order. Using this observation, one can show by induction that the local graph structure at Q_j of the surface graph after the first $j-1$ transformations ($j > 1$) looks as depicted in Fig. 6.

Hence, if the intersection of Q_j with preceding quadrilaterals is just a single edge shared with Q_a, we can apply transformation (e) of Fig. 3 using the facet Q_j and the adjacent quadrilaterals of the surface graph which has been created in the a-th step. If the intersection are two edges shared with Q_a and Q_b, we apply transformation (a) of Fig. 3 using the facet Q_j and the two adjacent quadrilaterals at the common corner of Q_a and Q_b in the current surface graph. Similarly, if the intersection contains three edges, we apply transformation (d) of Fig. 3. Note finally, that for any shelling order it is impossible that Q_j is completely surrounded by preceding quadrilaterals. The resulting sequence of graph operations transforms G' to G, i. e. it inserts a dual cycle. Reversing the order of transformations and each single operation yields the claimed sequence of transformations. $\square$

The shelling order for the enclosed quadrilaterals can be found in linear time using an algorithm of Danaraj and Klee [9]. Basically, this algorithm adds a quadrilateral to the initial shelling sequence in a greedy fashion if the shelling conditions are fulfilled. Danaraj and Klee's algorithm can be simplified in the special case of quadrilateral cells as we can check in constant time whether a quadrilateral can be added to an initial shelling sequence or not.

Fig. 7 shows a small example of a dual cycle which encloses four quadrilaterals and the corresponding graph transformations of a (re)insertion ("from left to right") or an elimination ("from right to left").

Note that in the example of Fig. 5 we have taken the nine quadrilaterals on the top facet as the set of quadrilaterals enclosed by the eliminated cycle. The quadrilaterals on the opposite side of the cycle are also connected with respect to the dual graph. However, if we take this other set of quadrilaterals as the basis for our sequence of graph transformations this would lead to a different hex complex.

Perfect cycle elimination schemes. A *perfect cycle elimination scheme* of a dual graph G^* of a quadrilateral surface mesh is an order $C_1, C_2, \ldots, C_k$ of its

M. Müller-Hannemann, *Shelling Hexahedral Complexes*, JGAA, 5(5) 59–91 (2001)73

k canonical cycles such that the first $k - 3$ cycles with respect to this order can be eliminated one after another in a feasible way, and such that the remaining 3 cycles form the dual surface graph of a single hexahedron. In particular, this means that the last 3 cycles are simple and cross pairwise exactly twice (two cycles *cross* if they share a common node).

The core algorithm. Whenever we know a perfect cycle elimination scheme for some graph G^*, we can iteratively apply Lemma 3.3 in reversed elimination order to give an explicit construction of a hex complex compatible to the prescribed surface mesh. This yields an algorithm for combinatorial meshes of topological balls which runs in two phases: In the first phase we determine a cycle elimination scheme, and in the second phase we build up a hex complex by adding sheets in reversed order of the elimination.

Example: the "double fold" problem revisited. We conclude this section with the example of the so-called "double fold" problem to illustrate how our core algorithm works. The same example has been used to study the whisker weaving algorithm [34]. The geometry of this example is simply a cube, but the surface mesh is quite irregular (several nodes of degree five and three), see Fig. 8. The Figs. 9 to 12 show the successive elimination of dual cycles from the original surface mesh to the mesh corresponding to a single hexahedron, and the construction phase. The first cycle to be eliminated encloses 9 dual nodes, the second 3, the third 2, and the fifth and sixth only one node. Hence, in the reversed order, the series of hexahedra to add to the first one is 1,1,2,3,9.

Figure 8: The "doublefold" example (the middle picture is the left cube with the front faces removed so you can see through to the back faces from the inside). The hex complex consists of 17 hexahedra.

M. Müller-Hannemann, *Shelling Hexahedral Complexes*, JGAA, 5(5) 59–91 (2001)74

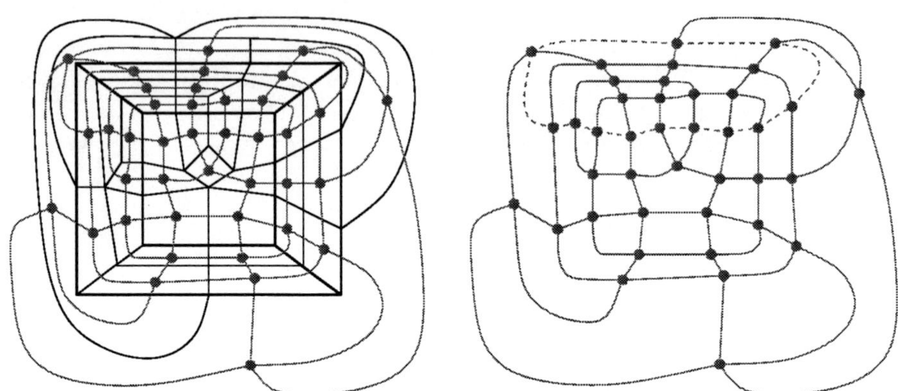

Figure 9: The primal and dual graph of the surface mesh of the "double fold" example, dual nodes are drawn as solid circles (left); the first dual cycle of a perfect elimination scheme is dashed, it encloses 9 dual nodes (right).

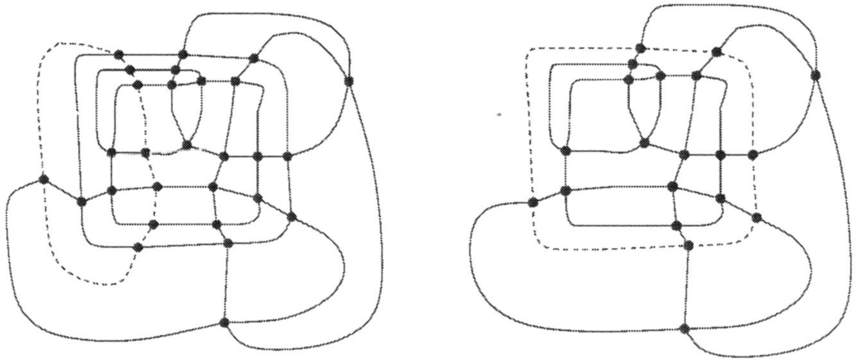

Figure 10: The dual graph after the first and second elimination. The cycle to be eliminated next is again dashed. The dashed cycle on the left side encloses 3 dual nodes, on the right side 2 dual nodes (on the unbounded side of this drawing!).

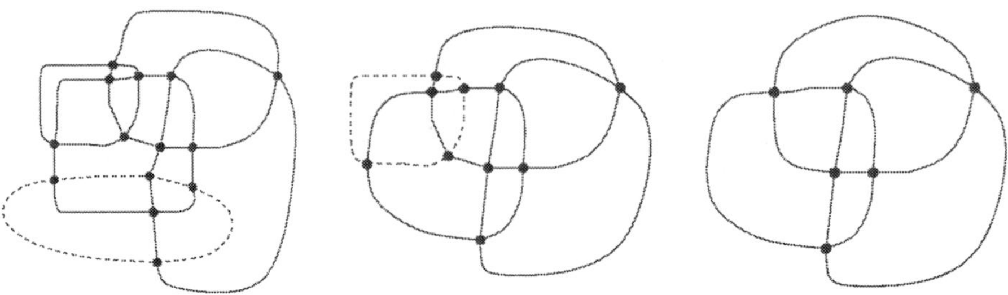

Figure 11: The dual graph after the third, forth and fifth elimination. The right figure shows final configuration of the remaining three cycles.

M. Müller-Hannemann, *Shelling Hexahedral Complexes*, JGAA, 5(5) 59–91 (2001)75

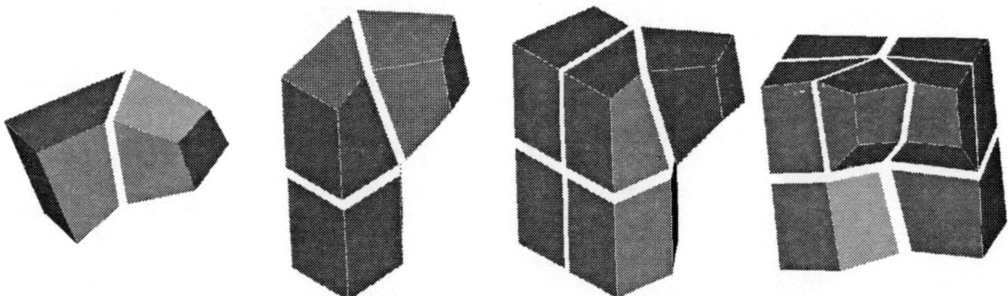

Figure 12: Construction of the hex complex for the "doublefold" example. The figures show from left to right the intermediate steps where hexahedra are added in reversed order of the elimination scheme.

4 Classes of Surface Meshes with a Perfect Elimination Scheme

As our approach relies on finding a perfect elimination scheme we would like to characterize the surface meshes of topological balls which admit such a scheme. While the decision problem whether a single cycle can be eliminated in a feasible way is solvable in polynomial time, the complexity status for the recognition of a perfect elimination scheme is still open (to our knowledge).

There are some obvious necessary conditions. By definition, existence of a perfect cycle elimination scheme requires that all dual cycles are simple. We will show in Section 5 how to guarantee this property for our surface meshes. Simplicity of a dual cycle implies even length, and all cycles being simple implies an even number of quadrilaterals. The latter is sufficient for the existence of a combinatorial mesh, but not for the existence of a perfect elimination scheme in general. To see a small counter-example, consider a 3-connected planar dual graph consisting of three canonical cycles which cross each other pairwise exactly four times (hence, we cannot reach the desired final state wherein three cycles cross pairwise twice).

We also note that non-shellable hex complexes exist. Furch's knotted hole ball is a well-known example of a hex complex (see Fig. 13) which is non-shellable [42]. However, it is insightful to check that the corresponding surface mesh has a perfect elimination scheme, and so can be decomposed into some hex complex, which is, of course, different from the non-shellable one. In other words, this shows that a fixed surface mesh may have different decompositions leading to shellable and non-shellable hex complexes.

Cycle arrangements. Consider a collection of k simple, closed Jordan curves in the plane such that no three curves meet in a point, which we call a *cycle arrangement*. We may associate in the obvious way a planar graph G^* to such a collection of curves (considered as a dual graph): The vertex set is formed by the intersection points between curves, and the edges are induced by the

M. Müller-Hannemann, *Shelling Hexahedral Complexes*, JGAA, 5(5) 59–91 (2001)76

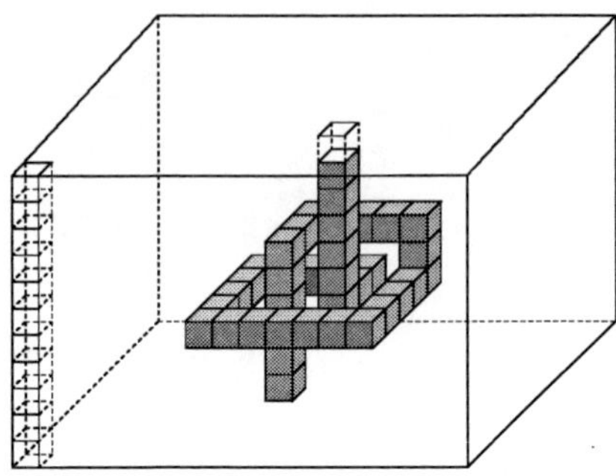

Figure 13: Furch's "knotted hole ball": a pile of $(k_1 \times k_2 \times k_3)$ hexahedra where hexahedra are missing along a knotted curve except for a single hexahedron at the top layer. This hex complex is non-shellable, but the surface mesh has a perfect elimination scheme.

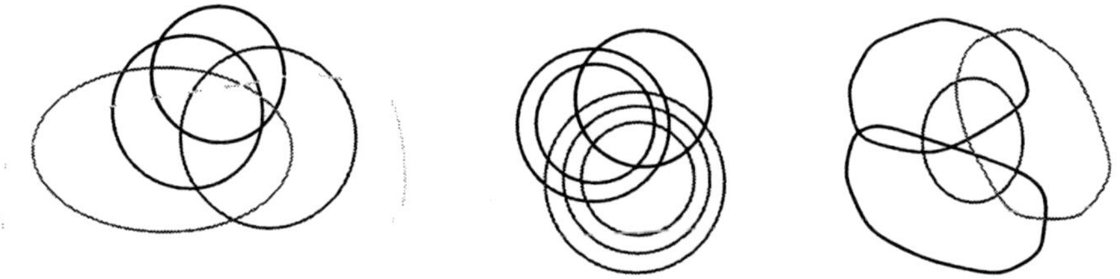

Figure 14: Examples of cycle arrangements of surface meshes: a zonotopal arrangement (left), a zonotopal arrangement with three equivalence classes of parallel neighbors (middle), and an arrangement of Class III (right). Note that the elimination order is not arbitrary in the latter example.

segments between intersection points. If G^* is simple and 3-connected, then so is the primal graph. As the degree of each dual vertex is four, such a primal graph corresponds to a quadrilateral surface mesh. Hence, any cycle arrangement such that the associated planar graph is simple and 3-connected is a *cycle arrangement of a surface mesh*.

Next we consider different classes of cycle arrangements which have perfect elimination schemes, examples are shown in Fig. 14.

Class I: "zonotopal cycle arrangements". Zonotopes are special polytopes which can be defined in many equivalent ways [41], for example, as the image of a higher-dimensional cube under an affine projection, or as the Minkowski sum of a set of line segments. It is well-known that zonotopes where all surface facets are quadrilaterals always have decompositions into hexahedra [33].

Moreover, such zonotopes constitute a class of polytopes for which every cycle order leads to a perfect elimination scheme (the "zones" of a zonotope

M. Müller-Hannemann, *Shelling Hexahedral Complexes*, JGAA, 5(5) 59–91 (2001)77

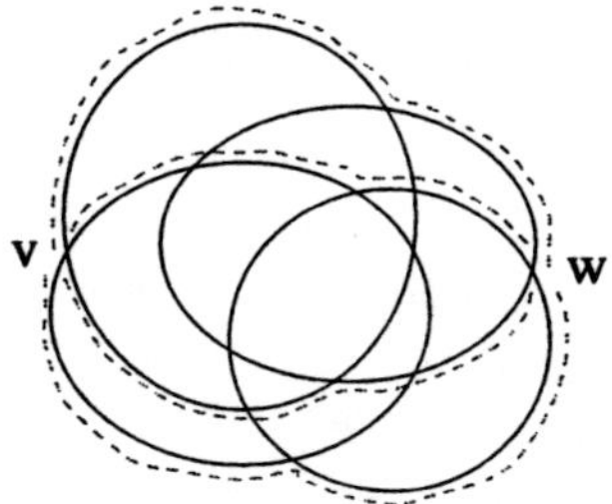

Figure 15: Zonotopal cycle arrangements are 4-connected: there exist four disjoint paths (dashed) between v and w.

correspond to the dual cycles). Certainly, for the existence of an elimination scheme it is sufficient that the surface graph of some mesh is combinatorially equivalent to that of a zonotope.

The dual cycles in graphs arising from zonotopes with $k+1$ zones have a nice combinatorial property: The dual cycles cross each other pairwise exactly twice, i. e. each dual cycle has length $2k$, and if a dual cycle is traversed, the first k intersections with other cycles appear in the same order as the second k intersections. We call an arrangement of three or more simple cycles which all have this ordering property a *zonotopal cycle arrangement*.

Lemma 4.1. *A zonotopal cycle arrangement has a perfect elimination scheme and every cycle order yields such a scheme.*

Proof: The only zonotopal cycle arrangement with three cycles is the arrangement of a single hexahedron, and so trivially has a perfect elimination scheme.

Any zonotopal cycle arrangement with more than three cycles is not only 3-connected (as we require for an elimination), it is even 4-connected. To see this, consider an arbitrary pair of vertices $v, w \in G^*$ and construct four internally vertex disjoint paths using the cycles through v and w as indicated in Fig. 15.

Hence, for a zonotopal cycle arrangement with $k+1$ cycles, we may select an arbitrary cycle C for elimination and the resulting graph remains 4-connected. For a feasible elimination we have still to check that, for some orientation of C, the dual vertex set $V_{C,\ell}^*$ on the left hand side from C is connected. To see this, consider any two dual vertices $v^*, w^* \in V_{C,\ell}^*$. Denote by C_{v^*} and C_{w^*} dual cycles going through (that is containing) v^* and w^*. If the two cycles are identical or intersect at some third vertex $s^* \in V_{C,\ell}^*$, there is a path from v^* to w^*. If, however, such an intersection between C_{v^*} and C_{w^*} does not exist within $V_{C,\ell}^*$, we get a contradiction to the order of the intersections with C in a zonotopal cycle arrangement, see Fig. 16. $\square$

Class II: "zonotopal cycle arrangements with parallel cycles". We say that a cycle C_1 is a *parallel neighbor* of another cycle C_2 of the same length in the graph G^* if their node sets are disjoint but for each node v^* of C_1 there is a node w^* of C_2 such that the edge (v^*, w^*) belongs to G^*.

M. Müller-Hannemann, *Shelling Hexahedral Complexes*, JGAA, 5(5) 59–91 (2001)78

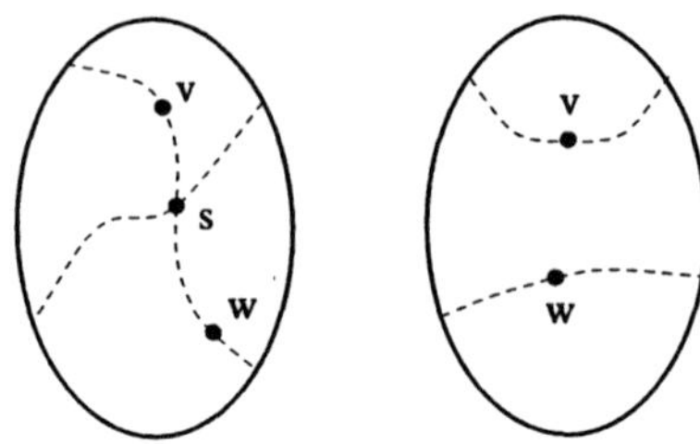

Figure 16: Vertices inside a cycle C of a zonotopal arrangement are always connected (left); otherwise the intersections with C appear in the wrong order (right).

Lemma 4.2. *If the dual cycles are partitioned into equivalence classes of parallel neighbors, then a mesh has a perfect elimination scheme if representing cycles of these equivalence classes form a zonotopal cycle arrangement. In this case, any elimination order is possible provided it keeps cycles from three different equivalence classes to the end.*

Proof: Just note that the class of zonotopal cycle arrangements with parallel neighbor cycles is closed under elimination of an arbitrary cycle (provided that the remaining arrangement has at least three equivalence classes). Then, basically the same arguments as in the proof of Lemma 4.1 show that such a cycle arrangement is 4-connected, and that the dual vertex set $V_{C,\ell}^*$ on the left hand side from an arbitrary cycle C is connected. $\qquad\square$

Note that although the order of eliminations is almost arbitrary, a different hexahedral mesh will result for different orders.

Class III: "cycles cross pairwise twice plus 3-connectedness". The condition "all cycles cross pairwise exactly twice" is to weak for a guarantee of a perfect elimination scheme, for an example see the picture on the right in Fig. 14 without the "inner cycle."

We consider the following recursively defined class of cycle arrangements, called *class III*. The only member of class III with three cycles is the arrangement of a hexahedron. A cycle arrangement with $k + 1$ cycles belongs to class III, if it can be derived from a member of class III with k cycles by the insertion of one simple cycle which crosses all others exactly twice and the resulting graph is 3-connected.

Lemma 4.3. *A cycle arrangement of class III has a perfect elimination scheme.*

Proof: By definition, a cycle arrangement which is member of class III can be constructed by the stepwise insertion of cycles. We claim that the reversed order of the insertion yields a perfect elimination scheme. Hence, if we eliminate a cycle, 3-connectivity is asserted. The only non-trivial fact to show is that either the vertex sets $V_{C,\ell}^*$ on the left hand side or on the right hand side $V_{C,r}^*$ are

M. Müller-Hannemann, *Shelling Hexahedral Complexes*, JGAA, 5(5) 59–91 (2001)79

connected. (It is not hard to construct an example where one of the two sets, say $V_{C,\ell}^*$, is disconnected.)

Assume that both sets are disconnected. This means that there are disconnected nodes $v, w \in V_{C,\ell}^*$ and $s, t \in V_{C,r}^*$. Denote by C_v, C_w, C_s, C_t cycles going through the vertices v, w, s, t. As v, w are not connected within $V_{C,\ell}^*$, the cycles C_v and C_w cannot intersect within $V_{C,\ell}^*$. Similarly, the cycles C_s and C_t do not cross within $V_{C,r}^*$. As cycles cross pairwise, this implies that C_v and C_w intersect in $V_{C,r}^*$, and C_s and C_t intersect in $V_{C,\ell}^*$.

Suppose that neither C_s nor C_t intersect C_v within $V_{C,\ell}^*$. This would imply that both cycles intersect C_v in $V_{C,r}^*$, in contradiction to the assumption that s and t are not connected in $V_{C,r}^*$. Hence, we may assume that at least one of both cycles, say C_s, intersects C_v in $V_{C,\ell}^*$. This in turn means that C_s crosses C_w in $V_{C,r}^*$. Now we get a contradiction. Either C_t crosses C_w in $V_{C,\ell}^*$ which would mean that v and w are connected in $V_{C,\ell}^*$ (via portions of C_v, C_s, C_t and C_w), or C_t crosses C_w in $V_{C,r}^*$, but then s and t are connected in $V_{C,r}^*$ (via portions of C_s, C_w and C_t). $\qquad\square$

5 Hexahedral Meshing for Arbitrary Domains

In this section we consider the meshing of arbitrary domains. We do not restrict our discussion to one of the many different CAD formats and possibilities to encode surfaces. However, we will assume that the CAD input model for our algorithm describes a solid body by polygonal surface patches.

The general algorithm. For arbitrary domains, we propose the following approach which consists of five major steps:

1. Decompose the whole domain into subdomains which are topologically balls and "almost convex."

2. Quadrangulate the surface mesh with half of the required density. Replace each quadrangle by four new ones.

3. Cancel self-intersecting dual cycles.

4. Do for each subdomain

    (a) Search for a perfect cycle elimination scheme.

    (b) Build up a hex complex.

5. Embed the hex complex and perform mesh smoothing.

In the following we will explain the steps of our general scheme.

Step 1: Decomposition into subdomains. In the very first step we have to decompose the given body into simpler subdomains which are topologically

M. Müller-Hannemann, *Shelling Hexahedral Complexes*, JGAA, 5(5) 59–91 (2001)80

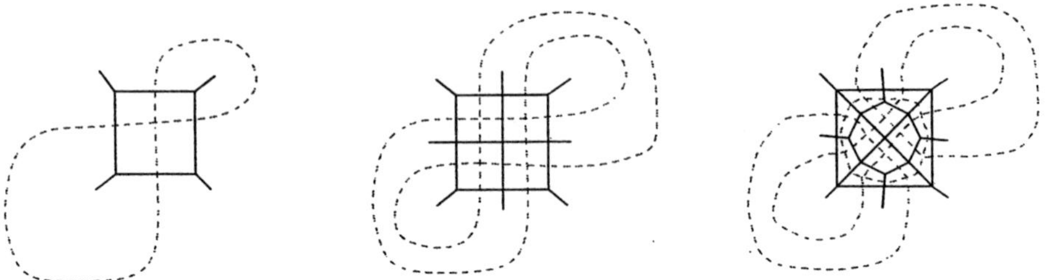

Figure 17: Getting rid of a self-intersecting dual cycle (left): the dual cycle is first duplicated (middle), and then the surface mesh is locally transformed (right).

balls and should be "almost convex." In general, this is a difficult problem which offers many degrees of freedom in how to subdivide a body. Solution methods depend very much on the used CAD data structures and geometric representations of the input model. The corresponding details go beyond the scope of this paper. We only mention that we add internal 2-manifolds using ideas of White et al. [40], but with weaker requirements for the resulting subregions. It is not necessary, although desirable, that these regions are "almost convex." Asking for a subdivision into convex regions in the usual mathematical sense would be too strict, as otherwise there would be no subdivision into finitely many regions for concave surfaces. So, roughly speaking, we mean by *almost convex* that a region should not deviate from a convex region by too much.

Step 2: Surface mesh quadrangulation. We can also be brief with remarks on the surface mesh generation, as any method which works consistently with branchings (because of the insertion of internal 2-manifolds) can be chosen. In our experiments, we used the surface meshing tool as described in [19, 22] without changes.

Step 3: Cancel self-intersecting dual cycles. We have to ensure that all dual cycles are simple as this is necessary for the existence of a perfect elimination scheme. Our preliminary paper [21] as well as Folwell & Mitchell [11] describe how to modify a quadrilateral surface mesh in order to remove self-intersections for a single domain. The main idea of the latter approach is to collapse a quadrilateral where a self-intersection occurs into two edges. This may result in degenerated vertices with degree two, but such a situation can be resolved by a so-called "pillowing" technique which places an additional ring of quadrilaterals around such a degenerated vertex. However, without a modification both methods have a serious drawback as they do not work in the presence of branchings. If branchings occur, the problem is that at least some dual cycles belong to several subdomains and cannot be modified independently. Hence, it may happen that self-intersections are removed in one subdomain but new self-intersections are created in some other subdomain.

M. Müller-Hannemann, *Shelling Hexahedral Complexes*, JGAA, 5(5) 59–91 (2001)81

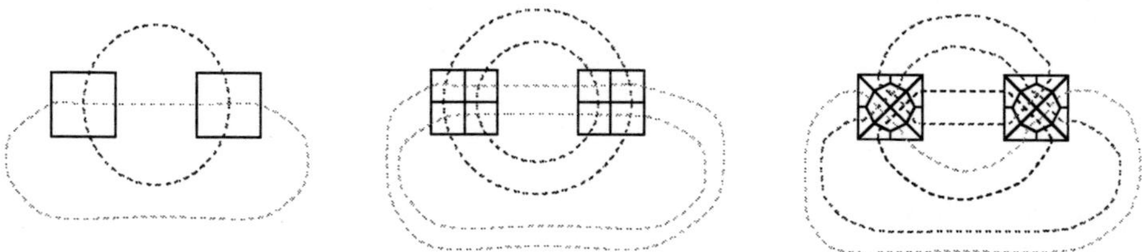

Figure 18: Simplicity of a pair of simple cycles (before duplication, left) is maintained if we apply the graph transformation at all crossings.

In this paper, we propose an alternative method which can also be applied in the presence of branchings. It works as follows: After decomposing the domain of our solid body into topological balls, we use the quadrilateral surface mesher with only half the required mesh density. Then we subdivide each quadrilateral into four new ones (by halving all edges) to meet the required mesh density.

This replacement duplicates all dual cycles. In particular, all quadrilaterals where self-intersections occur appear in pairs, see the middle part of Fig. 17. So it is possible to change the surface mesh locally at all such places, by replacing four quadrilaterals with 12 new ones, see the right part of Fig. 17. Obviously, the surface graph remains planar, simple and 3-connected, and we do not change the structure of other cycles than those going through such quadrilaterals. Most importantly, it serves its primary purpose to resolve each existing self-intersection. If a quadrilateral where a self-intersection occurred belongs to exactly one subdomain, the transformation cannot create a new self-intersection, otherwise it might do so with respect to the other subdomain. More precisely, if we apply the transformation to a quadrilateral where to simple dual cycles cross with respect to some subdomain (the common quadrilaterals of simple dual cycles are called *crossings*) then this transformation would create a self-intersection at one of the new quadrilaterals (if the two dual cycles are not changed in their remaining parts). Hence, we need some additional step for each pair of simple cycles affected by a transformation for some other subdomain. Fortunately, whenever necessary, we can avoid the complication of creating self-intersecting cycles if we apply the same local transformation at *all* four-tuples of quadrilaterals corresponding to crossings, for an example with two crossings see Fig. 18. (As the proposed transformation typically degrades the surface mesh quality we do not apply this transformation everywhere.) It is not hard but insightful to verify that this modification creates always simple cycles, for any number of crossings.

Step 4a: Search for a perfect elimination scheme. In principle, every perfect cycle elimination scheme allows us to build up a valid combinatorial mesh. Unfortunately we do not know an efficient algorithm to find such a scheme in the general case. Hence, we use a greedy strategy which iteratively eliminates some feasible cycle. However, the choice which cycle to eliminate next should not be arbitrary for several reasons. First, we note that, in general, different

M. Müller-Hannemann, *Shelling Hexahedral Complexes*, JGAA, 5(5) 59–91 (2001)82

elimination orders, lead to meshes of a different structure and size. Second, we also observe that two different geometric meshes can have the same combinatorial mesh, see Fig. 19. Hence, a careful cycle selection has to take the geometry of the surface into account.

Cycle selection. For that purpose, we determine for each dual edge the dihedral angle between the two quadrilateral faces which correspond to its endpoints. This, in turn, gives us a an initial classification for each primal edge as "sharp" or "plane" edges. A primal edge of the surface graph is a *sharp edge* if the dihedral angle is significantly smaller than 180 degrees.

For a simple dual cycle C, we use the term *neighboring primal cycles* to mean the two connected primal cycles induced by the union of all primal edges of the quadrilaterals corresponding to C which do not cross dual edges of C. We assign a *side elimination weight* to each dual cycle according to the number of sharp edges of the neighboring primal cycles, divided by the cycle length (number of edges). (In our implementation of side elimination weights, we classify a dihedral angle smaller than 120 degrees as sharp.) The weights can be used to define a preference order for dual cycles. A *first rule* is that a cycle should be preferred in the selection if it has a higher weight, i. e. if it has a higher quotient of sharp edges to the total number of cycle edges than some other cycle. We also keep track of the side for which the elimination weight has more sharp edges, as this side will be used as the enclosed set of quadrilaterals on which the construction phase adds hexahedra.

Moreover, we use a weight counting the number of sharp primal edges corresponding to edges of the dual cycle. A *second rule* says that one should eliminate a cycle only if this second weight is positive. The intuition behind this rule is that one should not eliminate a cycle which lies in a plane, as this may lead to a bad quality of the geometric mesh. (For that purpose, we use a different parameter classifying an edge as a plane edge, if the dihedral angle is larger than 170 degrees.)

After each cycle elimination we update this classification for all primal edges which are affected by the elimination. To be precise, if two primal edges become identified by contraction of a quadrilateral, the resulting edge is classified as a sharp edge if at least one of the edges was sharp before the identification. Hence, the weights of a dual cycle will change after eliminations.

Double cycle elimination. Suppose that some dual cycle C_m has two parallel neighbors, one on the right and one on the left side. Then it can certainly be eliminated in a feasible way. If, in addition, all edges of the corresponding left and right primal neighboring cycles are classified as sharp edges, the cycle C_m has a high preference to be selected for elimination. However, in such a situation it seems to be better to eliminate both the left and right neighboring parallel cycles simultaneously. In such a *double elimination* the set of enclosed quadrilaterals is just the dual cycle C_m in the middle. The elimination can be interpreted as removing a torus of hexahedra, see the left example of Fig. 19. Similarly to the statement of Lemma 3.3, one can prove that a double elimina-

M. Müller-Hannemann, *Shelling Hexahedral Complexes*, JGAA, 5(5) 59–91 (2001)83

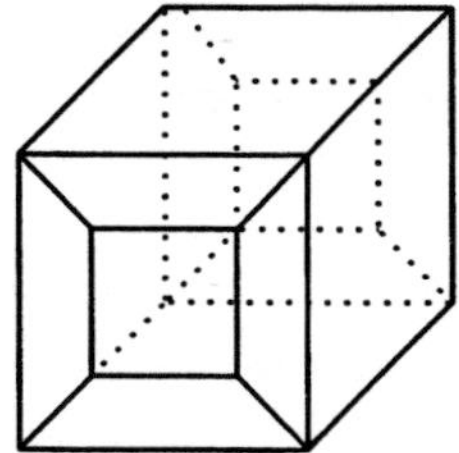 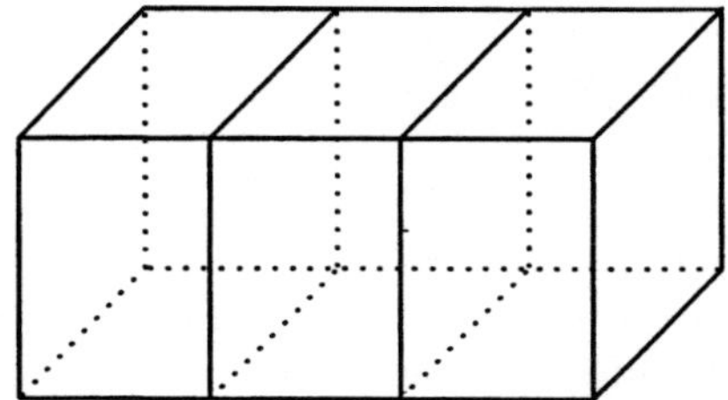

Figure 19: Examples of two solids with "obvious" decompositions into five and three hexahedra. Note that the surface meshes are combinatorially equivalent. Therefore, additional geometric information is necessary to yield the appropriate decompositions. For the left instance, a double cycle elimination is appropriate.

tion corresponds to a series of graph transformations from Fig. 3 (with $|V(C_m)|$ operations in total).

Adding new cycles. Of course, our greedy cycle selection can get stuck. We now sketch a strategy to resolve the situation that no cycle can be eliminated in a feasible way.

Even if no cycle can be eliminated, we can certainly still perform graph transformations (for which we have many degrees of freedom). Recall that all graph transformations of Fig. 3 can be reversed. Hence, through a series of such transformations we cannot only eliminate a dual cycle but also insert a new one. If no cycle can be eliminated in a feasible way, the idea is to add one or more new cycles in such a way to the current graph that at least one of the "old cycles" can be feasibly eliminated afterwards. We choose some cycle according to our selection criteria for elimination. The check for 3-connectivity after a testwise elimination tells not only a failure, it also gives us all 2-separators. To augment the local connectivity, a new cycle is inserted around a node of a 2-separator preferably in such a way that it crosses other cycles exactly twice. The latter heuristical placement is motivated by our characterization of shellable surface.

Step 4b: Building up the hex complexes. As soon as a perfect elimination scheme has been determined for some domain, we reverse the order of cycle eliminations and build up a hex complex layer by layer as described in Section 3.

Step 5: Embedding and smoothing. Up to this point we have only discussed how to find a combinatorial mesh. Embedding of a combinatorial hex complex into the prescribed surface such that all hexahedra are well-shaped is a non-trivial task. It is not even clear which conditions are sufficient for a "nice" embedding.

However, for convex domains we have been quite successful with a strikingly simple strategy, namely with a barycentric embedding algorithm (commonly also referred to as *Laplacian smoothing*).

This strategy has been inspired by a beautiful result of Tutte [38, 39] for

M. Müller-Hannemann, *Shelling Hexahedral Complexes*, JGAA, 5(5) 59–91 (2001)84

drawing planar graphs in the plane. Tutte showed that the barycentric straight-line embedding draws a 3-connected planar graph without crossings in the plane such that every facet is convex if the outer facet is prescribed as a convex polygon.

For our purposes, we consider a hex complex as a graph, fix the surface nodes according to the surface mesh, and for all other nodes the node position is determined as a convex combination (the barycenter) of the position of its neighbors. More precisely, the barycentric embedding determines vertex positions for all movable vertices from

$$p_v = \frac{1}{deg_v} \sum_{(v,w)\in E} p_w,$$

where we denote by p_v the position of vertex v in $\mathbb{R}^3$ and by deg_v the degree of vertex v. The barycentric embedding is the unique minimizer of the energy functional

$$E(p_1,\ldots,p_n) = \frac{1}{2} \sum_{(v,w)\in E} ||p_v - p_w||^2,$$

where $p_{k+1},\ldots,p_n$ are fixed positions of vertices on the surface. A straightforward Gauß-Seidel iteration on the system of equations for the node positions converges quickly and is computationally very fast. The barycentric embedding can be interpreted as a special case of force-directed algorithms in graph drawing, see for example [1, Chapter 10].

In the experiments with convex domains mentioned in this paper, the initial layout obtained from this method was already excellent. However, for this simple to implement and comparably fast algorithm it is well-known that it might fail to produce valid meshes (indeed it is quite likely to fail for non-convex domains). Therefore, we implemented additional embedding algorithms for optimized mesh smoothing and untangling, i.e., for getting rid of inverted hexahedra, as described in the pioneering work of Freitag and Knupp [12, 14, 15].

Example: Model of a shaft. We have implemented a first prototype for our approach in C++. To illustrate first experimental results we use the CAD model of a shaft with a flange shown in Fig. 20. Several features make this model useful for testing purposes: it contains 8 holes, a cylindrical inlet in the lower part (not visible in the hidden surface description of Fig. 20), and concave surface patches. The input model is axis-symmetric, and so is our output – although our algorithm has no tools incorporated to detect and to exploit symmetry. The complex input model has first been decomposed into topological balls, see the left part of Fig. 20. Then our quadrilateral meshing algorithm [22] has been run according to some specified uniform mesh density. The resulting surface mesh has the nice property to be free of self-intersecting dual cycles for all subregions. For all subregions, our algorithm was successful to find a perfect cycle elimination scheme. Even more, the embedding of the hex complexes has led also to geometric meshes of a very good quality. Figs. 21,

M. Müller-Hannemann, *Shelling Hexahedral Complexes*, JGAA, 5(5) 59–91 (2001)85

22, 23 and 24 show the resulting meshes for different parts of the model. In total, the hexahedral mesh for the complete model, depicted in the right part of Fig. 20, consists of 7488 hexahedra.

6 Summary and Future Work

In this paper, we proposed a new approach to hexahedral mesh generation. This approach combines techniques from several fields, namely decompositions of non-convex regions into convex pieces from computational geometry, minimum cost bidirected flows for surface meshing [20], shellings of cell complexes from topology, graph algorithms on planar graphs, and non-linear optimization methods for the embedding phase. Some main features are worth to be pointed out.

- The key idea is to separate the mesh generation process into a combinatorial part where an abstract complex of hexahedra is constructed from a surface mesh and a geometric embedding part where the exact positions of the mesh nodes are determined.

- Hexahedral mesh quality benefits from our surface mesh generator which yields very regular meshes (in contrast to paving or advancing front based methods).

- Successive elimination of dual cycles corresponds to the meshing of complete layers of hexahedra (sheets) in each step. Cycle selection can use global geometric information to find good elimination orders. Our strategy avoids all kind of combinatorial degeneracies rigorously. Problems with knives or wedges as known from the whisker weaving or plastering algorithms do not occur.

- The very core of our algorithm, and thus the computationally most expensive part, can be performed in parallel for each subdomain. Only the first three steps of our meshing procedure need global access to the complete set of data.

As mentioned, we have currently only implemented a first prototype for our algorithm and our experiences are still limited. First results are encouraging, but more testing has to be done to refine several parameters of our algorithm, for example the cycle selection rules. We would also like to improve our strategies for cycle insertion for those cases where the cycle elimination process gets stuck.

Self-intersections of dual cycles are a major source of difficulties within the mesh generation process. We developed a consistent strategy to remove all self-intersections, relying on the duplication of all dual cycles and a surface mesh modification. Note that this strategy is feasible if the mesh density requirements are not too strict. Problems may occur if the CAD input model contains very short edges. We are convinced that self-intersections should be avoided in the preprocessing part as far as possible. Future work should concentrate on how

M. Müller-Hannemann, *Shelling Hexahedral Complexes*, JGAA, 5(5) 59–91 (2001)86

to minimize the number of self-intersections of dual cycles in surface meshing algorithms in combination with rules for the subdivision into topological balls. Theoretical work should try to extend our characterization of classes of shellable surfaces.

Acknowledgments

The author wishes to thank G. Krause for providing us with the finite element preprocessor ISAGEN (which we used for our illustrations) and instances from the automobile industry.

References

[1] G. Di Battista, P. Eades, R. Tamassia, and I. G. Tollis, *Graph drawing: Algorithms for the visualization of graphs*, Prentice Hall, 1999.

[2] M. Bern and D. Eppstein, *Mesh generation and optimal triangulation*, Computing in Euclidean Geometry, 2nd Edition (D.-Z. Du and F. Hwang, eds.), World Scientific, Singapore, 1995, pp. 47–123.

[3] M. Bern and P. Plassmann, *Mesh generation*, chapter 6 of Handbook of Computational Geometry (J. Sack and J. Urrutia, eds.), North-Holland, Elsevier Science, 2000, pp. 291–332.

[4] T. D. Blacker and R. J. Meyers, *Seams and wedges in plastering: A 3D hexahedral mesh generation algorithm*, Engineering with Computers 9 (1993), 83–93.

[5] T. D. Blacker and M. B. Stephenson, *Paving: A new approach to automated quadrilateral mesh generation*, Int. J. Numer. Methods in Eng. **32** (1991), 811–847.

[6] N. A. Calvo and S. R. Idelsohn, *All-hexahedral element meshing by generating the dual mesh*, Computational Mechanics: New Trends and Applications (S. Idelsohn, E. Oñate, and E. Dvorkin, eds.), CIMNE, Barcelona, Spain, 1998.

[7] S. A. Canann, *Plastering: A new approach to automated, 3d hexahedral mesh generation*, Am. Inst. Aeronautics and Astronautics, Reston, VA (1992).

[8] W. A. Cook and W. R. Oakes, *Mapping methods for generating three-dimensional meshes*, Computers in Mechanical Engineering, CIME Research Supplement (1982), 67–72.

[9] G. Danaraj and V. Klee, *A representation of 2-dimensional pseudomanifolds and its use in the design of a linear-time shelling algorithm*, Annals of Discrete Mathematics **2** (1978), 53–63.

M. Müller-Hannemann, *Shelling Hexahedral Complexes*, JGAA, 5(5) 59–91 (2001)87

[10] D. Eppstein, *Linear complexity hexahedral mesh generation*, Computational Geometry — Theory and Applications **12** (1999), 3–16.

[11] N. T. Folwell and S. A. Mitchell, *Reliable whisker weaving via curve contraction*, Engineering with Computers **15** (1999), 292–302.

[12] L. A. Freitag and P. M. Knupp, *Tetrahedral element shape optimization via the jacobian determinant and condition number*, Proceedings of the 8th International Meshing Roundtable, South Lake Tahoe, CA, Sandia National Laboratories, Albuquerque, USA, 1999, pp. 247–258.

[13] J. E. Hopcroft and R. E. Tarjan, *Dividing a graph into triconnected components*, SIAM J. of Computing **2** (1973), 135–158.

[14] P. M. Knupp, *Matrix norms & the condition number: A general framework to improve mesh quality via node-movement*, Proceedings of the 8th International Meshing Roundtable, South Lake Tahoe, CA, Sandia National Laboratories, Albuquerque, USA, 1999, pp. 13–22.

[15] P. M. Knupp, *Achieving finite element mesh quality via optimization of the Jacobian matrix norm and associated quantities, part II — a framework for volume mesh optimization*, Int. J. Numer. Methods in Eng. **48** (2000), 1165–1185.

[16] T. S. Li, R. M. McKeag, and C.G. Armstrong, *Hexahedral meshing using midpoint subdivision and integer programming*, Computer Methods in Applied Mechanics and Engineering **124** (1995), 171–193.

[17] W. Min, *Generating hexahedron-dominant mesh based on shrinking-mapping method*, Proceedings of the 6th International Meshing Roundtable, Park City, Utah, Sandia National Laboratories, Albuquerque, USA, 1997, pp. 171–182.

[18] S. A. Mitchell, *A characterization of the quadrilateral meshes of a surface which admit a compatible hexahedral mesh of the enclosed volume*, Proceedings of the 13th Annual Symposium on Theoretical Aspects of Computer Science (STACS'96), Lecture Notes in Computer Science 1046, Springer, 1996, pp. 465–476.

[19] R. H. Möhring and M. Müller-Hannemann, *Complexity and modeling aspects of mesh refinement into quadrilaterals*, Algorithmica **26** (2000), 148–171.

[20] R. H. Möhring, M. Müller-Hannemann, and K. Weihe, *Mesh refinement via bidirected flows: Modeling, complexity, and computational results*, Journal of the ACM **44** (1997), 395–426.

[21] M. Müller-Hannemann, *Hexahedral mesh generation by successive dual cycle elimination*, Engineering with Computers **15** (1999), 269–279.

M. Müller-Hannemann, *Shelling Hexahedral Complexes*, JGAA, 5(5) 59–91 (2001)88

[22] M. Müller-Hannemann, *High quality quadrilateral surface meshing without template restrictions: A new approach based on network flow techniques*, International Journal of Computational Geometry & Applications **10** (2000), 285–307.

[23] P. Murdoch, *The spatial twist continuum: A dual representation of the all hexahedral finite element mesh*, Ph.D. thesis, Mechanical Engineering, Brigham Young University, Utah, 1995.

[24] P. Murdoch, S. E. Benzley, T. D. Blacker, and S. A. Mitchell, *The spatial twist continuum: a connectivity based method for representing and constructing all-hexahedral finite element meshes*, Finite Elements in Analysis and Design **28** (1997), 137–149.

[25] T. Nishizeki and N. Chiba, *Planar Graphs: Theory and Algorithms*, Annals of Discrete Mathematics, vol. 32, North–Holland, 1988.

[26] S. Owen, *Meshing research corner*, http://www.andrew.cmu.edu/user/sowen/mesh.html.

[27] M. A. Price and C. G. Armstrong, *Hexahedral mesh generation by medial surface subdivision: Part II, solids with flat and concave edges*, Int. J. Numer. Methods in Eng. **40** (1997), 111–136.

[28] M. A. Price, C. G. Armstrong, and M. A. Sabin, *Hexahedral mesh generation by medial surface subdivision: Part I, solids with convex edges*, Int. J. Numer. Methods in Eng. **38** (1995), 3335–3359.

[29] R. Schneiders, *Information on finite element mesh generation*, available online at http://www-users.informatik.rwth-aachen.de/~roberts/meshgeneration.html.

[30] R. Schneiders, *Open problem*, available online at http://www-users.informatik.rwth-aachen.de/~roberts/open.html, 1995.

[31] R. Schneiders, *A grid-based algorithm for the generation of hexahedral element meshes*, Engineering with Computers **12** (1996), 168–177.

[32] R. Schneiders, R. Schindler, and F. Weiler, *Octree-based generation of hexahedral element meshes*, Proceedings of the 5th International Meshing Roundtable, Sandia National Laboratories, Albuquerque, USA, 1996, pp. 205–215.

[33] G. C. Shephard, *Combinatorial properties of associated zonotopes*, Canadian Journal of Mathematics **26** (1974), 302–321.

[34] T. J. Tautges, T. Blacker, and S. A. Mitchell, *The whisker weaving algorithm: A connectivity-based method for constructing all-hexahedral finite element meshes*, Int. J. Numer. Methods in Eng. **39** (1996), 3327–3349.

M. Müller-Hannemann, *Shelling Hexahedral Complexes*, JGAA, 5(5) 59–91 (2001)89

[35] T. J. Tautges and S. A. Mitchell, *Whisker weaving: Invalid connectivity resolution and primal construction algorithm*, Proceedings of the 4th International Meshing Roundtable, Sandia National Laboratories, Albuquerque, USA, 1995, pp. 115–127.

[36] W. Thurston, *Hexahedral decomposition of polyhedra*, Posting to sci.math., 25 Oct., 1993, available online at http://www.ics.uci.edu/~eppstein/gina/Thurston-hexahedra.html.

[37] P. M. Tuchinsky and B. W. Clark, *The "HexTet" hex-dominant automesher: An interim progress report*, Proceedings of the 6th International Meshing Roundtable, Park City, Utah, Sandia National Laboratories, Albuquerque, USA, 1997, pp. 183–193.

[38] W. T. Tutte, *Convex representations of graphs*, Proceedings London Mathematical Society **10** (1960), 304–320.

[39] W. T. Tutte, *How to draw a graph*, Proceedings London Mathematical Society **13** (1963), 743–768.

[40] D. R. White, L. Mingwu, S. E. Benzley, and G. D. Sjaardema, *Automated hexahedral mesh generation by virtual decomposition*, Proceedings of the 4th International Meshing Roundtable, Sandia National Laboratories, Albuquerque, USA, 1995, pp. 165–176.

[41] G. M. Ziegler, *Lectures on polytopes*, revised ed., Graduate Texts in Mathematics, vol. 152, Springer-Verlag, New York, 1998.

[42] G. M. Ziegler, *Shelling polyhedral 3-balls and 4-polytopes*, Discrete & Computational Geometry **19** (1998), 159–174.

M. Müller-Hannemann, *Shelling Hexahedral Complexes*, JGAA, 5(5) 59–91 (2001)90

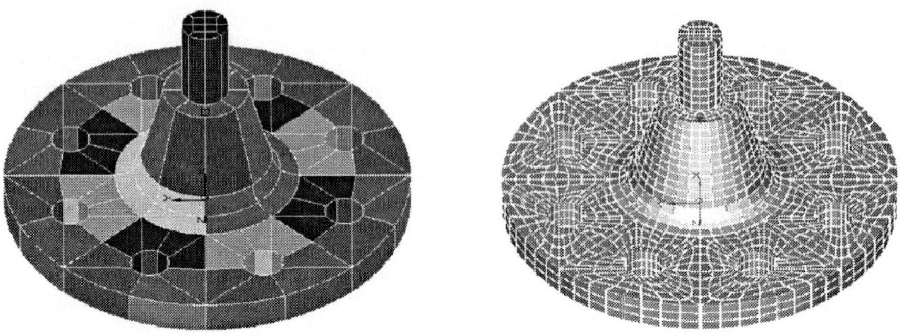

Figure 20: The model of a shaft with a flange which has been decomposed into simpler subdomains where different colors correspond to the subdomains (left), and the hexahedral mesh constructed by our algorithm (right).

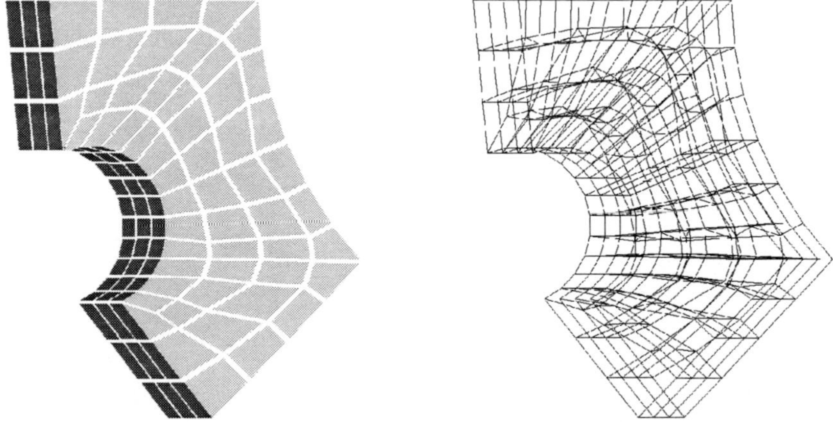

Figure 21: Hexahedral mesh for a part of the model in Fig. 20.

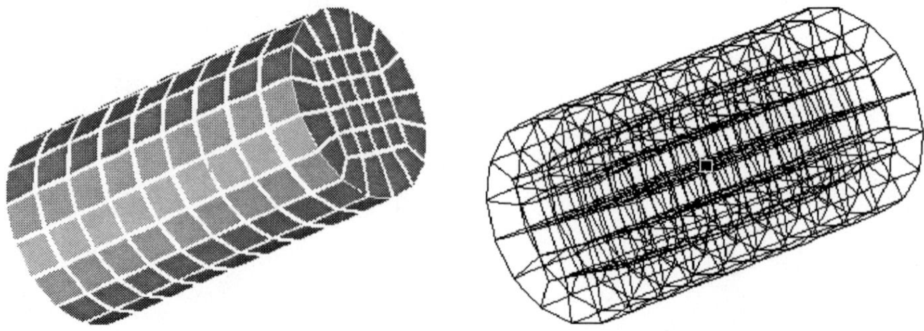

Figure 22: Hexahedral mesh for a cylinder (part of the model in Fig. 20). The cycle elimination scheme first selects parallel cycles corresponding to eight layers, and uses than a double elimination to remove the 16 hexahedra forming a torus on the last layer.

M. Müller-Hannemann, *Shelling Hexahedral Complexes*, JGAA, 5(5) 59–91 (2001)91

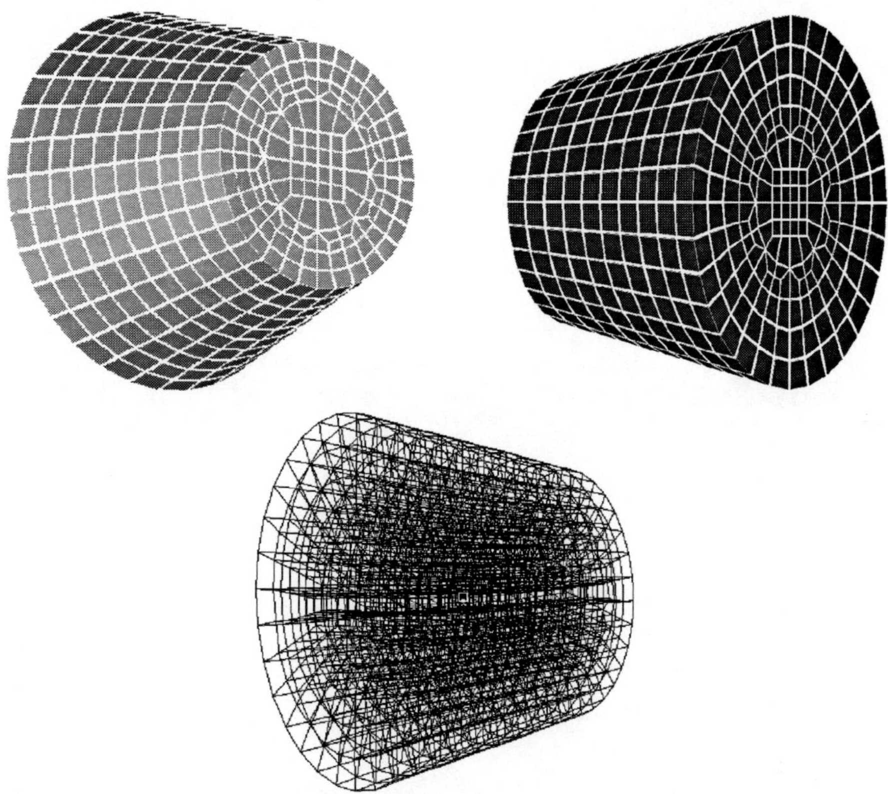

Figure 23: Hexahedral mesh for a truncated circular cone (also part of the model in Fig. 20). The meshing is non-trivial because the base cycles of the cone have slightly different surface meshes.

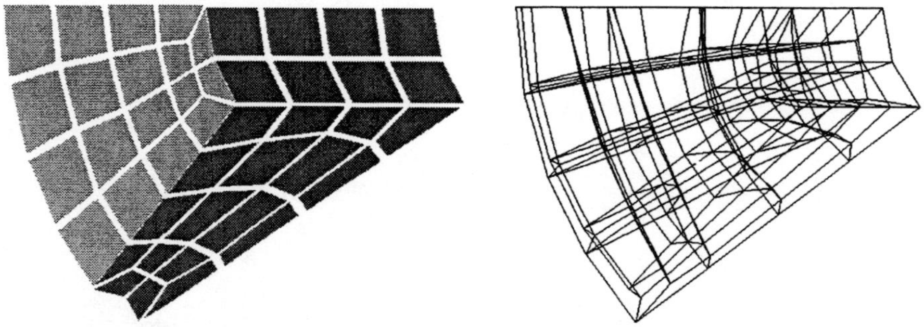

Figure 24: Hexahedral mesh for another part of the model in Fig. 20.

Journal of Graph Algorithms and Applications

http://www.cs.brown.edu/publications/jgaa/

vol. 5, no. 5, pp. 93–105 (2001)

Connectivity of Planar Graphs

H. de Fraysseix   *P. Ossona de Mendez*

CNRS UMR 8557
E.H.E.S.S.
54 Bd Raspail
75006 Paris
France
http://www.ehess.fr/centres/cams/
hf@ehess.fr   pom@ehess.fr

Abstract

We give here three simple linear time algorithms on planar graphs:
a 4-connexity test for maximal planar graphs, an algorithm enumerating
the triangles and a 3-connexity test. Although all these problems got
already linear-time solutions, the presented algorithms are both simple
and efficient. They are based on some new theoretical results.

Communicated by T. Nishizeki, R. Tamassia and D. Wagner:
submitted February 1999; revised April 2000.

Fraysseix, O. de Mendez, *Connectivity of Planar Graphs*, JGAA, 5(5) 93–105 (2001)94

1 Introduction

The study of graphs by means of special orientations is relatively recent. For instance, bipolar orientations became a basic tool in many graph drawing problems. We give here an example of relations between orientation and topological properties. Constrained orientations (i.e. orientations with bounded indegrees) lead to new characterizations on connexity of planar undirected graphs. Although usual 3-connexity testing of planar graphs are heavily related to planarity testing algorithms (see [10][17] and PQ-tree algorithms), the algorithm we present here assume that a graph is already embedded in the plane and a the problem drastically reduces to the acyclicity testing of a particular orientation. Concerning the 4-connexity testing of a maximal planar graph, the use of an indegree bounded orientation was already used in [2] to enumerate triangles. Here, the use of a specific orientation allows a further simplification of the algorithm. The 4-connexity test itself also reduces to an acyclicity test. It should be noticed that no special data structure is used for these algorithms as, in the planar case, the acyclicity of an orientation may be efficiently tested using a dual topological sort.

2 Preliminaries

In the following we consider plane graphs, that is planar graphs embedded in the plane. Each connected component of the complement in the plane of the vertex and edge sets is a *face region* of the graph. The *external face region* of G is the unbounded one. A *face* is the clockwise walk of the boundary of a face region. When considering an orientation of a graph, such walks also define a *dual orientation* of the dual graph: the outgoing edges of a vertex f of the dual are those traversed according to their orientation in a clockwise walk of the face corresponding to f.

If G is a graph, $V(G)$ and $E(G)$ denote the vertex set and the edge set of G, respectively. We denote G_A the subgraph of G induced by a subset A of vertices. We denote $d_G^-(x)$ the indegree of the vertex x in the graph G.

Let X and $\overline{X}$ be two complementary subsets of the vertices of an oriented graph. The *cocycle* $\omega(X)$ is the pair $(\omega^+(X), \omega^-(X))$ of the set $\omega^+(X)$ of edges oriented from X to $\overline{X}$ and the set $\omega^-(X)$ of edges oriented from $\overline{X}$ to X. A cocycle $\omega(X)$ is *elementary* if G_X and $G_{\overline{X}}$ are connected. Obviously, any cocycle is the disjoint union of elementary cocycles. A cocycle $\omega(X)$ is a *positive cocircuit* if $\omega^-(X)$ is empty, that is if no edge is directed from $\overline{X}$ to X.

Lemma 2.1 *Let X be a subset of $V(G)$. Then $\omega(X)$ is a positive cocircuit if and only if*

$$|E(G_X)| = \sum_{x \in X} d_G^-(x)$$

$\square$

Fraysseix, O. de Mendez, *Connectivity of Planar Graphs*, JGAA, 5(5) 93–105 (2001)95

A *cycle* γ is an Eulerian partial subgraph (i.e. with even vertices only). A cycle is *elementary* (or a *polygon*) if it is connected and 2-regular. A cycle γ is a *circuit* if each of its vertices has in γ an indegree equal to its outdegree. An elementary cycle γ defines a bipartition of the remaining vertices and edges of the graph as *internal* and *external* elements.

Two consecutive edges in the clockwise order at a vertex define an *angle* of the graph. The angle is *lateral* if one of the two edges is incoming and the other is outgoing; otherwise, the angle is *extremal*. The *angle graph* $A(G)$ of a 2-connected plane graph G is the incidence graph of the vertex and face sets of G (the *V-vertices* and *F-vertices* of $A(G)$). The edges of $A(G)$ correspond to the angles of G and their number is twice the number of edges of G. The graph $A(G)$ is maximal bipartite planar. Any embedding of G canonically defines an embedding of $A(G)$, where the faces correspond to the edges of G.

A *k-connected* graph is a graph with at least $k + 1$ vertices, such that the deletion of any subset of $k - 1$ vertices does not disconnect the graph. A *separating cycle* is an elementary cycle whose vertex set removal disconnects the graph.

Lemma 2.2 *Let X be a vertex subset of plane graph G. If $G_{\overline{X}}$ is connected, then $\overline{X}$ belongs to a same face region of G_X.*

Proof: Assume that two vertices u, v of $\overline{X}$ do not belong to a same face region of G_X. Then a path from u to v in $G_{\overline{X}}$ intersects the boundary of the face region and hence intersects X, which is a contradiction. $\square$

3 A 4-connexity test for maximal planar graphs

The algorithm is based on the following properties:

- A maximal planar graph is 4-connected if and only if it has no separating triangles, i.e. if each of its triangles is a face[19],

- Any maximal planar graph has an orientation where all the vertices (except the 3 external ones) have indegree 3 [3][14],

- In such an orientation, separating triangles corresponds to positive cocircuits (see Lemma 3.4).

An early linear-time algorithm may be found in [11], a more recent one, based on subgraph isomorphism detection, may also be found in [5].

Lemma 3.1 *Let G be a 3-connected planar graph and $\{x, y, z\}$ a cutset of G. Then, $G - \{x, y, z\}$ has 2 connected components.*

Proof: The graph $G - \{x, y, z\}$ has at least 2 connected components, as $\{x, y, z\}$ is a cutset. Assume $G - \{x, y, z\}$ has 3 connected components H_1, H_2, H_3 and let a_1, a_2, a_3 be vertices of H_1, H_2, H_3, respectively. As G is 3-connected, for

Fraysseix, O. de Mendez, *Connectivity of Planar Graphs*, JGAA, 5(5) 93–105 (2001)96

any $i \neq j$ in $\{1, 2, 3\}$, there exist three internally disjoint paths linking a_i and a_j [18] and these paths respectively include x, y and z. Hence, there exists in G 3 internally paths linking a_1 (resp. a_2, a_3) to x, y, z and whose internal vertices belong to H_1 (resp. H_2, H_3). Thus, a_1, a_2, a_3, x, y, z and these nine paths form a subdivision of $K_{3,3}$, which contradicts the planarity of G. $\qquad\square$

Lemma 3.2 *A triangle of a maximal planar graph is a separating triangle if and only if it is not a face.*

Proof: If a triangle is not a face, it separates its interior and exterior vertices. Conversely, assume a face $\{x, y, z\}$ is a separating triangle. A vertex may be added in this face, adjacent to x, y, z, while preserving the planarity. Then, $G - \{x, y, z\}$ has at least 3 components, what contradicts Lemma 3.1. $\qquad\square$

Lemma 3.3 (see [19]) *A maximal planar graph G is 4-connected if and only if its has no separating triangle, i.e. a cutset which is the vertex set of a triangle.*
$$\square$$

Lemma 3.4 *Let G be a maximal planar graph (with at least 5 vertices), which is oriented in such a way that all its vertices have indegree 3, except the 3 vertices of the external face which have indegree 1.*

Then, G is 4-connected if and only if it has only one positive cocircuit, namely the one defined by the vertex-set of its external face.

Proof: Let V_0 be the vertex set of the external face. Let us prove that the graph G has a cocircuit different from $\omega(V_0)$ if and only if G has a triangle which is not a face (this is equivalent to the G not being 4-connected, according to Lemma 3.3 and Lemma 3.2):

Algorithm 1 A 4-connexity test for a maximal planar graph G

Require: G is a maximal planar graph
Ensure: *IsFourConnected*=true if and only if G is 4-connected
1: **if** G has less than 6 vertices **then**
2:    *IsFourConnected* $\leftarrow$ false
3: **else**
4:    $G' \leftarrow G$
5:    $r_1, r_2, r_3 \leftarrow$ the vertices of some face of G'
6:    Orient G' in such a way that every vertex has indegree 3 (except r_1, r_2, r_3 which have indegree 1)
7:    Remove the vertices r_1, r_2, r_3
8:    Compute the oriented dual H of G'
9:    **if** the orientation of H is acyclic **then**
10:      *IsFourConnected* $\leftarrow$ true
11:    **else**
12:      *IsFourConnected* $\leftarrow$ false
13:    **end if**
14: **end if**

Fraysseix, O. de Mendez, *Connectivity of Planar Graphs*, JGAA, 5(5) 93–105 (2001)97

- Let $\omega(X)$ be an elementary positive cocircuit. The sum of the indegrees of the vertices of X is at least $3|X| - 6$, since only 3 vertices have indegree 1. Hence, according to Lemma 2.1, G_X has at least $3|X| - 6$ edges and then has exactly $3|X| - 6$ edges, is maximal planar and contains the vertices of the external face. Thus, according to Lemma 2.2, $\overline{X}$ belongs to a bounded face region of G_X and then is internal to some triangle of G. Thus, either X is the vertex set of the external face of G (i.e. V_0) or G has a triangle which is not a face.

- Let T be a triangle of G which is not a bounded face and let $\overline{X}$ be the set of the vertices internal to T. As G_X is maximal planar and contains r_1, r_2 and r_3, according to Lemma 2.1, the cocycle $\omega(X)$ is a cocircuit. Hence, $\omega(V_0)$ is a cocircuit and any triangle which is not a face defines a cocircuit (different from $\omega(V_0)$).

$\square$

Theorem 3.5 *Algorithm 1 tests in linear time whether a maximal planar graph is 4-connected or not.*

Proof: First notice that no 4-connected maximal planar graph has less than 6 vertices. Hence, the preliminary test at line 1: is valid and we may restrict ourselves to the case where G has at least 6 vertices.

The copy of the graph G into a graph G' may be performed in linear time. The orientation of G' performed at line 6: may be computed in linear time [3, 14].

Then, G is 4-connected if and only if G' has only one positive cocircuit, namely the one defined by $\{r_1, r_2, r_3\}$. After the deletion of r_1, r_2, r_3 at line 7:, we get that the graph G is 4-connected if and only if G' has no cocircuit, that is, if and only if its oriented dual H (which is computed in linear time at line 8:) has no circuit. This test (line 9:) can be done in linear time using a topological sort. $\square$

4   Enumerations of the triangles of a planar graph

Linear time algorithms enumerating the triangles of planar graphs may be found in [1] (using tree decompositions) or in [2] (using indegree bounded orientations).

The algorithm we present here has been optimized using Schnyder's decompositions, the definition of which we shall recall here:

Definition 4.1 (Schnyder, [14]) *Let G be a maximal planar graph and $\{r_1, r_2, r_3\}$ one of its faces. A Schnyder decomposition relative to $\{r_1, r_2, r_3\}$ is a tricoloration of the edges of G, each color $1 \le i \le 3$ forming a directed tree Y_i rooted at r_i such that there exists three total orders $<_1, <_2, <_3$ on the vertex set of G satisfying:*

- *If the arc (u, v) belongs to Y_i then $(u <_j v) \iff (j \ne i)$,*

Fraysseix, O. de Mendez, *Connectivity of Planar Graphs*, JGAA, 5(5) 93–105 (2001)98

- *If $\{x, y\}$ is an edge of G, then*

$$\forall u \notin \{x, y\}, \exists 1 \le i \le 3, \quad u >_i x \text{ and } u >_i y$$

Definition 4.2 *Let G be a planar graph on $n \le 3$ vertices and let r_1, r_2, r_3 be 3 vertices of G.*

A parent triplet (π_1, π_2, π_3) *relative to* $\{r_1, r_2, r_3\}$ *is a triplet of functions from $V(G)$ to $V(G) \cup \{0\}$, such that there exists a triangulation H of G and a Schnyder decomposition of H relative to $\{r_1, r_2, r_3\}$ which satisfies: $\pi_i(v)$ is either the parent of the vertex v if these vertices are adjacent in G, or 0 (otherwise).*

Algorithm 2 Computation of a parent triplet

Require: G is a planar graph with at least 4 vertices
Ensure: π_1, π_2, π_3 are Schnyder parent functions for G relative to r_1, r_2, r_3
 1: $H \leftarrow G$
 2: Triangulate H and mark the added edges
 3: $r_1, r_2, r_3 \leftarrow$ the vertices of some face of H
 4: Compute a Schnyder orientation of H as parent functions π_1, π_2, π_3 (extended with $\pi_i(0) = 0$)
 5: $\pi_1(r_2) \leftarrow r_1$
 6: $\pi_2(r_3) \leftarrow r_2$
 7: $\pi_3(r_1) \leftarrow r_3$
 8: **for all** marked edge $e = \{u, v\}$ **do**
 9:    **for all** $i \in \{1, 2, 3\}$ **do**
10:      **if** $u = \pi_i(v)$ **then**
11:        $\pi_i(v) \leftarrow 0$
12:      **else if** $v = \pi_i(u)$ **then**
13:        $\pi_i(v) \leftarrow 0$
14:      **end if**
15:    **end for**
16: **end for**

Theorem 4.1 *A parent triplet of a planar graph G may be computed in linear time using algorithm 2.*

Proof: A triangulation is easily performed in linear time. A Schnyder decomposition may also be computed in linear time [15], using the packing algorithm described in [3]. The modification we perform on the functions π is obviously linear. $\square$

Lemma 4.2 *When reversing the orientation of the edges of color i, the graph becomes acyclic.*

Fraysseix, O. de Mendez, *Connectivity of Planar Graphs*, JGAA, 5(5) 93–105 (2001)99

Proof: According to the definition, if there exists a directed path from x to y in the obtained orientation, then $x <_i y$. Thus, the orientation is acyclic. $\quad\square$

Lemma 4.3 *A triangle of G is a circuit if and only if it is 3-colored; otherwise, it is 2-colored*

Proof: The proof of the lemma will be a consequence of Lemma 4.2:

If a triangle is 2-colored, it does not become a circuit when reversing the orientation of the edges of the third color. Hence, it is not a circuit.

If a triangle is 3-colored, it does not become a circuit when reversing the orientation of any of its edges. Hence, it is a circuit. $\quad\square$

Theorem 4.4 *Algorithm 3 enumerates the triangles of a planar graph in linear time.*

Proof: Associate to each triangle of G either the sink of the triangle if it is acyclic, or the head of the edge colored 1 if it is a circuit. This way, to each triangle is associated exactly one vertex. The algorithm is then an obvious application of Lemma 4.3. $\quad\square$

Lemma 4.5 *Let (a, b, c, d) be a C_4. Then, it is not possible that (a, b) and (c, d) shall be both arcs of the same tree Y_i.*

Proof: Assume such a C_4 exists.

Considering the edge $\{b, c\}$ and according to the definition of a Schnyder decomposition, there exists j such that $a >_j b$ and $a >_j c$. As (a, b) belongs to Y_i, j shall only be equal to i. Hence, $a >_i c$.

Algorithm 3 Enumeration of the triangles of a planar graph

Require: G is a planar graph with at least 4 vertices
Ensure: *NumberOfTriangles* is the number of triangles of G
 1: Compute the Schnyder parent functions π_1, π_2, π_3 of G
 2: *NumberOfTriangles* $\leftarrow 0$
 3: **for all** vertex v **do**
 4:     **for all** $(i, j) \in \{1, 2, 3\}^2, i \neq j$ **do**
 5:         **if** $(\pi_i(v) \neq 0)$ and $(\pi_j(v) \neq 0)$ and $(\pi_i(\pi_j(v)) = \pi_i(v))$ **then**
 6:             *NumberOfTriangles* $\leftarrow$ *NumberOfTriangles* $+ 1$
 7:         **end if**
 8:     **end for**
 9:     **if** $(\pi_1(v) \neq 0)$ and $(\pi_2(\pi_1(v)) \neq 0)$ and $(\pi_3(\pi_2(\pi_1(v))) = v)$ **then**
10:         *NumberOfTriangles* $\leftarrow$ *NumberOfTriangles* $+ 1$
11:     **end if**
12:     **if** $(\pi_1(v) \neq 0)$ and $(\pi_3(\pi_1(v)) \neq 0)$ and $(\pi_2(\pi_3(\pi_1(v))) = v)$ **then**
13:         *NumberOfTriangles* $\leftarrow$ *NumberOfTriangles* $+ 1$
14:     **end if**
15: **end for**

Fraysseix, O. de Mendez, *Connectivity of Planar Graphs*, JGAA, 5(5) 93–105 (2001)100

Similarly, considering the edge $\{a,d\}$ and the vertex c, we get $c >_i a$ and are led to a contradiction. $\qquad\square$

Theorem 4.6 *Algorithm 4 enumerates in linear time the triangles of a planar graph.*

Proof: Algorithm 4 is a reorganized version of Algorithm 3 taking into account some exclusiveness in the cases. The only non-trivial exclusiveness used is that we cannot have simultaneously: $\pi_i(\pi_j(v)) = \pi_i(v)$ and $\pi_k(\pi_j(\pi_i(v))) = v$ (where none of the values taken by the π functions are 0). Otherwise, we would have a C_4: $(\pi_j(v), v, \pi_j(\pi_i(v)), \pi_i(v))$ with arcs $(\pi_j(v), v)$ and $(\pi_j(\pi_i(v)), \pi_i(v))$ colored j, which contradicts Lemma 4.5. $\qquad\square$

Remark 4.7 *Algorithm 4 obviously gives the upper bound of $3n - 8$ (1 in the bloc starting at line 12:, and $n - 3$ times 3 in the loop at line 17:) for the number of triangles of a planar graph having at least 4 vertices.*

Remark 4.8 *This algorithm may be modified to enumerate the separating triangles of 3-connected planar graphs, by enumerating the triangles which are not faces.*

Algorithm 4 Optimized enumeration of the triangles of a planar graph

---

Require: G is a planar graph with at least 4 vertices
Ensure: *NumberOfTriangles* is the number of triangles of G
  1: Compute the Schnyder parent functions π_1, π_2, π_3 of G and the roots r_1, r_2, r_3
  2: **if** $\pi_1(r_2) \neq 0$ and $\pi_2(r_3) \neq 0$ and $\pi_3(r_1) \neq 0$ **then**
  3:     *NumberOfTriangles* $\leftarrow 1$
  4: **else**
  5:     *NumberOfTriangles* $\leftarrow 0$
  6: **end if**
  7: **for all** vertex v different from r_1, r_2, r_3 **do**
  8:     $p_1 \leftarrow \pi_1(v), \quad p_2 \leftarrow \pi_2(v), \quad p_3 \leftarrow \pi_3(v)$
  9:     **if** $p_1 \neq 0$ **then**
 10:         **if** $(p_2 \neq 0)$ and $(\pi_2(p_1) = p_2)$ or $(\pi_1(p_2) = p_1)$ or $(\pi_3(\pi_2(p_1)) = v)$ **then**
 11:             *NumberOfTriangles* $\leftarrow$ *NumberOfTriangles* $+ 1$
 12:         **end if**
 13:         **if** $(p_3 \neq 0)$ and $(\pi_3(p_1) = p_3)$ or $(\pi_1(p_3) = p_1)$ or $(\pi_2(\pi_3(p_1)) = v)$ **then**
 14:             *NumberOfTriangles* $\leftarrow$ *NumberOfTriangles* $+ 1$
 15:         **end if**
 16:     **end if**
 17:     **if** $(p_3 \neq 0)$ and $(\pi_3(p_2) = p_3)$ or $(p_2 \neq 0)$ and $(\pi_2(p_3) = p_2)$ **then**
 18:         *NumberOfTriangles* $\leftarrow$ *NumberOfTriangles* $+ 1$
 19:     **end if**
 20: **end for**

Fraysseix, O. de Mendez, *Connectivity of Planar Graphs*, JGAA, 5(5) 93–105 (2001)101

5 A 3-connexity test for planar graphs

The algorithm is based on the following properties we shall prove later:

- A 2-connected planar graph is 3-connected if and only if each of the C_4 of its angle-graph is a face,

- Any planar quadrangulation has an orientation where all the vertices have indegree 2, except the 4 external ones, which have indegree 1.

- In such an orientation, the C_4 which are not faces correspond to positive cocircuits.

Algorithm 5 3-connexity test for a 2-connected planar graph G

Require: G is a 2-connected planar graph
Ensure: x=true if and only if G is 3-connected
 1: **if** G has less than 4 vertices **then**
 2:     $x \leftarrow$ false
 3: **else**
 4:     $H \leftarrow \mathcal{A}(G)$
 5:     $b_1, w_1, b_2, w_2 \leftarrow$ the vertices of some face of H
 6:     H is oriented in such a way that every vertex (except b_1, b_2) has 2 incoming edges
 7:     Remove the vertices b_1, w_1, b_2, w_2
 8:     $D \leftarrow$ oriented dual of H
 9:     **if** D is connected and its orientation is acyclic **then**
10:       $x \leftarrow$ true
11:     **else** $\{D$ has a directed circuit$\}$
12:       $x \leftarrow$ false
13:     **end if**
14: **end if**

Definition 5.1 *A* 2-articulated subgraph *of a 2-connected graph G is a connected proper induced subgraph H with at least 3 vertices, which may be disconnected from the remaining of the graph by the deletion of two vertices, the* articulation pair *of H.*

Lemma 5.1 *Let G be a 2-connected planar graph. Then G is 3-connected if and only if each C_4 of $A(G)$ is a face.*

Proof: Let γ be a C_4 of $A(G)$ which is not a face and let u, v be its V-vertices. As γ is not a face, there exists at least one vertex of $A(G)$ inside and outside γ. If the only vertices of $A(G)$ inside (resp. outside) γ where F-vertices, the faces inside (resp. outside) γ would correspond to multiple edges of G. Hence, $A(G)$ has at least one V-vertex internal to γ and one V-vertex external to γ. The subgraph H of G induced by the vertices corresponding to u, v and the

Fraysseix, O. de Mendez, *Connectivity of Planar Graphs*, JGAA, 5(5) 93–105 (2001)102

V-vertices of $A(G)$ inside γ meets then the requirement of the definition of a 2-articulated subgraph. Thus, G is not 3-connected.

Conversely, if G is not 3-connected, it has a 2-articulated subgraph H with articulation pair u, v. Let f_1 and f_2 be two faces of G adjacent to u and v, such that f_1 does not include the edge $\{u, v\}$ (if this edge exists). Then, f_1, u, f_2, v is not a face of $A(G)$ as it does not correspond to an edge of G. $\qquad\square$

Remark 5.2 *There will be no linear-time algorithm to enumerate the C_4 of 3-connected planar graphs, as this number may be quadratic (any double-wheel will do), although it is possible to "implicitly" enumerate them in linear time [1][4].*

Lemma 5.3 *Let G be a 2-connected planar graph with at least 4 vertices and let $A(G)$ its angle graph, oriented in such a way that each of its vertices have indegree 2, except the vertices of the external faces which have indegree 1.*

Then, the graph G is 3-connected if and only if $A(G)$ has only one positive cocircuit, namely the one defined by the vertex-set of its external face.

Proof: Let V_0 be the vertex set of the external face. Let us prove that the graph G has a cocircuit different from $\omega(V_0)$ if and only if $A(G)$ has a C_4 which is not a face (this is equivalent to the 3-connexity of G, according to Lemma 5.1):

- Let $\omega(X)$ be an elementary positive cocircuit. The sum of the indegrees of the vertices of X is at least $2|X| - 4$, since only 4 vertices have indegree 1. Hence, according to Lemma 2.1, G_X has at least $2|X| - 4$ edges and then has exactly $2|X| - 4$ edges, is a planar quadrangulation and contains the vertices of the external face. Thus, according to Lemma 2.2, $\overline{X}$ belongs to a bounded face region of G_X and then is internal to some C_4 of G. Thus, X is the vertex set of the external face (i.e. V_0) or G has a C_4 which is not a face.

- Let C be a C_4 of G which is not a bounded face and let $\overline{X}$ be the set of the vertices internal to C. As G_X is a planar quadrangulation and contains the vertices of the external face, according to Lemma 2.1, the cocycle $\omega(X)$ is a cocircuit. Hence, $\omega(V_0)$ is a cocircuit and any C_4 which is not a face defines a cocircuit (different from $\omega(V_0)$).

$\qquad\square$

Definition 5.2 *An e-bipolar orientation is an acyclic orientation with exactly one source s and one sink t linked by the edge e. Such an orientation may be computed in linear time [16, 8, 9].*

Lemma 5.4 *Let G be a 2-connected plane graph and let e_0 be an edge of G. Let $\{r_1, r_2, r_3, r_4\}$ be the face of $A(G)$ corresponding to e_0, where r_1 and r_3 are V-vertices.*

Fraysseix, O. de Mendez, *Connectivity of Planar Graphs*, JGAA, 5(5) 93–105 (2001)103

Algorithm 6 Optimized 3-connexity test for a 2-connected planar graph G

Require: G is a 2-connected planar graph
Ensure: x=true if and only if G is 3-connected
1: **if** G has less than 4 vertices **then**
2:    $x \leftarrow$ false
3: **else**
4:    $e_0 \leftarrow$ some edge of G
5:    $S \leftarrow \emptyset$ (empty stack)
6:    Compute a minimal e_0-bipolar orientation of G [8]
7:    **for all** edge e of G **do**
8:      $d[e] \leftarrow$ number of invertible angles at e
9:      **if** $d[e] = 0$ **then**
10:        Push e in the stack S
11:        Mark all the angles incident to e
12:      **end if**
13:    **end for**
14:    **while** S is not empty **do**
15:      Pop e from the stack S
16:      **for all** the neighbor edges e' of e **do**
17:        Decrement $d[e']$
18:        **if** $d[e'] = 0$ **then**
19:          Push e' in the stack S
20:          Mark all the angles incident to e'
21:        **end if**
22:      **end for**
23:    **end while**
24:    Mark all the angles incident to an edge adjacent to e_0
25:    Mark all the angles incident to an edge is a same face than e_0
26:    **if** all the angles are marked **then**
27:      $x \leftarrow$ true
28:    **else**
29:      $x \leftarrow$ false
30:    **end if**
31: **end if**

Any orientation of G defines an orientation of $A(G)$: an edge of $A(G)$ is directed from its incident V-vertex to its incident F-vertex if the corresponding angle of G is extremal.

If G is e_0-bipolarly oriented, then the induced orientation of $A(G)$ is such that every vertex has indegree 2, except r_1 and r_3 which are sources.

Proof: The poles have no lateral angles, any other vertex has at least two lateral angles and each face has at least two extremal angles. As $A(G)$ has $2|E(G)| = 2|F(G)| + 2(|V(G)| - 2)$ edges, the V-vertices different from the poles and the F-vertices have two incoming edges. $\qquad\square$

Fraysseix, O. de Mendez, *Connectivity of Planar Graphs*, JGAA, 5(5) 93–105 (2001)104

Theorem 5.5 *Algorithm 5 tests in linear time whether a 2-connected planar graph is 3-connected or not.*

Proof: A bipolar orientation of G will induce, according to Lemma 5.4, an orientation of $A(G)$ such that all the vertices of $A(G)$ (except the V-vertices incident to e_0) have indegree 2. Then, the validity of Algorithm 5 follows from Lemma 5.3. $\square$

Remark 5.6 *Using a particular e_0-bipolar orientation [8], we can ensure that all the circuits of the angle-graph are clockwise (the external face corresponding to e_0). Then, as the vertices and edges of the dual of the angle-graph are nothing but the edges and the angles of the original graph, Algorithm 5 may be translated on the original graph itself. Using the property of the particular e_0-bipolar orientation, we obtain (optimized) Algorithm 6.*

References

[1] N. Chiba and T. Nishizeki, *Arboricity and subgraph listing algorithms*, SIAM J. Computing vol. 14 (1985), 210–223.

[2] M Chrobak and D. Eppstein, *Planar orientations with low out-degree and compaction of adjacency matrices*, Theoretical Computer Science vol. 86 (1991), 243–266.

[3] H. de Fraysseix, J. Pach, and R. Pollack, *Small sets supporting Fary embeddings of planar graphs*, Twentieth Annual ACM Symposium on Theory of Computing, 1988, pp. 426–433.

[4] D. Eppstein, *Arboricity and bipartite subgraph listing algorithms*, IPL (1994), no. 51, 207–211.

[5] ———, *Subgraph isomorphism in planar graphs and related problems*, 6th ACM-SIAM Symp. Discrete Algorithms (San Francisco), 1995, pp. 632–640.

[6] ———, *Subgraph isomorphism in planar graphs and related problems*, J. Graph Algorithms and applications vol. 3 (1999), no. 3, 1–27.

[7] H. de Fraysseix and P. Ossona de Mendez, *On topological aspects of orientations*, Proc. of the Fifth Czech-Slovak Symposium on Combinatorics, Graph Theory, Algorithms and Applications, Discrete Math., (to appear).

[8] H. de Fraysseix, P. Ossona de Mendez, and J. Pach, *A left-first search algorithm for planar graphs*, Discrete Computational Geometry vol. 13 (1995), 459–468.

[9] H. de Fraysseix, P. Ossona de Mendez, and P. Rosenstiehl, *Bipolar orientations revisited*, Discrete Applied Mathematics vol. 56 (1995), 157–179.

Fraysseix, O. de Mendez, *Connectivity of Planar Graphs*, JGAA, 5(5) 93–105 (2001)105

[10] J.E. Hopcroft and R.E. Tarjan, *Dividing a graph into triconnected components.*, SIAM Journal on Computing (1973).

[11] Jean-Paul Laumond, *Connectivity of plane triangulations*, vol. 34 (1990), no. 2, 87–96.

[12] T Matsumoto, *Orientations contraintes*, Ph.D. thesis, Ecole des Hautes Etudes en Sciences Sociales, Paris, 1997.

[13] P. Ossona de Mendez, *Orientations bipolaires*, Ph.D. thesis, Ecole des Hautes Etudes en Sciences Sociales, Paris, 1994.

[14] W. Schnyder, *Planar graphs and poset dimension*, Order vol. 5 (1989), 323–343.

[15] ———, *Embedding planar graphs in the grid*, First ACM-SIAM Symposium on Discrete Algorithms, 1990, pp. 138–147.

[16] R.E. Tarjan, *Depth-first-search and linear graph algorithm*, SIAM J. Comp. vol. 2 (1972), 146–160.

[17] ———, *Testing graph connectivity*, Conference Record of Sixth Annual ACM Symposium on Theory of Computing (Seattle, Washington), 30 April–2 May 1974, pp. 185–193.

[18] H. Whitney, *Congruent graphs and the connectivity of graphs*, AM; J. Math. vol. 54 (1932), 150–168.

[19] D. Woods, *Drawing planar graphs*, Ph.D. thesis, Stanford University, 1982, Tech. Rep. STAN-CS-82-943.